T0310601

Dynamics of Galaxies

Second Edition

Our understanding of galaxies, the building blocks of the Universe, has advanced significantly in recent years. New observations from ground- and space-based telescopes, the discovery of dark matter, and new insights into its distribution have been instrumental in this. This textbook provides graduate students with a modern introduction to the gravitationally determined structure and evolution of galaxies. Readers will also benefit from detailed discussions of the issues involved in the process of modeling complex stellar systems. Additionally, the text provides an accessible framework for interpreting observations and devising new observational tests.

Based on the author's extensive teaching experience, this second edition features an up-to-date view of basic phenomenology, a discussion of the structure of dark halos in galaxies, the dynamics of quasi-relaxed stellar systems and globular clusters, galaxies and gravitational lensing, and an introduction to self-gravitating accretion disks. Extended problem sets are available from the accompanying resources website: www.cambridge.org/9781107000544.

Giuseppe Bertin is professor of physics at the University of Milan, Italy; previously, he was on the faculty of the Scuola Normale Superiore, Pisa. He has also held several positions at the Massachusetts Institute of Technology and has been a member of the Kavli Institute for Theoretical Physics, University of California, Santa Barbara (in 2006 and 2009). He is also author of *Spiral Structure in Galaxies: A Density Wave Theory*, with C. C. Lin, and editor of the proceedings of a series of three international workshops on plasmas in the laboratory and in the Universe. He was elected to Italy's Accademia Nazionale dei Lincei in 2013, the same year he received its Premio Nazionale del Presidente della Repubblica in Science.

Dynamics of Galaxies
Second Edition

GIUSEPPE BERTIN
Università degli Studi di Milano

CAMBRIDGE
UNIVERSITY PRESS

University Printing House, Cambridge CB2 8BS, United Kingdom

One Liberty Plaza, 20th Floor, New York, NY 10006, USA

477 Williamstown Road, Port Melbourne, VIC 3207, Australia

314-321, 3rd Floor, Plot 3, Splendor Forum, Jasola District Centre, New Delhi - 110025, India

79 Anson Road, #06-04/06, Singapore 079906

Cambridge University Press is part of the University of Cambridge.

It furthers the University's mission by disseminating knowledge in the pursuit of
education, learning and research at the highest international levels of excellence.

www.cambridge.org
Information on this title: www.cambridge.org/9781107000544

© Cambridge University Press 2000
© Giuseppe Bertin 2014

This publication is in copyright. Subject to statutory exception
and to the provisions of relevant collective licensing agreements,
no reproduction of any part may take place without the written
permission of Cambridge University Press.

First published 2000
Second edition 2014

A catalogue record for this publication is available from the British Library

Library of Congress Cataloging in Publication data
Bertin, G. (Giuseppe), author.
Dynamics of galaxies / Giuseppe Bertin, Università degli Studi di Milano. – Second edition.
    pages   cm
Includes bibliographical references and index.
ISBN 978-1-107-00054-4 (hardback)
1. Galaxies.   I. Title.
QB857.B47   2014
523.1´12–dc23   2013030425

ISBN 978-1-107-00054-4 Hardback

Additional resources for this publication at www.cambridge.org/9781107000544

Cambridge University Press has no responsibility for the persistence or
accuracy of URLs for external or third-party internet websites referred to in
this publication, and does not guarantee that any content on such websites is,
or will remain, accurate or appropriate.

Contents

Preface to the Second Edition

Since the year of publication of the first edition of *Dynamics of Galaxies*, the field has developed significantly. This book keeps the structure of the first edition but includes much new material. In general terms, it contains an up-to-date view of the basic phenomenology (in the eyes of a theorist), a new discussion of the properties of dark halos in galaxies in the light of recent results and projects, an extended description of the dynamics of quasi-relaxed stellar systems in connection with the dynamics of globular clusters, an introduction to gravitational lensing, and a primer on the properties of self-gravitating accretion disks. Motivation to undertake the writing of the new edition has come not only from personal research in the above-mentioned topics but also especially from the teaching experience in graduate and undergraduate courses (in Milan and Pisa); therefore, in this new edition the set of problems has increased considerably, especially in relation to Parts I and II.

The details of additions that characterize this edition are the following. Two new chapters have been placed at the end of the book, in Part V. Chapter 26, "Galaxies and Gravitational Lensing," provides the basic elements of the theory of gravitational lensing and a summary of some interesting astrophysical applications; an incentive to write such a chapter was given by the fact that in the last few years the combined use of stellar dynamics and gravitational lensing has proved to be an excellent diagnostic of the mass distribution in distant self-gravitating systems. Chapter 27, "Self-Gravitating Accretion Disks," describes gravitational collective effects applicable to a variety of contexts, ranging from protostellar nebulae, the sites of star and planet formation, to the outskirts of accretion disks around compact objects, such as active galactic nuclei; these collective effects are primarily related to density waves, and thus the chapter can be seen as an extension of the dynamics of galaxy disks, described in Part III, to the dynamics of disks on a much smaller scale. Furthermore, new material has often been incorporated in the form of new sections or subsections in already existing chapters: a more extended description of the global scaling laws (Chapter 4); dynamical friction in galaxies and relaxation processes in globular clusters (Chapter 7); the resolution of the Oort density discrepancy in the solar neighborhood, fully self-consistent models of disk-halo systems, and the problem of extraplanar gas (Chapter 14); a synthetic description of alternative scenarios for the interpretation of the large-scale spiral structure observed in galaxies (Chapter 17); a brief discussion of studies of spiral structure by means of numerical simulations and current studies of the large-scale shock scenario as a signature of quasi-stationary spiral structure, new observations of spiral structure in galaxies, and the issue of large-amplitude coherent patterns in the outer gaseous disk (Chapter 18); an improved description of the decomposition of rotation curves for spiral galaxies and some issues about the properties of dark halos inspired by cosmological arguments (Chapter 20); the construction of triaxial, tidally distorted and axisymmetric, rigidly or differentially rotating self-consistent models of ellipsoidal stellar systems (Chapter 22); a modern view of the problem of dark matter in elliptical galaxies and clusters of galaxies, also in relation to systems at distances of cosmological interest (Chapter 24); and the gravothermal catastrophe, some interesting properties of physically justified families of self-consistent models of elliptical galaxies, and a brief discussion of the homology of elliptical galaxies in view of possible scenarios for their formation and evolution (Chapter 25).

The main guideline followed in the writing of this book is the same as in the first edition, that is, to focus on theoretical issues inspired by the observations and to focus on theoretical models and results for which observational tests can be performed, at least in principle, if not with currently available telescopes and instruments.

Preface to the First Edition, 2000

This book is addressed to a readership of beginning graduate students and up. In contrast with other monographs on related subjects, it emphasizes some physical and methodological aspects. Two distinctive features are the emphasis on the similarities between the dynamics of stellar systems and the collective behavior of electromagnetic plasmas and the focus on a few important dynamical issues that are raised by observations. In the adopted semiempirical approach, the study of dynamics thus develops under the guidance and inspiration of empirical facts. Therefore, the book, although rich in quantitative analysis, reduces the mathematical discussion to its essentials. The level of presentation is not excessively technical. There is no intention to give a full account of all aspects (many of which are of fundamental importance) of interest in this broad research area; the reader is referred to the many authoritative books already available. Instead, this book tries to capture a lively synthesis to arouse curiosity in a reader who is not already professionally involved in this area of celestial physics.

From the methodological point of view, the book focuses on the general use of asymptotic methods, and it gives in detail the mathematical structure of some derivations when they may be useful for more general purposes. Asymptotic methods are flexible tools to obtain approximate solutions that give priority to the richness of ingredients in a physical problem; technically speaking, these methods recognize the importance of singular perturbations in the realization of physical processes. These aspects are often overlooked in toy models with exact solutions. The mathematical description developed here is not meant to provide a set of "recipes" for astronomical applications; rather, it serves to develop the reader's physical intuition and understanding of the processes involved.

On the one hand, the book is a synthesis of two graduate courses (one given at the Massachusetts Institute of Technology in the fall term of 1985 and another given regularly over the past fifteen years at the Scuola Normale, Pisa) addressed to students interested in undertaking research in a currently advanced and hot area of astrophysics or else to other students (sometimes already active in other research areas, such as plasma physics, geophysics, or applied mathematics) who wish to acquire an overall view of an interesting research field. In these courses it was assumed that students already had a reasonable background in classical physics, but otherwise, there were no special prerequisites. On the other hand, this book contains material expanded from recent extended review articles written by the author, so to astronomers it is also aimed at offering a comprehensive description of current research in galactic dynamics from a physical viewpoint that stresses general concepts, methods, the latest progress, and current unanswered questions.

Part III is devoted to the dynamics of spiral galaxies. The problem of spiral structure in galaxies has been treated recently in a separate book, *Spiral Structure in Galaxies: A Density Wave Theory*, written by the author in collaboration with C. C. Lin (MIT Press, 1996), in which the focus is on the general physical issues raised by observations of large-scale regular spiral structures; in that book, care was taken to keep the mathematical discussion of dynamical mechanisms to a minimum. In this respect, the analysis presented in Part III complements that already published; in addition, other dynamical issues, different from the development of spiral structure, are also addressed here.

The material contained in this book may serve as a basis for a basic course on normal galaxies. Roughly speaking, Chapters 1 through 9 might each correspond to approximately two hours of class; each of Chapters 10 through 25 may well require four hours of class. Thus such a course would actually exceed the normal length of a graduate course. However, various units are sufficiently independent of each other that the material of secondary interest to a given audience can be easily dropped.

The two initial parts define the astrophysical problems and the basic issues involved in the modeling process. Although the dynamical tools are introduced and developed, it is stressed that a crucial stage is reached when the dynamicist decides which equations and which parameter regimes may give a fair representation of the physical ingredients relevant to a given phenomenon being researched. In this sense, the concept of modeling goes beyond the more common definition of a simplified tool to frame the observations because it requires an evaluation of the physics involved in a given complex macroscopic phenomenon. Therefore, Part II, although covering general methodological aspects, is not an abstract discussion of various techniques but is aimed directly, especially by means of concrete examples and cross-references, at the points addressed in Parts III and IV. Part V briefly touches on a number of interesting points and questions that lie beyond the main scope of the book.

The set of problems provided in this book is primarily meant to offer the opportunity for additional discussion of a few selected points that are judged to be interesting. In many cases, the resolution is very simple, and the problem is an excuse to emphasize the scales associated with some important quantities. In other cases, the resolution may require a nontrivial analytical discussion or a simple numerical investigation. Sometimes the problem offers a way to focus on some derivation aspects that are thought to be instructive. In only a few cases are the problems plain exercises that involve a routine application of formulas provided in the main text.

A common practice, especially by the authors of astronomical literature, is to provide references with the purpose of establishing credit. Deciphering who has been the real initiator of a concept, a method, or a significant result is a very difficult task for which there is sometimes no clear solution, even when one looks at research in which one has actively participated. The many references provided in this book should not be interpreted as an attempt at a thorough investigation of all the relevant papers related to the various topics covered; thus the history of the results shown is not discussed. Many important articles probably have been missed, and others may not be properly emphasized. The references here are meant to provide the reader with a rather rich framework of research papers within which the issues that are discussed only briefly in this book are discussed in much greater detail than is possible here. They generally reflect personal experience. In this sense, it should be clear at the outset that the bibliography may be rather incomplete.

Acknowledgments

It would be impossible to thank properly all the scientists who have contributed to the ideas and results presented here. Special thanks are due to, among many others, A. Agnello, T. S. van Albada, R. J. Allen, N. C. Amorisco, S. E. Arena, C. V. Barbieri, R. H. Berman, F. Bertola, P. Bianchini, D. L. Block, S. V. Bulanov, S. Casertano, A. Cava, P.-H. Chavanis, D. F. Chernoff, L. Ciotti, M. Cipollina, B. Coppi, M. Crézé, C. De Boni, M. Del Principe, F. De Micheli, F. Fraternali, A. M. Fridman, K. Ganda, G. Gilmore, R. Giovanelli, C. Grillo, P. Grosbøl, J. Haass, D. C. Heggie, I. R. King, L. V. E. Koopmans, Y. Y. Lau, F. Leeuwin, C. C. Lin, T. V. Liseikina, G. Lodato, M. Lombardi, P. Londrillo, R. V. E. Lovelace, S. A. Lowe, D. Lynden-Bell, J. W.-K. Mark, S. Migliuolo, C. Nipoti, J. P. Ostriker, R. F. Pannatoni, M. Pasquato, F. Pegoraro, S. Pellegrini, E. Pignatelli, R. Pozzoli, L. A. Radicati, M. S. Roberts, W. W. Roberts, M. Romaniello, M. Romé, A. B. Romeo, F. Rubini, D. D. Ryutov, R. P. Saglia, R. Sancisi, C. L. Sarazin, F. H. Shu, A. Sonnenfeld, M. C. Sormani, M. Stiavelli, L. Sugiyama, M. Tazzari, R. P. Thurstans, T. Toniazzo, M. Trenti, T. Treu, A. L. Varri, M. A. W. Verheijen, E. Vesperini, K. B. Westfall, L. Woltjer, and A. Zocchi.

Part I

Basic Phenomenology

1 Scales

A piece of writing that best captures the beauty and mystery surrounding the dynamics of galaxies can be found in chapter 7 of the first volume of *The Feynman Lectures on Physics*.[1] There, in a few simple sentences and by means of a picture of a globular cluster and a spiral galaxy, we are rapidly brought to the scales and structure of the systems involved and to the underlying limits that our physical knowledge must recognize. We are told that gravitation is the prime actor, stars rarely collide, and angular momentum leads to a contraction in a plane, but we are also reminded of the facts that "of course we cannot prove that the law here is precisely inverse square," "what determines the shape of these galaxies has not been worked out," and that we are dealing with systems that are enormously complex.

Approximately four decades earlier, through a decisive distance measurement, some nebulae had been recognized as huge systems millions of light years away; those were nearby galaxies, and galaxies were thus found to be the visible building blocks of the Universe. If we examine that discovery,[2] we see that modern cosmology and quantum mechanics were developed at approximately the same time and well after the formulation of general relativity;[3] they were also developed well after the basic equations that govern the motion of a self-gravitating stellar system were studied by Jeans.[4]

Thus many of the objects that had been studied and cataloged for more than a century, through the patient work of Messier, the Herschels, and Dreyer, turned out to be galaxies.[5] Now, more than eight decades after their discovery, great progress has been made, and galaxies have been observed out to distances that are measured in billions of light years, almost at the end of the Universe (or, rather, at its beginning; Fig. 1.1).[6]

Typical masses of galaxies are in the range 10^9 to $10^{12} M_\odot$, where one solar mass is $1 M_\odot \approx 2 \times 10^{33}$ g. Their linear scales are in the range 1 to 100 kpc (kpc is a kiloparsec), where 1 kpc $\approx 3.1 \times 10^{21}$ cm. Velocities inside a galaxy are in the range 50 to 500 km s^{-1}, so typical dynamical time scales are 10^8 to 10^9 yr (note that 1 km s$^{-1} \times 10^9$ yr ≈ 1.02 kpc). We recall that the Hubble time, that is, the age of the Universe, is $\approx 10^{10}$ yr. The Milky Way galaxy has a mass estimated to be close to $2 \times 10^{11} M_\odot$. The Sun is located at ≈ 8 kpc from the galactic center, and the rotation velocity around the galactic center in the solar neighborhood is ≈ 220 km s^{-1}. The mass density in the solar neighborhood is ≈ 6 protons per cubic centimeter, whereas the local density of the Galaxy disk is $\approx 50 M_\odot$ pc^{-2}, which is close to that of an ordinary sheet of paper. The local energy densities (e.g., turbulent, magnetic, etc.) are close to 1 eV cm^{-3}, except for the overall rotation-energy density, which is close to 1 keV cm^{-3}; we recall that 1 eV $\approx 1.6 \times 10^{-12}$ erg.

Fig. 1.1. The Hubble Ultra Deep Field [credit: NASA, ESA, S. Beckwith (STScI), and the HUDF Team; see also Beckwith, S. V. W., Stiavelli, M., et al. 2006. *Astron. J.*, **132**, 1729]. This image was taken by observation with the *Hubble Space Telescope* in a 1-million-second exposure and contains an estimated 10,000 galaxies in a region one-tenth the diameter of the full Moon, with objects detected down to the 30th magnitude. A large fraction of these galaxies are at a redshift larger than unity, some at $z > 7$, as confirmed by spectroscopic observations with large telescopes from the ground; thus some galaxies existed earlier than 1 billion years after the Big Bang.

All these should be taken as scales for objects called *galaxies*. In reality, gravitation acts and isolates clumps of matter on all scales. Thus we can recognize well-defined subsystems inside galaxies; the largest of these subsystems are globular clusters and giant molecular clouds (with masses in the range 10^4 to $10^6 M_\odot$) down to open clusters and multiple stellar systems. In our Galaxy there are more than 150 globular clusters with a spheroidal space distribution. Elliptical galaxies may host thousands of globular clusters.[7]

On larger scales, galaxies are often found in groups or in clusters of galaxies. The Local Group contains more than thirty galaxies (M31, i.e., the Andromeda galaxy,[8] and the Milky Way galaxy[9] are the largest) spread out on the scale of a few million light years. The Large

Fig. 1.2. Gravitational lensing by an intervening cluster of galaxies [credit: NASA, ESA, Richard Ellis (Caltech), and Jean-Paul Kneib (Observatoire Midi-Pyrenees, France)]. The mass of Abell 2218 distorts the images of distant galaxies into arcs and arclets.

Magellanic Cloud is at a distance of ≈ 50 kpc, and M31 is at a distance of ≈ 800 kpc. The Local Group lies at the outskirts of the Virgo cluster, which contains more than 2,000 galaxies; the center of the Virgo cluster is usually associated with the giant elliptical galaxy M87, ≈ 15 Mpc away from us.[10] Clusters of galaxies thus contain large amounts of matter, which often acts as a gravitational lens for the light coming from more distant galaxies[11] (Fig. 1.2); a large fraction of their visible mass is in the form of a hot intergalactic medium that is confined in the potential well of the cluster and observed by X-ray telescopes.[12] Clusters are thus a very active environment in which galaxy-galaxy and galaxy-intracluster medium interactions participate in their long-term evolution. A striking case is that of a distant galaxy cluster that turned out to be the brightest X-ray cluster known (with $L_X \approx 2 \times 10^{46}$ erg s^{-1}) at the time of its discovery,[13] for which the energy loss that is due to X-rays was thought to induce an inflow of intergalactic matter toward its central regions at a rate estimated to exceed $3 \times 10^3 \, M_\odot$ yr^{-1}.

The visible mass in galaxies is primarily in the form of stars. Gas, even in the case of the so-called late-type spirals that have the largest gas content, does not contribute more than a few percent of the total mass. However, because the various forms of matter have different spatial distributions, it should be kept in mind that, at some locations, the gas content can be significant. For example, in the outer disks of spiral galaxies, the gas-to-star surface density ratio can easily exceed 15 percent. In the case of elliptical galaxies, traditionally described as gas-free systems, there are examples, such as NGC 4636, in which almost $10^{11} \, M_\odot$ (i.e., ≈ 10 percent of the estimated stellar mass) is in the form of hot, X-ray-emitting plasma; this hot corona occupies a volume larger than that of the luminous component. This point about the forms of mass distribution is further complicated by the fact that we have evidence for the presence of dark

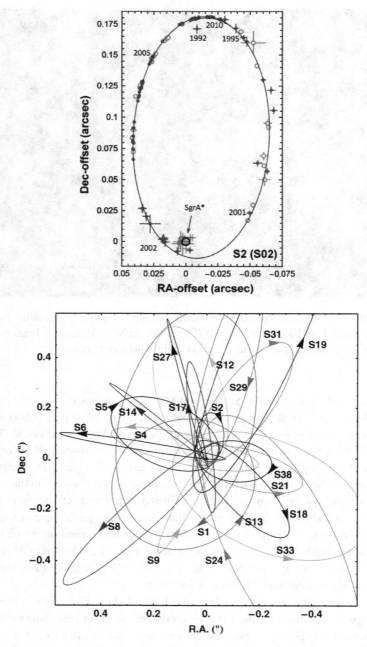

Fig. 1.3. *Top*: The elliptical orbit of the star S2 in the vicinity of Sgr A⋆ at the Galactic Center (Genzel, R., Eisenhauer, F., Gillessen, S. "The Galactic Center massive black hole and nuclear star cluster," 2010. *Rev. Mod. Phys.*, **82**, 3121; © 2010 by the American Physical Society). The positions in the sky were obtained in the period 1992–2010 by combining data from NTT/VLT with others from *Keck*. The pericenter of S2 is ≈ 125 AU (astronomical units), that is, ≈ 17 light hours from a compact object with mass ≈ 4 × 10⁶ $M_\odot$. *Bottom*: A set of twenty star orbits in the vicinity of Sgr A⋆ at the Galactic Center (Gillessen, S., Eisenhauer, F., et al. 2009. "Monitoring stellar orbits around the massive black hole in the Galactic Center," *Astrophys. J.*, **692**, 1075; reproduced by permission of the AAS).

matter, which may be in a form that is neither stellar nor gaseous. Conservative estimates suggest that typically the amount of dark matter in galaxies is close to the amount of visible matter, if referred to the volume in which the visible matter is observed. On bigger volumes, the relative amount of dark matter may be significantly larger. The appropriate numbers for these distributions are very important for proper modeling, especially because the stellar component and, most likely, the dark-matter component are collisionless and dissipationless, whereas the gas is a dissipative subsystem. The cold gas in galaxy disks is dissipative because of inelastic cloud-cloud collisions that occur on a very short time scale, whereas the hot gas in bright elliptical galaxies dissipates by means of radiative cooling.

A systematic study of galactic nuclei and cores has led to dramatic evidence for the presence of supermassive black holes even in inactive galaxies,[14] with masses in the range 10^6 to $10^9 M_\odot$. In fact, a beautiful study of stellar orbits[15] has shown that our Galaxy hosts a central black hole with a mass of $\approx 4 \times 10^6 M_\odot$ (Fig. 1.3). To what extent the dynamics of the central black holes is related to the large-scale dynamics of galaxies has yet to be clarified.[16]

From the dynamical point of view, a simple distinction can be made between rotation-supported (or cool) systems and pressure-supported (or hot) systems. Clearly, disks are the coldest systems (typical random-to-ordered kinetic energy ratios for stars or gas clouds of the thin disk in the solar vicinity are below 1 percent). However, even bulges and many low-luminosity spheroidal galaxies are to be considered cool to the extent that they have significant amounts of energy stored in the form of mean motions. In contrast, many bright ellipticals appear to be fully pressure-dominated.

When we talk about galaxies, in terms of either observations or dynamical theories, we naturally tend to refer to bright objects. Still we should keep in mind that there are many dim galaxies and that there are many unresolved problems associated with them. Much as for the luminosity distribution of stars,[17] the faint end of the luminosity distribution of galaxies is hard to determine empirically and may hide large numbers of objects. Some studies have tried to assess this distribution inside nearby clusters.[18] A commonly used distribution[19] is $f(L) \sim (L/L_\star)^\alpha \exp[-(L/L_\star)]$. The characteristic luminosity is found to correspond to an absolute magnitude $M_B^\star \approx -21$, which may reach approximately -22 if the cD galaxies are counted. The characteristic exponent is found to be $\alpha \approx -1$ for galaxies in the field and $\alpha \approx -1.25$ for cluster galaxies.[20] If a universal luminosity function indeed could be shown to underlie these data, this would be of high theoretical interest in relation to the processes of galaxy formation.[21] These considerations become especially important when we consider the distribution of galaxies at large distances and how this distribution may be affected by evolutionary processes (among which are galaxy-galaxy encounters and merging).

In this and the following three chapters the stage is set, within the proper empirical context, for a number of important dynamical issues that will be introduced starting with Chapter 5. There is clearly no intention here to provide an exhaustive discussion of the many studies that define galactic and extragalactic astronomy. For a thorough introduction to the subject, the reader is referred to monographs and review articles.[22]

Notes

1. Feynman, R. P., Leighton, R. B., Sands, M. 1963. *The Feynman Lectures on Physics*, Addison-Wesley, Reading, MA.

2. Hubble, E. 1925. *Astrophys. J.*, **62**, 409. A thorough account of the discovery made by Hubble and of the debate on the nature of the nebulae is given by A. Sandage (1961) in the introduction to his *Hubble Atlas of Galaxies*, Publ. 618, Carnegie Institution of Washington, Washington, DC.

3. We recall that the matrix mechanics and wave mechanics that are generally associated with the foundation of quantum mechanics start with Heisenberg, W. 1925. *Z. Phys.*, **33**, 879; and Schrödinger, E. 1925. *Ann. Phys.*, **79**, 361.

4. Jeans, J. H. 1915. *Mon. Not. Roy. Astron. Soc.*, **76**, 70.

5. One major catalog of galaxies is the RC3 catalog: de Vaucouleurs, G., de Vaucouleurs, A., et al. 1991. *Third Reference Catalogue of Bright Galaxies*, Springer-Verlag, New York.

6. Relatively normal galaxies have been found at redshifts larger than unity. The Hubble Deep Fields, obtained from the *Space Telescope* by observing nearly continuously for periods of ≈ 10 days in one dark spot in the constellation of Ursa Major and, later, in a region of Tucana, are populated with thousands galaxies, many of them at redshifts larger than unity; Williams, R. E., Blacker, B., et al. 1996. *Astron. J.*, **112**, 1335; Williams, R. E., Baum, S., et al. 2000. *Astron. J.*, **120**, 2735. The following Hubble Ultra Deep Field, obtained from the *Space Telescope* with a 1-million-second-long exposure in the direction of Fornax, taken at the end of 2003 and the beginning of 2004, is so far the deepest visible-light view of the cosmos, with nearly 10,000 galaxies imaged; Beckwith, S. V. W., Stiavelli, M., et al. 2006. *Astron. J.*, **132**, 1729.

7. Ashman, K., Zepf, S. E. 1998. *Globular Cluster Systems*, Cambridge University Press, Cambridge, UK.

8. Hodge, P. 1992. *The Andromeda Galaxy*, Kluwer, Dordrecht, The Netherlands.

9. Bok, B. J., Bok, P. F. 1974. *The Milky Way*, 4th ed., Harvard University Press, Cambridge, MA.

10. Setting the exact value of the distance to M87 and the Virgo cluster is important in order to calibrate the distance scale in cosmology; lower values of the distance are generally favored by studies that suggest a relatively high value for the Hubble constant. See, e.g., Tammann, G. A., Sandage, A., Reindl, B. 2008. *Astron. Astrophys. Rev.*, **15**, 289; Freedman, W. L., Madore, B. F. 2010. *Annu. Rev. Astron. Astrophys.*, **48**, 673.

11. See Schneider, P., Ehlers, J., Falco, E. E. 1992. *Gravitational Lenses*, Springer-Verlag, Heidelberg.

12. Rosati, P., Borgani, S., Norman, C. 2002. *Annu. Rev. Astron. Astrophys.*, **40**, 539.

13. Schindler, S., Hattori, M., et al. 1997. *Astron. Astrophys.*, **317**, 646; the cluster is at redshift $z = 0.45$.

14. For the initial stages of these studies, see Crane, P., Stiavelli, M., et al. 1993. *Astron. J.*, **106**, 1371; Stiavelli, M., Møller, P., Zeilinger, W. W. 1993. *Astron. Astrophys.*, **277**, 421; Harms, R. J., Ford, H. C., et al. 1994. *Astrophys. J. Lett.*, **435**, L35; Miyoshi, M., Moran, J., et al. 1995. *Nature*, **373**, 127; Eckart, A., Genzel, R. 1997. *Mon. Not. Roy. Astron. Soc.*, **284**, 576; Marconi, A., Axon, D. J., et al. 1997. *Mon. Not. Roy. Astron. Soc.*, **289**, L21; van der Marel, R. P., de Zeeuw, P. T., et al. 1997. *Nature*, **385**, 610; van der Marel, R. P., de Zeeuw, P. T., Rix, H.-W. 1997. *Astrophys. J.*, **488**, 119; Kormendy, J., Bender, R., et al. 1997. *Astrophys. J. Lett.*, **482**, L139.

15. Schödel, R., Ott, T., et al. 2002. *Nature*, **419**, 694; Gillessen, S., Eisenhauer, F., et al. 2009. *Astrophys. J.*, **692**, 1075; Genzel, R., Eisenhauer, F., Gillessen, S. 2010. *Rev. Mod. Phys.*, **82**, 3121.

16. See Ciotti, L. 2009. *La Rivista del Nuovo Cimento*, **32**, 1, and references therein.

17. But see Gould, A., Bahcall, J. N., Flynn, C. 1997. *Astrophys. J.*, **482**, 913.

18. For the Virgo cluster, Sandage, A., Binggeli, B., Tammann, G. A. 1985. *Astron. J.*, **90**, 1759, go down to the faint end of $M_B \approx -14$ (total absolute luminosity in the B band). For the Ursa Major cluster, see Verheijen, M. A. W. 1997. Ph.D. thesis, University of Groningen, The Netherlands.

19. Schechter, P. 1976. *Astrophys. J.*, **203**, 297.

20. See Binggeli, B., Sandage, A., Tammann, G. A. 1988. *Annu. Rev. Astron. Astrophys.*, **26**, 509; see also Dressler, A. 1978. *Astrophys. J.*, **223**, 765; Lugger, P. 1986. *Astrophys. J.*, **303**, 535.

21. Zwicky, F. 1942. *Phys. Rev.*, **61**, 489; 1957. *Morphological Astronomy*, Springer-Verlag, Berlin; Press, W. H., Schechter, P. 1974. *Astrophys. J.*, **187**, 425.

22. See Mihalas, D., Binney, J. J. 1981. *Galactic Astronomy*, Freeman, San Francisco; Binney, J., Merrifield, M. 1998. *Galactic Astronomy*, Princeton University Press, Princeton, NJ. For a review of the general properties of normal galaxies along the Hubble sequence, see Roberts, M. S., Haynes, M. P. 1994. *Annu. Rev. Astron. Astrophys.*, **32**, 115.

2 Observational Windows

When we make observations, we rely on a number of spectral windows for which we have instruments to study the radiation that reaches the Earth. Obviously, astronomy is centered around optical radiation. After a long period of work with photographic techniques, most of the current optical studies are based on charge-coupled-device (CCD) detectors, which are solid-state devices based on two-dimensional arrays of detection elements with a wide range of capabilities.[1] The visible light extends from the ultraviolet (UV) region (UV radiation is associated with wavelengths[2] in the range 100 to 4,000 Å) to the near-infrared (near-IR) region (IR radiation has wavelengths between 7,500 Å and the millimeter range). Observations from the ground are limited by the way the light is transmitted by the atmosphere. For example, much of the near-IR radiation is absorbed by the atmosphere, and observations beyond 1 μm are best performed at high altitude (where significant transmission occurs in correspondence with the standard IR filters J, H, and K) or directly from space.

New observational windows have been opened, especially in the second part of past century, and these have dramatically changed our view of the Universe (Fig. 2.1). A major step forward has been identification of the 21-cm line (at a frequency of 1,420 MHz; a hyperfine transition in atomic hydrogen associated with the spin-flip in the electron-proton pair[3]), which has given an enormous boost to radioastronomy.[4] Studies of the kinematics of atomic hydrogen, especially those of the 1970s and the early 1980s, have provided the decisive evidence for the existence of dark matter. Radio studies have marked significant breakthroughs in extragalactic astronomy. They have also been the starting point for the discovery of multiple images by gravitational lensing[5] and have led to what is probably the best evidence to date for the existence of massive black holes at the centers of external galaxies.[6] In different contexts, the radio window has led to major discoveries, in particular, the cosmic microwave background radiation[7] and pulsars.[8]

The possibility of using telescopes from space opened the way for observational windows that have no counterpart from the ground. The X-ray Universe has thus unfolded, starting with observations from balloons and rockets in the 1960s,[9] followed by many missions, with a major impact made by the *Uhuru* satellite[10] and the *Einstein* observatory, launched in the late 1970s,[11] and later by ROSAT (*Röntgen Satellit*) in the 1990s. This led to the discovery of hot, diffuse intergalactic (intracluster) matter and to the identification of some important iron lines[12] and of the electron-positron annihilation line at 0.511 MeV. The X-ray window turned out to be very important for our understanding of normal galaxies;[13] in the past decade, great progress has been made based on X-ray telescopes of a new generation, in particular, XMM (*X-Ray Multi-Mirror Mission*) and *Chandra*. At the same time, at lower energies, the domain of UV light was explored.[14]

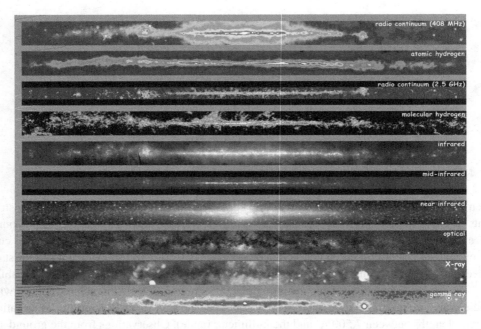

Fig. 2.1. Multiwavelength Milky Way (credit: Astrophysics Data Facility and Astronomical Data Center of the Goddard Space Flight Center, NASA). The various strips represent how the Milky Way looks within various wave bands. *From top to bottom*: Radio continuum (408 MHz), atomic hydrogen, radio continuum (2.5 GHz), molecular hydrogen (inferred from CO), IR (12–100 μm), mid-IR (6.8–10.8 μm) , near IR (1.25–3.5 μm), optical, X-ray (0.25–1.25 keV), gamma ray.

At higher energies, the *Compton Gamma Ray Observatory* had a big impact. In 1997, results from the SAX (*Satellite per Astronomia X*) satellite allowed us to pinpoint the position of the long mysterious gamma-ray bursts and thus identify them optically.[15] The study of the gamma-ray sky is now carried out by INTEGRAL (*INTErnational Gamma-Ray Astrophysics Laboratory*), *Swift* (a multiwavelength observatory dedicated to the study of gamma-ray bursts), AGILE (*Astrorivelatore Gamma a Immagini LEggero*), and *Fermi*.

IR astronomy from space [with IRAS (*Infrared Astronomical Satellite*), COBE (*Cosmic Background Explorer*), ISO (*Infrared Space Observatory*), WMAP (*Wilkinson Microwave Anisotropy Probe*), and *Spitzer*] has led to great advances in not only the cosmological context but also in new perceptions of normal galaxies;[16] the most exciting results are now coming from the data being acquired by *Herschel* and *Planck*, launched with the same rocket in 2009.

Even for the observational windows fully open from the ground, astronomy from space has led to an extraordinary set of new results. Among the many accomplishments of HST (*Hubble Space Telescope*), we should mention the discovery of Cepheid variables in the Virgo cluster of galaxies, the finding of a large number of gravitational lenses, the identification of the most distant galaxies (up to $z \approx 5$ and beyond), and the finding that many quasars are associated with relatively normal galaxies.[17] HST and the *Hipparcos* satellite have led to a much sharper determination of the relevant distance scales. Work based on HST observations of distant supernovae has led to the discovery[18] that the Universe is accelerating, leading to the current concordance model characterized by low mass density ($\Omega_m < 1$) and nonvanishing gravitational constant ($\Lambda \neq 0$).

In the meantime, new large telescopes from the ground [the *Keck* telescopes and others, such as the VLT (*Very Large Telescope*) of the European Southern Observatory] allow us to study the spectroscopic properties of the faintest sources. The use of active and adaptive optics now overcomes some of the limits imposed by atmospheric turbulence on the images (the related seeing spreads the image of a point source well beyond the nominal diffraction limit).

The revolution in extragalactic astronomy that we are witnessing is going to be dwarfed soon by major new initiatives that are currently in progress. Among these, we should mention ALMA (*Atacama Large Millimeter/Sub-millimeter Array*), currently under completion on the Chajnantor plain of the Chilean Andes, 5,000 m above sea level; JWST (*James Webb Space Telescope*), new interferometric missions, such as GAIA (*Global Astrometric Interferometer for Astrophysics*) and SIM Lite (*Space Interferometry Mission*), conceived with the capability of a few microarcsecond resolution in the optical;[19] E-ELT (*European Extremely Large Telescope*), a 42-m fully adaptive optical-IR telescope; and SKA (*Square Kilometer Array*), which should give a greatly improved sensitivity, by two orders of magnitude, with respect to any existing radiotelescope, with the possibility of studying kinematics and morphology of gas-rich galaxies out to $z \approx 1$ and beyond.

The impact on studies of normal galaxies of each observational window develops through the two natural elements of imaging and spectroscopy. Imaging emphasizes morphological characteristics that result from the intrinsic structure of the observed object, whereas spectroscopy offers quantitative information on kinematics and chemical composition. In practice, what can be inferred about internal structure, dynamics, and intrinsic physical characteristics is model-dependent. Note that imaging in different wave bands can also provide some information that would be best derived spectroscopically, with the advantage that images can be gathered more easily than full spectra. For example, a first and generally accurate estimate of the redshift of distant galaxies in the Hubble Deep Field has been obtained from the images and later confirmed spectroscopically.[20]

In the following sections, a schematic and qualitative description is given of the main characteristics of the various observational windows in relation to the prevalent applications of the dynamics of normal galaxies. These should be taken as highlights only; a thorough review is clearly beyond the scopes of this chapter because it would require covering most of modern astrophysics.[21] Note that the fact that we are now observing galaxies at cosmological redshifts, even beyond $z = 1$, produces significant mismatches between the range of frequencies characteristic of the source and the frequencies at which they are detected by our telescopes through the available observational windows. The description here is given mostly in view of the case where such mismatch is not significant.

2.1 Radio

In the context of the large-scale structure of normal galaxies, two important sources of radio waves[22] are abundant atomic hydrogen, with its 21-cm line emission, and high-energy electrons that provide synchrotron continuum emission based on the interstellar magnetic fields. Part of the continuum emission is thermal (free-free) and is largely contributed to by the HII regions (regions of warm ionized gas). In addition, we should also keep in mind the emission from the nuclear region, which is quite common among ellipticals and S0 galaxies.

Atomic hydrogen is very important in galactic dynamics.[23] It is an excellent kinematic tracer of the underlying gravitational field, as will be better described in Chapter 4. Physically, its presence in spiral galaxies makes the disk cooler from the dynamical point of view, thus favoring the conditions for Jeans instability. Atomic hydrogen tends to have a diffuse distribution, thus effectively increasing the length scale of the mass distribution of the disk (with respect to that determined by the stars alone). The spatial distribution of atomic hydrogen is clumpy,[24] being associated with clouds with sizes up to 100 pc and characterized by random velocities of a few kilometers per second with respect to streaming motions. Cloud-cloud inelastic collisions are rather frequent and make the cold-gas medium a dissipative component. These phenomenological aspects will play an important role in the discussion of Chapter 16. Atomic hydrogen is present in only very small amounts in elliptical galaxies. In many ellipticals, it is not even detected.

Continuum emission has been used to diagnose the overall structure of the interstellar magnetic fields and has shown interesting connections with the large-scale spiral structure, for example,[25] in M51. The amount of mass involved in the source particles is very small, and thus this component is not expected to play an active role in the overall dynamics of the galaxy. In some cases, spectacular phenomena such as radio jets do occur on the largest scales, but these appear to be phenomena where self-gravity plays a secondary role. Even for these, gravity may be the ultimate source of energy (at the center of active galactic nuclei). There are very interesting problems to be studied in this general area,[26] but these are not covered in the treatment of the dynamics of normal galaxies, as provided in Parts III and IV.

2.2   Millimetric

The interstellar medium in spiral galaxies has been found, starting with the work of the 1960s, to be rich in molecules.[27] Dozens of molecules have been identified, and in many cases this identification is by means of millimetric radiation. Probably the most important source of this type of radiation[28] is that of CO (in particular, the 2.6-mm emission line of ^{12}CO), which is generally used as a tracer of a much more abundant molecule, that of H_2 (H_2 has no radiofrequency spectrum; it has been detected directly by UV absorption in diffuse clouds).[29] The study of the radiation processes involved and the physical characteristics of the interstellar medium[30] makes it likely that significant errors can be introduced when the easily observed CO emission is converted, as is often done, into H_2 content by means of a simple proportionality factor.[31] In some cases there is more excitation in CO that is due to interstellar light, and the H_2 column densities involved need not be large. However, it may be that large amounts of cold H_2 are currently escaping our detection. A clarification of these points would have a major impact on the dynamics of spiral galaxies because this dynamically cold material adds up to what is already known to be present in significant amounts in the form of atomic hydrogen.

2.3   Infrared

The near-IR radiation (in particular, the $\approx$ 2-μm radiation of the K and K' bands) is dominated by the light of evolved red giant stars and thus traces well the "underlying massive stellar disk" of spiral galaxies[32] (Fig. 2.2). Other optical radiation is absorbed and (especially) scattered more efficiently by interstellar dust.[33] The extinction due to interstellar dust can be enormous. For

Fig. 2.2. NGC 1300, imaged in IR light with the HAWK-I camera on VLT at Paranal Observatory in Chile. The field of view of the image is ≈ 6.4 arcminutes (credit: ESO/P. Grosbøl).

example, in the direction of the Galactic Center, the estimated total visual extinction is up to 30 magnitudes, whereas in the near IR it is much smaller. Some of the observed near-IR radiation may come from red supergiants, if present, and thus may be associated with star-formation events characteristic of the Population I disk. In practice, the overall near-IR emission appears to be generally only little contaminated by this effect.

At longer wavelengths, in the far-IR region, there is thermal emission radiated by the dust grains themselves or by processes in the central regions of active galactic nuclei. Bright far-IR (40 to 500 μm) sources may be associated with exceptionally high star-formation events, interacting galaxies, and active galactic nuclei.[34]

2.4 Optical

The optical radiation that we observe is dominated by stellar light. This is produced by the combination of many stars because these stars are usually not resolved separately inside external galaxies. In the simplest case, for the luminosity produced by the many stars contained in a portion of a galaxy, one may imagine that the light observed derives from a distribution of stars of approximately the same age that have evolved from an initial mass function[35] plus additional contributions from more recent star-formation events. Young stars, such as OB types, when present, tend to dominate the optical picture because they are exceptionally bright. Thus optical pictures often may leave the incorrect impression of high-density contrasts (e.g., in spiral arms) that do not correspond to the actual density distributions.

Colors and spectra are indicative of the stellar population involved. However, there is a kind of degeneracy that makes it hard to disentangle whether given spectral characteristics are due to the age or to the chemical composition (often referred to in terms of metallicity) of the under-lying stellar population.[36] A clarification of these issues is important from the dynamical point of view. On the one hand, we may be led to identify interesting scenarios for galaxy formation and for star-formation history. On the other hand, finding which stellar population is involved would actually determine what the relevant mass-to-light ratio to be used is. In practice, given the difficulty in ascertaining the number of low-luminosity stars present, the mass-to-light ratio is best constrained by dynamics, and this information is then used to select the relevant *population synthesis models*.[37] Various theoretical models of chemical evolution of galaxies have been developed[38] with the exciting possibility of interpreting observed properties of galaxies at cosmological distances.[39]

Some of the light also shines from regions of ionized gas, the HII regions, found frequently in spiral galaxies. Elliptical galaxies often show some light associated with warm ionized gas. The stellar light has spectra marked by absorption features produced in the stellar atmospheres, whereas warm gas is associated with emission lines, such as Hα (in the Balmer series of the hydrogen atom, the transition from the third level to the second level is at $\approx 6,563$ Å) and Hβ (fourth to second level, at $\approx 4,861$ Å), which are of considerable interest for spectroscopic observations.

2.5   Ultraviolet

UV radiation[40] is a good tracer of the hottest and youngest stars, so its emission in spirals[41] is taken to mark the present star-formation rate. For example, evolutionary synthesis models[42] indicate that the mean age of stars that contribute to the extreme ultraviolet (EUV) spectrum below 912 Å (the limit of the Lyman series of the hydrogen atom, i.e., the transitions to the ground state) is ≈ 2 Myr. One major spectroscopic feature in the UV spectrum is the Lyman α line of the hydrogen atom (the transition from the second to the first level) at $\approx 1,216$ Å; for a source at redshift $z > 2.5$, the line is brought into the visible part of the spectrum, and this has wide applications within the cosmological context.

In elliptical galaxies, a curious aspect found in recent years has been the realization that even the most normal objects show an unexpected UV excess below 2,000 Å, which has sparked the interest of experts in population synthesis models.[43]

One important feature of the UV part of the spectrum is that it is associated with a deep minimum for the UV-optical-IR frequency range in the night-sky background radiation (at 1,600 to 2,400 Å), making it an ideal window for low-surface-brightness observations. In contrast, dust extinction[44] is largest, with a characteristic signature at 2,175 Å.

Quasars and active galactic nuclei have a large UV flux. UV radiation is often used to discriminate between a nucleus made of stars and a gaseous accretion disk at the center of some galaxies.

Fig. 2.3. The Bullet Cluster (1E 0657-56) in Carina, as viewed by *Chandra* and *Magellan*. The picture is ≈ 6 arcminutes across. The cluster is at a distance of ≈ 3.8 billion light years (credit: X-ray: NASA/CXC/CfA/M. Markevitch et al.; Optical: NASA/STScI; Magellan/U.Arizona/ D. Clowe et al.).

2.6 X-Rays

In the context of galactic dynamics, X-rays are associated with two main types of sources:[45] the hot interstellar medium, which often continues into the intergalactic, intracluster gas, with temperatures of ≈ 1 keV, and discrete sources, such as X-ray binary stars, which are usually associated with a harder spectrum. Studies of nearby galaxies are able to assess the relative share between these sources. In X-ray bright elliptical galaxies, the emission appears to be dominated by the hot coronal gas. In spiral galaxies, some of the soft X-ray emission comes from hot shocked gas in the interstellar medium.[46] Note that X-rays are absorbed by cold interstellar hydrogen.

The emission of X-rays can be very powerful, and in some cases we can estimate that the time scale for radiation cooling is comparable with the relevant dynamical time scale. In these cases, the hydrostatic equilibrium condition that we may imagine cannot be maintained, and the hot gas is probably undergoing a massive cooling flow, with matter falling toward the center of the potential well, at a rate of a few solar masses per year. This is a small-scale version, at the galactic level, of the more powerful processes occurring on the scale of galaxy clusters.[47] The cooling-flow paradigm has been challenged by modern observations, which have led to the discovery of a number of interesting phenomena in the intracluster medium (Fig. 2.3).

The brightest X-ray ellipticals are generally cluster members, and the X-ray emission merges with that of the cluster. From the dynamical point of view, we should also consider the possibility of outflows of interstellar gas toward the cluster and, for the cases for which such an outflow does not occur, the possibility that the hot gas in a galaxy is partly confined by the intracluster medium itself.

2.7 Other Sources

Most of the gamma rays observed at energies above 100 MeV are produced by collisions of cosmic rays with hydrogen nuclei in gas clouds; in this respect, they may offer interesting constraints on the overall distribution of the elusive molecular hydrogen.[48] Other potential windows (such as neutrinos,[49] cosmic rays,[50] and gravitational waves,[51] when detected) so far have less to offer in relation to the dynamics of galaxies. In general, we should keep an open mind for any kind of data available because in astrophysics we often obtain important clues from apparently unrelated sources of information. After all, we are trying to decipher an extremely complex puzzle, and we should be ready to use any empirical suggestion that may be offered.

The empirical physical framework outlined in Part I is obtained by means of observations and by careful modeling, which is required for interpreting the observations. By themselves, observations contribute very little. But, of course, without observations, models and theories would be mere speculations, and some of them would never be born.

Notes

1. These devices were introduced in the mid-1970s; see Kitchin, C. R. 1984. *Astrophysical Techniques*, Hilger, Bristol, UK (5th ed. published in 2009 by CRC Press, Taylor & Francis Group, Boca Raton, FL); Mackay, C. D. 1986. *Annu. Rev. Astron. Astrophys.*, **24**, 255.
2. We recall that the wavelength of a 1 eV photon is $\approx 1.2398 \times 10^4 \text{Å} = 1.2398 \times 10^{-4} \text{ cm} = 1.2398 \, \mu\text{m}$.
3. van de Hulst, H. C. 1945. *Ned. Tijdschr. Natuurkd.*, **11**, 210.
4. Starting with the first detection of the 21-cm emission by Ewen, H. I., Purcell, E. M. 1951. *Nature (London)*, **168**, 356; Muller, C. A., Oort, J. H. 1951. *Nature (London)*, **168**, 357; Christiansen, W. N., Hindman, J. V. 1951. *Nature (London)*, **168**, 358. A first study of spiral structure in the atomic hydrogen disk of the Milky Way galaxy was made by van de Hulst, H. C., Muller, C. A., Oort, J. H. 1954. *Bull. Astron. Inst. Neth.*, **12**, 117, and a full map of the radio emission was produced by Oort, J. H., Kerr, F. J., Westerhout, G. 1958. *Mon. Not. Roy. Astron. Soc.*, **118**, 379. The analogous transition in atomic deuterium, occurring at 92 cm (at a frequency of 327 MHz), has been marginally detected by Chengalur, J. N., Braun, R., Burton, B. W. 1997. *Astron. Astrophys. Lett.*, **318**, L35; see also Blitz, L., Heiles, C. 1987. *Astrophys. J. Lett.*, **313**, L95; and the following study by Rogers, A. E. E., Dudevoir, K. A., Bania, T. M. 2007. *Astron. J.*, **133**, 1625.
5. Walsh, D., Carswell, R. F., Weymann, R. J. 1979. *Nature (London)*, **279**, 381, selected a source from a radio survey at 966 MHz, finding two optical counterparts with the same redshift $z \approx 1.4$; subsequently, the galaxy responsible for the lensing was detected at a redshift $z_d \approx 0.36$; see also Moran, J. M., Hewitt, J. N., Lo, K. Y. 1989. In Vol. 330 of the Springer-Verlag *Lecture Notes in Physics Series*, Springer-Verlag, Berlin.
6. Miyoshi, M., Moran, J., et al. 1985. *Nature (London)*, **373**, 127; their work is based on the water-vapor maser transition at $\lambda \approx 1.35$ cm.
7. Penzias, A. A., Wilson, R. W. 1965. *Astrophys. J.*, **142**, 419.
8. Hewish, A., Bell, S. J., et al. 1968. *Nature (London)*, **217**, 709.

9. The first X-ray extrasolar source and the X-ray extragalactic background were discovered during a rocket flight; see Giacconi, R., Gursky, H., et al. 1962. *Phys. Rev. Lett.*, **9**, 439.

10. Giacconi, R., Murray, S., et al. 1972. *Astrophys. J.*, **178**, 281.

11. Jones, C., Mandel, E., et al. 1979. *Astrophys. J. Lett.*, **234**, L21.

12. Mitchell, R. J., Culhane, J. L., et al. 1976. *Mon. Not. Roy. Astron. Soc.*, **176**, 29p.

13. Fabbiano, G. 1989. *Annu. Rev. Astron. Astrophys.*, **27**, 87.

14. For example, see Kondo, Y., Boggess, A., Maran, S. P. 1989. *Annu. Rev. Astron. Astrophys.*, **27**, 397.

15. See Costa, E., Frontera, F., et al. 1997. *Nature (London)*, **387**, 783; van Paradijs, J., Groot, P. J., et al. 1997. *Nature (London)*, **386**, 686.

16. See many articles in Block, D. L., Greenberg, J. M., eds. 1996. *New Extragalactic Perspectives in the New South Africa*, Kluwer, Dordrecht, The Netherlands.

17. For the last point, see Bahcall, J. N., Kirhakos, S., et al. 1997. *Astrophys. J.*, **479**, 642.

18. See Perlmutter, S. Gabi, S., et al. 1997. *Astrophys. J.*, **483**, 565; Perlmutter, S., Aldering, G., et al. 1998. *Nature (London)*, **391**, 51; Garnavich, P. M., Kirshner, R. P., et al. 1998. *Astrophys. J. Lett.*, **493**, L53.

19. See the report of the Space Interferometry Mission Science Working Group given by Allen, R. J., Peterson, D., Shao, M. 1997. *Soc. Photo.-Opt. Instrum. Eng.*, **2871**, 504; for GAIA, see Lindegren, L., Perryman, M. A. C. 1996. *Astron. Astrophys. Suppl.*, **116**, 579.

20. Hogg, D. W., Cohen, J. G. et al. 1998. *Astron. J.*, **115**, 1418. One record case, apparently confirmed by ground-based spectroscopy, is that of a galaxy in the Hubble Deep Field at $z \approx 8.6$; Lehnert, M. D., Nesvadba, N. P. H. et al. 2010. *Nature (London)*, **467**, 940. Another interesting case is that of a candidate galaxy at $z \approx 10$; Bouwens, R. J., Illingworth, G. D., et al. 2011. *Nature (London)*, **469**, 504.

21. For a general introduction to radiation mechanisms relevant to the interstellar medium, see Spitzer, L., Jr. 1978. *Physical Processes in the Interstellar Medium*, Wiley, New York; and Rybicki, G. B., Lightman, A. P. 1979. *Radiative Processes in Astrophysics*, Wiley, New York. Another interesting general reference is Harwit, M. 2006. *Astrophysical Concepts*, 4th ed., Springer, New York. An overview of many topics associated with the physics of galaxies is given by Fabbiano, G., Gallagher, J. S., Renzini, A., eds. 1990. *Windows on Galaxies*, Kluwer, Dordrecht, The Netherlands.

22. See Condon, J. J. 1992. *Annu. Rev. Astron. Astrophys.*, **30**, 575; see also van der Kruit, P. C., Allen, R. J. 1976. *Annu. Rev. Astron. Astrophys.*, **14**, 417.

23. Among the finest results, we should recall the detailed study of the spiral structure in M81 by Visser, H. C. D. 1977. Ph.D. thesis, University of Groningen, The Netherlands. See also the map of atomic hydrogen for the entire group of galaxies associated with M81 by Yun, M. S., Ho, P. T. P., Lo, K. Y. 1994. *Nature (London)*, **372**, 530. Among the most recent surprising results we mention the discovery of slowly rotating extra-planar gas in spiral galaxies, the detection of prominent grand-design spiral structure well outside the bright optical disk, and the discovery of faint but regular gaseous disks in elliptical galaies; these topics will be addressed in Parts III and IV.

24. Hartmann, D., Burton, W. B. 1997. *Atlas of Galactic Neutral Hydrogen*, Cambridge University Press, Cambridge, UK.

25. Mathewson, D. S., van der Kruit, P. C., Brouw, W. N. 1972. *Astron. Astrophys.*, **17**, 468.

26. See Peterson, B. M. 1997. *An Introduction to Active Galactic Nuclei*, Cambridge University Press, Cambridge, UK.

27. In the context of the dynamics of galaxies, see the many articles in Combes, F., Casoli, F., eds. 1991. *Dynamics of Galaxies and Their Molecular Cloud Distributions*, Kluwer, Dordrecht, The Netherlands. Dame, T. M., Ungerechts, H., et al. 1987. *Astrophys. J.*, **322**, 706, give a CO survey for the entire Milky Way galaxy.

28. Combes, F. 1991. *Annu. Rev. Astron. Astrophys.*, **29**, 195; Young, J. S., Scoville, N. Z. 1991. *Annu. Rev. Astron. Astrophys.*, **29**, 581.

29. Carruthers, G. P. 1970. *Astrophys. J. Lett.*, **161**, L81

30. One important aspect is the physics of photodissociation regions; see Crawford, M. K., Genzel, R., et al. 1985. *Astrophys. J.*, **291**, 755; Tielens, A. G. G. M., Hollenbach, D. 1985. *Astrophys. J.*, **291**, 722.

31. See the critical analysis by Allen, R. J. 1996. In *New Extragalactic Perspectives in the New South Africa*, eds. D. L. Block, J. M. Greenberg. Kluwer, Dordrecht, The Netherlands, p. 50.

32. See Rix, H.-W. 1993. *Publ. Astron. Soc. Pacific*, **105**, 999; Rix, H-W., Rieke, M. J. 1993. *Astrophys. J.*, **418**, 123.

33. See Witt, A. N. 1988. In *Dust in the Universe*, eds. M. E. Bailey, D. A. Williams. Cambridge University Press, Cambridge, UK, p. 1; Whittet, D. C. B. 1992. *Dust in the Galactic Environment*, Institute of Physics Publishing, Bristol, UK.

34. Soifer, B. T., Houck, J. R., Neugebauer, G. 1987. *Annu. Rev. Astron. Astrophys.*, **25**, 187; Sanders, D. B., Mirabel, I. F. 1996. *Annu. Rev. Astron. Astrophys.*, **34**, 749; Devereux, N. 1996. In *New Extragalactic Perspectives in the New South Africa*, eds. D. L. Block, J. M. Greenberg. Kluwer, Dordrecht, The Netherlands, p. 357.

35. From a study of the solar neighborhood, Salpeter, E. E. 1955. *Astrophys. J.*, **121**, 161, proposed a power-law distribution $M^{-2.35}$ for the range 0.4 to $10 M_\odot$. More recent studies suggest different options; see Miller, G. E., Scalo, J. M. 1979. *Astrophys. J. Suppl.*, **41**, 513; Rana, N. C. 1991. *Annu. Rev. Astron. Astrophys.*, **29**, 129; Kroupa, P. 2001. *Mon. Not. Roy. Astron. Soc.*, **322**, 231; Chabrier, G. 2003. *Publ. Astron. Soc. Pacific*, **115**, 763; Treu, T., Auger, M. W., et al. 2010. *Astrophys. J.*, **709**, 1195, and references therein.

36. Worthey, G. 1994. *Astrophys. J. Suppl.*, **95**, 107.

37. See, e.g., Peletier, R. 1989. Ph.D. thesis, University of Groningen, The Netherlands; Bruzual, G., Charlot, S. 2003. *Mon. Not. Roy. Astron. Soc.*, **344**, 1000; Maraston, C. 2005. *Mon. Not. Roy. Astron. Soc.*, **362**, 799.

38. See, e.g., Greggio, L., Renzini, A. 1990. *Astrophys. J.*, **364**, 35; Bressan, A., Chiosi, C., Tantalo, R. 1996. *Astron. Astrophys.*, **311**, 425; Matteucci, F. 1996. *Fund. Cosmic Phys.*, **17**, 283; Pagel, B. E. J. 1998. *Nucleosynthesis and Chemical Evolution of Galaxies*, Cambridge University Press, Cambridge, UK; Matteucci, F. 2001. *The Chemical Evolution of the Galaxy*, Kluwer, Dordrecht, The Netherlands.

39. See Madau, P., Ferguson, H. C., et al. 1996. *Mon. Not. Roy. Astron. Soc.*, **289**, 1388; Madau, P., Pozzetti, L., Dickinson, M. 1998. *Astrophys. J.*, **498**, 106.

40. A good review is given by O'Connell, R. W. 1990. In *Windows on Galaxies*, eds. G. Fabbiano, J. S. Gallagher, A. Renzini. Kluwer, Dordrecht, The Netherlands, p. 39.

41. See the interesting near-UV image of M81 by Hill, J. K., Bohlin, R. C., et al. 1992. *Astrophys. J. Lett.*, **395**, L37.

42. Lequeux, J., Maucherat-Joubert, M., et al. 1981. *Astron. Astrophys.*, **103**, 305; Larson, R. B., Tinsley, B. M. 1978. *Astrophys. J.*, **219**, 46.

43. See Greggio, L., Renzini, A. 1990. *op. cit.*

44. See Whittet, D. C. B. 1992. *op. cit.*

45. See Fabbiano, G. 1989. *op. cit.*

46. Snowden, S. L., Freyberg, M. J., et al. 1995. *Astrophys. J.*, **454**, 643, give the maps of the soft-X-ray diffuse background of the ROSAT survey.

47. Sarazin, C. L. 1988. *X-Ray Emissions from Clusters of Galaxies*, Cambridge University Press, Cambridge, UK.

48. Hunter, S. D., Bertsch, D. L., et al. 1997. *Astrophys. J.*, **481**, 205.

49. We recall that several neutrinos from the supernova SN1987A in the Large Magellanic Cloud were detected by Kamiokande-II (Hirata, K., Kajita, T., et al. 1987. *Phys. Rev. Lett.*, **58**, 1490) and the Irvine-Michigan-Brookhaven detector (Bionta, R. M., Blewitt, G., et al. 1987. *Phys. Rev. Lett.*, **58**, 1494).

50. The Pierre Auger Collaboration has announced observing a correlation of the highest-energy cosmic rays with the positions of nearby active galactic nuclei; Abraham, J., Abreu, P., et al. 2007. *Science*, **318**, 938.

51. So far the best evidence for gravitational waves is an indirect detection; see Hulse, R. A., Taylor, J. H. 1975. *Astrophys. J.*, **195**, L51; Weisberg, J. M., Taylor, J. H. 2005. *ASP Conference Series*, **328**, 25.

3 Classifications

Jeans wrote: "The great nebulae exhibit an enormous difference of structural detail, but Hubble, who has devoted much skill and care to their classification, finds that most of the observed forms can be reduced to law and order."[1] This general statement applies in even stronger terms today, now that several decades of work have confirmed Hubble's intuition.[2] In fact, "The conclusion is that the modern classification indeed describes a true order among the galaxies, an order not imposed by the classifier."[3]

The efforts to sharpen the empirical morphological classification[4] are extremely important for our knowledge of the dynamics of normal galaxies. Indeed, "The ultimate purpose of the classification is to understand galaxy formation and evolution."[5] Thus the very existence of the morphological classification scheme proves that the observed morphology reflects a few intrinsic characteristics that vary with continuity along the Hubble sequence and that the overall structure is likely to be quasi-stationary.[6] This plays a central role in the development of a dynamical framework for the classification of disk galaxies,[7] for which the spiral structure is the most spectacular morphological property considered, as is outlined in Part III.

It has been noted that morphology changes significantly with the wave band of observation, with redder images generally found to be characterized by a higher degree of smoothness and regularity.[8] For a dynamical theory aimed at bringing out the role of gravity in the classification, the recent near-infrared (near-IR) studies that probe the underlying evolved stellar disk thus become of primary importance, especially when they may show a contrast with the morphology based on standard optical images.[9] The reason is that many features of the optical images may result from the exaggerated influence of the so-called Population I objects, which are known to be associated with more transient dynamical behavior.

Given the emphasis on intrinsic characteristics, continuity, and quasi-stationarity established empirically by the value of the morphological classification system, some interesting dynamical processes that are known to be associated with fast evolution (such as galaxy-galaxy mergers) are probably to be considered the exception rather than the rule in determining the general morphology of normal galaxies.

Somehow disk galaxies present a richer morphology that allows us to set up a better defined intrinsic empirical classification. For elliptical galaxies instead, much of what we see is probably determined by projection effects. However, we will see later that there may be clues to intrinsically different categories of elliptical galaxies. Besides normal elliptical galaxies, one should also consider other ellipsoidal stellar systems that may be characterized by completely different dynamics and formation history. These are the "dwarf ellipticals" and the "dwarf spheroidals" (dSph; low-surface-brightness objects quite common in the Local Group and in nearby clusters[10]), "compact ellipticals" (small objects with high surface brightness, such as M32, close to the Andromeda spiral galaxy), and cD galaxies (giant galaxies, generally found in the central parts of clusters of galaxies, with excess luminosity in their outer parts). Most of what will be discussed in this book refers to the case of normal elliptical galaxies.

3.1   Hubble Classification

The traditional Hubble classification distinguishes two main types of galaxies, elliptical (E) and spiral (S) galaxies. Elliptical galaxies (Fig. 3.1) are usually denoted by En, with n running from 0 to 7, based on the flattening of the observed image (0 indicates a round elliptical, whereas 7 corresponds to an observed aspect ratio of 3:1). Elliptical galaxies most frequently appear as E2 galaxies.[11] The Hubble tuning-fork diagram separates spirals in two linear sequences, normal S galaxies (also denoted as SA galaxies), which lack a prominent bar, and barred SB galaxies, which display a prominent bar. The two sequences run from early-type objects (denoted by a) through the b types to late-type galaxies denoted by c. The terms *early* and *late* should not be taken in the sense of time evolution. The key parameter controlling the sequences from early- to late-type spirals is probably the gas content (discussed later).

Fig. 3.1. Giant elliptical galaxy NGC 4486 (= M87) in the nearby Virgo cluster (credit: Anglo-Australian Observatory/D. Malin). Many globular clusters can be recognized as prominent small clumps.

The main improvement with respect to the original scheme, but following the needs already pointed out by Hubble himself, has been to establish the role of the S0 class (disks without spiral structure, with or without a prominent bar) as an element of continuity between elliptical and spiral galaxies. In addition, for spiral galaxies, the modern classification better recognizes the continuity among the various categories (by the introduction of suitable intermediate types); extends the spiral sequences beyond the *c* class (to *d* and *m* types, which partly replace the so-called irregular spirals); incorporates additional information relative to the *luminosity classes*,[12] with a roman numeral from I to V (from luminous galaxies with long, well-developed filamentary arms to fainter objects with fuzzier, broader, less-organized optical arms); and better characterizes the type of spiral structure involved by the specification of indicators such as *r* (for ring) or *s*. Thus a galaxy such as NGC 5236 (following the denomination of the *New General Catalogue*;[13] the same galaxy is also called M83 in the Messier catalog[14]) is classified as SB*c*(*s*)II, a galaxy such as NGC 5364 as S*c*(*r*)I, and a galaxy such as NGC 1566 as S*bc*(*s*)I.

The most important factors that correlate and define the sequences from *a* to *d* are gas content (increasing), bulge size (decreasing), pitch angle of spiral arms (from tight to open), size of HII regions of ionized gas (increasing), and nature of spiral arms. Early-type spirals are often more massive than late-type spirals. Other quantities, such as integrated color and current star-formation rate, also correlate along the sequence.[15] Further details on the classification scheme are not provided here. The reader can find a thorough discussion with the necessary illustrations in the monographs quoted earlier, in other books, and in recent articles.[16]

In what follows, we focus on only a few selected morphological aspects that are naturally related to some dynamical questions addressed later in this book, especially in Parts III and IV.

3.2 Morphology of Elliptical Galaxies

Probably the first question that comes to mind about an observed elliptical galaxy is this: What does the image correspond to in terms of intrinsic three-dimensional shape of the galaxy? This question has no easy answer, as we will further discuss in Part IV. From an empirical point of view, we may try to gather clues from the orientation of dust lanes[17] or from the structure and kinematics of ionized gas[18] that is sometimes present. But, of course, assumptions have to be made as to whether the dust or the gas has settled in one of the principal planes of the potential well.

The most convincing evidence for the presence of triaxiality (at least in some cases) is derived from the (not infrequent) observation of a significant twisting of the isophotes[19] (Fig. 3.2). Studies have also addressed the issue of the dependence of ellipticity on galactocentric radius, but no general systematic trends seem to emerge for such ellipticity profiles.[20]

There is some evidence that departures of the isophotes from a pure ellipse, either in the boxy or the disky direction, may be characteristic of two distinct categories of elliptical galaxies.[21] In support of this conclusion, an interesting correlation has been found between the boxy/disky character and the X-ray luminosity.[22] Given the frequent evidence for a disk component, attempts have been made for a disk-spheroid decomposition based on both kinematic and photometric data.[23]

Another interesting morphological feature that is sometimes detected is the presence of shells and ripples.[24]

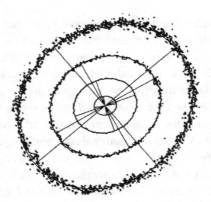

Fig. 3.2. Twist of isophotes in the elliptical galaxy NGC 1549 (from Jedrzejewski, R. J. 1987. In IAU Symposium: *Structure and Dynamics of Elliptical Galaxies*, ed. T. de Zeeuw. Reidel, Dordrecht, The Netherlands, p. 37; © 1987 by the International Astronomical Union; used with kind permission from Springer Science and Business Media B.V.).

3.3 Morphology of Spiral Galaxies

Much of the morphology of spiral galaxies derives from a complex interplay between what is sometimes called the *Population I disk* (young stars, cold interstellar medium, molecules, and dust) and the *Population II disk*. (Here by Population II disk stars we mean just the older stars of the stellar disk that are characterized by relatively large epicycles; in the astronomical terminology, various populations are often considered, and Population II is frequently used with a more specific meaning.) The former component is dynamically colder, being associated with very small deviations from circular motions, and thinner, whereas the latter is warmer and thicker. The morphology is further traced by star formation, especially through HII regions of ionized gas, which takes place in the Population I disk. For grand-design spiral galaxies, star-formation events mark the arms as "beads on a string" or as "whitecaps of ocean waves."[25] We saw briefly in Chapter 2 which observational windows are best suited to probe the different components involved.

Spiral structure may occur in a variety of forms. In many cases, a grand design is displayed (Fig. 3.3), sometimes "extending over the whole galaxy, from the nucleus to its outermost part, and consisting of two arms starting from diametrically opposite points."[26] Such a grand design may be present or absent within the same Hubble type. For example, NGC 2841 and M81 are similar objects, both classified as Sb, but the former lacks a grand design and is considered a prototype of a flocculent galaxy, whereas the latter is one of the nicest examples of a regular, bisymmetric, normal (i.e., unbarred), grand-design spiral. Given the apparent decoupling of regularity from the criteria that define the Hubble sequence, a regularity classification has been proposed[27] to quantify this important morphological characteristic. Twelve different levels of regularity have been considered (the scheme introduced probably overemphasizes bisymmetry; in principle, one might have a highly regular three-armed grand design).

Studies in the near IR now demonstrate[28] that the grand-design character is associated mainly with the older stellar disk, whereas the Population I component is more active in defining less regular spiral features. The near-IR surveys show that the underlying grand-design structure is

Fig. 3.3. Spiral galaxy NGC 4321 (= M100) in the nearby Virgo cluster, imaged in IR light with the HAWK-I camera on VLT at Paranal Observatory in Chile (credit: ESO/P. Grosbøl).

very common and is mostly bisymmetric;[29] a significant role is also played by the $m = 1$ character.[30] In turn, a multiple-armed structure is associated mainly with the Population I disk and tends to disappear in the near IR. These important findings are discussed further in Chapter 18 in light of various theoretical models.

A regular, large-scale spiral structure is generally associated with the main optical disk and stops inside the bulge, if present (as in M81 or NGC 2997). However, in grand-design spirals there are indications that long spiral arms may continue in the interstellar medium, possibly at much smaller amplitudes, all the way to the center[31] and outward too, well beyond the bright optical disk[32] (as in M101, NGC 6946, and NGC 628).

A curious and sometimes spectacular morphological feature is that of rings, often found to be associated with the cold gaseous component, especially in barred galaxies.[33] Rings turn out to be important tracers of the dynamics of the disk and are presumably associated with special resonances.

The disk of spiral galaxies is often found to be warped on large scales[34] (Fig. 3.4). The typical structure exhibits a large-scale $m = 1$ bending of the disk. For our Galaxy, for which the required observation can be made much more easily, there is also evidence for high-m corrugations.[35] The phenomenon is associated mainly with the gaseous outer disk, but there are also indications of deviations from the planar symmetry in the outer stellar disk.[36] Another curious case is that of the warped dust disk in the galaxy ESO 510-G13. These morphological features are the focus of a dynamical analysis in Chapter 19.

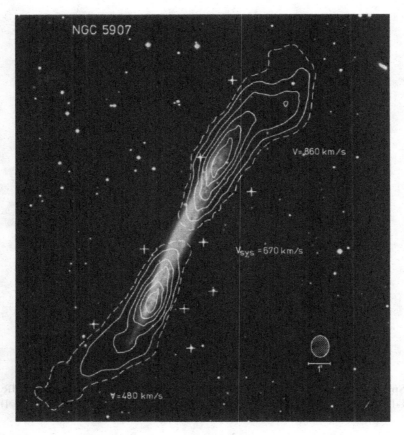

Fig. 3.4. Warped galaxy NGC 5907 (from Sancisi, R. 1976. *Astron. Astrophys.*, **53**, 159; reproduced with permission, © ESO). The HI radio map is superimposed on the optical image of the galaxy, which is seen edge-on.

The structure of cold atomic hydrogen reveals other morphological features for which a dynamical explanation is only partly available. Here we may recall the general issue of "high-velocity clouds,"[37] which might trace a general accretion process of extragalactic material onto the disk. High-velocity material also has been detected around external galaxies,[38] such as NGC 6946. An HI halo has been shown to exist, with different kinematical properties with respect to the disk,[39] in the edge-on galaxy NGC 891. In the galaxy M101, a giant bubble in atomic hydrogen[40] may correspond to a highly energetic event that occurred in the disk. The cold gaseous disk, either by means of accretion (in a quasi-steady manner or by means of episodic events) or through ejections and outflows, thus should be thought of as a lively component of spiral galaxies.[41]

In meteorology, hurricanes and other low-pressure systems generally occur as large-scale quasi-stationary structures with a grand-design morphology, often characterized by a prominent $m = 1$ component (Fig. 3.5). Regular one-armed cyclonic patterns also have been observed in the Mars atmosphere (in particular, from the *Viking* orbiter and later by the *Hubble Space Telescope*).

Fig. 3.5. *Top:* GOES-13 satellite view of Hurricane Irene in 2011 (credit: NASA/NOAA GOES Project). *Bottom:* Image of a low-pressure system over Iceland taken by *Aqua/MODIS* in 2004 (credit: Jacques Descloitres, MODIS Rapid Response Team, NASA/GSFC).

Notes

1. Jeans, J. 1929. *Astronomy and Cosmogony*, Cambridge University Press, Cambridge, UK (reprinted in 1961 by Dover, New York).
2. Hubble, E. 1926. *Astrophys. J.*, **64**, 321; 1936. *The Realm of the Nebulae*, Yale University Press, New Haven, CT.

3. The words that seem to complete Jeans's statement are from Sandage, A., Bedke, J. 1994. *The Carnegie Atlas of Galaxies*, Publ. 638, Carnegie Institution of Washington, Washington, DC. The atlas illustrates 1,168 galaxies and most of the 1,246 galaxies contained in the original catalog of Shapley, H., Ames, A. 1932. *Ann. Harvard College Obs.*, **88**, no. 2. Sandage and Bedke describe their atlas as a "textbook in classification." Their work completes a long-term project with major steps marked by Sandage, A. 1961. *The Hubble Atlas of Galaxies*, Publ. 618, Carnegie Institution of Washington, Washington, DC; Sandage, A., Tammann, G. A. 1981 (2nd ed. 1987). *A Revised Shapley-Ames Catalog of Bright Galaxies*, Publ. 635, Carnegie Institution of Washington, Washington, DC; see also Sandage, A., Bedke, J. 1988. *Atlas of Galaxies Useful for Measuring the Cosmological Distance Scale*, NASA SP-496, Washington, DC.

4. Two classical articles dealing with the problem of classification are de Vaucouleurs, G. 1959. In *Handbook der Physik*, Vol. 53, Springer-Verlag, Berlin, p. 275; Sandage, A. 1975. In *Galaxies and the Universe*, eds. A. Sandage, M. Sandage, J. Kristian. University of Chicago Press, Chicago, p. 1.

5. The citation is also from Sandage, A., Bedke, J. 1994. *op. cit.*

6. See also Roberts, M. S. 1963. *Annu. Rev. Astron. Astrophys.*, **1**, 149.

7. Bertin, G., Lin, C. C. 1996. *Spiral Structure in Galaxies: A Density Wave Theory*, MIT Press, Cambridge, MA.

8. Zwicky, F. 1957. *Morphological Astronomy*, Springer-Verlag, Berlin; see also the 1-μm study by Elmegreen, D. M. 1981. *Astrophys. J. Suppl.*, **47**, 229.

9. Block, D. L., Wainscoat, R. J. 1991. *Nature (London)*, **353**, 48; Block, D. L., Bertin, G., et al. 1994. *Astron. Astrophys.*, **288**, 365. Block, D. L., Puerari, I. 1999. *Astron. Astrophys.*, **342**, 627, propose a classification system based on the pitch angle of spiral arms and the lopsidedness or bisymmetry of spiral structure, as diagnosed in the near-IR. A comparison between optical and near-IR morphology is given by Eskridge, P. B., Frogel, J. A., et al. 2002. *Astrophys. J. Suppl.*, **143**, 73. An interesting classification study in the mid-IR (at 3.6 μm) has been based on the data acquired by *Spitzer*; see Buta, R. J., Sheth, K., et al. 2010. *Astrophys. J. Suppl.*, **190**, 147.

10. Dwarf spheroidals around our Galaxy are often regarded as the stellar systems that are most dominated by dark matter. Currently, they are studied with great attention, especially in view of the problem that cosmological simulations of galaxy formation predict the existence of large numbers of substructures that are apparently not observed; Diemand, J., Moore, B., Stadel, J. 2005. *Nature (London)*, **433**, 389; Springel, V., White, S. D. M., et al. 2005. *Nature (London)*, **435**, 629.

11. From a study of the shapes of galaxies in the *Sloan Digital Sky Survey*, the distribution of the observed flattening of elliptical galaxies appears to be peaked in correspondence of E2; see fig. 5 in Alam, S. M. K., Ryden, B. S. 2002. *Astrophys. J.*, **570**, 610.

12. van den Bergh, S. 1960. *Astrophys. J.*, **131**, 215, and 558.

13. Dreyer, J. L. E. 1888. *Mem. Roy. Astron. Soc.*, **49**, 1.

14. Messier, C. 1781. In *Connaissance des Temps pour 1784*, p. 227.

15. A quantitative account of the trends along the Hubble sequence can be found in Roberts, M. S., Haynes, M. P. 1994. *Annu. Rev. Astron. Astrophys.*, **32**, 115.

16. A simplified but rather complete introduction is given in the monograph by Bertin, G., Lin, C. C. 1996. *op. cit.*; the best description is probably the one given by Sandage, A., Bedke, J. 1994. *op. cit.*, fully devoted to the problem of classification. See also de Vaucouleurs, G., de Vaucouleurs, A., et al. 1991. *Third Reference Catalogue of Bright Galaxies*, Springer-Verlag, New York; Buta, R. J., Corwin, H. G., Oderwahn, S. C. 2007. *The de Vaucouleurs Atlas of Galaxies*, Cambridge University Press, Cambridge, UK. Extensive classification studies have been made on the data from the *Sloan Digital Sky Survey*; see Fukugita, M., Nakamura, O., et al. 2007. *Astron. J.*, **134**, 579; Nair, P. B., Abraham, R. G. 2010. *Astrophys. J. Suppl.*, **186**, 427. Lintott, C. J., Schawinski, K., et al. 2008. *Mon. Not. Roy. Astron. Soc.*, **389**, 1179, provide morphological classification for nearly 1 million galaxies.

17. Bertola, F., Galletta, G. 1978. *Astrophys. J. Lett.*, **226**, L115; Sadler, E. M., Gerhard, O. E. 1985. *Mon. Not. Roy. Astron. Soc.*, **214**, 177.

18. Bertola, F., Bettoni, D., et al. 1991. *Astrophys. J.*, **373**, 369; Buson, L. M., Sadler, E. M., et al. 1993. *Astron. Astrophys.*, **280**, 409.

19. Carter, D. 1978. *Mon. Not. Roy. Astron. Soc.*, **182**, 797; Bertola, F., Galletta, G. 1979. *Astron. Astrophys.*, **77**, 363; Lauer, T. R. 1985. *Astrophys. J. Suppl.*, **57**, 473; Jedrzejewski, R. J. 1987. *Mon. Not. Roy. Astron. Soc.*, **226**, 747; Peletier, R. F., Davies, R. L., et al. 1989. *Astron. J.*, **100**, 1091.

20. di Tullio, A. 1979. *Astron. Astrophys. Suppl.*, **37**, 591; Bender, R., Döbereiner, S., Möllenhoff, C. 1988. *Astron. Astrophys. Suppl.*, **74**, 385; Caon, N., Capaccioli, M., Rampazzo, R. 1990. *Astron. Astrophys. Suppl.*, **86**, 429; Sparks, W. B., Wall, J. V., et al. 1991. *Astrophys. J. Suppl.*, **76**, 471.

21. Jedrzejewski, R. J. 1987. *op. cit.*; Bender, R., Möllenhoff, C. 1987. *Astron. Astrophys.*, **177**, 71; Carter, D. 1987. *Astrophys. J.*, **312**, 514; Bender, R. 1988. *Astron. Astrophys. Lett.*, **193**, L7; Nieto, J.-L., Bender, R., et al. 1991. *Astron. Astrophys. Lett.*, **244**, L25; Nieto, J.-L., Bender, R., Surma, P. 1991. *Astron. Astrophys. Lett.*, **244**, L37.

22. Bender, R., Surma, P., et al. 1989. *Astron. Astrophys.*, **217**, 35.

23. Scorza, C., Bender, R. 1995. *Astron. Astrophys.*, **293**, 20.

24. Malin, D. F. 1979. *Nature (London)*, **277**, 279; Schweizer, F., Seitzer, P., et al. 1990. *Astrophys. J. Lett.*, **364**, L33.

25. We may refer once more to the monograph by Bertin, G., Lin, C. C. 1996. *op. cit.*, where the various morphological aspects involved are introduced with the help of a wide set of illustrations.

26. Oort, J. H. 1962. In *Interstellar Matter in Galaxies*, ed. L. Woltjer. Benjamin, New York, p. 234. In his formulation of the problem of spiral structure in galaxies, Oort makes a clear distinction between grand-design spiral structure and other less regular structure, made mostly of small-scale spiral features, which often coexists with the large-scale regular pattern.

27. Elmegreen, D. M., Elmegreen, B. G. 1982. *Mon. Not. Roy. Astron. Soc.*, **201**, 1021.

28. See Block, D. L., Bertin, G., et al. 1994. *op. cit.*, and papers in the volume Block, D. L., Greenberg, J. M., eds. 1996. *New Extragalactic Perspectives in the New South Africa*, Kluwer, Dordrecht, The Netherlands.

29. One issue often addressed is that of the frequency of barred structure, which may be significantly different if studied in the optical or the near IR; see Eskridge, P. B., Frogel, J. A., et al. 2000. *Astron. J.*, **119**, 536; Menéndez-Delmestre, K., Sheth, K., et al. 2007. *Astrophys. J.*, **657**, 790; Sheth, K., Elmegreen, D. M., et al. 2008. *Astrophys. J.*, **675**, 1141; Nair, P. B., Abraham, R. G. 2010. *Astrophys. J. Lett.*, **714**, L260; Cameron, E., Carollo, C. M., et al. 2010. *Mon. Not. Roy. Astron. Soc.*, **409**, 346.

30. See also Baldwin, J. E., Lynden-Bell, D., Sancisi, R. 1980. *Mon. Not. Roy. Astron. Soc.*, **193**, 313; Richter, O.-G., Sancisi, R. 1994. *Astron. Astrophys. Lett.*, **290**, L9.

31. As beautifully shown in M51; see Zaritsky, D., Rix, H.-W., Rieke, M. J. 1993. *Nature (London)*, **364**, 313.

32. Shostak, G. S., van der Kruit, P. C. 1984. *Astron. Astrophys.*, **132**, 20; Dickey, J. M., Hanson, M. M., Helou, G. 1990. *Astrophys. J.*, **352**, 522; Kamphuis, J. 1993. Ph.D. thesis, Groningen University, The Netherlands; Meurer, G. R., Carignan, C., et al. 1996. *Astron. J.*, **111**, 1551; Begum, A., Chengalur, J. N., Karachentsev, I. D. 2005. *Astron. Astrophys.*, **433**, L1. A particularly impressive example is given by the case of NGC 6946; Boomsma, R. 2007. Ph.D. thesis, Groningen University, The Netherlands; Boomsma, R., Oosterloo, T. A., et al. 2008. *Astron. Astrophys.*, **490**, 555.

33. See Buta, R., Combes, F. 1996. *Fund. Cosmic Physics*, **17**, 95, and the many references therein.

34. Sancisi, R. 1976. *Astron. Astrophys.*, **53**, 159; Bosma, A. 1978. Ph.D. thesis, University of Groningen, The Netherlands. For the Milky Way galaxy, a warp had been detected much earlier; see Burke, B. F. 1957. *Astron. J.*, **62**, 90; Kerr, F. J. 1957. *Astron. J.*, **62**, 93; Kerr, F. J. 1969. *Annu. Rev. Astron. Astrophys.*, **7**, 39.

35. Kulkarni, S. R., Heiles, C., Blitz, L. 1982. *Astrophys. J. Lett.*, **259**, L63; Blitz, L., Fich, M., Kulkarni, S. 1983. *Science*, **220**, 1233.

36. For the case of NGC 4762, see Tsikoudi, V. 1977. Ph.D. thesis, University of Texas, Austin: Strom, S. E., Strom, K. M. 1979. *Scientific American*, April, 72. For NGC 4244, NGC 4565, and NGC 5907, see van der Kruit, P. C. 1979. *Astron. Astrophys. Suppl.*, **38**, 15.

37. Giovanelli, R. 1980. *Astron. J.*, **85**, 1155.

38. Kamphuis, J., Sancisi, R. 1993. *Astron. Astrophys. Lett.*, **273**, L31.

39. Swaters, R. A., Sancisi, R., van der Hulst, J. M. 1997. *Astrophys. J.*, **491**, 140; Oosterloo, T., Fraternali, F., Sancisi, R. 2007. *Astron. J.* , **134**, 1019; for NGC 2403, see Fraternali, F., van Moorsel, G., et al. 2002. *Astron. J.*, **123**, 3124.

40. Kamphuis, J., Sancisi, R., van der Hulst, T. 1991. *Astron. Astrophys. Lett.*, **244**, L29.

41. For a recent review, see Sancisi, R., Fraternali, F., et al. 2008. *Astron. Astrophys. Rev.*, **15**, 189.

4 Photometry, Kinematics, and Dark Matter

Normal galaxies have a variety of sizes and morphological categories. In this chapter their main empirically established structural characteristics are summarized schematically,[1] and the problem of dark matter in the general cosmological context is briefly introduced. The problem of dark matter in galaxies[2] will be better formulated in Chapter 5, among other broad issues addressed in this book, and will be discussed separately for spiral galaxies (Chapter 20) and for elliptical galaxies (Chapter 24) after the relevant dynamical tools have been properly developed.

Our world of galaxies is biased in favor of bright objects and against objects with low surface brightness.[3] However, the tail of the galaxy luminosity function is widely populated by low-surface-brightness galaxies.[4] We should thus be aware that our knowledge of the structure of galaxies and the discussion of the dynamics of normal galaxies that have been developed so far have focused mainly on bright galaxies; much remains to be done for fainter systems.

From astronomical observations, we ideally would wish to extract information on mass distributions and overall kinematics (mean flow motions and velocity dispersions). In practice, we must interpret images and spectroscopic information, gathered through various observational windows, and we must mediate the process by simplified models. Obviously, a major source of uncertainty is the a priori unknown three-dimensional structure of the object under investigation because the data give us quantities projected along the line of sight.

The large-scale structure of galaxies is characterized mainly by the luminosity profiles (surface brightness as a function of radius), which tell us how stars are spatially distributed, and by the rotation curves and velocity-dispersion profiles (velocities as a function of radius) derived from spectroscopic studies of Doppler-induced line shifts and linewidths. The extraction of these (one-dimensional) profiles implicitly assumes the existence of axisymmetry, which is often only crudely justified by the data. However, the interpretation of these functions is obviously more straightforward than the inspection of the two-dimensional images or the two-dimensional kinematics contour maps.

For spiral galaxies, rotation curves give a rather accurate direct measure, at least sufficiently far from the bulge region, of the force field associated with the mean gravitational field as a function of galactocentric radius. Inclination effects usually can be taken into account rather easily. For elliptical galaxies, the observed Doppler profiles resulting from projection along the line of sight are related to the intrinsic kinematics (mean flow motions and velocity-dispersion ellipsoid as a function of position in the galaxy) in a complex manner; in this case, the relation with the underlying force field associated with the mean gravitational potential depends significantly on the model that is used.

4.1   Luminosity Profiles

Because normal galaxies are primarily stellar systems, it is natural to imagine that the luminosity profiles, especially those obtained in the near infrared (IR), trace the visible mass. However, it is well known that significant amounts of cold gas exist in spiral galaxies, for which the local gas-to-star density ratio increases steadily with radius. Correspondingly, for bright ellipticals, one finds that relatively large amounts of hot gas exist in the form of a diffuse X-ray halo. In other words, other observational windows should be taken into account to complete the picture of the visible galaxy, which is only partly determined by the luminosity profiles derived by means of optical telescopes.

The fact that color gradients and metallicity gradients inside individual galaxies are relatively small is often taken as an empirical justification for converting luminosity profiles into mass profiles for the luminous component by means of a constant mass-to-light ratio[5] (the mass-to-light ratio may vary from galaxy to galaxy). In general, there are no good reasons to choose different values of the mass-to-light ratio for the disk and the bulge, if present, so the most straightforward disk-bulge decompositions are often carried out with only one mass-to-light ratio. In reality, the data available and the consideration of two separate components with possibly different star-formation histories would often leave considerable leverage for the conversion of luminosity profiles into mass profiles. Note that, in practice, even if color gradients can be considered to be small in relation to the mass-to-light ratio assumptions, observations in different bands may give different parameters for the scales that define the luminosity profiles (either h or R_e; see Subsections 4.1.1 and 4.1.2).

4.1.1   Exponential Disks

The luminosity profiles of galaxy disks are approximately exponential (Fig. 4.1), being reasonably well fitted by the law

$$I(R) = I_0 \exp(-R/h), \tag{4.1}$$

which has two scale parameters, the central brightness I_0 and the exponential length h. The associated total luminosity is given by $L = 2\pi h^2 I_0$; half of it is derived from the disk inside the half-light radius $R_e \approx 1.678h$. Expressed in standard astronomical units (magnitudes per seconds of arc squared), the exponential law becomes

$$\mu(R) = \mu_0 + 1.086\frac{R}{h}. \tag{4.2}$$

In practice, the visible disk is associated with the R_{25} or the Holmberg radius (i.e., the location where, in the relevant band, the surface brightness drops to 25 or 26.5 mag arcsec^{-2}), so the disk can be considered as almost terminating at $R = 4h$ to $5h$. There have been efforts to see whether the luminous disk actually terminates abruptly (i.e., whether the exponential is truncated[6]). This might have interesting dynamical consequences.[7] However, at those locations, the contribution of the cold gas to the local disk density is generally important. The optical truncation thus might mark an abrupt change in the properties of the disk, in the sense that the formation of stars from the gas, for some reason, has been and is ineffective without a significant truncation in the density profile.

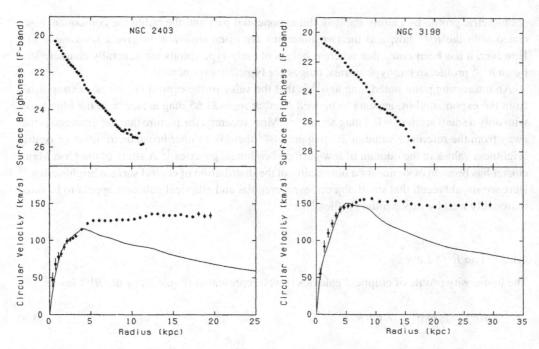

Fig. 4.1. The exponential luminosity profile of the spiral galaxies NGC 2403 (*left*) and NGC 3198 (*right*) and the related radio-determined rotation curves (from Sancisi, R., van Albada, T. S. 1987. In IAU Symposium: *Dark Matter in the Universe*, eds. J. Kormendy, G. R. Knapp. Reidel, Dordrecht, The Netherlands, p. 67; © 1987 by the International Astronomical Union, with kind permission from Springer Science and Business Media B.V.). Solid curves in the lower panels show the rotation curves expected for models with a constant (maximum) mass-to-light ratio for the stellar disk and no dark matter.

The observed brightness profiles often show significant deviations from the ideal exponential behavior, part of which may be associated with the structure of the inner disk, in which a bulge may be present, especially in early-type spirals. A concentrated component is called a *nucleus*, whereas a flat profile in the disk is sometimes called a *lens*. Two main types of deviations from the exponential profile, which is best set by the brightness profile of the outer disk, have been noted.[8] In one type of deviation, the observed profile always stays above the fitted exponential, and the deviation becomes systematically larger when we compare the observed profile with the fitted exponential at small radii; in contrast, in the second type of deviation, at small radii we first see a depression with respect to the exponential, and then a steep luminosity increase toward the nucleus, which may overshoot with respect to the extrapolated exponential.

The bulge is often considered a small elliptical and thus is modeled as a typically slightly oblate three-dimensional system, characterized by an $R^{1/4}$ law (see Subsection 4.1.2). In practice, the three-dimensional structure of the central region is hard to distinguish from the data (especially in the generic case in which the galaxy is not seen edge-on). It is also unclear whether, when a three-dimensional component exists, the bulge coexists with the (approximately exponential) disk or replaces the disk. Obviously, the two possibilities correspond to two significantly different dynamical systems. In many cases we simply proceed by decomposing an observed photometric profile by calling the disk the exponential part and the bulge the component associated with the $R^{1/4}$ law, and the two functions are often sufficient to give a reasonable fit.[9] However, it has been noted that whereas bulges of early-type spirals are generally characterized by an $R^{1/4}$ profile, in late-type spirals, bulges are typically exponential.[10]

An interesting point noted long ago[11] is that the value of the central brightness, extrapolated from the exponential fit, appears to be well fixed at $\mu_0 \approx 21.65$ mag arcsec^{-2} in the blue band, with only a small scatter (≈ 0.3 mag arcsec^{-2}). More recently, the picture that has emerged is that away from the reference value of 20 mag arcsec^{-2} there is a rather broad distribution of central brightness values in the domain of low-surface-brightness galaxies.[12] A study of the Ursa Major cluster has brought evidence for a bimodality in the distribution of central surface brightnesses.[13] Here we should recall that small objects, even irregular and elliptical galaxies, appear to be fitted reasonably well by exponential profiles.[14]

4.1.2   The $R^{1/4}$ Law

The luminosity profile of elliptical galaxies is well represented (Fig. 4.2) by the $R^{1/4}$ law:[15]

$$I(R) = I_0 \exp[-7.67(R/R_e)^{1/4}]. \tag{4.3}$$

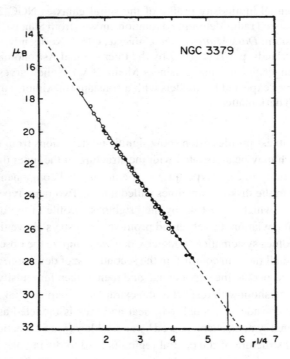

Fig. 4.2. The $R^{1/4}$ luminosity profile of the elliptical galaxy NGC 3379 (from de Vaucouleurs, G., Capaccioli, M. "Luminosity distribution in galaxies. I: The elliptical galaxy NGC 3379 as a luminosity distribution standard," 1979. *Astrophys. J. Suppl.*, **40**, 699; reproduced by permission of the AAS).

The factor 7.67 guarantees that half the total luminosity associated with $I(R)$ is contained in the disk of (projected) radius R_e, often called the *effective radius* of the galaxy. Expressed in units of magnitudes per seconds of arc squared, the $R^{1/4}$ law becomes

$$\mu(R) = \mu_0 + 8.325 \left(\frac{R}{R_e}\right)^{1/4}. \tag{4.4}$$

The extrapolated central value is attained in a small region usually affected by atmospheric seeing that smears out the innermost profile. The surface-brightness scale of the profile is often given in terms of the value of the surface brightness at R_e, which is formally related to μ_0 by $\mu_e = \mu_0 + 8.325$. Note that a typical value of $\mu(2R_e)$ is 24 to 25 mag arcsec^{-2} in the blue band (i.e., often more than 3 magnitudes fainter than the sky level). The most impressive, radially extended confirmation of the accuracy of the $R^{1/4}$ law is probably the case of NGC 3379, for which the law has been tested over a range of approximately 11 magnitudes,[16] from $\approx 0.01R_e$ to $\approx 7R_e$. The deviations have systematic trends within ± 0.1 magnitudes. When isophotes are not round, the radius to be used in the profile is the circularized radius ($\sqrt{ab}$ for an ellipse of semiaxes a and b). The value of R_e determined by the observations is found to depend significantly on the radial range within which the photometric data are fitted (with variations of up to 50 percent).

There has been some debate as to whether ellipticals are indeed characterized by fully homologous profiles, that is, whether the $R^{1/4}$ law is indeed universal. Many astronomers now carry out their studies with an open mind for laws of the $R^{1/n}$ type

$$I(R) = I_0 \exp\left[-b(n)(R/R_e)^{1/n}\right], \tag{4.5}$$

where n is a free parameter.[17] As for the $R^{1/4}$ law, the parameter $b(n)$ is chosen in such a way that half the total luminosity associated with $I(R)$ is contained in the disk of (projected) radius R_e. An asymptotic expansion of $b(n)$ has been derived[18]

$$b(n) \sim 2n - \frac{1}{3} + \frac{4}{405n} + \frac{46}{25515n^2} \tag{4.6}$$

and shown to be accurate with relative error smaller than 10^{-6} for $n \geq 1$.

So far, significant deviations from $n = 4$ appear to be justified for only low-luminosity galaxies, best fitted with n in the 1 to 2 range.[19] Some interesting exceptions, such as NGC 4374 and NGC 4552, which are best fitted with $n \approx 10$, are noted.[20]

4.1.3 Azimuthal Structure

Some of the morphological aspects mentioned in Chapter 3, in particular, properties of spiral arms and bar structure in galaxy disks and isophotal twisting, diskiness, and boxiness in elliptical galaxies, can be quantified in terms of suitable profiles that can be derived from the photometric studies of galaxy images. A detailed description of these quantitative tools developed to characterize the observed luminosity distributions would bring us beyond the scope of this chapter.

4.2   Doppler Line Shifts and Linewidths

The reconstruction of kinematic properties of galaxies relies on a suitable interpretation of observed line shifts and linewidths induced by the standard Doppler effect. The case of the Milky Way galaxy is special because of our unique experience of being inside the system for which we would like to make the measurements; here kinematic studies have developed somewhat independently. The discovery of differential rotation in our Galaxy, confirming the hypothesis put forward by Lindblad, has been a landmark achievement in galactic astronomy.[21] The main aspects of the kinematics of our Galaxy are briefly recalled in Part III (see Chapters 13 and 14). Here it is mentioned only that reconstruction of the rotation curve for our Galaxy, especially outside the solar circle, is a difficult task for which several open problems remain.[22] The data are consistent with a rather flat rotation velocity $V \approx 220$ km s^{-1} from $0.5R_\odot$ to $\approx 2.5R_\odot$, where $R_\odot$ is the radius of the solar circle. Measurement of the velocity-dispersion ellipsoid (i.e., of the relevant pressure tensor) in the solar neighborhood shows that the energy contained in random motions is small and that the related pressure is definitely anisotropic. An accurate measurement of the velocity-dispersion profile (in particular, of the radial velocity dispersion as a function of galactocentric radius) is very important from a dynamical point of view.[23]

Progress in measuring the kinematics of external galaxies had been made initially, up until the mid-1970s, exclusively for spiral galaxies. The measurements were based on the use of strong emission lines, either the 21-cm line for atomic hydrogen or the emission lines from ionized gas in star-forming regions across the disk. One simple expectation, that in the outer disk the rotation should show the signature of a Keplerian decline ($V \sim R^{-1/2}$), was soon proved to be incorrect,[24] but the potential implications of this fact were not recognized immediately. Eventually this led to the discovery of dark halos, initiating the quest for dark matter in astrophysics. The fact that kinematical tracers similar to those readily available in spiral galaxies are not systematically present in sufficient amounts in elliptical galaxies left the kinematics of ellipticals essentially unexplored until the first measurements[25] in the mid-1970s, which were based on stellar absorption lines (produced in the atmospheres of the stars that provide the observed luminosity), showed dramatically that the flattening of elliptical galaxies is statistically not associated with rotation.

4.2.1   Rotation Curves of Spiral Galaxies

The main kinematic characteristic of the disk of spiral galaxies is that of rotation. The velocity field in the disk plane is given approximately by $\mathbf{V} = r\Omega(r)\mathbf{e}_\theta$; here r and θ are standard polar coordinates, $\mathbf{e}_\theta$ is the unit vector in the tangential direction, and $\Omega(r)$ is the differential rotation. The departures from circular motions of the individual (microscopic) mass elements are generally very small, so that the radial momentum balance for a given component, either atomic hydrogen or disk stars, is dominated by rotation (the pressure gradients would contribute only a few percent):

$$\frac{V^2}{r} \sim \frac{\partial \Phi}{\partial r}. \tag{4.7}$$

The quantity Φ represents the total mean gravitational potential. Thus the observed rotation curve $V(r)$ for a given component is often confused with the circular velocity, that is, the velocity for which the preceding equation holds exactly.

The gas, especially molecular gas, is the coldest component (which is characterized by a very small turbulent speed of a few kilometers per second). Stars, especially those of the old disk, are warmer, being characterized by sizable epicyclic motions, although they are also a rotation-dominated system.[26]

The atomic hydrogen emission is ideally suited for determination of the rotation curves of spiral galaxies because atomic gas is cold (so its mean flow motion gives an accurate direct measurement of the gradient of the mean gravitational potential, much like that for single particles in circular motion) and because it often extends well outside the bright optical disk while still retaining a regular disk shape. There may be signs of nonaxisymmetric structure and warps of the gaseous disk, which raise interesting dynamical questions, but there are many galaxies for which the atomic hydrogen disk is remarkably flat, smooth, and axisymmetric. The latter cases are best suited for probing the underlying gravitational field in its overall properties, especially in relation to the problem of the amount and distribution of dark matter. After the first indications that indeed the rotation curves of spiral galaxies are, except for the rise in the central regions, rather flat,[27] a remarkable series of Ph.D. theses produced at the Kapteyn Astronomical Institute in Groningen has been devoted to the study of accurate and radially extended rotation curves in spiral galaxies[28] (Fig. 4.3). The most rapidly rotating disk galaxy known (with measured

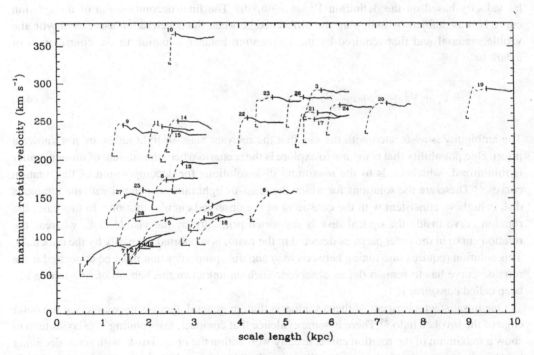

Fig. 4.3. Systematic trends for a set (twenty-eight spiral galaxies) of high-quality HI rotation curves (from Casertano, S., van Gorkom, J. H. "Declining rotation curves: The end of a conspiracy?" 1991. *Astron. J.*, **101**, 1231; reproduced by permission of the AAS). For each curve, rescaled here in arbitrary units, the origin gives the optical exponential scale length (horizontal axis) and the maximum rotation velocity (vertical axis). Each rotation curve is drawn with a thick line outside a reference radius taken to be two-thirds of the optical radius of the galaxy. The dashed part of the rotation curves may be unreliable.

rotation velocity slightly below 500 km s^{-1}) is UGC 12591.[29] An unusual galaxy characterized by similarly high rotation (very close to the center, the rotation curve is unresolved and reaches ≈ 490 km s^{-1}, but then it rapidly declines to a fairly constant asymptotic value of ≈ 240 km s^{-1}) is UGC 4458 (NGC 2599).[30]

There are already some impressive cases from optical rotation curves, derived from emission lines in the ionized gas, in which the rotation curve stays flat in the disk out to very large radii,[31] in galaxies such as UGC 2885 or NGC 801. Yet, as is discussed in Chapter 20, decisive evidence for the presence of dark halos is derived only from radio rotation curves extending well beyond the optical radius. The rotation curve of NGC 3198 remains flat,[32] within less than 10 km s^{-1}, at $V \approx 150$ km s^{-1} from $R \approx 2h$ out to $R \approx 11h$.

When a given rotation curve is measured, there are various options for its decomposition, that is, for determination of the relative share in the contributions of disk, halo, and other components (such as gas or bulge) to the total gravitational field. Under the hypothesis of a constant mass-to-light ratio, the photometric profiles of the disk I_d and the bulge I_b can be converted into the corresponding density profiles $\sigma_\star$ and ρ_b. Similarly, under suitable assumptions on the molecular gas distribution, the atomic hydrogen density profile can be converted into a gas density profile σ_g. Each of these three density components is thus expected to generate a contribution $\partial \Phi_i / \partial r$ to the total gravitational acceleration, which is usually expressed in terms of a circular velocity based on the definition $V_i^2 = r \partial \Phi_i / \partial r$. The final decomposition of the rotation curve is thus defined by assigning the difference between the acceleration associated with the visible material and that required by the momentum balance equation to the contribution of a dark halo

$$V^2 = V_\star^2 + V_b^2 + V_g^2 + r \frac{\partial \Phi_h}{\partial r}. \tag{4.8}$$

The ambiguity is associated with the fact that the relevant mass-to-light ratios are not known a priori. One possibility that is natural to explore is the scenario where the amount of unseen matter is minimized, which leads to the maximum disk solutions for decomposition of the rotation curves.[33] These are the solutions for which the mass-to-light ratio associated with the luminous disk is highest, consistent with the constraint of the observed circular motions. In this case, the rotation curve inside the optical disk is supported primarily by the visible disk, whereas the rotation curve in the outer parts, as detected in the radio, is supported primarily by the dark halo. This solution requires fine-tuning between halo and disk properties that must be concerted if the rotation curve has to remain flat as observed. Such an unknown mechanism of fine-tuning has been called *conspiracy*.

One unsatisfactory aspect of these studies is that, in general, there is no way to see the outer edge of the invoked halo.[34] There is some evidence that compact, fast-rotating[35] galaxies tend to show a maximum of the rotation curve, usually well within the optical disk, with some declining trend in the outer parts, but even two objects with a significant decline outside the optical disk, NGC 2683 and NGC 3521, are found to require the presence of a dark halo of which the data are unable to probe the full radial extent.[36] In turn, for small late-type spirals, rotation curves are generally found to rise out to the outermost kinematic point.[37] Low-surface-brightness galaxies probably do not conform to the maximum-disk hypothesis because, for these galaxies, a maximum-disk decomposition appears to require a value of mass-to-light ratio for the disk that is judged to be excessive.[38]

Attempts have been made to characterize systematic trends in the shape of the rotation curves (in particular, the deviations from a flat profile), especially as functions of Hubble type and of size or luminosity parameters. A general idea is to look for universal properties by a suitable rescaling of the observed rotation curves.[39] There also have been claims that the large-scale properties of rotation curves would correlate not so much with Hubble type but rather with the environment;[40] here the argument is that radially extended dark halos would have to compete with cluster dynamics if the galaxy is not in the field.

Rotation curves also have been measured in galaxy disks on the basis of stellar absorption lines, following the methods developed for elliptical galaxies. One curious phenomenon revealed by the observations is that there may be cases where two stellar disks are counterrotating, that is, rotating around the galaxy in opposite directions. Great excitement was generated by the kinematic study[41] of the E7/S0 galaxy NGC 4550. If confirmed, systems of this type would raise interesting dynamical questions as to their formation and their stability.[42]

Given the coldness of the disks, it is not so easy to obtain measurements of the relevant velocity dispersions for the various components, especially for the stars, to the accuracy level desired by dynamical theories, but general trends can be established. For the cold gas, there are several cases with interesting quantitative determinations.[43] For the stars, some tools have been developed, but they give results mostly for the inner parts of the stellar disks,[44] and the final numbers are affected by sizable uncertainties.

4.2.2 Kinematic Profiles of Elliptical Galaxies

Spectroscopic studies of absorption lines associated with the stars give radial velocity profiles or line-of-sight velocity curves [usually indicated by $V(R)$, in kilometers per second] and velocity-dispersion profiles [usually indicated by $\sigma(R)$, in kilometers per second] that correspond to the observed line shifts and linewidths[45] (Fig. 4.4). As has already been emphasized, the conversion

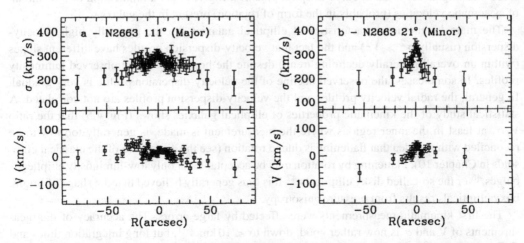

Fig. 4.4. Kinematics of the elliptical galaxy NGC 2663 (from Carollo, C. M., Danziger, I. J. "Colours Line Strengths and Stellar Kinematics of NGC 2663 and NGC 5018," 1994. *Mon. Not. Roy. Astron. Soc.*, **270**, 523; by permission of Oxford University Press, on behalf of the Royal Astronomical Society). The measurements extend out to $\approx 1.8 R_e$.

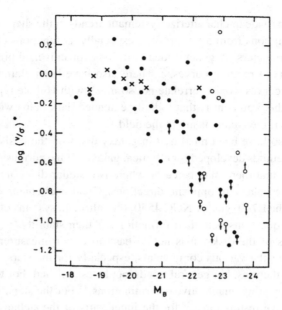

Fig. 4.5. The plot shows the logarithm of the ratio of the observed systematic velocity to velocity dispersion normalized to the value expected for oblate isotropic rotators versus the galaxy absolute magnitude. Filled circles are ellipticals, crosses are bulges, open circles are cD and brightest-cluster galaxies (from Davies, R. L. 1987. In *Structure and Dynamics of Elliptical Galaxies*, ed. T. de Zeeuw. Reidel, Dordrecht, The Netherlands, p. 63; © 1987 by the International Astronomical Union; used with kind permission from Kluwer Academic).

of these profiles into intrinsic kinematic profiles is model-dependent. Here the function $V(R)$ is not simply related to the local gravity field at distance R from the center and does not have the same meaning as the rotation curve of spiral galaxies; it gives only an indication of the amount of systematic velocities (probably in the form of rotation) present in the galaxy.

The main kinematic characteristics of elliptical galaxies are the generally high velocity-dispersion (usually $\sigma^2 \gg V^2$) and the fact that velocity-dispersion profiles have different slopes (within an overall generally declining trend) despite the homology of the observed luminosity profiles. In some cases, the observed decline of the velocity-dispersion profile is very gradual. In general, the radial velocity profiles and the velocity-dispersion profiles are not correlated. A statistical study of the kinematic properties of elliptical galaxies shows (Fig. 4.5) that the ratio V/σ, at least in the inner regions where the measurement is made, is generally too low to be reconciled with the idea that flattening is due to rotation (see the discussion of the classical ellipsoids in Chapter 10). Flattening by rotation may be acceptable for only low-luminosity ellipticals, bulges,[46] or the so-called disky ellipticals.[47] It is thus generally believed that the shape of bright, boxy ellipticals results from pressure anisotropy.

The first kinematic measurements were affected by large errors. The accuracy of the measurements of V and σ is now rather good, down to $< 10\,\mathrm{km\,s^{-1}}$, but long integration times and careful analyses[48] are required for deriving accurate extended profiles beyond R_e. In the outer parts, the radial velocity profiles may rise significantly, showing a behavior that resembles that found for S0 galaxies.[49] A point of major interest in this context, which will be further described in Chapter 24, is the development of kinematic studies based on the diagnostics provided by planetary nebulae.[50]

Further indications that there is a nontrivial relation between the observationally determined projected kinematic profiles and the intrinsic kinematic profiles are derived from the rather frequent observation, especially in the central regions, of curious features such as counterrotation or minor-axis rotation that pose challenging interpretation problems.[51] A kinematic tool that has been the focus of recent interest is determination of the shape of the Doppler distortion of the absorption lines, that is, of the so-called line-of-sight velocity profiles; in most cases, only modest distortions from Gaussian profiles are observed.[52]

4.3 Global Scaling Laws

One of the hopes of studies of galaxies is to establish the existence of global relations involving physically natural quantities (such as total mass M, total gravitational energy E_{tot}, or total angular momentum J_{tot}). For example, some cosmological theories for which the rotation of galaxies is acquired by means of tidal torques between neighboring protogalaxies during hierarchical formation[53] predict the value of a rotation parameter, defined as

$$\lambda = \frac{J_{tot}|E_{tot}|^{1/2}}{GM^{5/2}}, \tag{4.9}$$

which should be close[54] to 0.05. Unfortunately, the data provide only incomplete determinations of the physically interesting parameters. For example, large amounts of angular momentum may be stored in the outer parts of elliptical galaxies, in which the kinematic profiles are less known and the density is likely to decline rather slowly. If we refer to the total mass of a galaxy, our inability to probe the outer boundary of the halo properly automatically implies our failure in making such a measurement; still, a quantity such as the square of a typical velocity times a typical radius is certainly related to the mass. We thus may be content with finding correlations involving physically interesting quantities, as defined empirically, although the final translation into the desired physical variables is generally bound to be model-dependent. The most important steps in establishing the laws that will be outlined later are a proper definition of the quantities involved from an operational point of view (i.e., what is meant by total luminosity, effective radius, central velocity dispersion, or maximum rotational velocity) and the choice of optimal variables from the point of view of physical interest. The reader is referred to the articles cited for a discussion of these important aspects.

Some global relations would be especially exciting because they would break the scale-free character of pure dynamics. Here, as an example (see Part IV), we may recall that self-consistent stellar dynamical models can be constructed with realistic density profiles for the description of either globular clusters or elliptical galaxies. However, those models have free scales (which can be taken to match the observed scales) that the dynamical theory does not take into account; the mass scale and the length scale can be set independently of each other, and the virial theorem relation only requires that the velocity scale be consistent with them. Another way to state this is that in a purely dynamical study of stellar systems there is no way to set a preferred mass scale.

However, physical processes other than pure gravitational forces during formation[55] and possibly during evolution may act to enforce some scaling laws. Empirically established global scaling relations thus would provide clues on the formation and evolution of galaxies. In addition, because dimensional relations are involved, the scaling laws would have natural applications in cosmology, especially to the problem of distance determination.[56]

One of the first attempts[57] in the direction of global scaling laws was to propose a relation between mass and radius in the form

$$M \sim R^2. \tag{4.10}$$

We will see later that, to some extent, under a suitable assumption on the relevant mass-to-light ratio, the main scaling laws established for spiral and elliptical galaxies may be connected with a relation of this form.

One important aspect of scaling laws is that they generally involve quantities such as the total luminosity that are determined by only the visible mass and others such as the rotation velocity that depend on the potential well and thus on the presence of dark matter. The small scatter in the scaling laws is further evidence for the so-called conspiracy problem.

4.3.1 Luminosity-Velocity Relation for Spiral Galaxies

For spiral galaxies, a very important relation has been found between total luminosity and (suitably defined) rotation velocity in the form[58]

$$L \propto V^p. \tag{4.11}$$

The value of the parameter p is known to be wave-band-dependent. It is natural to anticipate that if it traces an intrinsic property of the mass distribution, the law should be best derived, with smallest scatter, in the near IR,[59] where the parameter p is found to be close to 4. The scatter in velocity at a given value of luminosity is ≈ 5 percent. A relatively recent study, using $B, R, I,$ and K' surface photometry, of a volume-limited complete sample of spiral galaxies in the Ursa Major cluster[60] (Fig. 4.6) shows that with a proper definition of the rotation velocity (the amplitude of the outer flat parts of the rotation curve, measured in HI), the tightest correlation is found for the absolute magnitude in K' (2.1 μm) versus $\log V$, with "a slope -10.3 ± 0.4 and a total observed scatter of 0.29 mag, consistent with no intrinsic scatter" (we recall that -10 corresponds to $p = 4$).

A break in the relation, with a significant increase in the parameter p, has been noted for low-velocity, gas-rich galaxies ($V < 90$ km s^{-1}). The continuity in the slope of the relation is restored if the luminosity of the galaxy is replaced by the baryonic mass, defined as the sum of the stellar mass, obtained from the luminosity on the basis of reasonable assumptions on the stellar populations responsible for the optical emission, and of the gas mass, obtained from the HI data.[61]

Several projects are under way, aimed at studying the evolution of this scaling law with redshift, but nontrivial problems make it difficult to obtain well-established trends.[62]

4.3.2 The Rising Part of the Rotation Curves of Spiral Galaxies

The general statement that the inner parts of the rotation curve can be explained in terms of the visible matter alone, with no need to invoke a dark halo (this point will be addressed

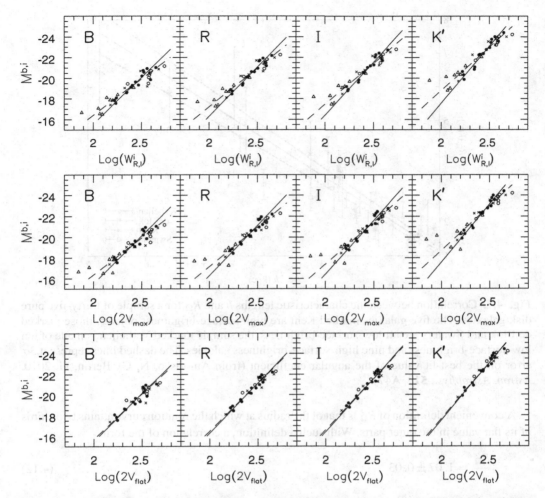

Fig. 4.6. Luminosity–velocity relation for a sample of spiral galaxies in the Ursa Major cluster (from Verheijen, M. A. W. 1997. Ph.D. thesis, University of Groningen, The Netherlands, p. 204), based on different passbands (*from left to right*) and different definitions of the relevant velocity variable (*top to bottom*), based on *Westerbork* radio observations.

further in Chapter 20), or, in other words, that a maximum-disk decomposition of rotation curves is generally viable requires that the rising part of the rotation curve be characterized by a scale length R_Ω well in tune with the scale length of the visible matter. For simple disk galaxies, that is, spiral galaxies with no bulge and negligible amounts of gas, we thus would expect (see Subsection 14.3.1) that the exponential scale h of the optical photometric profile correlates with R_Ω. Given the ubiquity and major role of dark matter in galaxies and the different formation histories of dark halos and visible-matter components, such correlation is surprising and defines one additional aspect of the conspiracy problem mentioned earlier.

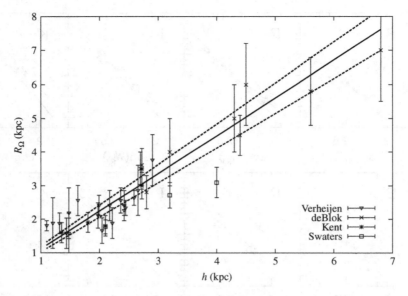

Fig. 4.7. Correlation between the characteristic lengths h and R_Ω for a sample of thirty-five pure disk galaxies. The five galaxies marked Kent are high-surface-brightness objects, those marked de Blok and Swaters are all low-surface-brightness, the sample marked Verheijen is made of ten low-surface-brightness and nine high-surface-brightness galaxies. The dashed lines represent 3σ error on the best-fit value of the angular coefficient (from Amorisco, N. C., Bertin, G. 2010. *Astron. Astrophys.*, **519**, A47).

A convenient definition of R_Ω is that of the radius at which the rotation curve attains two-thirds of its flat value in the outer parts. With such a definition, a correlation of the form

$$\frac{R_\Omega}{h} = 1.07 \pm 0.03 \tag{4.12}$$

has indeed been found[63] for a sample of thirty-five spiral galaxies, twenty-one of which are classified as low-surface-brightness galaxies (Fig. 4.7). This suggests that the distribution of dark and visible matter in pure disk galaxies is homologous, that is, characterized by similar global relations in different galaxies, even if significantly different in size and type. To some extent, the physical meaning of this scaling law is contained in the finding that if radii are expressed in optical-disk scale-lengths, the rotation curves of the low-surface-brightness galaxies and those of high-surface-brightness galaxies often have almost identical shape.[64]

4.3.3   The Fundamental Plane of Elliptical Galaxies

A relation similar to Eq. (4.11) has been proposed and studied for elliptical galaxies:[65]

$$L \propto \sigma^4. \tag{4.13}$$

Here σ denotes the central velocity dispersion measured along the line of sight. The observed correlation presents a relatively large scatter.[66]

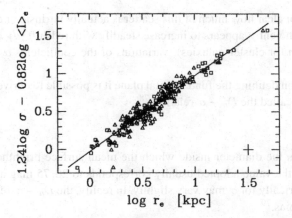

Fig. 4.8. The fundamental plane (seen edge on) for a sample of 226 ellipticals and S0s in ten clusters of galaxies (from Jørgensen, I., Franx, M., Kjærgaard, P. "The fundamental plane for cluster E and S0 galaxies," 1996. *Mon. Not. Roy. Astron. Soc.*, **280**, 167; by permission of Oxford University Press, on behalf of the Royal Astronomical Society).

The scaling law just proposed reflects the existence of a better correlation, known as the *fundamental plane*,[67] which is often given in the form

$$R_e \propto \sigma^{1.35} I_e^{-0.84}. \tag{4.14}$$

Here R_e is the effective radius (measured in kiloparsecs) and I_e is the surface brightness at the effective radius. The exponents found in the correlation typically vary by ± 0.15 depending on the data set[68] and on the procedure used for their determination. For a sample of 226 Es and S0s in ten clusters of galaxies,[69] the exponents found for σ and I_e are 1.24 and -0.82, respectively, with a scatter of ≈ 0.07 in $\log R_e$ or ≈ 15 percent in R_e (Fig. 4.8); when used for distance determinations, the scatter corresponds to less than 20 percent uncertainty on the distance estimated for a galaxy.

An alternative, often-used representation is

$$\log R_e = \alpha \log \sigma_0 + \beta SB_e + \gamma. \tag{4.15}$$

Here σ_0 is defined[70] as the velocity dispersion referred to an aperture radius of $R_e/8$. The mean surface brightness is defined as $SB_e = -2.5 \log \langle I \rangle_e$, where $\langle I \rangle_e = L/(2\pi R_e^2)$. The coefficients α, β, and γ depend slightly on the photometric band considered. By measuring R_e in kiloparsecs, σ_0 in kilometers per second, and $SB_e = 42.0521 - 2.5 \log [L/(2\pi R_e^2)]$ in magnitudes per seconds of arc squared, where L is expressed in units of the solar blue luminosity, reported[71] values are $\alpha = 1.25 \pm 0.1$, $\beta = 0.32 \pm 0.03$, $\gamma = -8.895$ (note that this value of γ refers to a long cosmological distance scale, i.e., a Hubble constant $H_0 = 50$ km s^{-1} Mpc^{-1}).

Several studies have addressed the issue of reducing the scatter by modifying the definition of σ in the presence of significant values of rotation velocity and including a metallicity term.[72] There also may be systematic differences between cluster galaxies and field galaxies.[73]

Great progress has been made at measuring the properties of the fundamental plane out to cosmological distances at relatively high redshifts (up to $z \approx 1.1$) for cluster galaxies[74] and in the field,[75] with a slight increase of scatter with redshift, which is estimated to be ≈ 23 percent in R_e at $z = 1$. It is not clear how much of this scatter is actually intrinsic or due to observational precision.[76] The coefficient γ appears to increase steadily with redshift ($d\gamma/dz \approx 0.58$, possibly with slower evolution for cluster galaxies); variations of the coefficients α and β appear to be less significant.[77]

From the correlation defining the fundamental plane it is possible to derive another often-used scaling law, which is called the $D_n - \sigma$ relation:[78]

$$D_n \propto \sigma^{1.35}. \tag{4.16}$$

Here D_n is defined as the diameter inside which the mean surface brightness equals a chosen reference value $\langle I \rangle_n$; the value adopted usually corresponds to 20.75 mag arcsec^{-2} (in B). The exponent found empirically for σ may vary slightly. In reality, the $D_n - \sigma$ relation is affected by a surface-brightness bias.[79]

Finally, another empirical scaling relation that involves only photometric quantities is the $I_e - R_e$ relation,[80] with an indication of a change of character[81] in the properties of small and large ellipticals (with respect to a reference value of R_e).

The fundamental plane relation appears to extend in the direction of both small stellar systems and large self-gravitating systems.[82] For globular clusters, the fundamental plane may hold with similar coefficients;[83] curiously, in the standard fundamental plane coordinates, the set of points is found to occupy a rather slim, axisymmetric, cylindrical region of parameter space, which suggests that the relevant scaling relation might be around a line rather than a plane.[84] In the resulting $\log R_e - M_V$ photometric plane, globular clusters appear to have a distinct behavior with respect to that of dwarf spheroidals.[85] The fundamental plane relation also has been studied for clusters of galaxies.[86]

4.3.4   The Mass of the Central Black Hole – Velocity Dispersion Relation for Spheroids

As noted in Chapter 1, one of the most exciting discoveries of the last decades has been the finding that even nonactive galaxies generally host a central supermassive black hole, with mass $M_\bullet$ in the range 10^6 to 10^9 $M_\odot$. It has soon become apparent that the mass of the central black hole correlates with other global properties of the spheroid in which it resides.[87] The most interesting correlation is that with the (projected) central velocity dispersion σ_0 of the hosting spheroid

$$M_\bullet \propto \sigma_0^q. \tag{4.17}$$

The commonly accepted value for the exponent is $q \approx 4$; the scatter around this relation is very small.[88] A related correlation is with the estimated stellar mass of the hosting spheroid ($M_\bullet \sim 1.4 \times 10^{-3} M_\star$).[89] Other studied correlations are those with the index n (see Subsection 4.1.2) of the photometric profile[90] and with the number of globular clusters[91] of the host spheroid.

4.4   **Dark Matter and Cosmology**

Many of the currently hot topics and unresolved puzzles in astrophysics can be traced to the issue of dark matter.[92] As outlined in Chapter 2, sophisticated telescopes and instrumentation from the ground and from space allow us to explore the Universe in several windows across the whole spectrum of electromagnetic waves – from radio waves to X-rays. In a number of different contexts (in galactic and extragalactic astronomy), it is now generally accepted that dark matter (i.e., matter that is not observed directly through its electromagnetic radiation) must be present in significant amounts.[93]

Yet no clear-cut answer is available as to what makes up dark matter, whether it is made of low-luminosity celestial objects of the type we know (such as cold gas, planets, or dwarf stars, collectively referred to as *baryonic dark matter*), or of black holes, or (as is by now generally believed) of some so-far-undetected elementary particles that may be the relic of the early Big Bang (in this latter case we often talk about *nonbaryonic dark matter*). Several projects are under way with the goal of identifying possible candidates of such unseen material.[94]

The frustration arising from our inability to identify what makes up dark matter is so great that some scientists have come to believe that the laws of physics, as discovered on the laboratory scale, no longer hold or rather should be revised when astrophysical scales are involved (see Section 20.5). In this case, dark matter would be like cosmic ether, something that we use to hide our ignorance of the laws of physics. In fact, even by simply looking at pictures of a globular cluster or a galaxy, we have direct evidence that gravitation dominates, but "of course we cannot prove that the law here is precisely inverse square, only that there is still an attraction, at this enormous dimension, that holds the whole thing together."[95] Indeed, the measurements that currently demand the existence of unseen material are precisely the experiments that should test the validity of the laws of physics, and especially the validity of the law of gravitation, on the large scale of galactic and intergalactic distances.

It should be recognized, however, that the laws of physics learned from the laboratory and a small number of general principles of symmetry and simplicity have led to the Big Bang cosmological framework, which has been rewarded by remarkable cases of observational confirmation (such as measurement of the cosmic background radiation) despite the persistence of many unresolved mysteries.[96] Strengthened by such success, for a long time cosmological theorists have provided significant pressure in the direction of a scenario in which $\Omega_m = \rho_0/\rho_{crit} = 1$ (defined later), thus contributing much of the excitement and motivation to focus on the issue of dark matter in astrophysics. The main relation between cosmology and the problem of dark matter, as described in the context of the dynamics of galaxies (see Chapters 20 and 24), is through the comparison between the cosmologically interesting density ρ_{crit} and the actual density ρ_0 as determined from observations. It has long been evident that the amount of mass associated with visible material yields a very small value of ρ_0. The need for dark matter so clearly established by observations on the galactic scale soon became a strong motivation to allow for the possibility that ρ_0 is actually much higher than that directly observed. However, it should be stressed that the dark halos that are inferred to be around galaxies have proven to be insufficient from a cosmological point of view; in particular, they might contribute and bring the present density up to the value of ρ_{crit} only if they extend out to galactocentric radii $r_{halo} \approx 2.5h^{-1}$ Mpc (the definition of the dimensionless parameter h, of common usage in the cosmological context, is given later). In this respect, although much work through the 1970s and 1980s strengthened the case for dark matter around galaxies, the gap between the dynamically determined mass density for the Universe and the value of ρ_{crit} has become sharper and sharper.

Within the classical homogeneous Friedmann-Lemaître cosmological models, an empirical determination of the type of Universe we are in requires the measurement of three parameters: the Hubble constant H_0, the density parameter Ω_m, and the cosmological constant Λ. The three terms appear in the expansion equation, which for a matter-dominated Universe can be written as

$$H^2 \equiv \left(\frac{\dot{a}}{a}\right)^2 = H_0^2[\Omega_m(1+z)^3 + \Omega_R(1+z)^2 + \Omega_\Lambda], \qquad (4.18)$$

where $\Omega_m + \Omega_R + \Omega_\Lambda = 1$ and $\Omega_\Lambda \equiv \Lambda/(3H_0^2)$, and scale differently with cosmic time, so the relative ordering of gravity, curvature, and cosmological constant contributions is rapidly changing if we move backward to early epochs (i.e., to large values of the cosmological redshift z), when the gravity term is bound to dominate. Similarly, the cosmological acceleration equation for the universal expansion factor $a(t)$ can be written as

$$\frac{\ddot{a}}{a} = H_0^2\left[\Omega_\Lambda - \frac{\Omega_m(1+z)^3}{2}\right]. \qquad (4.19)$$

If $\Omega_\Lambda > 0$, in the course of evolution, the Universe may change from a state of deceleration to a state of acceleration.

Until the end of last century, the most interesting astrophysical constraints were on the first two parameters, and the cosmological constant was often tacitly neglected. The Hubble constant is usually written as $H_0 = 100h$ km s^{-1} Mpc^{-1}, and many independent astronomical studies had come to the conclusion that h is in the range between 0.5 and 0.9. A low value of the Hubble constant corresponds to a long distance scale and to a long time scale for the age of the Universe. The density parameter is proportional to the current mean density of the Universe ρ_0, being defined as $\Omega_m = \rho_0/\rho_{\text{crit}} = 8\pi G\rho_0/(3H_0^2)$. The Einstein–de Sitter model considers Λ as negligible and sets $\Omega_m = 1$, with a specific relation between Hubble time ($1/H_0 = 0.98 \times 10^{10}h^{-1}$ yr) and age of the Universe t_0, given by $H_0t_0 = 2/3$. Note that the critical density that is associated with the Einstein–de Sitter solution is $\rho_{\text{crit}} = 1.88 \times 10^{-29}h^2$ g cm^{-3}. Especially as a result of a long-term project based on the study of Cepheids in the Virgo cluster, the accepted value of the cosmological distance scale is now converging[97] to $h \approx 0.7$.

At the end of the last century, some interesting projects based on observation of Type Ia supernovae at high redshifts[98] showed that the Universe is accelerating and thus led to a revival of interest in the role of the cosmological constant and its physical interpretation as manifestation of dark energy. At present, a vast body of measurements, which include redetermination of the Hubble constant,[99] study of the baryon acoustic oscillations in the galaxy distribution,[100] and mapping of the cosmic microwave background radiation,[101] is leading to a concordance cosmological model in which the age of the Universe is $t_0 \approx 13.75 \times 10^9$ yr, the matter density parameter is $\Omega_m \approx 0.27$ (with dark to baryonic matter densities approximately in the ratio 5:1), and the dark-energy parameter is $\Omega_\Lambda \approx 0.73$.

We should recall that the observed degree of homogeneity of the Universe appears to be hard to reconcile within the classical Friedmann-Lemaître cosmological models with the existence of the horizon problem (parts of the Universe that are not in causal contact appear to have relaxed into a common state). The currently accepted theoretical framework to resolve this difficulty is that of the inflationary scenario,[102] which leads to a Universe with negligibly small space curvature.

In the early 1980s, much interest grew in the possibility that dark matter might be made of massive neutrinos,[103] especially after some experiments initially indicated a neutrino mass of the right order of magnitude.[104] Massive neutrinos are often referred to as *hot dark matter* because at the time of decoupling, when during the cooling of the Universe the plasma combines into atomic hydrogen and matter is released from the radiation drag, these particles would be characterized by high relativistic velocities. In the mid-1980s, much attention was directed toward the possibility of *cold dark matter* (particles more massive than 1 keV), with the great disadvantage that the candidate particles of this type have to be taken within a large set of exotic theoretical options that so far have not been detected in the laboratory (such as photinos,[105] usually called *WIMPs*, for weakly interacting massive particles). The current concordance cosmological model is based on the cold-dark-matter hypothesis.

Notes

1. The reader is referred to other books and reviews that address these issues thoroughly. For the general astronomical framework, see Mihalas, D., Binney, J. 1982. *Galactic Astronomy*, Freeman, San Francisco; Binney, J., Merrifield, M. R. 1998. *Galactic Astronomy*, Princeton University Press, Princeton, NJ. The surface photometry of elliptical galaxies has been reviewed by Kormendy, J., Djorgovski, S. 1989. *Annu. Rev. Astron. Astrophys.*, **27**, 235. Important aspects of the structure and kinematics of spiral galaxies are discussed in great detail by Verheijen, M. A. W. 1997. Ph.D. thesis, University of Groningen, The Netherlands. A recent summary is provided by Blanton, M. R., Moustakas, J. 2009. *Annu. Rev. Astron. Astrophys.*, **47**, 159, and by van der Kruit, P. C., Freeman, K. C. 2011. *Annu. Rev. Astron. Astrophys.*, **49**, 301.
2. For a review of the problem of dark matter in galaxies, see Ashman, K. M. 1992. *Publ. Astron. Soc. Pac.*, **104**, 1109; Sancisi, R. 2004. In *Dark Matter in Galaxies*, eds. S. D. Ryder, D. J. Pisano et al. Astronomical Society of the Pacific, San Francisco, p. 233. A modern general discussion of the problem is given by Sanders, R. H. 2010. *The Dark Matter Problem: A Historical Perspective*, Cambridge University Press, Cambridge, UK.
3. Disney, M. J., Phillips, S. 1983. *Mon. Not. Roy. Astron. Soc.*, **205**, 1023; Binggeli, B., Sandage, A., Tammann, G. A. 1988. *Annu. Rev. Astron. Astrophys.*, **26**, 509.
4. Binggeli, B., Sandage, A., Tarenghi, M. 1984. *Astron. J.*, **89**, 64; Binggeli, B., Sandage, A., Tammann, G. A. 1985. *Astron. J.*, **90**, 1681; Impey, C., Bothun, G. 1997. *Annu. Rev. Astron. Astrophys.*, **35**, 267; Zhong, G. H., Liang, Y. C., et al. 2008. *Mon. Not. Roy. Astron. Soc.*, **391**, 986.
5. Peletier, R. F. 1989. Ph.D. thesis, University of Groningen, The Netherlands.
6. van der Kruit, P. C., Searle, L. 1981. *Astron. Astrophys.*, **95**, 105 and 116; 1982. *Astron. Astrophys.*, **110**, 61; Pohlen, M., Dettmar, R.-J., Lütticke, R. 2000. *Astron. Astrophys.*, **357**, L1; Kregel, M., van der Kruit, P. C., de Grijs, R. 2002. *Mon. Not. Roy. Astron. Soc.*, **334**, 646; Pohlen, M., Trujillo, I. 2006. *Astron. Astrophys.*, **454**, 759.
7. See the model of NGC 5907 by Casertano, S. 1983. *Mon. Not. Roy. Astron. Soc.*, **203**, 735. See also van der Kruit, P. C. 2007. *Astron. Astrophys.*, **466**, 883.
8. Freeman, K. C. 1970. *Astrophys. J.*, **160**, 811.
9. For example, see Shaw, M. A., Gilmore, G. 1989. *Mon. Not. Roy. Astron. Soc.*, **237**, 903.
10. Courteau, S., de Jong, R. S., Broeils, A. H. 1996. *Astrophys. J.*, **457**, L73; Carollo, C. M., Stiavelli, M., et al. 2001. *Astrophys. J.*, **546**, 216; Carollo, C. M., Stiavelli, M., et al. 2002. *Astron J.*, **123**, 159; Seigar, M., Carollo, C. M., et al. 2002. *Astron J.*, **123**, 184.
11. Freeman, K. C. 1970. *op. cit.*

12. McGaugh, S. S., Bothun, G. D., Schombert, J. M. 1995. *Astron. J.*, **110**, 573; McGaugh, S. S. 1996. *Mon. Not. Roy. Astron. Soc.*, **280**, 337; de Jong, R. S. 1995. Ph.D. thesis, University of Groningen, The Netherlands; de Jong, R. S. 1996. *Astron. Astrophys.*, **313**, 45.

13. Tully, R. B., Verheijen, M. A. W. 1997. *Astrophys. J.*, **484**, 145; they formally split low-surface-brightness galaxies from high-surface-brightness galaxies by using the reference value $\mu_0 = 18.5$ mag arcsec^{-2} in the K' band (2.1 μm).

14. Binggeli, B., Sandage, A., Tarenghi, M. 1984. *op. cit.*

15. de Vaucouleurs, G. 1948. *Ann. Astrophys.*, **11**, 247; 1953. *Mon. Not. Roy. Astron. Soc.*, **113**, 134. See also Kormendy, J. 1977. *Astrophys. J.*, **218**, 333; King, I. R. 1978. *Astrophys. J.*, **222**, 1; Schweizer, F. 1979. *Astrophys. J.*, **233**, 23.

16. de Vaucouleurs, G., Capaccioli, M. 1979. *Astrophys. J. Suppl.*, **40**, 699.

17. Sérsic, J.-L. 1968. *Atlas de Galaxias Australes*, Observatorio Astronomico de Córdoba, Córdoba, Argentina; Ciotti, L. 1991. *Astron. Astrophys.*, **249**, 99; Caon, N., Capaccioli, M., D'Onofrio, M. 1993. *Mon. Not. Roy. Astron. Soc.*, **265**, 1013.

18. Ciotti, L., Bertin, G. 1999. *Astron. Astrophys.*, **352**, 447.

19. Binggeli, B., Cameron, L. M. 1991. *Astron. Astrophys.*, **252**, 27; see fig. 3 in Prugniel, P., Simien, F. 1997. *Astron. Astrophys.*, **321**, 111. For an application to bulges, see Andredakis, Y. C., Peletier, R. F., Balcells, M. 1995. *Mon. Not. Roy. Astron. Soc.*, **275**, 874.

20. Caon, N., Capaccioli, M., Rampazzo, R. 1990. *Astron. Astrophys. Suppl.*, **86**, 429; Caon, N., Capaccioli, M., D'Onofrio, M. 1994. *Astron. Astrophys. Suppl.*, **106**, 199; Bertin, G., Ciotti, L., Del Principe, M. 2002. *Astron. Astrophys.*, **386**, 149.

21. See Oort, J. H. 1927. *Bull. Astron. Inst. Netherlands*, **3** (120), 275, and references therein.

22. See Gilmore, G., Wyse, R. F. G., Kuijken, K. 1989. *Annu. Rev. Astron. Astrophys.*, **27**, 555; Fich, M., Tremaine, S. 1991. *Annu. Rev. Astron. Astrophys.*, **29**, 409.

23. For this, very interesting indications come from the work of Lewis, J. R., Freeman, K. C. 1989. *Astron. J.*, **97**, 139, who find an approximate exponential decline for the radial velocity dispersion in the old stellar disk out to $\approx 2R_\odot$.

24. See the data for M31 by van de Hulst, H. C., Raimond, E., van Woerden, H. 1957. *Bull. Astron. Inst. Netherlands*, **14**, 1.

25. Starting with Bertola, F., Capaccioli, M. 1975. *Astrophys. J.*, **200**, 439; Illingworth, G. D. 1977. *Astrophys. J. Lett.*, **218**, L43.

26. For a discussion of the kinematic aspects that differentiate gas from stars (which plays a major role in the dynamics of spiral structure), see the introductory chapters of Bertin, G., Lin, C. C. 1996. *Spiral Structure in Galaxies: A Density Wave Theory*. MIT Press, Cambridge, MA.

27. See Roberts, M. S. 1976. *Comm. Astrophys.*, **6**, 105; Rubin, V. C. 1978. In *The Large-Scale Characteristics of the Galaxy*, ed. W. B. Burton. Reidel, Dordrecht, The Netherlands, p. 211; Faber, S. M., Gallagher, J. S. 1979. *Annu. Rev. Astron. Astrophys.*, **17**, 135, and the many references cited.

28. Among others, we should mention the theses by Bosma, A. 1978; Begeman, K. G. 1987; Broeils, A. H. 1992; de Blok, W. J. G. 1997; Verheijen, M. A. W. 1997; Swaters, R. 1999 (late-type, low-luminosity galaxies); Noordermeer, E. 2006 (early-type, high-luminosity galaxies).

29. Giovanelli, R., Haynes, M. P., et al. 1986. *Astrophys. J. Lett.*, **301**, L7.

30. Noordermeer, E., van der Hulst, J. M., et al. 2007. *Mon. Not. Roy. Astron. Soc.*, **376**, 1513.

31. Rubin, V. C., Ford, W. K., Thonnard, N. 1978. *Astrophys. J. Lett.*, **225**, L107. A systematic study of optical rotation curves in spiral galaxies along the Hubble sequence has been performed by Rubin, V. C., Ford, W. K., Thonnard, N. 1980. *Astrophys. J.*, **238**, 471; Rubin, V. C., Ford, W. K., et al. 1982. *Astrophys. J.*, **261**, 439; Rubin, V. C., Burstein, D., et al. 1985. *Astrophys. J.*, **289**, 81. See also Rubin, V. C. 1987. In *Dark Matter in the Universe*, eds. J. Kormendy, G. R. Knapp. Reidel, Dordrecht, The Netherlands, p. 51.

32. van Albada, T. S., Bahcall, J. N., et al. 1985. *Astrophys. J.*, **295**, 305.

33. van Albada, T. S., Sancisi, R. 1986. *Phil. Trans. Roy. Soc. London, A*, **320**, 447.

34. Interesting constraints on the actual size of dark halos have been derived by means of galaxy-galaxy lensing studies; see Casertano, S., Griffiths, R. E., Ratnatunga, K. U. 2001. In *Gravitational Lensing: Recent Progress and Future Goals*, eds. T. G. Brainerd, C. S. Kochanek. Astronomical Society of the

Pacific, San Francisco, p. 387; Hoekstra, H., Franx, M., et al. 2003. *Mon. Not. Roy. Astron. Soc.*, **340**, 609.

35. Above ≈ 150 km s^{-1}. But see the interesting case of the large galaxy UGC 12591 with a rather flat rotation curve at ≈ 500 km s^{-1} from 10 to 40 kpc; see Rubin, V. C. 1987. *op. cit.*

36. Casertano, S., van Gorkom, J. H. 1991. *Astron. J.*, **101**, 1231.

37. Carignan, C., Freeman, K. C. 1985. *Astrophys. J.*, **294**, 494.

38. de Blok, W. J. G., McGaugh, S. S. 1996. *Astrophys. J.*, **469**, 89; de Blok, W. J. G., McGaugh, S. S., van der Hulst, T. 1996. *Mon. Not. Roy. Astron. Soc.*, **283**, 18; Verheijen, M. A. W. 1997. *op. cit.*; see also Swaters, R. 1999. Ph.D. thesis, University of Groningen, The Netherlands; Swaters, R. A., Sancisi, R., et al. 2011. *Astrophys. J.*, **729**, id.118.

39. Burstein, D., Rubin, V. C., et al. 1982. *Astrophys. J.*, **253**, 70; Persic, M., Salucci, P. 1991. *Astrophys. J.*, **368**, 60. In the article by Persic, M., Salucci, P., Stel, F. 1996. *Mon. Not. Roy. Astron. Soc.*, **281**, 27, such a universal rotation curve is determined empirically by proper averaging over a sample of 714 optical rotation curves and 33 HI rotation curves; an accurate study of a complete sample of galaxies in the Ursa Major cluster shows that about one-third of the objects defy the proposed universal rotation curve; see Verheijen, M. A. W. 1997. *op. cit.*

40. Burstein, D., Rubin, V. C., et al. 1986. *Astrophys. J. Lett.*, **305**, L11.

41. Rubin, V. C., Graham, J. A., Kenney, J. D. P. 1992. *Astrophys. J. Lett.*, **394**, L9; Rix, H.-W., Franx, M., et al. 1992. *Astrophys. J. Lett.*, **400**, L5.

42. Bertin, G., Cava, A. 2006. *Astron. Astrophys.*, **459**, 333.

43. In particular, see Dickey, J. M., Hanson, M. M., Helou, G. 1990. *Astrophys. J.*, **352**, 222.

44. For example, see Bottema, R., van der Kruit, P. C., Freeman, K. C. 1987. *Astron. Astrophys.*, **178**, 77; Bottema, R. 1995. Ph.D. thesis, University of Groningen, The Netherlands.

45. Several techniques have been devised to derive this information, either by cross-correlation or by Fourier quotient, based on comparison of the spectra with those of template stars. See Tonry, J. L., Davis, M. 1979. *Astron. J.*, **84**, 1511; Sargent, W. L. W., Schechter, P. L., et al. 1977. *Astrophys. J.*, **212**, 326; Bender, R. 1990. *Astron. Astrophys.*, **229**, 441; Rix, H.-W., White, S. D. M. 1992. *Mon. Not. Roy. Astron. Soc.*, **254**, 389.

46. Davies, R. L., Efstathiou, G., et al. 1983. *Astrophys. J.*, **266**, 41. A very recent analysis of the available data about slow and fast rotators is given by Emsellem, E., Cappellari, M., et al. 2011. *Mon. Not. Roy. Astron. Soc.*, **414**, 888.

47. Bender, R. 1988. *Astron. Astrophys.*, **193**, L7.

48. Davis, R. L., Birkinshaw, M. 1988. *Astrophys. J. Suppl.*, **68**, 409; Saglia, R. P., Bertin, G., et al. 1993. *Astrophys. J.*, **403**, 567; Bertin, G., Bertola, F., et al. 1994. *Astron. Astrophys.*, **292**, 381; Carollo, C. M., de Zeeuw, P. T., et al. 1995. *Astrophys. J. Lett.*, **441**, L25.

49. For the case of NGC 3115, see Capaccioli, M., Cappellaro, E., et al. 1993. *Astron. Astrophys.*, **274**, 69.

50. See Napolitano, N. R., Romanowsky, A. J., et al. 2011. *Mon. Not. Roy. Astron. Soc.*, **411**, 2035, and references therein.

51. Bender, R. 1988. *Astron. Astrophys.*, **202**, L5; Jedrzejewski, R. J., Schechter, P. L. 1988. *Astrophys. J. Lett.*, **330**, L87; Franx, M., Illingworth, G. D. 1988. *Astrophys. J. Lett.*, **327**, L55.

52. Franx, M., Illingworth, G. D. 1988. *op. cit.*; Bender, R. 1990. *op. cit.*; Rix, H.-W., White, S. D. M. 1992. *op. cit.*; Winsall, M. L., Freeman, K. C. 1993. *Astron. Astrophys.*, **268**, 443; Gerhard, O. 1993. *Mon. Not. Roy. Astron. Soc.*, **265**, 213; van der Marel, R., Franx, M. 1993. *Astrophys. J.*, **407**, 525; Kuijken, K., Merrifield, M. R. 1993. *Mon. Not. Roy. Astron. Soc.*, **264**, 712; Bender, R., Saglia, R. P., Gerhard, O. 1994. *Mon. Not. Roy. Astron. Soc.*, **269**, 785; Carollo, C. M., de Zeeuw, P. T., et al. 1995. *op. cit.*; Gerhard, O., Kronawitter, A., et al. 2001. *Astron. J.*, **121**, 1936.

53. Strömberg, G. 1934. *Astrophys. J.*, **79**, 460; Hoyle, F. 1949. In *Problems in Cosmical Aerodynamics*, eds. J. M. Burgers, H. C. van de Hulst. Central Air Documents Office, Dayton, OH, p. 195; Peebles, P. J. E. 1969. *Astrophys. J.*, **155**, 393. See also Fall, S. M., Efstathiou, G. 1980. *Mon. Not. Roy. Astron. Soc.*, **193**, 189.

54. See the simulations by Barnes, J., Efstathiou, G. 1987. *Astrophys. J.*, **319**, 575.

55. For example, see the arguments provided by Silk, J. 1977. *Astrophys. J.*, **211**, 638; Rees, M. J., Ostriker, J. P. 1977. *Mon. Not. Roy. Astron. Soc.*, **179**, 541.

56. For example, see Jacoby, G. H., Branch, D., et al. 1992. *Publ. Astron. Soc. Pacific*, **104**, 599; Giovanelli, R., Haynes, M. P., et al. 1997. *Astrophys. J. Lett.*, **477**, L1; Bender, R., Saglia, R. P., et al. 1998. *Astrophys. J.*, **493**, 529.

57. Fish, R. 1964. *Astrophys. J.*, **139**, 284; see also Kaastra, J. S., van Bueren, H. G. 1981. *Astron. Astrophys.*, **99**, 7, for the context of clusters of galaxies.

58. Tully, R. B., Fisher, J. R. 1977. *Astron. Astrophys.*, **54**, 661.

59. Aaronson, M., Huchra, J., et al. 1982. *Astrophys. J. Suppl.*, **50**, 241.

60. Verheijen, M. A. W. 1997. *op. cit.*; 2001. *Astrophys. J.*, **563**, 694.

61. This corrected relation is referred to as the *baryonic Tully-Fisher relation*; see McGaugh, S. S., Schombert, J. M., et al. 2000. *Astrophys. J. Lett.*, **533**, L99; McGaugh, S. S. 2012. *Astron. J.*, **143**, 40, and references therein.

62. For example, see Conselice, C. J., Bundy, K., et al. 2005. *Astrophys. J.*, **628**, 160; Fernández Lorenzo, M., Cepa, J., et al. 2010. *Astron. Astrophys.*, **521**, A27; Miller, S. H., Bundy, K., et al. 2011. *Astrophys. J.*, **741**, 115; Miller, S. H., Ellis, R. S., et al. 2012. *Astrophys. J.*, **753**, id.74, and references therein.

63. Amorisco, N. C., Bertin, G. 2010. *Astron. Astrophys.*, **519**, A47. Another scaling law related to the rising part of the rotation curve has been noted recently; see Lelli, F., Fraternali, F., Verheijen, M. 2013. *Mon. Not. Roy. Astron. Soc.*, **433**, L30. The interest in the important role of the gradient in the rising part of the rotation curve has often been emphasized by Renzo Sancisi; see also last comment at the end of Subsection 20.1.2 and the related reference.

64. See fig. 4 in Swaters, R. A., Madore, B. F., Trewhella, M. 2000. *Astrophys. J.*, **531**, L107.

65. Faber, S. M., Jackson, R. E. 1976. *Astrophys. J.*, **204**, 668; Schechter, P. L., Gunn, J. E. 1979. *Astrophys. J.*, **229**, 472; Tonry, J. L., Davis, M. 1981. *Astrophys. J.*, **246**, 680; Dressler, A. 1984. *Astrophys. J.*, **281**, 512.

66. See also Davies, R. L., Efstathiou, G., et al. 1983. *op. cit.*

67. Dressler, A., Lynden-Bell, D., et al. 1987. *Astrophys. J.*, **313**, 42; Djorgovski, S., Davis, M. 1987. *Astrophys. J.*, **313**, 59. See also Bender, R., Burstein, D., Faber, S. M. 1992. *Astrophys. J.*, **399**, 462.

68. An important database used in several analyses is the so-called 7 Samurai sample of 321 elliptical galaxies; see Faber, S. M., Wegner, G., et al. 1989. *Astrophys. J. Suppl.*, **69**, 763.

69. Jørgensen, I., Franx, M., Kjærgaard, P. 1996. *Mon. Not. Roy. Astron. Soc.*, **280**, 167, for photometry in Gunn r. These authors actually use the mean surface brightness inside the effective radius $\langle I \rangle_e$; for an $R^{1/4}$ law, there is a proportionality relation, $\langle I \rangle_e \approx 3.61 I_e$, with the quantity used in relation (4.14).

70. Jørgensen, I., Franx, M., Kjærgaard, P. 1993. *Astrophys. J.*, **411**, 34.

71. Bender, R., Saglia, R. P., et al. 1998. *Astrophys. J.*, **493**, 529.

72. See Guzmán, R., Lucey, J. R., et al. 1992. *Mon. Not. Roy. Astron. Soc.*, **257**, 187; Djorgovski, S., Santiago, B. X. 1993. In *Structure, Dynamics, and Chemical Evolution of Elliptical Galaxies*, eds. I. J. Danziger, W. W. Zeilinger, K. Kjär. European Southern Observatory Publications, Garching, Germany, p. 59; Jørgensen, I., Franx, M., Kjærgaard, P. 1993. *Astrophys. J.*, **411**, 34; see also van Albada, T. S., Bertin, G., Stiavelli, M. 1995. *Mon. Not. Roy. Astron. Soc.*, **276**, 1255.

73. de Carvalho, R. R., Djorgovski, S. 1992. *Astrophys. J. Lett.*, **389**, L49.

74. van Dokkum, P. G., Franx, M. 1996. *Mon. Not. Roy. Astron. Soc.*, **281**, 985; Kelson, D. D., van Dokkum, P. G., et al. 1997. *Astrophys. J. Lett.*, **478**, L13; Wuyts, S., van Dokkum, P. G., et al. 2004. *Astrophys. J.*, **605**, 677; van der Wel, A., Franx, M. 2005. *Astrophys. J.*, **631**, 145, and references therein.

75. Treu, T., Stiavelli, M., et al. 1999. *Mon. Not. Roy. Astron. Soc.*, **308**, 1037; Treu, T., Stiavelli, M., et al. 2005. *Astrophys. J.*, **633**, 174, and references therein.

76. Auger, M. W., Treu, T., et al. 2010. *Astrophys. J.*, **724**, 511, estimate the intrinsic scatter to be as low as 11 percent.

77. Treu, T., Stiavelli, M., et al. 2005. *op. cit.*

78. Dressler, A., Lynden-Bell, D., et al. 1987. *op. cit.*; Lynden-Bell, D., Faber, S. M., et al. 1988. *Astrophys. J.*, **326**, 19.

79. See Lynden-Bell, D., Faber, S. M., et al. 1988. *op. cit.*; and the study of the Coma cluster by Lucey, J. R., Bower, R. G., Ellis, R. S. 1991. *Mon. Not. Roy. Astron. Soc.*, **249**, 755; Lucey, J. R., Guzmán, R., et al. 1991. *Mon. Not. Roy. Astron. Soc.*, **253**, 584; see also van Albada, T. S., Bertin, G., Stiavelli, M. 1993. *Mon. Not. Roy. Astron. Soc.*, **265**, 627.

80. Kormendy, J. 1977. *Astrophys. J.*, **218**, 333; Hamabe, M., Kormendy, J. 1987. In *Structure and Dynamics of Elliptical Galaxies*, ed. P. T. de Zeeuw. Reidel, Dordrecht, The Netherlands, p. 379.

81. Capaccioli, M., Caon, N., D'Onofrio, M. 1992. *Mon. Not. Roy. Astron. Soc.*, **259**, 323.

82. See Burstein, D., Bender, R., et al. 1997. *Astron. J.*, **114**, 1365.

83. Djorgovski, S. 1995. *Astrophys. J.*, **438**, L29; McLaughlin, D. E. 2000. *Astrophys. J.*, **539**, 618; Barmby P., McLaughlin, D. E., et al. 2007. *Astron. J.*, **133**, 2764; Pasquato, M., Bertin, G. 2008. *Astron. Astrophys.*, **489**, 1079.

84. Bellazzini, M. 1998. *New Astron.*, **3**, 219; Pasquato, M., Bertin, G. 2008. *op. cit.*

85. Pasquato, M., Bertin, G. 2010. *Astron. Astrophys.*, **512**, A35; van den Bergh, S. 2008. *Mon. Not. Roy. Astron. Soc.*, **390**, L51.

86. Schaeffer, R., Maurogordato, S., et al. 1993. *Mon. Not. Roy. Astron. Soc.*, **263**, L21; Adami, C., Mazure, A., et al. 1998. *Astron. Astrophys.*, **331**, 493; Lanzoni, B., Ciotti, L., et al. 2004. *Astrophys. J.*, **600**, 640.

87. For a review, see Ciotti, L. 2009. *La Rivista del Nuovo Cimento*, **32**, 1.

88. Ferrarese, L., Merritt, D. 2000. *Astrophys. J.*, **539**, L9; Gebhardt, K., Bender, R., et al. 2000. *Astrophys. J.*, **539**, L13; Merritt, D., Ferrarese, L. 2001. *Astrophys. J.*, **547**, L140; Tremaine, S., Gebhardt, K., et al. 2002. *Astrophys. J.*, **574**, 740. A steepening of the relation up to $q \approx 5$ has been noted for the most massive galaxies as a result of the inclusion in the relevant models of the role of the dark-matter halo in the host galaxy; see Gebhardt, K., Thomas, J. 2009. *Astrophys. J.*, **700**, 1690; Saglia, R. P. 2012. Private communication; Rusli, S. P., Thomas, J., Saglia, R. P., et al. 2013. *Astron. J.*, **166**, id.45.

89. Magorrian, J., Tremaine, S., et al. 1998. *Astron. J.*, **115**, 2285; Marconi, A., Hunt, L. K. 2003. *Astrophys. J.*, **589**, L21; Häring, N., Rix, H.-W. 2004. *Astrophys. J.*, **604**, L89.

90. Graham, A. W., Erwin, P., et al. 2001. *Astrophys. J.*, **563**, L11; Graham, A. W., Driver, S. P. 2007. *Astrophys. J.*, **655**, 77.

91. Spitler, L. R., Forbes, D. A. 2009. *Mon. Not. Roy. Astron. Soc.*, **392**, L1; Burkert, A., Tremaine, S. 2010. *Astrophys. J.*, **720**, 516; Harris, G. L. H., Harris, W. E. 2011. *Mon. Not. Roy. Astron. Soc.*, **410**, 2347; Snyder, G. F., Hopkins, P. F., Hernquist, L. 2011. *Astrophys. J. Lett.*, **728**, id.L24.

92. Parts of this discussion and of Chapters 20 and 24 are taken from "Materia oscura," by Bertin, G., van Albada, T. S. 1998. In *Enciclopedia del Novecento*, **XI**, 149, Istituto della Enciclopedia Italiana fondata da Giovanni Treccani, Roma.

93. In 1908, while addressing in his essay "Science et Méthode" the issue of estimating the size and mass of the Milky Way, Henri Poincaré wrote: "... mais ne pourrait-il y avoir des astres obscurs qui circuleraient dans les espaces interstellaires et dont l'existence pourrait rester longtemps ignorée? Mais alors, ce que nous donnerait la méthode de lord Kelvin, ce serait le nombre total des étoiles, en y comprenant les étoiles obscures."

94. One interesting study is the DAMA (*DArk MAtter*) experiment under way at Gran Sasso, Italy; see Bernabei, R., Belli, P., et al. 2011. *Can. J. Phys.*, **89**, 11, and references therein. High expectations are based on the forthcoming results from LHC (the Large Hadron Collider) in operation at CERN.

95. Feynman, R. P., Leighton, R. B., Sands, M. 1963. *The Feynman Lectures on Physics*, Vol. 1. Addison-Wesley, Reading, MA, p. 7–7.

96. Peebles, P. J. E. 1993. *Principles of Physical Cosmology*, Princeton University Press, Princeton, NJ, gives an interesting description of the many subtle aspects involved.

97. See Freedman, W., Madore, B., et al. 1994. *Nature (London)*, **371**, 757; Sandage, A., Saha, A., et al. 1996. *Astrophys. J. Lett.*, **460**, L15; Tammann, G. A., Sandage, A., Reindl, B. 2008. *Astron. Astrophys. Rev.*, **15**, 289; Freedman, W. L., Madore, B. F. 2010. *Annu. Rev. Astron. Astrophys.*, **48**, 673. For determination of the Hubble constant, the study of delays in multiple images produced by strong lensing has been shown to lead to competitive results with respect to more traditional methods; see Suyu, S. H., Marshall, P. J., et al. 2010. *Astrophys. J.*, **711**, 201.

98. Perlmutter, S., Gabi, S., et al. 1997. *Astrophys. J.*, **483**, 565; Perlmutter, S., Aldering, G., et al. 1998. *Nature (London)*, **391**, 51; Garnavich, P. M., Kirshner, R. P., et al. 1998. *Astrophys. J. Lett.*, **493**, L53.

99. Riess, A. G., Macri, L., et al. 2009. *Astrophys. J.*, **699**, 539; Riess, A. G., Macri, L., et al. 2011. *Astrophys. J.*, **730**, 119. See also Pietrzyński, G., Graczyk, D., et al. 2013. *Nature (London)*, **495**, 76. For the use of time delays in gravitational lenses to measure the Hubble constant, see Chapter 26.

100. Percival, W. J., Reid, B. A., et al. 2010. *Mon. Not. Roy. Astron. Soc.*, **401**, 2148.

101. See Komatsu, E., Smith, K. M., et al. 2011. *Astrophys. J. Suppl.*, **192**, 18. The newly released data from the *Planck* mission appear to reveal some significant discrepancies, in particular, pointing to a relative low value of the Hubble constant $H_0 = 67.3 \pm 1.2$ km s^{-1} Mpc^{-1}; see Ade, P. A. R., Aghanim, N., et al. 2013. arXiv:1303.5076.

102. See Olive, K. A. 1990. *Phys. Rep.*, **190**, 307; Linde, A. D. 1990. *Particle Physics and Inflationary Cosmology*, Harwood, New York.

103. See Sciama, D. W. 1993. *Modern Cosmology and the Dark Matter Problem*, Cambridge University Press, Cambridge, UK.

104. Lubimov, V. A., Novikov, E. G., et al. 1980. *Phys. Letters*, **94B**, 266.

105. See Kolb, E. W., Turner, M. S. 1990. *The Early Universe*, Addison-Wesley, Redwood City, CA.

5 Basic Questions, Semiempirical Approach, and the Dynamical Window

The discovery of dark halos, the spectroscopic evidence that elliptical galaxies are dominated by collisionless dynamics (which rules out the applicability of simple fluid models for their description) and the opening of new observational windows (especially in the near infrared) able to provide direct information on the underlying stellar mass distribution in galaxies have significantly changed, in the third quarter of last century, our perception of the internal structure of galaxies. As will be pointed out in general terms in Part II, the modeling tools and the theories developed to explain many interesting observations (from the study of global spiral and bar modes of galaxy disks to the construction of self-consistent anisotropic collisionless models to explain the universality of the luminosity profile of elliptical galaxies) present many analogies to parallel work in the physics of electromagnetic plasmas. At the frontier of current research in extragalactic astrophysics, the *Hubble Space Telescope* and new large telescopes from the ground are giving us a view of the early dynamical stages of galaxies, providing interesting insights into aspects of galactic formation and evolution, and, on the small scale for relatively nearby galaxies, are offering unprecedented accurate data on their structure and kinematics.

Within the same basic framework of dynamical studies, very different approaches can be taken. Dynamics is a powerful tool that allows us not only to interpret the data and to form a better picture of the internal structure of galaxies but also to develop and test physical scenarios. As such, its role is not limited to the descriptive stage. Two extremes are often taken in the study of the dynamics of galaxies.

Some dynamical investigations aim at staying as close as possible to the most direct interpretation of the data available, with little or no physical input, generally reducing the assumptions of the modeling to a few geometric hypotheses. These studies are a priori descriptive and leave many interesting questions raised by the observations without an answer. Here the proposed strategy is to postpone the physical discussion to a later stage, when it is hoped that all the structural parameters have been secured directly from the data with the desired accuracy. The underlying motivation is that, in principle, if we knew the current conditions of a galaxy with sufficient precision, dynamics, possibly with the help of a sufficiently powerful computer, should allow us to deduce the evolution of the galaxy under investigation. In a sense, the spirit in which dynamics is taken here is that of weather forecasting. In many studies it is even implied that from a sufficiently accurate set of observed conditions we should be able to reconstruct the past history of the galaxy; in other words, dynamics should allow us, after the descriptive stage is completed, to determine the past. This approach suffers from the fact that the observational windows alone are unable to give us a sufficiently sharp picture of the current structure of galaxies and that the evolution of complex systems is too sensitive to initial conditions to be derived from a limited set of observational constraints. Furthermore, it is well known that for dissipative systems or for systems with an effective relaxation, going backward in time is not feasible, in contrast to the case of single-particle orbits.

At the other extreme, some scientists focus on dynamical mechanisms in a deductive approach. These are best studied in relatively simple physical and mathematical contexts, in which the rules are set at the beginning by our choice of models and dynamical equations. Here we can safely proceed, in principle, through an infinite number of cases, and because of the interest in the mechanisms per se, we often postpone a confrontation with the specific astrophysical issues raised by the observations. The hope is that, eventually, we may point to data and to observed phenomena for which the theoretically investigated dynamical mechanisms can be argued to be relevant. A difficulty with this approach, at least in terms of astrophysical feasibility, is that we may easily be misled into dynamical studies that have little contact with real galaxies.

Somewhere in the middle between the preceding extremes, we can take a semiempirical approach, in which the mechanisms studied are those inspired by the observations. This is the approach that is preferred in this book. Here we proceed by formulating physically plausible scenarios and then checking the dynamical consequences of the adopted hypotheses in relation to the observations. This approach is inherently predictive, in the sense that it generally calls for improved observations to confirm or dismiss the physical scenario that is proposed and used. Such a semiempirical approach is difficult to use, but for the cases for which predictions turn out to be sufficiently well confirmed by the observations, it is highly rewarding. When such dynamical theories can be developed, the view of galactic structure and evolution is significantly sharpened by the dynamical window because the data coming from the observational windows can be placed into a unified interpretation.

5.1   Structure

Probably the most important question that we would like to answer is, What is the internal structure of galaxies? Some classification systems have been developed (see Chapter 3), and these point to an underlying law and order that the dynamicist should try to explain. There is thus a need for a dynamical classification of galaxies as a theoretical counterpart to the empirical classification schemes.

What are the basic symmetries of galaxies? This question obviously refers to an idealization process that leads to the identification of a basic state (see Chapter 9). Most or all quantitative conclusions that can be drawn on the physical state of galaxies depend on such identification. Many spiral galaxies are probably well described as approximately axisymmetric. How adequate is this description in the case of disks with well-developed arms and, especially, with well-developed bars? What can we say about the intrinsic symmetry of elliptical galaxies? Should they be considered to be generically triaxial? How can the various shapes be supported in terms of distributions of stellar orbits? What determines, in spiral and elliptical galaxies, the amount of pressure anisotropy present in the collisionless stellar component? Can a dynamically meaningful classification of elliptical galaxies be proposed?

Dynamics deals with models for the mass distributions, whereas telescopes observe radiation. A major source of uncertainty is the connection between light and mass. To what extent does light trace mass? How much of the mass is contained in the cold interstellar medium? A sizable amount of gas, in the form of cold molecular material, may escape most observations and still have a strong impact on the dynamics of galaxies. How frequent is the presence of a light regular HI disk in elliptical galaxies? What is the origin of the extraplanar gas found in several spiral galaxies? How can we explain the observed kinematic profiles? Dark halos are required for explaining the observed rotation curves of spiral galaxies. Does the halo have a significant impact on the dynamics of the inner visible part of the galaxy, or is its role mostly confined to the outer regions? Are the halos round or flat? Can they be considered as basically axisymmetric? How far out do the halos extend? Given the established presence of dark matter around spiral galaxies, what can we say about the amount and distribution of dark matter in elliptical galaxies? Is it reasonable for bright ellipticals to assume that the mass-to-light ratio is approximately constant inside the visible galaxy? What about low-luminosity ellipsoidal stellar systems? Do low-luminosity ellipticals possess significant halos, and what is their dark-matter distribution? Why, in this respect, do globular clusters and dwarf spheroidals appear to have so different characteristics?

The dynamics of galaxies provides us with the experiments that test the validity of the laws of mechanics on the kiloparsec scale. Given the elusive nature of dark matter, do we need to revise our basic physical laws?

On the smallest scale of galaxy nuclei we also have a problem of unseen mass. Are supermassive black holes present at the centers of most galaxies, only there, and why? Can this be unequivocally established by dynamical studies? Given the interesting correlations found (see Subsection 4.3.4), is there a direct relation between large-scale dynamics of a galaxy and the dynamical activity in its nuclear region?

5.1.1 Morphology and Pattern Formation in Complex Systems

The most spectacular morphological properties of galaxies include spiral structure (with or without a bar), warps of galaxy disks, polar rings, shells, and isophotal twisting in elliptical galaxies. How do we explain the observed morphologies? How can large-scale coherent structures be supported in differentially rotating disks? Are they associated with quasi-stationary perturbations, or are the structures caught by the observations in the process of rapid evolution? How much of the large-scale spiral structure depends on the dissipative interstellar medium, and how much depends on the collisionless stellar component? It appears that for disk galaxies, the morphology results from a subtle balance and regulation between the two components. Near-infrared (near-IR) observations now demonstrate that large-scale spiral structures are present as density perturbations in the evolved stellar disk, occur frequently, and are characterized by a small number of arms (typically the structure is bisymmetric, but a one-armed structure is also often observed).

Why do certain galaxies have a grand design and others do not? Why do some galaxies have a prominent bar and others do not? Can we provide an explanation for the observed large-scale morphologies in terms of intrinsic dynamics, or do we have to resort to galaxy-galaxy interactions as a general mechanism for pattern formation? How frequently do galaxy-galaxy interactions have a dominant impact on the observed morphologies? In Part III it is shown that a study based on a quasi-stationary spiral structure supported by intrinsic global modes gives a natural description of the observed large-scale patterns and leads to a successful unified framework for interpretation of the morphological categories of spiral galaxies.

Starting from a given interpretation of the observed morphological properties, what is the expected long-term evolution of the galaxy morphology? Here dynamical studies are still in the speculative stage because processes of nonlinear evolution are difficult and have been only partially explored.

In a semiempirical approach, it is interesting to combine a physical picture for the explanation of some morphological features with the available observational data in order to obtain a dynamical measurement of the properties of the basic state. In particular, this procedure often may be used to set additional constraints on the amount and distribution of unseen matter.

5.2   Formation and Evolution

Some structural properties immediately bring us to consider formation scenarios. In particular, the existence of universal luminosity profiles (either the $R^{1/4}$ law for bright elliptical galaxies or the exponential law for galaxy disks) suggests a common formation process for objects characterized by similar luminosity profiles. As will be shown in Part IV, for bright quasi-spherical galaxies, the empirical $R^{1/4}$ law can be explained successfully within a framework in which formation has proceeded by means of a basically collisionless collapse. Still, for low-luminosity ellipticals and for significantly nonspherical galaxies, satisfactory dynamical arguments and models are not yet available. The regularity and smoothness of flat rotation curves in many spiral galaxies (see Chapter 20) and the corresponding regularity of the total density profiles of distant bright ellipticals (see Chapter 24) suggest a strange dynamical tuning between visible matter and dark halo, which, if true, may have been established in the distant past. The regularity and thinness of many galaxy disks suggest that minor and major mergers, as a rule, have not disturbed the recent evolution of many spiral galaxies.

Empirical global scaling laws, in particular, the luminosity-velocity relation for spirals and the fundamental plane for ellipticals establish the rules by which physical scales are realized in galaxies. This is very important and allows astronomers, from a (distance-independent) measurement of their internal kinematics, to use galaxies effectively as standard candles or as standard rods. Galaxies then can be used as distance indicators and as tools to probe the geometry of the Universe (i.e., to set constraints on the cosmological parameters). Most likely these scaling laws are the relic of the relevant formation processes. However, dynamical evolution also may drive galaxies in the direction of some scaling relations. Therefore, even apart from the important applications that can be made, the global scaling laws raise interesting questions and provide important clues about understanding the mechanisms of galaxy formation and galaxy evolution. The observed properties of the fundamental plane of early-type galaxies out to $z \approx 1$ suggest that in the last several billion years they have evolved mostly passively (i.e., following the natural aging of their stellar populations).

We may note that within the context of laboratory physics, scaling laws are often studied and developed for different goals (e.g., designing and planning new machines in physical regimes not yet explored in the laboratory). In practice, some of the questions addressed present analogies to questions addressed in the study of scaling laws mentioned earlier.

Can we set up a global mass-luminosity relation for galaxies or, at least, for classes of galaxies? How much of the current structure of galaxies depends on a continuous shaping induced by galaxy-galaxy or galaxy-cluster interactions, and how much instead is genetically frozen in the initial conditions associated with the formation processes in the distant past? Are galaxies, as dynamical systems, still undergoing rapid and significant evolution? Deducing the evolution of complex dynamical systems is the hardest and boldest step to take, but can we make use of observations of galaxies at high redshifts at least to set empirically the stage for reasonable evolution models? Can we actually separate the impact of evolutionary processes from that of initial conditions? A similar discussion on the relative share of initial conditions and evolution mechanisms in the determination of the parameters that characterize the observed objects today is carried out in the analysis of the correlations found for the properties of globular cluster systems in galaxies.

One interesting aspect of the study of the evolution of stellar systems, which makes it different from the study of electromagnetic plasmas, is that by suitable spectroscopic analyses we may determine the actual age of their constituents, that is, of the relevant stellar populations. The reason for this is that because the evolution of individual stars is largely well understood, stars carry a clock aboard ready for us to read. To be sure, currently observed stellar populations may include stars born before or after the actual epoch of galaxy assembly. In a given galaxy, by suitable spectroscopic study we may tell that some components (e.g., the globular cluster system or the thin disk or the bulge) must have had different formation histories. Obviously, clues of this type are not available in the context of electromagnetic plasmas and make the study of galaxy formation and evolution much more interesting than a simple reconstruction of the time evolution of a pure dynamical system.

5.2.1 The Cosmological Context

As already noted in each of the previous chapters, much of the interest in the physics of galaxies is driven by cosmological questions. In turn, we may also argue that every breakthrough in understanding the dynamics of galaxies is also a step forward in making a clear picture of the Universe as a whole and thus can be considered as a breakthrough in cosmology. These comments apply especially to the issues of galaxy formation and evolution briefly described earlier.

Here we should recall that in recent years great efforts have been made in the direction of studying the problem of structure formation in the evolving Universe by means of dedicated simulations. These assign a dominant role to the dynamics and evolution of dark matter in the cosmological context, starting from initial conditions set to represent the current concordance cosmological model, and then specify a number of properties of the visible matter (i.e., the baryonic component) during evolution, often by means of simple semianalytical recipes, so as to meet the main empirical constraints posed by the observations. Dark matter is generally treated as a purely collisionless component, whereas visible matter includes a number of dissipative effects. There is a general optimistic expectation that this procedure, properly refined and updated by the inclusion of the latest observational constraints and by the development of the most powerful and accurate numerical tools, will soon allow us to predict the properties, down to the kiloparsec scale or less, of the galaxies observed in the nearby Universe from first principles, starting from the initial conditions of the Big Bang as set for example, by the observed properties of the cosmic microwave background radiation.

This point of view will not be followed in this book, for at least three reasons. As noted earlier in this chapter, a semiempirical approach, not a deductive approach, is preferred throughout this book. The focus will be on studying physical mechanisms suggested by specific phenomena observed in galaxies and not on devising simple recipes to summarize the available astronomical information. In addition, if we note how fast our general cosmological perceptions have been changing in the last few years, it would appear as premature to follow all the consequences of such a deductive approach. Finally and most important, the experience from other fields of research devoted to the study of complex macroscopic systems in the laboratory suggests that great caution should be taken in studying the evolution of nonlinear structures beyond a few dynamical time scales and over an excessive range of spatial scales.

5.3 Modeling and Some Fundamental Questions

There are also several very interesting issues that are inspired by the study of the dynamics of galaxies, which, because of their generality, transcend the immediate astrophysical applications. We will soon see in Part II that many of the dynamical tools of investigations that are developed refer to well-defined mathematical models for which some basic questions remain unresolved. For example, the characterizations of integrability of a dynamical system, of what determines whether a potential is endowed with regular or irregular orbits, and of what happens to orbits when a given symmetry is slightly perturbed are all interesting and fundamental questions that are met frequently in the construction of self-consistent models of stellar systems. Yet many of these issues, for example, those related to the asymptotic long-term behavior of orbits in the future, may refer to conditions (in particular, to time scales) well beyond any sensible astrophysical interest.

Probably the most interesting and difficult problem in this context of fundamental issues is that of building a satisfactory statistical mechanics theory of self-gravitating systems. Here we must find a way to overcome the stumbling blocks associated with the loss of additivity (with respect to more standard thermodynamical systems) and with the singularities that accompany complete relaxation. Related to this is the need for a more thorough evaluation of the relevant relaxation processes (see Chapters 7 and 8). Even for the conceptually simple context of quantifying the role of star-star encounters, a quantitative theory to describe transport processes inside inhomogeneous gravitating systems is still unavailable. In addition the continuum limit of a system of N stars, with $N \to \infty$, which leads to the collisionless Boltzmann equation for the distribution function in the six-dimensional phase space, would require a more solid justification.

Similarly, many of the dynamical problems explored in this book turn out to involve various mechanisms that would deserve a deeper theoretical investigation for their own sake, independent of the possible astrophysical applications that may be imagined. In this respect, some points of interest are identified when the opportunity arises, even though more fundamental questions are not further elaborated.

5.4 Relation to Other Branches of Astrophysics

Besides the clear analogies to other fields of research in physics and applied mathematics (especially hydrodynamics, geophysics, and plasma physics) set by the questions raised and by the methods used in the study of the dynamics of galaxies, what is learned about the topics covered in this book has important consequences relevant to other branches of astrophysics; in turn, the study of the dynamics of galaxies depends crucially on the progress made in these separate fields. Research areas in astrophysics with which this mutual exchange is most beneficial are stellar evolution, chemical evolution and nucleosynthesis, astrophysics of the interstellar medium, star formation, gravitational lensing, and cosmology. Occasionally, some of the relations involved are pointed out, but in general, the focus is on the dynamical aspects of the physics of galaxies.

Part II
Physical Models

6 Self-Gravity and Relation to Plasma Physics

The dynamics of galaxies is to some extent similar to that of electromagnetic plasmas. This is not at all a new way of looking at stellar systems. In fact, the work of Jeans,[1] whose goal was to describe the motion of a large collection of stars under the influence of their mutual gravitational attractions (see Chapter 8), and the work of Chandrasekhar,[2] in setting out the equations for the effects of collisions in a stellar system (see Chapter 7), precede much of the work related to the Vlasov equation[3] and the various transport processes in electromagnetic plasmas.[4] Thus it is no surprise that many scientists have contributed, either explicitly or indirectly, to both fields of electromagnetic and gravitational plasmas.[5] Several articles emphasize the common concepts at the roots of the two fields.[6] To be sure, the field of stellar dynamics is much less diverse than that of electromagnetic plasmas. This is probably more the result of the several cancellations related to the fact that the charge-to-mass ratio equals unity for the gravitational charge, which limits the number of relevant frequency windows, rather than the effect of the absence of the complexity of the full set of Maxwell's equations (with classical gravity, we need to keep track of only the Poisson equation).

What are the key analogies and differences? To answer this question, we distinguish among fundamental aspects, basic modeling tools, specific mechanisms, and general approach. In this book, the focus is not on the subtle issues involved in a rigorous derivation of the basic equations for idealized self-gravitating systems based on a systematic study of statistical mechanics. These issues would deserve a monograph of their own; instead, we are interested in moving on quickly to some interesting results about the dynamics of galaxies. Therefore, some fundamental aspects of the modeling process, for which comparison with plasma physics could be made in great detail, are summarized only briefly in Chapters 7 and 8. When the relevant fluid or stellar dynamics equations are introduced, it is made clear that, for most practical applications, they should be thought of as model equations.

The similarity of tools and specific mechanisms is stressed in this book quite often by frequent use of examples from the theory of electromagnetic plasmas when the opportunity for an analogy arises (e.g., in relation to the guiding-center orbit theory in Chapter 13 or the results on bending waves on a current sheet in Chapter 19). However, an effort is made not to overemphasize the various technical aspects involved that often exist and in many cases contribute to excite the researcher when an analogy is made (e.g., see the comparison between density waves in a stellar disk and some kinetic waves in a magnetized plasma in Subsection 15.1.1). In this respect, it is sometimes amazing to find that certain results can be literally translated from one field to the other by means of a suitable dictionary (see comments in Section 6.3).

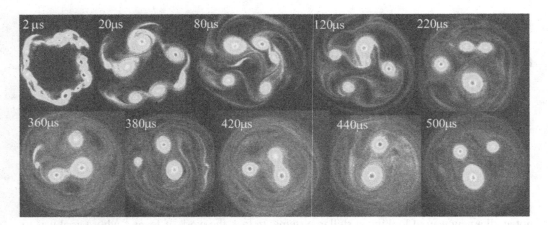

Fig. 6.1. A typical electron plasma evolution from experimental data obtained in the Malmberg-Penning trap ELTRAP. The first frame (corresponding to a trapping time of 2 μs) reflects the initial conditions. The highly nonlinear evolution rapidly leads to the formation of several small vortices, which then interact through close encounters resulting in merger events and emission of vorticity filaments (from Bettega, G., Pozzoli, R., Romé, M. "Multiresolution analysis of the two-dimensional free decaying turbulence in a pure electron plasma," 2009. *New J. Phys.*, **11**, 053006; © IOP Publishing, Ltd., and Deutsche Physikalische Gesellschaft; published under a CC BY-NC-SA license).

What it is hoped will emerge most sharply from the presentation of this book is an overall similarity of approaches.[7] In both research areas, the major struggle is to find the most appropriate equations to describe inherently complex macroscopic systems governed by collective behavior in the presence of long-range forces (Fig. 6.1). For these systems, geometry, parameter regimes, inhomogeneities, and boundary conditions give rise to a fantastically diverse set of phenomena and evolutionary possibilities. There is no way to derive by a deductive approach all the possible interesting cases. Instead, a semiempirical approach, based on clues and facts from specific observations or from laboratory experiments, helps us to narrow down the focus of interest and to form physically plausible pictures. These are best formulated and investigated if they can be made quantitative and predictive. At the end of Chapter 8 it is argued that for realistic applications, asymptotic methods are thus bound to give us the richest picture from the physical point of view. To stress the importance of some methodological aspects, a rather long digression is made on the basic perception of separating the study of equilibrium and stability (Chapter 9), with the help of a classical example (Chapter 10). Then we take an excursion through the dynamics of dispersive waves (Chapter 11) so as to better appreciate the role of self-gravity in the simplest context (Chapter 12), with a hint at the difficulties involved when we move from homogeneous to inhomogeneous media (either kinetic or fluid). The discussion of Part II is thus meant as a methodological prerequisite, in which the general approach is defined, before we arrive at the specific problems relevant to the dynamics of galaxies. Even if the large-scale structure of normal galaxies appears to be shaped by gravity rather than by electromagnetic forces, there are also several interesting problems of genuine plasma physics in the galactic context, for example, those related to the large-scale magnetic configuration in spiral galaxies, the acceleration and confinement of cosmic rays, or the generation of large-scale radio and optical jets and radio lobes in active galaxies. Not much attention is devoted here to these interesting issues, but this is one more reason to stress the interest in plasma physics for anyone who approaches the study of the dynamics of galaxies.

6.1 Gravity and Self-Gravity

Gravity is a long-range force that tends to dominate the scene on a large scale. We will find detailed evidence for this statement in Chapter 7, in which it is shown that the cumulative effects of distant stars on the motion of a test star are dominant and that their contribution to the scattering cross section formally diverges.

The importance of gravity in astronomy is ubiquitous, from the classical examples of celestial mechanics to the more modern studies of the powerful accretion disks, for which material is captured by a massive central object. Yet it is less frequently appreciated that large-scale distributed masses give rise to a mean gravitational field, generated self-consistently by the parts that make up the system, that naturally dictates a global behavior. Thus the way spiral galaxies rotate around their center departs significantly from the Keplerian behavior that is most intuitive from experience with the solar system or with planetary rings. It so happens that with respect to the galactocentric radius, the mean potential in a bright elliptical galaxy or in the equatorial plane within the disk of a spiral galaxy is approximately logarithmic. The orbits in the mean gravitational field can be very unusual, especially when departures from axisymmetry occur.

Because of self-gravity, stellar systems have to face a natural tendency to collapse. In the absence of dissipation, one may assume equilibrium configurations in which random motions are able to oppose the tendency to collapse. In practice, this effect eventually fails at sufficiently long scales, at which the system has to become inhomogeneous (see Chapter 12). However, on the equatorial plane of a rotating system, rotation turns out to be able to oppose the tendency toward collapse on long scales.

There is obviously no way to screen the gravitational force, and inhomogeneity is the rule for stellar systems. This also implies the loss of additivity, thereby undermining one of the basic assumptions of standard thermodynamical theories. In addition, there is no such a thing as an isolated galaxy. Galaxies interact with each other all the time through tidal forces that tend to tear them apart. One interesting problem thus is to ascertain how much of the observed phenomena in normal galaxies depends on their intrinsic dynamics and how much is induced by the tidal forces of neighboring objects.

In contrast, the existence of opposite charges and the sign difference in the force between equal charges make it possible for electromagnetic plasmas to reach a quasi-neutrality condition. Nonetheless, even in this somewhat simpler environment, we realize that global properties, geometry, boundary conditions, and nonthermodynamical behavior are key aspects of dynamics.

6.2 Collective Behavior

Even when the importance of the action of a mean field is taken into account, we should distinguish between the dynamics of individual orbits, which is rather intuitive, and the collective dynamics of the system. For example, encouraged by well-known problems of celestial mechanics, and given the convenience from the numerical point of view, we may simplify the description of some processes in terms of the dynamics of test stars.[8] In fact, it may be rather easy to visu-alize the behavior of single-particle orbits that are part of a specific mechanism. However, we should be aware that the collective behavior often defies simple intuition and introduces subtle aspects that are easily missed.

One point sometimes overlooked in the discussion of single-particle dynamics is the fact that the center of mass of the whole stellar system cannot accelerate. Apparently, this has sometimes led to wrong conclusions even in the relatively simpler context of the dynamics of planetary rings.

Judging from the study of individual orbits around corotation in the presence of a rigidly rotating spiral field, we see a tendency for the mass to accumulate out of phase with respect to the imposed nonaxisymmetric field (see Chapter 13). Many papers, based on the construction of thousands of individual orbits, have tried to determine the conditions for compatibility with the imposed gravitational field. These studies are interesting because they show in detail the difference between individual and collective behavior. However, they also show the limits of a single-particle description. As for the simple processes of sound waves in a gas, we thus need new tools better suited to describe the collective behavior. Indeed, a breakthrough of the 1960s was the introduction, within the context of stellar dynamics, of methods able to deal directly with wave collective processes (see the description of density waves given in Part III). These methods take advantage of a continuum description (see Chapter 8), together with the tools developed to describe dispersive waves (see Chapter 11).

Similarly, we may try to build nontrivial equilibrium configurations for collisionless stellar systems by direct orbit superposition. Study of the collisionless Boltzmann equation and use of the Jeans theorem provide powerful tools that go well beyond such a scheme (see Chapter 8 and Part IV).

6.3   A Concise Dictionary

The similarity in the long-range behavior of electrostatic and gravity forces and the strict analogy between the Coriolis force in a rotating frame and the Lorentz force for a charged particle in the presence of a magnetic field make it possible to set up a dictionary for the translation of several quantities and concepts commonly used in one field into those of the other. Because of the similarity in the role of collisions, high-temperature plasmas are found to be close in their behavior to collisionless stellar systems (see Chapter 7). Some of the graininess effects due to the presence of discrete particles are treated in the two fields of research by similar tools, with similar options of a fluid or a kinetic description in the relevant continuum limit (see Chapter 8).

There is a formal analogy between the plasma frequency and the natural frequency for a self-gravitating system given by $(4\pi G\rho)^{1/2}$. The analysis that brings out such similarity (see Chapter 12) also draws a correspondence between Debye length and Jeans length, although physically different phenomena are involved. At a more subtle level, basic processes, such as Landau damping (see Chapter 12), instabilities due to pressure anisotropies (see Chapter 23), negative-energy modes (see Section 15.5), and dissipation-driven instabilities (see Chapter 10), are found in both fields of research.

Pressure versus rotation support has a counterpart in plasma physics, in that magnetically confined plasmas, with a relatively small β parameter (which measures the ratio of thermal to magnetic pressure), are similar in many respects to galaxy disks, which are rotation-supported (i.e., they are high-t systems, in the language of Chapter 10). Under these conditions, the orbital description can be developed with parallel tools, especially the physically intuitive guiding-center description (see Chapter 13); in particular, the epicyclic frequency corresponds to the Larmor gyration frequency of magnetized plasmas. The shear flow present in a differentially rotating disk presents interesting analogies to the dynamics of sheared magnetic configurations.

Pressure-supported stellar systems share with high-β plasmas a general difficulty that is met in the orbital description and in the stability analysis.

In making use of the dictionary, we should always be aware that we are comparing systems characterized by tremendously different scales. The Alfvén time scale in a magnetically confined toroidal plasma, for devices devoted to controlled thermonuclear research, may be of the order of 10^{-5} s, whereas the typical dynamical time in a galaxy is $\approx 10^8$ yr. Some plasma experiments thus run on a time scale that can be several orders of magnitude longer than the relevant Alfvén time scale, whereas the Hubble time for galaxies, although long, is typically smaller than 100 dynamical time scales. The toroidal plasma may have 10^{15} particles per cubic centimeter and a Debye length smaller than a tenth of a millimeter, whereas in the solar vicinity there is less than one star per cubic parsec, and the typical Jeans scale is ≈ 1 kpc. Another nontrivial source of different dynamical behavior is that in many laboratory plasmas, much like in many experiments in fluid dynamics, controlled boundary conditions are set by the machine, whereas galaxies are associated with free boundaries, so it is generally unrealistic to consider stellar systems with a finite edge.

Notes

1. Jeans, J. H. 1915. *Mon. Not. Roy. Astron. Soc.*, **76**, 70.
2. Chandrasekhar, S. 1942. *Principles of Stellar Dynamics*, University of Chicago Press, Chicago (reprinted in 1960 by Dover, New York).
3. Vlasov, A. A. 1945. *J. Phys. USSR*, **9**, 25.
4. See Braginskii, S. I. 1965. *Rev. Plasma Phys.*, **1**, 205.
5. As an example, the topic of weakly collisional stellar systems is well introduced in a book by Spitzer, L. 1987. *Dynamical Evolution of Globular Clusters*, Princeton University Press, Princeton, NJ; one of the first standard references in plasma physics is by the same author, Spitzer, L. 1962. *Physics of Fully Ionized Gases*, Wiley, New York.
6. An interesting introduction is given by Lynden-Bell, D. 1967. In *Relativity Theory and Astrophysics 2: Galactic Structure*, Vol. 9 of *Lectures in Applied Mathematics*, ed. J. Ehlers. American Mathematical Society, Providence, RI, p. 131. See also Bertin, G. 1980. *Phys. Rep.*, **61**, 1.
7. In the last decades, several schools and workshops held in Varenna and Como (Italy) have brought together plasma physicists and astrophysicists to share their results in the study of collective phenomena in complex macroscopic systems. On these occasions, many contributions have addressed the dynamics of self-gravitating systems, treated as gravitational plasmas. The proceedings of the last three workshops are Bertin, G., Farina, D., Pozzoli, R., eds. 2004. *Plasmas in the Laboratory and in the Universe: New insights and New Challenges*, AIP Conference Proceedings, Vol. 703, Melville, NY; Bertin, G., Pozzoli, R., et al. eds. 2007. *Collective Phenomena in Macroscopic Systems*, World Scientific, Singapore; Bertin, G., De Luca, F., et al. eds. 2010. *Plasmas in the Laboratory and in the Universe: Interactions, Patterns, and Turbulence*, AIP Conference Proceedings, Vol. 1242, Melville, NY.
8. For example, see the interesting study of galaxy-galaxy interactions by means of a restricted three-body code by Toomre, A., Toomre, J. 1972. *Astrophys. J.*, **178**, 623.

7 Relaxation Times, Absence of Thermodynamical Equilibrium

Empirical evidence proves that stellar systems do not obey the simple laws of equilibrium thermodynamics. For example, in the solar neighborhood, the data show that the distribution of stellar orbits is such that the associated pressure tensor is definitely anisotropic, which is a primary indication of insufficient relaxation; the axis ratios of the velocity ellipsoid for old disk stars[1] are approximately $c_r: c_\theta: c_z = 39: 23: 20$ (the pressure ratios go with the square of these figures).

In this chapter an elementary discussion is given of why large stellar systems are characterized by a very low degree of collisionality, and then the issue of how such collisionless stellar systems can still be subject to relaxation processes that play a significant role in their evolution is briefly addressed. Thus, even if stellar encounters act on a very long time scale, arbitrary distributions of stellar orbits are probably not realized in nature. It appears that, also with the help of subtle mechanisms of collisionless relaxation, most normal galaxies are in a slowly evolving state of incomplete relaxation.

7.1 Two-Star Relaxation Times

The following discussion summarizes well-known results that apply to stellar systems and ionized gases.[2] The two-star relaxation times quantify the effects of star-star collisions in changing the orbit of a star with respect to that determined by the smooth mean field generated by the whole stellar system. Here the stars are considered as point masses interacting with each other according to Newton's law. The relaxation times thus measure the effects of the discreteness of the mass distribution in a stellar system. Different relaxation times can be defined relative to different aspects of the scattering process. Their definition is that of a time scale beyond which the relevant aspect of the scattering process becomes significant in the course of the orbit of a reference star under consideration.

We can quantify the relaxation times by considering a test star of mass m_t with initial velocity v_∞ incident on a set of many field stars of mass m_f, initially at rest, and with mean density (number of field stars per unit volume) n. Each encounter of the test star with a field star is treated as a separate two-body problem. Then we sum over all encounters as if their global effect could be obtained by simple superposition.

7.1.1 Heuristic Analysis

Let us focus on the deflection induced in the stellar orbit of the test star. Then the property of the scattering process involved is the gain of velocity $\Delta v_\perp$ transverse to the original direction of motion. Statistically, the deflections due to separate encounters will average out, but the sum of $(\Delta v_\perp)^2$ will keep increasing. It is natural to define a relaxation time relative to deflection T_D as the time scale beyond which the cumulative value of $(\Delta v_\perp)^2$ becomes comparable with the original specific kinetic energy of the star under consideration. Then the relaxation time scale T_D could be defined by the integral relation

$$v_\infty^2 = T_D \int (\Delta v_\perp)^2 2\pi (v_\infty n) b \, db. \tag{7.1}$$

Here the integral over the impact parameter b replaces the sum that would be more appropriate for a discrete set of stars.

For a quick estimate of the quantity T_D just defined, an additional simplification can be made. We may focus on small scattering angles, approximate each discrete interaction by means of the *unperturbed straight orbit* of the test star, and estimate the relevant interaction time to be limited to $2b/v_\infty$. In reality, because the two-body problem can be solved exactly, there is really no need to make these approximations (see below). Then, if we take, for simplicity, $m_f = m_t = m$, we find that

$$\Delta v_\perp \approx \frac{Gm}{b^2} \Delta t \approx \frac{Gm}{b^2} \frac{2b}{v_\infty} = \frac{2Gm}{bv_\infty}, \tag{7.2}$$

which, when inserted in Eq. (7.1), yields

$$v_\infty^2 \approx 8\pi n \frac{G^2 m^2}{v_\infty} T_D \ln (b_{max}/b_{min}). \tag{7.3}$$

This leads to the desired estimate for T_D. If we take a representative value of 10 for the factor $\ln (b_{max}/b_{min})$, we get

$$T_D \approx 2 \times 10^8 \frac{v_\infty^3}{m^2 n} \text{ yr}, \tag{7.4}$$

where now v_∞ is measured in kilometers per second, m in solar masses, and n is the number of stars per cubic parsec. Note that the dependence on v_∞ (third power) dominates that on the density n.

7.1.2 Full Expressions for the Relaxation Times

The divergence of the integral at small impact parameters, which is resolved in the heuristic derivation by the introduction of a cutoff at b_{min}, shows only that our assumption of small scattering angles does not carry through, and therefore, for small impact parameters, we need to refer to the exact scattering formulas. The divergence at large impact parameters is instead genuine and reflects the long-range nature of the gravitational interaction. In fact, the exact calculation gives

$$\Delta v_\perp = \frac{2m_f v_\infty}{(m_t + m_f)} \frac{(b/b_0)}{1 + (b/b_0)^2}, \tag{7.5}$$

where b_0 is an appropriate scale for the impact parameter:

$$b_0 = \frac{G(m_t + m_f)}{v_\infty^2}. \tag{7.6}$$

Thus a calculation of T_D from Eq. (7.1) based on the exact orbits would give

$$T_D = \frac{v_\infty^3}{8\pi n G^2 m_f^2 I_D},\tag{7.7}$$

where the integral over impact parameters can be approximated as

$$I_D = \int_0^{b_{\max}/b_0} \frac{t^3}{(1+t^2)^2}\, dt = \frac{1}{2}\left[\ln(1+\Lambda^2) - \frac{\Lambda^2}{1+\Lambda^2}\right] \approx \ln\Lambda.\tag{7.8}$$

Here the Coulomb logarithm, defined as

$$\ln\Lambda = \ln(b_{\max}/b_0),\tag{7.9}$$

is anticipated to be large.

Two other relaxation times also may be introduced, the energy equipartition time T_{eq} relative to the energy exchange ΔE and the dynamical friction time T_{fr} relative to the specific momentum loss $\Delta v_\parallel$ parallel to the original direction of motion. These relaxation times turn out to be related to T_D:

$$T_{eq} = \frac{m_f}{m_t} T_D,\tag{7.10}$$

$$T_{fr} = \frac{2m_f}{m_t + m_f} T_D.\tag{7.11}$$

We can recover these expressions by using the following exact formulas for a classical two-body encounter:

$$(\Delta v_\perp)^2 = \frac{4\mu^2}{m_t^2} v_\infty^2 \sin^2(\theta/2)\cos^2(\theta/2),\tag{7.12}$$

$$\Delta E = -\frac{2\mu^2}{m_f} v_\infty^2 \sin^2(\theta/2),\tag{7.13}$$

$$\Delta v_\parallel = -\frac{2\mu}{m_t} v_\infty \sin^2(\theta/2).\tag{7.14}$$

Here $\mu = m_t m_f/(m_t + m_f)$ is the reduced mass and θ is the deflection in the center-of-mass frame of reference, which is related to the impact parameter

$$\sin^2(\theta/2) = \frac{1}{1 + (b/b_0)^2}.\tag{7.15}$$

Final expressions (7.10) and (7.11), together with Eqs. (7.7) and (7.8), refer to the leading-order estimates, under the assumption that $\Lambda \gg 1$.

For comparison, we record here the formula[3] for the reference time scale $1/v_0$ for Coulomb collisions in an ionized gas for test particles of charge q_t on field particles of charge q_f:

$$\frac{1}{v_0} = \frac{m_t^2 v_\infty^3}{4\pi q_t^2 q_f^2 n \ln\Lambda}.\tag{7.16}$$

The charge-to-mass cancellations account for the simpler expressions in the gravitational case. The Rutherford cross section

$$\frac{d\sigma}{d\Omega} = -\frac{b}{\sin\theta}\frac{db}{d\theta} = \frac{q_i^2 q_f^2}{4\mu^2 v_\infty^4 \sin^4\theta/2},$$

(7.17)

leads to the counterintuitive property that the relaxation rate decreases when the relative speed v_∞ increases.

7.1.3 Setting the Parameter Dependence

For an estimate of the relaxation times in specific cases, we should decide what the appropriate values of v_∞ and b_{max} are to be used in the preceding formulas. As to the value of v_∞, a calculation based on Maxwellian velocity distributions for the field and the test particles shows that for many applications, one should refer to the relevant velocity dispersion, that is, to the appropriate thermal speed characterizing the system.[4] If the system is made of more than one component with separate kinematic properties (e.g., a galaxy made of stars and a population of globular clusters or a plasma of ions and electrons), it is clear that several processes can be considered, and different values of v_∞ should be used to represent them.[5]

For the plasma context, this is treated systematically.[6] Here it is interesting to note that the resistivity in an ionized gas[7] scales as $T^{-3/2}$ because of the dependence of the collision rate on the thermal speed, so high-temperature plasmas are better conductors. In fact, the term *high-temperature plasma* is often synonymous with collisionless plasma. (A related interesting point made in the study of electromagnetic plasmas is the possibility of slide-away or runaway tails in the distribution functions that may be generated in the presence of electric fields because the accelerated particles are less influenced by friction.)

The choice of b_{max} is somewhat controversial because there are arguments to take it as small as the star-star distance $n^{-1/3}$ or as large as the natural size of the system.[8] However, the discussion on b_{max} only affects a logarithmic term, so the two extreme choices mentioned here lead to results within a factor of 2. In modeling of the relaxation processes in an inhomogeneous stellar system there are other important limitations that also should be addressed.

7.1.4 Relation to Other Time Scales

For a finite, one-component, nonrotating stellar system of size R made of N stars of mass m, the natural velocity dispersion is given by the virial theorem $c^2 \approx GNm/R$, and the natural crossing time is $\tau_d = R/c$. Therefore, from Eq. (7.7), we find the scaling

$$\frac{T_D}{\tau_d} \sim \frac{N}{\ln N},$$

(7.18)

which readily shows why large stellar systems are to be considered collisionless.[9] In fact, typical numbers for conditions applicable to elliptical galaxies easily give $T_D \approx 10^{14}$ yr. For disk galaxies, even in the central regions, the relaxation time typically exceeds 10^{11} yr. Chapter 8 introduces the tools that describe a collisionless stellar system.

For comparison, here we recall the main time scales that characterize a normal galaxy. The shortest time in a galaxy is generally the time scale for a light signal to cross the system τ_ℓ, often

$\approx 10^5$ yr. The lifetime of the brightest stars can be 10^6 to 10^7 yr; in a spiral galaxy this also covers the range for the period of vertical oscillations in the main disk and for the dissipation time associated with the cold interstellar medium. The dynamical time τ_d, either meant as typical crossing time or revolution time, is often $\approx 10^8$ yr. The long-term evolution of a galaxy is expected to occur on the time scale of $\approx 10^9$ yr. The ages of galaxies are comparable with the Hubble time, of the order of 10^{10} yr.

7.2   Collisionality and Dynamical Friction in Inhomogeneous Systems

The preceding analysis deals with an idealized homogeneous environment and straight orbits for the test particles through the field. Dynamical friction is often visualized in terms of a wake of particles created by scattering behind the test object.[10] In an inhomogeneous system, with the test object moving along a quasi-periodic orbit, the qualitative picture of wake-particle interaction is bound to change significantly. Fortunately, the order-of-magnitude estimates of the relaxation times obtained when typical values are taken for n and v_∞ show that large stellar systems are well within the collisionless regime, so the limitations of the idealized model used and the exact numerical factors to be determined are probably unimportant for most applications. However, not only from the conceptual point of view but especially because there are processes for which the effects of collisionality have interesting astrophysical consequences (see Subsections 7.2.2 through 7.2.4), it would be desirable to test the accuracy of the predictions that can be made based on the formulas for the relaxation times just recorded.

One case in which collisionality becomes important is the motion of one massive object (e.g., a globular cluster or a small satellite galaxy) through a field of stars so that $m_t \gg m_f$ (this condition has been jokingly compared to the passing of a truck through a field of bicycles). For a globular cluster inside a stellar system, the mass ratio may be easily 10^5. Therefore, even when T_D for the stars exceeds the Hubble time, the dynamical friction relaxation time $T_{fr} \approx 2(m_f/m_t)T_D$ can produce interesting effects on a short time scale. The dynamical friction equation, with the description of Section 7.1, becomes

$$\frac{d\mathbf{v}}{dt} = -\frac{\mathbf{v}}{T_{fr}} = -\frac{4\pi G^2 m_t \rho_f F(v) \ln \Lambda}{v^3} \mathbf{v}, \tag{7.19}$$

where $\rho_f = m_f n$ is the mass density through which the massive object moves, and the factor $F(v) \leq 1$ takes into account the velocity distribution of field stars in relation to the velocity of the test object; the function $F(v)$ is usually taken to be the fraction of field stars with velocity smaller than the velocity v of the massive object.[11] Equation (7.19) should give a quantitative description of the sinking of a satellite toward the center of a galaxy as a result of dynamical friction.

7.2.1   Adequacy of the Classical Local Description

The study of relaxation by means of N-body simulations is a vast field of research.[12] In particular, the process of a sinking satellite has been studied by means of numerical simulations for a number of different conditions.[13]

In systems with a relatively broad core (e.g., polytropic models), numerical experiments confirm the adequacy of the dynamical friction formulas, although questions remain as to the importance of the additional effects due to the collective response of the system through which the object sinks[14] and to the related possibility of some effective friction even in vacuum[15] [where formula (7.19) would predict no orbital decay].

We may leave the $\ln \Lambda$ term as a free parameter to be fitted by the simulation

$$\ln \Lambda = -v \left(\frac{dE_t}{dt} \right) \frac{1}{4\pi G^2 m_t^2 \rho_f F(v)}, \qquad (7.20)$$

where E_t and m_t are the energy and mass of the sinking satellite, and ρ_f is the mass density of the host galaxy; the right-handside of the equation can be easily monitored during the simulation. Curiously, even when the simple relaxation formula just given is claimed to be basically adequate, the empirical values of $\ln \Lambda$ thus determined turn out to be embarassingly small (especially for soft, extended satellites), sometimes even smaller than unity. The latter outcome is theoretically unsatisfactory; in any case, for possible astrophysical applications, it implies slower relaxation than that naively anticipated. In addition, the sinking process is found to be significantly model-dependent. In concentrated models, different values of $\ln \Lambda$ have to be taken at different galactocentric radii;[16] for King models, the dependence has been found to be approximately $\ln \Lambda \propto r$. Some suggestions have been made for a choice of b_{max}, or a suitable modification of the classical Coulomb logarithm, able to reproduce the observed behavior.[17]

In cases in which the sinking satellite reaches the central regions of models with an extended homogeneous core, dynamical friction has been found to practically stop. This latter effect has been related to the possible action of a resonance condition[18] that would make dynamical friction vanish. In approximately harmonic potentials, the phase at which the wake of scattered particles is created may turn out to be compatible with no net friction.

Other studies have addressed the issue of the dependence of dynamical friction on the finite size or finite mass of the sinking satellite.[19] Simulations have shown that dynamical friction is only little influenced by the velocity-dispersion anisotropy possibly present in the host galaxy.[20] Current attention is focused mainly on the way dynamical friction and overall evolution proceed in the context in which the satellite is treated as a live stellar system, with the possible formation of tidal tails and the possibility that the satellite is eventually disrupted by tidal interactions in the host galaxy.[21] Obviously, as the simulations become more and more realistic, the number of parameters introduced makes it more and more difficult to extract a general statement on the ways dynamical friction proceeds.

Another interesting aspect of the problem of a sinking satellite is related to the possibility that dynamical friction circularizes the orbit of the captured satellite.[22] Simulations show that circularization occurs in models with a broad core, whereas in concentrated models eccentric orbits are hardly circularized.

7.2.2 Slow, Nonadiabatic Evolution of Elliptical Galaxies Induced by Dynamical Friction

Because of the large values of the relevant relaxation times, elliptical galaxies are generally modeled as collisionless equilibrium stellar systems (in particular, as will be described in Part IV, in terms of a time-independent distribution function in the six-dimensional phase space). However, in real galaxies, a number of physical effects are expected to drive a slow evolution of the system. Among the possible sources of evolution we may consider the exchange of energy and momentum with a large population of globular clusters (e.g., we recall that the galaxy M87 hosts more than 10,000 globular clusters) that are slowly dragged inward by dynamical friction toward the center. A similar effect also might be induced by the frequent capture of small satellites (minimergers) impinging on the galaxy along randomly distributed orbits. The question is whether this slow evolution can be properly modeled and, in particular, whether it can be treated as adiabatic. This general issue is interesting not only within the strict dynamical context but also because the paradigm of adiabatic growth has received great attention within cosmological scenarios of structure formation, for which it has been proposed as a natural framework to calculate the effects of infall of baryonic matter into the potential wells dominated by dark-matter halos.[23]

A problem of this type can be formulated in a relatively simple way, for example, by considering a spherical system of stars hosting a spherically symmetric distribution of heavy clumps undergoing dynamical friction.[24] We may then ask if evolution tends to make the stellar system more (or less) concentrated in the course of time and whether such evolution is accompanied by the development of pressure anisotropy (the concept of pressure anisotropy will be introduced in Chapter 8). We also may ask the same questions for the case in which some spherically distributed matter is adiabatically (but arbitrarily) moved toward the center of the host galaxy and compare the two cases; this latter process can be worked out analytically (with methods similar to those presented in Subsection 22.2.2), but the action of dynamical friction is best followed by means of simulations. While the sinking of a single satellite, especially if its mass is nonnegligible (e.g., one-tenth the mass of the galaxy), is characterized by a relatively rough interaction, the approximate spherical symmetry of the distribution of many sinking clumps (e.g., 100 clumps each weighing only one-thousandth the mass of the galaxy) makes the simulations much smoother so that it is much easier to properly monitor the effects to be ascribed to dynamical friction.

The results of several dedicated simulations show that dynamical friction tends to lead to a softening of the initial density profile of the host galaxy, which becomes less concentrated in the course of time (Fig. 7.1a). Because this occurs while the mass of the sinking objects is brought to the center, we may ask what the resulting evolution is of the total (galaxy plus sinking clumps) density profile. For models that start with more concentrated density profiles, the softening also occurs for the total density profile. For sinking clumps spiraling in on quasi-circular orbits, this density evolution is accompanied by the development of significant tangential anisotropy. In contrast, in the adiabatic case, evolution leads to an overall contraction (Fig. 7.1b), that is, to a steepening of the density profile; in turn, the development of tangential pressure anisotropy is similar to the case of dynamical friction. It is not surprising that in the adiabatic case and in the case of dynamical friction, evolution leads to different end states because we know that the microscopic mechanisms at the basis of dynamical friction (scattering of orbits in a discrete picture or wave-particle resonance in a continuum description) are, despite their slowness, inherently nonadiabatic. Spherical adiabatic evolution is characterized by the detailed conservation of angular momentum and radial action; these conservations do not apply to the process of dynamical friction.

For a single sinking satellite, numerical experiments suggest that the induced evolution in the host galaxy reflects the initial conditions adopted for the captured satellite. Satellites spiraling in on quasi-circular orbits tend to modify the pressure tensor of the host galaxy in the tangential direction; in addition, the orbital angular momentum of the satellite has been observed to

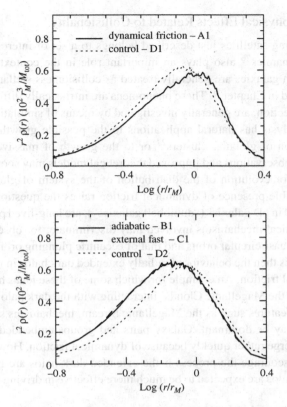

Fig. 7.1. Slow evolution of the host galaxy density profile induced by the fall of a shell of twenty satellites by dynamical friction (*a*; *top*) compared with the evolution due to the infall of a comparable amount of matter in a purely adiabatic case (*b*; *bottom*). In the two frames, solid lines are the final density profiles; dotted lines, the initial density profiles; as indicated on the vertical axes, the plotted quantity represents the volume-weighted density $r^2 \rho(r)$ to better bring out the shift of mass associated with evolution (from Arena, S. E., Bertin, G., et al. 2006. *Astron. Astrophys.*, **453**, 9).

be converted into an overall quasi-solid-body rotation of the host galaxy. In turn, satellites captured along quasi-radial orbits tend to induce pressure anisotropy in the radial direction. Whereas satellites captured along quasi-circular orbits make a galaxy change its shape from spherical to oblate, satellites captured along quasi-radial orbits tend to induce a shape in the host galaxy of the prolate type. The induced softening in the density distribution and the changes in the pressure anisotropy are less pronounced for systems that start out as well concentrated and quasi-isotropic. In particular, systems starting with an $R^{1/4}$ profile tend to remain with the $R^{1/4}$ profile after the capture of a single sinking satellite. This result suggests that one important factor in determining the shape of early-type galaxies may just be the characteristics of occasional mergers. In contrast, in the past it was often argued[25] that the shape of early-type galaxies would result from the effectiveness of the radial-orbit instability (see Chapter 23) during the process of formation via collisionless collapse.

7.2.3   Other Astrophysical Effects Related to Collisionality

The problem of sinking satellites just described belongs to a set of interesting phenomena for which collisional dynamics[26] also plays an important role in the context of the dynamics of galaxies, even though galaxies are generally treated as collisionless stellar systems, with tools that will be introduced in Chapter 8. These phenomena are intrinsically difficult to model and, as noted earlier in this section, are generally investigated by means of simulations.

Orbital decay analysis has natural applications to the possible growth of small bulges by capture and disruption of globular clusters[27] or to the growth of massive central galaxies in rich clusters.[28] New observations and improved theoretical models may soon lead to a consistent dynamical scenario for evolution of the distribution of the system of globular clusters inside galaxies.[29] The possible presence of dynamical friction raises the question of whether double nuclei, often observed in cD galaxies ("dumbbell galaxies"), are long-lived physical associations; a study of the dynamical mechanisms involved may discriminate, for observed configurations, between the case of quasi-circular orbits and that of eccentric plunging orbits.[30]

One general issue is then the behavior of radially extended dark halos in relation to tidal interactions and dynamical friction. An example for which some of these ideas have long been tested and debated is that of the Magellanic Clouds' interacting with the dark halo of our Galaxy,[31] but for some interesting features such as the Magellanic Stream, mechanisms strictly separate from dynamical friction may be dominant. Galaxy pairs and groups embedded in very large halos might be driven to merge rather quickly because of dynamical friction. However, the factor $F(v)$ of Eq. (7.19) may discourage the process if the extended dark halos are made of fast-moving particles. Thus cold halos are expected to be much more effective in driving the merging than hot halos.

Therefore, a broad category of phenomena in which dynamical friction is expected to play a role is that of merging of stellar systems[32] (Fig. 7.2), which has been postulated as a primary mechanism for the formation of elliptical galaxies. The point is interesting, but there are arguments against it.[33] To support the picture that merging is natural and ubiquitous, it is often noted that ellipticals are the dominant population in the central regions of clusters; however, the conditions of galaxy-galaxy interaction in the central regions of rich clusters are not so favorable to merging because galaxies close to the center are likely to be in fast relative motion (because they are at the bottom of a deep potential well). Merging takes place as a result of a combination of relaxation processes in addition to dynamical friction; in particular, it may be accompanied by mechanisms associated with the dissipative interstellar medium.[34] Curiously, it may be dominated by a form of collisionless relaxation (see Section 7.3).

Galaxy-galaxy interactions, galaxy-intergalactic gas interactions, and gas stripping are interesting complex processes expected to shape the evolution of clusters of galaxies. Simulations indicate[35] that because of these interactions significant numbers of stars should be shed in the intergalactic medium. The search for such intergalactic stellar debris is in progress, and some evidence for it is being collected.[36]

7.2.4   Relaxation Processes in Globular Clusters

In practice, collisional dynamics is primarily of interest for the internal dynamics of stellar systems much smaller than galaxies, in particular, for globular clusters,[37] and, as such, will find

Fig. 7.2. A classical case of interacting galaxies, the Antennae (NGC 4038/4039) (credit: B. Whitmore, NASA, ESA, and the Hubble Heritage Team).

little application to the issues addressed in Parts III and IV. It would be impossible to review the subject properly in only a few pages. Here we mention only some topics of general interest.[38]

The dense environment of globular clusters introduces the possibility of significant effects due to multiple interactions[39] and direct physical encounters, with tidal and dissipative processes beyond the simple two-body scattering. This can lead to the formation of binaries and even a change in the structural characteristics of the stars involved (e.g., the formation of exotic objects, such as the stars called *blue stragglers*). Actually, a large fraction of binaries in globular clusters may just be of primordial origin.[40]

Even within the simpler framework of star-star scattering processes quantified earlier in this chapter, a number of interesting phenomena can take place, among which we mention evaporation, mass segregation and equipartition, and core collapse. Evaporation can be studied in a simple spherical spatially truncated model (such as spherical King models; see Section 22.3) by arguing that when collisions bring stars above a certain energy level, the stars will escape. It also can be studied in a more realistic environment, in which the boundary of the system is defined by the three-dimensional structure of the so-called Roche lobe associated with a tidal interaction. It appears that mass loss from globular clusters proceeds primarily by means of stars along radial orbits, so the outermost parts of the system tend to develop tangential pressure anisotropy.[41] Another important process related to collisionality is the one that leads to mass segregation, that is, a preferential concentration toward the center of heavier stars. In principle, full relaxation would be expected to lead to energy equipartition, with the square of the relevant velocity dispersion inversely proportional to the star mass.[42] In recent years, it has been realized that this process may be significantly altered in the presence of binaries or of a central massive black hole.[43]

Finally, for concentrated stellar systems collisionality is expected to trigger an instability called *gravothermal catastrophe* leading to a collapse of the core, which is expected to be ultimately halted by binaries. This general process was studied initially in the context of gas spheres[44] and later investigated for stellar systems.[45] We will briefly return to this mechanism in Chapter 22 in the discussion of stellar dynamical models of elliptical galaxies and in Chapter 25.

7.3   Collisionless Relaxation Processes

A system in which two-body encounters play a negligible role may find other means of relaxation. One general process is related to phase mixing and refers to the fact that our macroscopic description basically operates by coarse-grained distribution functions.[46] In the vicinity of dynamical equilibrium, some relaxation is effectively induced by collective modes.[47] When the system starts away from equilibrium, violent relaxation can take place[48] by means of rapid fluctuations in the gravitational field. Clearly, the approach to equilibrium of a collisionless stellar system is intertwined with the general problem of the statistical mechanics of self-gravitating systems and the problem of galaxy formation and evolution, which will be further addressed in Part V. Here it is stressed that because of the efficiency of collisionless relaxation processes, stellar systems are not expected to be characterized by arbitrary distribution functions. Still, it may be difficult in general to assess whether a given distribution function is realistic. Some of these issues have important consequences for our understanding of elliptical galaxies and will be addressed in detail in Part IV.

7.3.1   Phase Mixing

Collisionless relaxation by phase mixing is well illustrated by the evolution of a set of oscillators[49] with different energies E and period $\tau = \tau(E)$. If the system begins with a certain distribution of relative phases, even without any interaction among the oscillators, just because of kinematics, the information on the relative phases is bound to be lost in the course of time, unless we use a very fine-grained description (Fig. 7.3).

7.3.2   Resonances and Collective Modes

Relaxation by means of collective modes is also known to occur and is a familiar process in plasma physics. A prime example is that of Landau damping, which occurs as a result of wave-particle resonance (see Chapter 12 for the context of the Jeans instability). Resonances are associated with an effective collisionality through rather subtle mechanisms,[50] with several confirmations in the laboratory. Collective modes often remove unfavorable gradients in the distribution function that act as sources of instability.

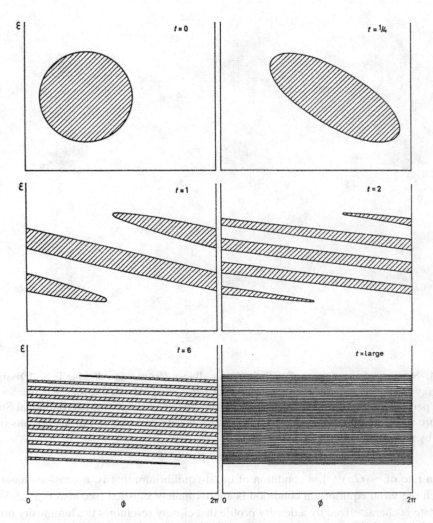

Fig. 7.3. Illustration of the process of collisionless relaxation by means of phase mixing (from Lynden-Bell, D. "Statistical mechanics of violent relaxation in stellar systems," 1967. *Mon. Not. Roy. Astron. Soc.*, **136**, 101; by permission of Oxford University Press, on behalf of the Royal Astronomical Society).

7.3.3 Violent Relaxation

For a system away from equilibrium and collapsing under its own gravity, the rapid fluctuations in the gravitational potential and the clumpiness generated by instabilities during collapse can provide a source of violent relaxation.[51] These ideas have generated numerical experiments aimed at investigating whether elliptical galaxies can be viewed as end products of collisionless collapse[52] (Fig. 7.4). The following discussion has led to further progress and to new questions.[53]

This chapter closes with a brief mention of the basic features found in numerical experiments of collisionless collapse. From a variety of irregular initial conditions, systems starting from a small virial ratio (i.e., sufficiently cold systems) are found to reach, in a few crossing times

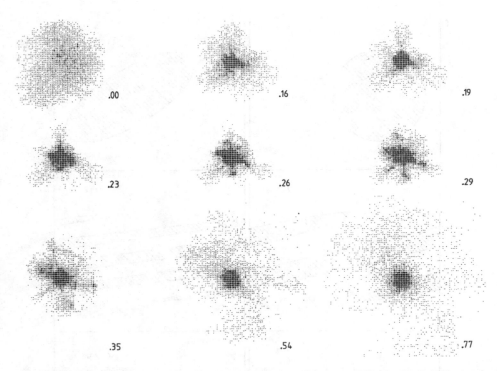

Fig. 7.4. Numerical simulation of collisionless collapse (from van Albada, T. S. "Dissipationless galaxy formation and the R to the 1/4-power law," 1982. *Mon. Not. Roy. Astron. Soc.*, **201**, 939; by permission of Oxford University Press, on behalf of the Royal Astronomical Society). The figure shows the projected density of particles during collapse at various instants of time, marked by numbers in appropriate units.

[i.e., at a rate of $\sim (G\rho)^{1/2}$], a condition of quasi-equilibrium, that is, a quasi-stationary state for which the virial equilibrium condition is approximately satisfied (see also Fig. 25.3). Such a final state is characterized by a density profile that closely resembles the luminosity profile of elliptical galaxies. The final state is characterized by a distribution function that is approximately an isotropic Maxwellian in the inner regions (where the rapid fluctuations have had time to act efficiently), whereas it is strongly anisotropic in the envelope, which is populated mostly by radial orbits. These experiments show that violent relaxation operates differentially, being almost complete in only the inner regions. In contrast, the system is left far from thermodynamical equilibrium in the outer parts, where relaxation remains incomplete. These findings form the basis for interesting theoretical developments that will be described in Part IV.

Notes

1. Wielen, R. 1974. *Highlights of Astronomy*, Vol. 3, ed. G. Contopoulos. Reidel, Dordrecht, The Netherlands, p. 395; for a review of related issues, see Gilmore, G., Wyse, R. F. G., Kuijken, K. 1989. *Annu. Rev. Astron. Astrophys.*, **27**, 555. Similarly, direct empirical evidence that the solar wind is associated with a collisionless plasma comes from in situ measurements that indicate that the relevant ion and electron distribution functions exhibit strong deviations from a standard Maxwellian; see the *Helios*

data described by Marsch, E. 1984. In *Plasma Astrophysics*, ESA Publications SP-207, European Space Agency, Paris, France, p. 33.

2. The presentation is based mainly on the lecture by Woltjer, L. 1967. Vol. 9 of AMS *Lectures in Applied Mathematics Series*, ed. J. Ehlers. American Mathematical Society, Providence, RI, p. 1. Two classical references are Chandrasekhar, S. 1960. *Principles of Stellar Dynamics*, Dover, New York (originally published in 1942); and Braginskii, S. I. 1965. *Rev. Plasma Phys.*, **1**, 205. Interesting discussions are given by Hénon in Contopoulos, G., Hénon, M., Lynden-Bell, D. 1973. *Dynamical Structure and Evolution of Stellar Systems*, eds. L. Martinet, M. Mayor. Geneva Observatory, Sauverny, Switzerland; and Spitzer, L., Jr. 1987. *Dynamical Evolution of Globular Clusters*, Princeton University Press, Princeton, NJ.

3. Huba, J. D. 2002. *NRL Plasma Formulary*, Naval Research Laboratory, Washington, DC, p. 31.

4. Rosenbluth, M. N., MacDonald, W. M., Judd, D. L. 1957. *Phys. Rev.*, **107**, 1; see the basic references mentioned earlier in this chapter.

5. A recent study addresses the problem of estimating the relevant relaxation times for the case in which the masses of the field stars are distributed with a mass spectrum; Ciotti, L. 2010. In *Plasmas in the Laboratory and in the Universe: Interactions, Patterns, and Turbulence*, eds. G. Bertin, F. De Luca, et al. AIP Conference Proceedings, Vol. 1242, Melville, NY, p. 117.

6. Huba, J. D. 2002. *op. cit.*, gives all the formulas applicable to a plasma of ions and electrons characterized by different temperatures for various transport processes.

7. Spitzer, L., Jr., 1962. *Physics of Fully Ionized Gases*, Wiley Interscience, New York.

8. The scale of an inhomogeneous stellar system is its typical Jeans length (see Chapter 12). This choice would make the case analogous to the plasma context, within which we generally set the maximum impact parameter to be the Debye length by invoking the collective screening effects of the plasma. See Persico, E. 1926. *Mon. Not. Roy. Astron. Soc.*, **86**, 93; Cohen, R. S., Spitzer, L., Jr., Routly, P. McR. 1950. *Phys. Rev.*, **80**, 230.

9. For conditions applicable to a globular cluster, referred to its half-mass radius, the representative time ratio would be $T_{Dh}/\tau_{dh} = N/[11 \ln(0.4N)]$; see formula 4-1 in the monograph by Spitzer, L., Jr. 1987. *op. cit.*

10. Mulder, W. A. 1983. *Astron. Astrophys.*, **117**, 9; Tremaine, S., Weinberg, M. D. 1984. *Mon. Not. Roy. Astron. Soc.*, **209**, 729; Palmer, P. L., Papaloizou, J. 1985. *Mon. Not. Roy. Astron. Soc.*, **215**, 691; Weinberg, M. D. 1986. *Astrophys. J.*, **300**, 93; 1989. *Mon. Not. Roy. Astron. Soc.*, **239**, 549; Bertin, G., Pegoraro, F., et al. 1994. *Astrophys. J.*, **434**, 94. See sect. 2 and fig. 2 in Bertin, G., Liseikina, T., Pegoraro, F. 2003. *Astron. Astrophys.*, **405**, 73.

11. Chandrasekhar, S. 1943. *Astrophys. J.*, **97**, 255.

12. For example, see Hohl, F. 1973. *Astrophys. J.*, **184**, 353; Huang, S., Dubinski, J., Carlberg, R. G. 1993. *Astrophys. J.*, **404**, 73; Theuns, T. 1996. *Mon. Not. Roy. Astron. Soc.*, **279**, 827.

13. Lin, D. N. C., Tremaine, S. 1983. *Astrophys. J.*, **264**, 364; White, S. D. M. 1983. *Astrophys. J.*, **274**, 53; Duncan, M. J., Farouki, R. T., Shapiro, S. L. 1983. *Astrophys. J.*, **271**, 22; Bontekoe, T. R., van Albada, T. S. 1987. *Mon. Not. Roy. Astron. Soc.*, **224**, 349; Bontekoe, T. R. 1988. Ph.D. thesis, University of Groningen, The Netherlands; Zaritsky, D., White, S. D. M. 1988. *Mon. Not. Roy. Astron. Soc.*, **235**, 289; Cora, S. A., Muzzio, J. C., Vergne, M. M. 1997. *Mon. Not. Roy. Astron. Soc.*, **289**, 253; Bertin, G., Liseikina, T., Pegoraro, F. 2003. *op. cit.*; Arena, S. E., Bertin, G., et al. 2006. *Astron. Astrophys.*, **453**, 9; Arena, S. E., Bertin, G. 2007. *Astron. Astrophys.*, **463**, 921.

14. See especially, Hernquist, L., Weinberg, M. D. 1989. *Mon. Not. Roy. Astron. Soc.*, **238**, 407; Weinberg, M. D. 1989. *op. cit.*; Maoz, E. 1993. *Mon. Not. Roy. Astron. Soc.*, **263**, 75.

15. See Lin, D. N. C., Tremaine, S. 1983. *op. cit.*; Prugniel, P., Combes, F. 1992. *Astron. Astrophys.*, **259**, 25; Leeuwin, F., Combes, F. 1997. *Mon. Not. Roy. Astron. Soc.*, **284**, 45.

16. Bontekoe, T. R. 1988. *op. cit.*

17. Hashimoto, Y., Funato, Y., Makino, J. 2003. *Astrophys. J.*, **582**, 196; Spinnato, P. F., Fellhauer, M., Portegies Zwart, S. F. 2003. *Mon. Not. Roy. Astron. Soc.*, **344**, 22; Just, A., Peñarrubia, J. 2005. *Astron. Astrophys.*, **431**, 861.

18. Tremaine, S., Weinberg, M. D. 1984. *op. cit.*; Weinberg, M. D. 1986. *op. cit.*

19. White, S. D. M. 1976. *Mon. Not. Roy. Astron. Soc.*, **174**, 467, provides a formula for the Coulomb logarithm applicable to the sinking of extended objects.

20. Arena, S. E., Bertin, G. 2007. *op. cit.*

21. For example, see Peñarrubia, J., Just, A., Kroupa, P. 2004. *Mon. Not. Roy. Astron. Soc.*, **349**, 747; Fuji, M., Funato, Y., Makino, J. 2005. *Publ. Astron. Soc. Jpn*, **58**, 743.

22. See section 4.1 in Bontekoe, T. R., van Albada, T. S. 1987. *op. cit.*

23. Blumenthal, G. R., Faber, S. M., et al. 1986. *Astrophys. J.*, **301**, 27; Mo, H. J., Mao, S., White, S. D. M. 1988. *Mon. Not. Roy. Astron. Soc.*, **295**, 319; Kochanek, C. S., White, M. 2001. *Astrophys. J.*, **559**, 531; Gnedin, O. Y., Kravtsov, A. V., et al. 2004. *Astrophys. J.*, **616**, 16.

24. Here we summarize the results obtained in a series of papers: Bertin, G., Liseikina, T., Pegoraro, F. 2003. *op. cit.*; Arena, S. E., Bertin, G., et al. 2006. *op. cit.*; Arena, S. E., Bertin, G. 2007. *op. cit.* The general trend in the direction of a softening of the density profile has also been noted in other papers, among which are El-Zant, A., Shlosman, I., Hoffman, Y. 2001. *Astrophys. J.*, **560**, 636; El-Zant, A., Hoffman, Y., et al. 2004. *Astrophys. J.*, **607**, L75; Ma, C.-P., Boylan-Kolchin, M. 2004. *Phys. Rev. Lett.*, **93**, 021301; Nipoti, C., Treu, T., et al. 2004. *Mon. Not. Roy. Astron. Soc.*, **355**, 1119; Lackner, C. N., Ostriker, J. P. 2010. *Astrophys. J.*, **712**, 88.

25. For example, see Aguilar, L., Merritt, D. 1990. *Astrophys. J.*, **354**, 33; Cannizzo, J. K., Hollister, T. C. 1992. *Astrophys. J.*, **400**, 58; Warren, M. S., Quinn, P. J., et al. 1992. *Astrophys. J.*, **399**, 405; Theis, C., Spurzem, R. 1999. *Astron. Astrophys.*, **341**, 361.

26. Chandrasekhar, S. 1943. *Rev. Mod. Phys.*, **15**, 1; Alladin, S. M. 1965. *Astrophys. J.*, **141**, 786; Alladin, S. M., Narasimhan, K. S. V. S. 1982. *Phys. Rep.*, **92**, 339; Saslaw, W. C. 1985. *Gravitational Physics of Stellar and Galactic Systems*, Cambridge University Press, Cambridge, UK.

27. This idea was put forward for M31 by Tremaine, S., Ostriker, J. P., Spitzer, L. 1975. *Astrophys. J.*, **196**, 407.

28. See Merritt, D. 1985. *Astrophys. J.*, **289**, 18.

29. See Harris, W. E. 1988. In *Globular Cluster Systems in Galaxies*, eds. J. E. Grindlay, A. G. Davis-Philip. Kluwer, Dordrecht, The Netherlands, p. 237. Harris, W. E., Harris, G. L. H., McLaughlin, D. E. 1998. *Astron. J.*, **115**, 1801, study in detail the case of the giant elliptical M87.

30. Tonry, J. L. 1984. *Astrophys. J.*, **279**, 13; 1985. *Astrophys. J.*, **291**, 45; Valentijn, E. A., Casertano, S. 1988. *Astron. Astrophys.*, **206**, 27.

31. See Fich, M., Tremaine, S. 1991. *Annu. Rev. Astron. Astrophys.*, **29**, 409.

32. Toomre, A. 1977. In *The Evolution of Galaxies and Stellar Populations*, eds. B. Tinsley, R. B. Larson. Yale University Press, New Haven, CT, p. 401; Schweizer, F. 1986. *Science*, **231**, 193; Barnes, J. F. 1988. *Nature (London)*, **338**, 123; Schechter, P. L. 1990. In *Dynamics and Interactions of Galaxies*, ed. R. Wielen. Springer-Verlag, Berlin, p. 508.

33. Ostriker, J. P. 1980. *Comments Astrophys.*, **38**, 1717.

34. See Kormendy, J., Sanders, D. B. 1992. *Astrophys. J. Lett.*, **390**, L53.

35. Starting with Toomre, A., Toomre, J. 1972. *Astrophys. J.*, **178**, 623.

36. See Theuns, T., Warren, S. J. 1997. *Mon. Not. Roy. Astron. Soc.*, **284**, L11.

37. Curiously, in his essay "Science et Méthode," Livre IV, *La Science Astronomique*, 1908, Henri Poincaré gives an extremely vivid and correct representation of the process of relaxation inside a globular cluster.

38. For the dynamical evolution of globular clusters, the reader is referred to Spitzer, L., Jr., 1987. *op. cit.*; Heggie, D., Hut, P. 2003. *The Gravitational Million-Body Problem: A Multidisciplinary Approach to Star Cluster Dynamics*, Cambridge University Press, Cambridge, UK.

39. For example, see Hut, P., Bahcall, J. N. 1983. *Astrophys. J.*, **268**, 319.

40. Hut, P., McMillan, S., et al. 1992. *Publ. Astron. Soc. Pacific*, **104**, 981.

41. See Giersz, M., Heggie, D. C. 1997. *Mon. Not. Roy. Astron. Soc.*, **286**, 709; Takahashi, K., Lee, H. M., Inagaki, S. 1997. *Mon. Not. Roy. Astron. Soc.*, **292**, 331; Aarseth, S. J., Heggie, D. C. 1998. *Mon. Not. Roy. Astron. Soc.*, **297**, 794; Giersz, M. 2001. *Mon. Not. Roy. Astron. Soc.*, **324**, 218; Baumgardt, H., Makino, J. 2003. *Mon. Not. Roy. Astron. Soc.*, **340**, 227; Lee, K. H., Lee, H. M., Sung, H. 2006. *Mon. Not. Roy. Astron. Soc.*, **367**, 646; Gieles, M., Baumgardt, H. 2008. *Mon. Not. Roy. Astron. Soc.*, **389**, L28; Vesperini, E., McMillan, S. L. W., Portegies Zwart, S. 2009. *Astrophys. J.*, **698**, 615.

42. Models of quasi-relaxed stellar systems will be described in Chapter 22. For a discussion of the properties of simple multicomponent models under equipartition and mass segregation, see Da Costa, G.

S., Freeman, K. C. 1976. *Astrophys. J.*, **206**, 128; Kondrat'ev, B. P., Ozernoy, L. M. 1982. *Astrophys. Sp. Sci.*, **84**, 431.

43. See Gill, M., Trenti, M., et al. 2008. *Astrophys. J.*, **686**, 303, and references therein.

44. Ebert, R. 1955. *Zeit. Astrophys.*, **37**, 217; Bonnor, W. B. 1956. *Mon. Not. Roy. Astron. Soc.*, **116**, 351; a generalization to the nonspherical geometry is given by Lombardi, M., Bertin, G. 2001. *Astron. Astrophys.*, **375**, 1091.

45. Antonov, V. A. 1962. *Vestnik Leningr. Univ.*, **19**, 96 (English transl.: 1986. In *Structure and Dynamics of Elliptical Galaxies*, ed. T. de Zeeuw. Reidel, Dordrecht, The Netherlands, p. 531); Lynden-Bell, D., Wood, R. 1968. *Mon. Not. Roy. Astron. Soc.*, **138**, 495; Lynden-Bell, D., Eggleton, P. P. 1980. *Mon. Not. Roy. Astron. Soc.*, **191**, 483; Bertin, G., Trenti, M. 2003. *Astrophys. J.*, **584**, 729.

46. Lynden-Bell, D. 1962. *Mon. Not. Roy. Astron. Soc.*, **124**, 279.

47. Kadomtsev, B. B., Pogutse, O. P. 1970. *Phys. Rev. Lett.*, **25**, 1155.

48. Lynden-Bell, D. 1967. *Mon. Not. Roy. Astron. Soc.*, **136**, 101.

49. Lynden-Bell, D. 1968. In *Astrophysics and Relativity*, Vol. 1, eds. M. Chrétien, S. Deser, J. Goldstein. Gordon & Breach, New York, p. 1.

50. The effective dissipation associated with the Alfvén resonance in an ideal MHD plasma is discussed by Bertin, G., Einaudi, G., Pegoraro, F. 1986. *Comments Plasma Phys. Cont. Fusion*, **10**, 173.

51. Lynden-Bell, D. 1967. *op. cit.*; Saslaw, W. C. 1970. *Mon. Not. Roy. Astron. Soc.*, **150**, 299; Shu, F. H. 1978. *Astrophys. J.*, **225**, 83.

52. van Albada, T. S. 1982. *Mon. Not. Roy. Astron. Soc.*, **201**, 939. See also Doroshkevich, A. G., Klypin, A. A. 1981. *Sov. Astron.*, **25**, 127.

53. See Luwel, M., Severne, G. 1985. *Astron. Astrophys.*, **152**, 305; Pfenniger, D. 1986. *Astron. Astrophys.*, **165**, 74; Stiavelli, M., Bertin, G. 1987. *Mon. Not. Roy. Astron. Soc.*, **229**, 61; Madsen, J. 1987. *Astrophys. J.*, **316**, 497; Ziegler, H. J., Wiechen, H. 1989. *Mon. Not. Roy. Astron. Soc.*, **238**, 1261; Londrillo, P., Messina, A., Stiavelli, M. 1991. *Mon. Not. Roy. Astron. Soc.*, **250**, 54; Hjorth, J., Madsen, J. 1991. *Mon. Not. Roy. Astron. Soc.*, **253**, 703; Nozakura, T. 1992. *Mon. Not. Roy. Astron. Soc.*, **257**, 455; Spergel, D., Hernquist, L. 1992. *Astrophys. J. Lett.*, **397**, L75. See also Trenti, M., Bertin, G., van Albada, T. S. 2005. *Astron. Astrophys.*, **433**, 57, and references therein.

8 Models

How can we formulate appropriate equations that are able to provide a realistic and quantitative description of the processes involved in the dynamics of galaxies? How should we model specific phenomena, such as the occurrence of spiral structure or the equilibrium configuration that underlies the observed photometric properties of elliptical galaxies? The modeling process can be articulated in two steps: (1) choice of equations and (2) identification of the relevant basic state and parameter characterizations. Because we are dealing with extremely complex and composite systems (which include different populations of stars, gas in various forms, magnetic fields, etc.), we should recognize from the beginning that no matter how detailed we try to be, we are introducing only model equations, that is, equations for ideal systems that have only some of the ingredients of the real systems we would like to describe. Once the equations are chosen, a second major aspect of the modeling process is that of choosing a good match to the parameters, boundary conditions, and overall characteristics that define the astrophysical phenomena to be studied.

There is a tradeoff between the number of ingredients that can be included in the equations and the model and what can be practically calculated. Special cases with fewer ingredients may be amenable to exact analytical solutions or to efficient numerical calculations, with the obvious advantages implied by such a possibility. However, experience in parallel fields, such as hydrodynamics, geophysics, and plasma physics, has taught us that great progress can be made by means of asymptotic studies, in which we can handle significantly more realistic models if we are satisfied with only approximate analytical solutions or numerical solutions to approximate equations. Numerical integrations then complete the investigation by providing detailed answers, tests, and inspiration. Because we are dealing with model equations, numerical simulations may provide precious insights but should not be thought of as true experiments. The contrast between exact models and asymptotic models may be traced to the contrast between a deductive approach and a semiempirical approach[1] (see Chapter 5).

8.1 Systems of Many Particles

A first natural description of a galaxy is as a system of a large number N of mutually interacting stars modeled as point masses. It soon became clear that a numerical integration of the equations of motion from given initial conditions could handle a relatively large number of particles sufficiently accurately as to give simulations relevant to the dynamics of galaxies.[2] Several kinds of codes have been developed with the general goal of simulating systems of many particles.

This important type of modeling is not described in this book. The interested reader is referred to monographs that give a wide coverage of the subject[3] and to the many articles that provide a review of the main themes involved in this rapidly growing line of research.[4] Note that the codes developed are generally dedicated to different needs; for stellar systems such as globular clusters, collisionality is the focus of the attention, whereas for large stellar systems, we would like to reproduce the collisionless aspects of galactic dynamics. In the cosmological context, other issues become important.[5] In particular, the geometry of the system to be simulated (e.g., disk geometry, spheroidal geometry, multicenter mass distributions) largely determines the optimal code to be used. It is obviously more difficult to simulate systems with multiple scales properly.

Broadly speaking, methods to simulate the dynamics of many particles can be divided in three main categories: (1) direct codes,[6] by which particle-particle interactions are calculated directly, (2) tree codes,[7] by which the interactions with distant particles are systematically simplified by proper grouping, and (3) particle-mesh or mean-field codes, by which the gravitational field is calculated on a grid of fixed points and interpolated from there.[8,9] In some cases, the potential associated with a distribution of particles is handled by means of suitable expansions, for example, in spherical harmonics, even without the use of a grid.[10]

The complexity of the code (i.e., its scaling with the number of particles) is different for the various categories, ranging from N^2 for direct codes, to $N \ln N$ for tree codes, to N for mean-field and particle-mesh schemes. Note that the nominal complexity is only one factor that determines the efficiency of the algorithm.

Because the number N that may be dealt with in a code is much smaller than the number of stars present in a galaxy, we should keep in mind that each particle is a superparticle; that is, it represents the mass of many stars, and these are unrealistically grouped at exactly the same location in phase space. This leads to unrealistic graininess and fluctuations of the field even for codes, such as mean-field codes, that are best suited for the simulation of collisionless stellar systems. Furthermore, strictly two-dimensional (2D) layers collisionally relax on the dynamical time scale,[11] which is again undesirable in the galactic context. Many codes, in particular, many direct codes, artificially smooth the interactions between close particles by effectively modifying Newton's law through a device called *softening*.[12] When this is done for the case of zero-thickness disk simulations, the softening may have the welcome role of simulating to simulate an effective thickness of the disk. Determination of the optimal recipe in the choice of softening and evaluation of how realistic the simulations are have to be scrutinized for each class of problems and models under investigation.[13] In general, we should be aware that the simulation may include undesired relaxation and may be intrinsically unsuitable for the study of some small-scale processes and long-term evolution.

Simulations are widely appreciated because they provide what observations and semianalytical theories hardly touch, that is, the time evolution of the system under investigation. In fact, a major motivation for a numerical modeling in terms of systems of many particles is the quest for experiments in which we would like to observe directly how galaxies evolve. In this respect, there are two opposite directions that may be taken. On the one hand, we would naturally try to make the simulations more realistic by including more physical ingredients (e.g., the dynamics of a dissipative gas component, star-formation processes, etc.); in turn, this is bound to strain the demands on the computer and therefore involve introducing devices (e.g., softening for the relatively simple problem of close encounters) that simplify but modify the actual physical interactions. This certainly leads to more lively simulations, but it also leads to results that are markedly different from experiments, as currently meant for laboratory physics. The adequacy of such physically complex numerical models in providing a fair representation of the actual physical state of galaxies becomes controversial.

On the other hand, we may wish to simplify the system of many particles even further, away from the complexities of galaxies, into test cases for the study of dynamical mechanisms. The simulation is thus conceived from the beginning to be an excellent discretized representation of a specific (simple) dynamical problem and is used, as a laboratory experiment, to check dependence on parameters, the role of nonlinearities, some aspects of evolution, and so on. This approach has less ambitious goals, but it can be very successful and leads to solid and inspiring results. The simulations are thus meant primarily to test elementary dynamical mechanisms (e.g., the role of resonances, the adequacy of the classical relaxation-rate formulas, or the occurrence of the two-stream instability) that are known to take part in the grand physical framework of galaxy evolution. This approach is very natural to physicists and plasma physicists,[14] although it deflects our attention somewhat away from the hot astrophysical questions that we would like to answer.

8.2   **Continuum Limit and Stellar Dynamics**

For the description of a system of many stars, given the results of Chapter 7 (which suggest that the relevant two-star relaxation times are very long for conditions applicable to the stellar component of galaxies), we may take the continuum limit and refer to the collisionless Boltzmann equation, which states that the one-particle distribution function $f(\mathbf{x}, \mathbf{v}, t)$ evolves in the six-dimensional phase space under the action of the mean-field potential $\Phi(\mathbf{x}, t)$ according to the continuity equation of an incompressible "fluid"

$$\frac{Df}{Dt} = \frac{\partial f}{\partial t} + \mathbf{v}\frac{\partial f}{\partial \mathbf{x}} - \frac{\partial \Phi}{\partial \mathbf{x}}\frac{\partial f}{\partial \mathbf{v}} = 0. \tag{8.1}$$

In the fully self-consistent case, in which the mean potential is completely determined by the stars described by f, the preceding relation is supplemented by the Poisson equation

$$\Delta\Phi = 4\pi G \int f \, d^3v, \tag{8.2}$$

where we have normalized f to the mass density. In the kinetic theory of gases, the Boltzmann equation is conceptually simple because it represents a linear hyperbolic equation. Here, in the self-consistent case, the equation traces a highly nonlinear problem because Φ is not to be considered as an assigned function but is instead determined by the distribution f itself. Much of Parts III and IV will be devoted to subtle and surprising aspects of such self-consistent problems. In plasma physics, Eq. (8.1) is generally known as the *Vlasov equation*,[15] whereas in the context of the dynamics of galaxies, the preceding equations are often referred to as the *fundamental equations of stellar dynamics*.[16]

If we express f as a function of generalized coordinates and momenta, in a standard Hamiltonian formulation, Eq. (8.1) can be conveniently rewritten in terms of the Poisson brackets much like the Liouville theorem (which generally applies only to the $6N$-dimensional phase space):

$$\frac{\partial f}{\partial t} + \{f, H\} = 0. \tag{8.3}$$

Here $H = p^2/2 + \Phi$ represents the one-particle Hamiltonian associated with the mean potential Φ (for simplicity, we are considering a one-component system of identical stars, and we refer to specific energy and momenta).

When weak collisionality is present, we may adopt the so-called Fokker–Planck description, which has on the right-hand side of Eq. (8.1) a collisional term, with the diffusion coefficients related to the relaxation times calculated in Chapter 7:[17]

$$\left(\frac{\partial f}{\partial t} \right)_c = \frac{\partial}{\partial \mathbf{v}} \left[\langle \frac{\Delta \mathbf{v}}{\Delta t} \rangle f \right] + \frac{1}{2} \frac{\partial^2}{\partial v_i \partial v_j} \left[\langle \frac{\Delta v_i \Delta v_j}{\Delta t} \rangle f \right]. \tag{8.4}$$

The Fokker–Planck equation corresponds to the leading terms of an expansion in the Coulomb logarithm, $\ln \Lambda \gg 1$. The first term represents dynamical friction, and the second is a velocity-diffusion term. The Fokker–Planck equation is a convenient model with many limitations.[18] It has an important application in the dynamics of globular clusters, in which the spherical geometry suggests a partitioning of phase space by use of energy and angular momentum so that friction and diffusion terms in velocity space are more conveniently reexpressed[19] in the variables E and J.

Many books[20] explain the meaning of the collisionless Boltzmann equation and how this equation describes the continuum limit of a system of many particles with $N \to \infty$. Therefore, the rather well-known comments and explanations that are commonly given in more deductive introductions to the subject are not repeated here. In plasma physics, the basic framework for a systematic justification of the collisionless Boltzmann equation is the so-called Bogoliubov–Born–Green–Kirkwood–Yvon (BBGKY) hierarchy;[21] this is essentially an expansion in the plasma parameter $g \ll 1$ that states that the number of particles in the Debye sphere is taken to be large [note that $\ln \Lambda \sim \ln (1/g)$]. However, even in the plasma case, the conditions that justify neglecting higher-order terms are not well demonstrated, and application of the BBGKY hierarchy to the gravity case is not straightforward.

We should be aware that despite the fact that the collisionless Boltzmann equation has had so wide an application in stellar dynamics, there are still fundamental aspects of it that are only partly understood. A real stellar system has a granularity that is generally missed in the continuum mean-field description of Eq. (8.1). We might ask how the role of fluctuations in the dynamics of stellar systems could be properly approached, and in general, we would like to develop a sounder statistical derivation of the relevant macroscopic equations.[22] Another interesting problem is whether the overall description may lead to a genuine thermodynamical limit.[23] For the goals of this book, these interesting themes are not addressed, and instead an applied approach to studying the dynamics of galaxies is followed. We thus make standard use of the collisionless Boltzmann equation to address specific issues for spiral and elliptical galaxies. Some important consequences of this description, in particular, some consequences of the Jeans theorem, will be outlined in Chapter 9. More subtle questions will be raised when demanded by the astrophysical context (e.g., some nontrivial aspects of the definition of integrals of the motion will be given in Part IV).

8.3 Fluid Limit and Fluid Models

From a kinetic description in terms of a distribution function we can easily construct some fluid quantities, that is, quantities from which the information in velocity space is integrated out. In particular, we can define a mass density $\rho(\mathbf{x}, t)$, a fluid velocity $\mathbf{u}(\mathbf{x}, t)$, and a pressure tensor $p_{ij}(\mathbf{x}, t)$:

$$\rho = \int f \, d^3v, \tag{8.5}$$

$$\mathbf{u} = \frac{1}{\rho} \int \mathbf{v} f \, d^3v, \tag{8.6}$$

$$p_{ij} = \int (v_i - u_i)(v_j - u_j) f d^3v. \tag{8.7}$$

The process of taking various moments of the collisionless Boltzmann equation relative to the velocity variables thus leads to a hierarchy of fluid equations. The information from the original kinetic equation is spread out on an infinite set of moment equations (see the example given in Subsection 8.3.1), of which the first can be identified with the continuity and Euler equations of fluid dynamics. However, there is generally no way to break the system into a finite, complete set of fluid equations because each equation that is constructed introduces new variables corresponding to higher moments (the first fluid variables are ρ, $\mathbf{u}$, and p_{ij}).

For a standard fluid, the pressure tensor is a scalar (i.e., $p_{ij} = p\delta_{ij}$). To define a finite, complete set of equations, we generally consider an equation of state, that is, a local relation between pressure and density, or an equation for the energy transport. This must be based on physical arguments that depend on the system under investigation. Even when thermodynamical concepts can be safely used, the molecular theory leading to a justification of the relevant fluid equations is nontrivial, especially when we are trying to obtain a realistic description of the basic transport properties (i.e., heat conduction and viscosity) for energy and momentum.[24] Thus, even for a rather normal fluid, it is not trivial to justify the final set of equations for its description. For a gas component in a galaxy (e.g., for the cold atomic hydrogen component in the disk of a spiral galaxy), we should also take into account the additional complication that its state is generally turbulent.

It is also well known, but often not sufficiently stressed, that the justification for the use of a given set of fluid equations may be different for an equilibrium state and for different types of perturbations on such a state. In the parallel field of plasma physics, it is common practice to consider the fluid limit as a rigorous set of conditions under which the use of fluid equations is justified. Within such a fluid limit, another case that has wide application is the magnetohydrodynamic (MHD) limit (the case of ideal MHD is that for which resistivity is taken to vanish), in which the dynamics of electrons and ions is considered synthetically through currents, fluid velocities, and plasma mass density.[25] In practice, too many interesting physical problems violate the assumptions that are required for a rigorous application of the relevant fluid limit. It may be useful, then, to work on fluid-model equations as a tool, for which it should be stressed from the beginning that the conclusions that can be derived should be supplemented by knowledge of effects that are calculated separately from the set of fluid equations used. We can make good progress by abandoning a strictly deductive approach if we find a way to retain the key physical ingredients in the model.

Similarly, for a stellar system, we may consider the conditions for the use of appropriate equations of stellar hydrodynamics[26] as the fluid limit of stellar dynamics (see the short description in Subsection 13.5.2). In general, though, we are left with equations that are not as simple as desired or too difficult to justify in practical applications. Thus we often resort to model equations within a semiempirical approach. Fluid models, much like MHD equations in plasma physics, if properly accompanied by information gathered from separate kinetic analyses, have proved to be a useful source of knowledge. This is especially true in the dynamics of spiral galaxies. A demonstration of the methodology involved will be given in Part III in the discussion of global spiral modes in self-gravitating disks. So far the use of fluid models has been less fruitful in the case of elliptical galaxies (but see the interesting use of the so-called Jeans equations mentioned in Part IV in Subsection 22.1.3).

8.3.1 An Example of Moment Equations

As an example of the structure of the fluid moment equations that can be derived from the collisionless Boltzmann equation, we consider the simple case of a two-dimensional (2D) distribution function suitable for describing an axisymmetrical equilibrium disk. If we use coordinates (R, θ) and conjugate momenta (p_R, J), the relevant Hamiltonian can be written as

$$H = \frac{1}{2} \left(p_R^2 + \frac{J^2}{R^2} \right) + \Phi(R), \tag{8.8}$$

and from Eq. (8.3), the distribution function $F(R, p_R, J)$ must satisfy the equation

$$\frac{\partial H}{\partial p_R} \frac{\partial F}{\partial R} - \frac{\partial H}{\partial R} \frac{\partial F}{\partial p_R} = p_R \frac{\partial F}{\partial R} + \left(\frac{J^2}{R^3} - \frac{d\Phi}{dR} \right) \frac{\partial F}{\partial p_R} = 0. \tag{8.9}$$

If we now perform a change of variables $(R, p_R, J) \rightarrow (r, v_r, v_\theta)$, with $r = R$, $v_r = p_R$, and $v_\theta = J/R$, the collisionless Boltzmann equation for the function $f(r, v_r, v_\theta) = F(R, p_R, J)$ becomes

$$v_r \frac{\partial f}{\partial r} - \frac{v_r v_\theta}{r} \frac{\partial f}{\partial v_\theta} + \frac{1}{r} \left[v_\theta^2 - (r\Omega)^2 \right] \frac{\partial f}{\partial v_r} = 0. \tag{8.10}$$

Here we have used the standard notation $r\Omega^2 = d\Phi/dr$. If we now multiply Eq. (8.10) by v_r and integrate over the 2D velocity space, we find the following equation for the momentum balance in the radial direction:

$$\frac{\partial}{\partial r} (r\sigma c_r^2) + \sigma \left[(r\Omega)^2 - u_\theta^2 \right] - \sigma c_\theta^2 = 0. \tag{8.11}$$

Here $\sigma = \int f \, dv_r dv_\theta$ is the disk mass density, and we have used the common notation $p_{rr} = \sigma c_r^2$, $p_{\theta\theta} = \sigma c_\theta^2$ to introduce the relevant velocity dispersions. The fluid equation thus obtained readily shows that in the presence of finite-pressure effects, the fluid velocity u_θ generally differs from the one-star rotation velocity $r\Omega$ along a circular orbit. This point will be discussed further in Chapter 14, in which the distribution functions for a cool disk will be described in detail.

8.3.2 Equations for a Zero-Thickness Barotropic Fluid Disk

We now record the complete set of equations for a one-component, zero-thickness inviscid fluid model of a disk embedded in a spheroidal bulge-halo in polar cylindrical coordinates (r, θ, z). These equations form the basis for many applications to the dynamics of spiral galaxies (see Part III). It is again stressed that these are model equations (in particular, they simplify and unify the dynamics of stars and gas present in the disk). Note, for example, that we will take a scalar pressure, whereas the stellar component of the disk is characterized by anisotropic velocity dispersion. In addition, the zero-thickness limit also should be taken in the spirit of a useful model. In fact, the zero-thickness limit of a 3D fluid layer is nontrivial; we will see, for example, that the effects of finite thickness can be very important in the dynamics of spiral density waves (see Subsection 15.1.4).

We denote the disk mass density by σ, the tangential and radial components of the disk velocity field by $v \equiv u_\theta$ and $u \equiv u_r$, respectively, and the total gravitational potential by Φ. We assume a barotropic equation of state so that the relevant pressure term can be derived from an enthalpy function h. Then the mass and momentum conservation equations are written as

$$\frac{\partial \sigma}{\partial t} + \frac{1}{r}\frac{\partial}{\partial r}(r\sigma u) + \frac{1}{r}\frac{\partial}{\partial \theta}(\sigma v) = 0, \tag{8.12}$$

$$\frac{\partial u}{\partial t} + u\frac{\partial u}{\partial r} + \frac{v}{r}\frac{\partial u}{\partial \theta} - \frac{v^2}{r} = -\frac{\partial}{\partial r}(\Phi + h), \tag{8.13}$$

$$\frac{\partial v}{\partial t} + u\frac{\partial v}{\partial r} + \frac{v}{r}\frac{\partial v}{\partial \theta} + \frac{uv}{r} = -\frac{1}{r}\frac{\partial}{\partial \theta}(\Phi + h), \tag{8.14}$$

the equation of state as

$$\frac{dp}{\sigma} = dh = c^2\frac{d\sigma}{\sigma}, \tag{8.15}$$

and the Poisson equation as

$$\frac{\partial^2 \Phi}{\partial r^2} + \frac{1}{r}\frac{\partial \Phi}{\partial r} + \frac{1}{r^2}\frac{\partial^2 \Phi}{\partial \theta^2} + \frac{\partial^2 \Phi}{\partial z^2} = 4\pi G\sigma\delta(z) + 4\pi G\rho_{BH}. \tag{8.16}$$

Here ρ_{BH} is the bulge-halo density distribution. These equations assume that the disk density distribution remains flat on the equatorial plane $z = 0$ in the course of time, and thus they are suitable for the description of density waves given in Chapters 15 through 17. In those chapters, the bulge-halo will generally be considered as *immobile*; to take into account the possibility of interaction between the disk and the bulge-halo components, the preceding equations should be supplemented by another set of equations defining the evolution of ρ_{BH}. Equation of state (8.15) is given in a form that can easily be applied to perturbations over a given basic state. In such a case, the equivalent acoustic speed c is one of the three key functions that characterize the equilibrium configuration of the disk (see Section 14.4).

8.4   Virial Equations

The development of models, from those in terms of many particles to those described by fluid equations, involves successive steps of averaging in phase space so that we get a description that is more and more macroscopic, losing information about the details on the orbits of the individual particles. The loss of information is compensated for by the simplicity of the model and by the focus on a few physical quantities that are usually those that are more easily constrained by observations.

At the end of this ladder we consider the virial equations. We obtain these by taking moments of the fluid equations with respect to the spatial coordinates and by integrating over the volume occupied by the system under investigation. For the relatively simple system of N particles in mutual gravitational interaction, we can derive the scalar virial equation in a straightforward way from the set of $\mathbf{F}_i = m_i\mathbf{a}_i$ equations, where $i = 1, \ldots, N$, by multiplying by (scalar product) the position vector of the ith particle $\mathbf{x}_i$ and by summing over i.

The scalar virial equation for a self-gravitating system is given by

$$\frac{1}{2}\frac{d^2 I}{dt^2} = W + 2K_{\text{tot}}, \tag{8.17}$$

where W is the total gravitational energy, K_{tot} is the total kinetic energy, and I is an appropriate moment of inertia. At equilibrium, we have the virial constraint $W + 2K_{\text{tot}} = 0$. The kinetic energy includes a part associated with the internal energy of random motions (i.e., with pressure) and a part associated with the kinetic energy of fluid motions. In Chapter 10, full definitions are given for the second-order virial equations in the time-dependent case (see Subsection 10.2.4; we would obtain higher-order virial equations by taking higher moments with respect to the spatial coordinates).

Virial Eq. (8.17) states what is, at equilibrium, the proper equipartition of energy in various forms. A cold system is one for which the balance is primarily between W and the kinetic energy of fluid motions, whereas a hot system is one for which the balance is mostly between W and the internal (kinetic) energy in the form of random motions. For structured systems (such as a galaxy model), the scalar virial equation provides only one global necessary constraint and, as such, has only limited predictive power. For highly idealized, homogeneous models (such as the classical ellipsoids; see Chapter 10), the first virial equations may summarize most of what is needed in a dynamical study.

Despite its extreme simplicity, the scalar virial constraint proves to be very useful and is often invoked in the formulation of interesting physical arguments. Because the total energy is given by $E_{\text{tot}} = W + K_{\text{tot}}$, the virial equilibrium equation requires $E_{\text{tot}} = -K_{\text{tot}}$. The latter relation contains the essence of a curious property of self-gravitating systems: They are basically systems with negative specific heat. In fact, if energy is supplied to the system so that E_{tot} increases, for equilibrium to be maintained, the total kinetic energy has to decrease. (A one-particle example of the same concept is that of a satellite on a quasi-circular geocentric orbit spiraling down because of energy loss by friction with the upper atmosphere: The satellite actually accelerates because quasi-circular orbits at smaller radii correspond to higher orbital speeds.)

In addition, the equilibrium virial condition is often used to test the quality of some results obtained by means of numerical integration (e.g., see Subsection 22.4.3). Note that in some numerical simulation schemes, for which the elementary force interactions are modified (e.g., when softening is introduced in N-body simulations), commonly used relations such as Eq. (8.17) no longer hold and have to be properly reformulated. (In this respect, other standard relations, such as total energy or angular-momentum conservation, that are commonly used to test the quality of N-body simulations in the course of time also have to be checked and reformulated.)

8.5 Asymptotics versus Toy Models

The most striking structural property of galaxies is that they are highly inhomogeneous. The surface brightness of the disk in spiral galaxies typically spans five length scales of an approximately exponential decline. In addition, spiral galaxies have a 3D structure with rapid variation of the density along the vertical direction; for early-type spiral galaxies, a bulge is present and, in the inner parts, accompanies or even replaces the disk. For some ellipticals, the $R^{1/4}$ luminosity profile can be followed over more than 10 magnitudes, with a brightness range of a factor of more than 10^4. The study of formation processes should aim at explaining the regularities observed in the overall star distributions. The most spectacular large-scale phenomena, such as spiral structure and warps, take place in an inhomogeneous environment that dictates well-defined boundary conditions at small and large galactocentric radii. Even for the case of electromagnetic plasmas, for which at least quasi-neutrality can guarantee the existence of a relatively uniform background for a number of interesting processes, macroscopic phenomena in laboratory experiments or in space plasmas are associated with an inhomogeneous environment.

In turn, the intrinsic inhomogeneities make it hard to devise models that can be solved exactly by means of analytical calculations. Analytical results have a clear priority, in that they can best clarify the various dynamical mechanisms involved. It is thus common practice to look for models, even if they may be special, that can be solved analytically. Unfortunately, these are either homogeneous or highly symmetric, sometimes singular models. For example, it is possible to work out most of the properties of the homogeneous classical ellipsoids analytically (see Chapter 10). The basic processes that lead to gravitational collapse are easily calculated in the homogeneous case (see the description of the Jeans instability in Chapter 12; severe complications arise in the relatively simple slab geometry, see Section 12.3). Later it is mentioned that the study of one key instability mechanism in a differentially rotating disk can be conveniently performed within the homogeneous shearing-sheet model (with a constant density and a constant Q parameter; see Section 15.6). In the context of elliptical galaxies, the special case of Stäckel potentials (see Section 21.3) leads the way to the study of interesting triaxial configurations. All these are keystones for which exact analytical equations and calculations have led to major progress by clarifying important dynamical mechanisms. But it must be recognized that all these are bound to remain, however interesting, just toy models, to the extent that they are not matched to a realistic inhomogeneous environment.

At first we might thus be brought to the pessimistic conclusion that only nontrivial numerical calculations or simulations are of any interest for the dynamics of galaxies. This would leave us with studying separately a plethora of different cases because we would lose much of the unifying power of analytical calculations.

Fortunately, there is a very broad approach that shares the merits of exact analytical calculations but opens the way to more realistic models, and this is the use of asymptotic methods of investigation.[27] The basic idea is, when relatively complex systems are approached, to identify physically relevant dimensionless parameters that, under realistic conditions, can be assumed to be small and to proceed by means of systematic expansions. A careful study may lead to tractable (approximate) equations and may allow us to derive approximate analytical solutions.

Naively, in the mathematical framework within which we look for convergent series solutions, we might be persuaded that the approximation tools developed are able to provide satisfactory answers only when the expansion parameter is very small (e.g., 0.001) and when convergence can be adequately established. In practice, asymptotic methods are often able to give accurate solutions even when the parameter involved is barely smaller than unity (e.g., 0.25) and even when there is no guarantee of convergence for the formal series resulting from the expansion. Numerous examples of this outcome are given by the asymptotic expressions that approximate some well-known special functions.[28] The accuracy of a given asymptotic result is usually checked a posteriori by comparison with exact numerical integrations (the range of applicability of the asymptotic tool depends on the problem considered and is generally difficult to ascertain a priori). One broad area in which the use of asymptotic methods allows us to deal successfully with the dynamics of inhomogeneous environments in many interesting cases is that of dispersive waves (see Chapter 11). For these, in addition to the numerical checks mentioned earlier, real laboratory experiments are often available to confirm the value of asymptotically based results (in particular, in hydrodynamics and plasma physics). These tools eventually may allow us to handle global-stability problems analytically (see Chapter 17).

A common misunderstanding is that the whole issue at stake when approximate solutions are looked for is that of *accuracy*. In reality, the introduction of a small factor ($\epsilon \neq 0$) may open the way to qualitatively new solutions that would be missed altogether if the simpler ($\epsilon = 0$) case is used. In a sense, there is a lack of continuity between toy models and more realistic configurations. A way to categorize this important concept is to distinguish regular from singular perturbations. There are many phenomena that could be mentioned to show the dramatic consequences involved in this distinction. Because in the following some interesting examples will emerge directly from the study of the dynamics of galaxies, here the simplest example is used for a quick demonstration.

Consider the quadratic algebraic equation

$$\epsilon x^2 - x - 1 = 0, \tag{8.18}$$

where $\epsilon \ll 1$. Equation (8.18) can be seen as a (singular) perturbation to the equation $x + 1 = 0$. Because the small parameter multiplies the higher power, a new root appears. To understand the character of the new root, we might try to balance the first and the last terms, adopting the ordering $x = O(\epsilon^{-1/2})$, but this would be inconsistent because the second term would dominate with respect to the two terms selected to balance each other. There remains one other possibility: that the balance occurs between first and second terms. This leads to the consistent ordering $x = O(\epsilon^{-1})$. In addition, we also have the regular modified solution characterized by $x = -1 + O(\epsilon)$. In contrast to other less trivial problems, here we have the opportunity to compare these conclusions directly with the behavior of the exact roots.

8.5.1 Matched Asymptotic Expansions

The preceding algebraic example has an important counterpart in a simple second-order ordinary differential equation that leads to the concept of boundary layers. Consider the equation

$$\epsilon \frac{d^2 u}{dx^2} + \frac{du}{dx} + u = 0, \tag{8.19}$$

to be solved in the interval $[0, 1]$, where ϵ is a small positive parameter, under the boundary conditions $u(0) = 0$ and $u(1) = 1$. Obviously, if we approximate Eq. (8.19) by dropping the term involving the second-order derivative, we would be unable to satisfy both boundary conditions, whereas a rapid change in a layer around $x = 0$ with a thickness of $O(\epsilon)$ would be obtained in the solution if we retain the first term. The origin of such a boundary layer is explicitly demonstrated when we work out the full analytical solution, which is easily obtained in this example (Fig. 8.1). In many interesting, conceptually similar cases, suggested by a variety of physical problems,

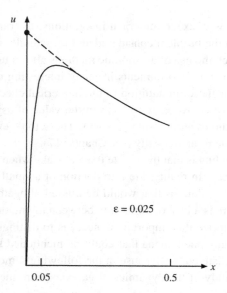

Fig. 8.1. Formation of a boundary layer. The solid curve demonstrates the rapid change of a solution in a small layer with thickness of order ε; the dashed curve displays the behavior of the outer solution for the same problem.

such an exact analytical solution is not available, yet we may resort to asymptotic methods and obtain an approximate analytical solution.

By trial and error we guess that a boundary layer occurs at $x = 0$. Then we consider an outer region where the approximation $du^{(\mathrm{out})}/dx + u^{(\mathrm{out})} \sim 0$ holds. Then, by the same argument mentioned in the ordering of the terms of Eq. (8.18) and by proper rescaling to a stretched coordinate $\zeta = x/\epsilon$, we find the appropriate equation $d^2 u^{(\mathrm{in})}/d\zeta^2 + du^{(\mathrm{in})}/d\zeta \sim 0$ for the inner region (the boundary layer). A matching of the solutions is then performed in an intermediate region, where the coordinate ζ can be considered to be large for $u^{(\mathrm{in})}$ and the coordinate x can be taken to be small for $u^{(\mathrm{out})}$. The matching eventually determines the relevant integration constants because $u^{(\mathrm{out})}$ carries the information about only the outer boundary condition (at $x = 1$), whereas $u^{(\mathrm{in})}$ has only one constant fixed by the boundary condition at $\zeta = 0$.

There are no rules that can guarantee the success of asymptotic approaches of this type, but there are many examples for which the art of asymptotics has led to remarkable results. Some of these will be given later in this book, especially in Part III, which is devoted to spiral galaxies.

Notes

1. It would be interesting to elaborate further on the concept of modeling in science, but this would soon carry us away from the main goals of this book. The reader may consult, among others, the interesting article by Kac, M. 1969. *Science*, **166**, 695.

2. See Holmberg, E. 1941. *Astrophys. J.*, **94**, 385; von Hoerner, S. 1958. *Z. Astrophys.*, **44**, 221; 1960. *Z. Astrophys.*, **50**, 184; 1963. *Z. Astrophys.*, **57**, 47; Lindblad, P. O. 1960. *Stockholm Obs. Ann.*, **21** (4).

3. Potter, D. 1973. *An Introduction to Computational Physics*, Wiley, London; Hockney, R. W., Eastwood, J. W. 1981. *Computer Simulations Using Particles*, McGraw-Hill, New York; Birdsall, C. K., Langdon, A. B. 1985. *Plasma Physics via Computer Simulation*, McGraw-Hill, New York; Tajima,

T. 1989. *Computational Plasma Physics, with Applications to Fusion and Astrophysics*, Addison-Wesley, Reading, MA; Aarseth, S. J. 2003. *Gravitational N-Body Simulations*, Cambridge University Press, Cambridge, UK; Heggie, D., Hut, P. 2003. *The Gravitational Million-Body Problem: A Multidisciplinary Approach to Star Cluster Dynamics*, Cambridge University Press, Cambridge, UK.

4. Buneman, O., Dunn, D. 1966. *Science J.*, **2**, 34 (July); Lecar, M., ed. 1972. *Gravitational N-Body Problem, IAU Colloquium*, Vol. 10, Reidel, Dordrecht, The Netherlands; Spitzer, L. 1975. In *IAU Symposium*, Vol. 69, ed. A. Hayli. Reidel, Dordrecht, The Netherlands, p. 3; Aarseth, S. J., Lecar, M. 1975. *Annu. Rev. Astron. Astrophys.*, **13**, 1; Dawson, J. M. 1983. *Rev. Mod. Phys.*, **55**, 403; van Albada, T. S. 1986. In *The Use of Supercomputers in Stellar Dynamics*, eds. P. Hut, S. McMillan. Springer-Verlag, Berlin, p. 23; Sellwood, J. A. 1987. *Annu. Rev. Astron. Astrophys.*, **25**, 151.

5. For example, see Bertschinger, E., Gelb, J. M. 1991. *Computers in Physics*, **5**, 164.

6. See Aarseth, S. J. 1971. *Astrophys. Space Sci.*, **13**, 324; **14**, 118; Ahmad, A., Cohen. L. 1973. *J. Comp. Phys.*, **12**, 389; Aarseth, S. J. 1985. In *Multiple Time Scales*, eds. J. U. B. Brackbill, B. I. Cohen. Academic, New York, p. 377.

7. See Appel, A. W. 1985. *SIAM J. Sci. Stat. Comp.*, **6**, 85; Barnes, J., Hut, P. 1986. *Nature (London)*, **324**, 446; 1989. *Astrophys. J. Suppl.*, **70**, 389; McMillan, S. L. W., Aarseth, S. J. A. 1993. *Astrophys. J.*, **414**, 200.

8. See Miller, R. H., Prendergast, K. H. 1968. *Astrophys. J.*, **151**, 699; Miller, R. H., Prendergast, K. H., Quirk, W. J. 1970. *Astrophys. J.*, **161**, 903; Miller, R. H. 1971. *Astrophys. Space Sci.*, **14**, 73; Birdsall, C. K., Fuss, D. 1969. *J. Comp. Phys.*, **3**, 494; Hohl, F., Hockney, R. W. 1969. *J. Comp. Phys.*, **4**, 306; Morse, R. L., Nielson, C. W. 1970. *Phys. Fluids*, **12**, 2418; Hohl, F. 1971. *Astrophys. J.*, **168**, 343; 1973. *Astrophys. J.*, **184**, 353; James, R. A., Sellwood, J. A. 1978. *Mon. Not. Roy. Astron. Soc.*, **182**, 331; Berman, R. H., Brownrigg, D. R. K., Hockney, R. W. 1978. *Mon. Not. Roy. Astron. Soc.*, **185**, 861; Miller, R. H., Smith, B. F. 1979. *Astrophys. J.*, **227**, 407; Hohl, F., Zang, T. A. 1979. *Astron. J.*, **84**, 585; Wilkinson, A., James, R. A. 1982. *Mon. Not. Roy. Astron. Soc.*, **199**, 171.

9. See Hénon, M. 1964. *Ann. Astrophys.*, **27**, 83; Hénon, M. 1973. *Astron. Astrophys.*, **24**, 229; van Albada, T. S., van Gorkom, J. H. 1977. *Astron. Astrophys.*, **54**, 121; van Albada, T. S. 1982. *Mon. Not. Roy. Astron. Soc.*, **201**, 939; Villumsen, J. V. 1982. *Mon. Not. Roy. Astron. Soc.*, **199**, 493; McGlynn, T. A. 1984. *Astrophys. J.*, **281**, 13; Allen, A. J., Palmer, P. L., Papaloizou, J. 1992. *Mon. Not. Roy. Astron. Soc.*, **242**, 576.

10. For example, see Trenti, M. 2005. Ph.D. thesis, Scuola Normale Superiore, Pisa, Italy; Trenti, M., Bertin, G., van Albada, T. S. 2005. *Astron. Astrophys.*, **433**, 57.

11. Rybicki, G. B. 1971. *Astrophys. Space Sci.*, **14**, 15; 1972. In *IAU Colloquium*, Vol. 10, ed. M. Lecar. Reidel, Dordrecht, The Netherlands, p. 22. But see Marcos, B. 2013. *Phys. Rev. E*, **88**, id.032112.

12. Aarseth, S. J. 1963. *Mon. Not. Roy. Astron. Soc.*, **126**, 223; Miller, R. H. 1970. *J. Comp. Phys.*, **21**, 400.

13. For example, see White, R. L. 1988. *Astrophys. J.*, **330**, 26; Hernquist, L., Barnes, J. E. 1990. *Astrophys. J.*, **349**, 562; Pfenniger, D., Friedli, D. 1993. *Astron. Astrophys.*, **270**, 561; Romeo, A. B. 1994. *Astron. Astrophys.*, **286**, 799.

14. See Birdsall, C. K., Langdon, A. B. 1985. *op. cit.* Many examples of interesting physical processes discovered by simple numerical experiments can be cited; e.g., see Berman, R. H., Tetrault, D. J., Dupree, T. H. 1983. *Phys. Fluids*, **26**, 2437.

15. Vlasov, A. A. 1945. *J. Phys. USSR*, **9**, 25.

16. Jeans, J. H. 1915. *Mon. Not. Roy. Astron. Soc*, **76**, 70; Chandrasekhar, S. 1942. *Principles of Stellar Dynamics*, University of Chicago Press (reprinted by Dover, New York, 1960). See also Hénon, M. 1982. *Astron. Astrophys.*, **114**, 211.

17. Chandrasekhar, S. 1943. *Rev. Mod. Phys.*, **15**, 1; Rosenbluth, M. N., MacDonald, W. M., Judd, L. D. 1957. *Phys. Rev.*, **107**, 1; Spitzer, L., Jr. 1987. *Dynamical Evolution of Globular Clusters*, Princeton University Press, Princeton, NJ.

18. A conceptually more satisfactory description of the effects associated with the discreteness of the system is known in plasma physics: Balescu, R. 1960. *Phys. Fluids*, **3**, 52; Lenard, A. 1960. *Ann. Phys. (N.Y.)*, **3**, 390; Rostoker, N., Rosenbluth, M. N. 1960. *Phys. Fluids*, **3**, 1; Klimontovich, Y. L. 1967. *The Statistical Theory of Nonequilibrium Processes in a Plasma*, MIT Press, Cambridge, MA.

19. Spitzer, L., Jr. 1987. *op. cit.*

20. For the context of stellar dynamics, the reader may refer, e.g., to Chandrasekhar, S. 1942. *op. cit.*; Ogorodnikov, K. F. 1965. *Dynamics of Stellar Systems*, Pergamon, New York; Saslaw, W. C. 1985. *Gravitational Physics of Stellar and Galactic Systems*, Cambridge University Press, Cambridge, UK; Palmer, P. L. 1994. *Stability of Collisionless Stellar Systems*, Kluwer, Dordrecht, The Netherlands; Ciotti, L., *An Introduction to Stellar Dynamics*, Cambridge University Press, Cambridge, UK (in press).

21. Named after N. N. Bogoliubov, M. Born, H. S. Green, J. G. Kirkwood, and J. Yvon. See Krall, N. A., Trivelpiece, A. W. 1973. *Principles of Plasma Physics*, McGraw-Hill, New York; Ichimaru, S. 1973. *Basic Principles of Plasma Physics: A Statistical Approach*, Benjamin, Reading, MA. The hierarchy follows from the Mayer cluster expansion, which introduces the various n-particle correlation functions; see Mayer, J. E., Mayer, M. G. 1940. *Statistical Mechanics of Fluids*, Wiley, New York. A similar systematic approach has had wide application in cosmology; see Peebles, P. J. E. 1980. *The Large-Scale Structure of the Universe*, Princeton University Press, Princeton, NJ.

22. Kandrup, H. E. 1980. *Phys. Rep.*, **63**, 1; 1981. *Astrophys. J.*, **244**, 316.

23. Interesting general aspects of this problem are discussed by Thirring, W. 1980. *Quantum Mechanics of Large Systems, A Course in Mathematical Physics*, Vol. IV, Springer-Verlag, New York.

24. For example, see Chapman, S., Cowling, T. G. 1952. *The Mathematical Theory of Non-Uniform Gases*, Cambridge University Press, Cambridge, UK; Batchelor, G. K. 1967. *Fluid Dynamics*, Cambridge University Press, Cambridge, UK.

25. Alfvén, H., Fälthammar, C.-G. 1963. *Cosmic Electrodynamics*, Oxford University Press, Oxford, UK; Chandrasekhar, S. 1961. *Hydrodynamic and Hydromagnetic Stability*, Oxford University Press, Oxford, UK (reprinted in 1981 by Dover, New York); Krall, N. A., Trivelpiece, A. W. 1973. *op. cit.*

26. See Hunter, C. 1970. *Studies Appl. Math.*, **49**, 59; Berman, R. H., Mark, J. W.-K. 1977. *Astrophys. J.*, **216**, 257; Hunter, C. 1979. *Astrophys. J.*, **227**, 73.

27. An excellent introduction to the methods and spirit of asymptotic analysis is given by Lin, C. C., Siegel, L. A. 1988. *Mathematics Applied to Deterministic Problems in the Natural Sciences*, Society for Industrial and Applied Mathematics, Philadelphia. Methods frequently used in Parts III and IV are thoroughly treated in more advanced books: Bender, C. M., Orszag, S. A. 1978. *Advanced Mathematical Methods for Scientists and Engineers*, McGraw-Hill, New York; Erdélyi, A. 1956. *Asymptotic Expansions*, Dover, New York; De Bruijn, N. G. 1970. *Asymptotic Methods in Analysis*, North-Holland, Amsterdam, The Netherlands; Heading, J. 1962. *An Introduction to Phase-Integral Methods*, Wiley, New York; Nayfeh, A. H. 1973. *Perturbation Methods*, Wiley, New York; van Dyke, M. 1975. *Perturbation Methods in Fluid Mechanics*, Parabolic Press, Stanford, CA.

28. See Abramowitz, M., Stegun, I. A. 1965. *Handbook of Mathematical Functions*, Dover, New York.

9 Equilibrium and Stability: Symmetry and Symmetry Breaking

Much of the study of the dynamics of collective systems proceeds through investigation of the properties of equilibrium configurations (often referred to as *basic states*) and analysis of their stability with respect to a variety of perturbations. The choice of the equilibrium state is primarily a choice of symmetry. The study of how perturbations evolve on such a basic state is thus a study of the mechanisms that lead to a breaking of the assumed symmetry (although we may consider perturbations that preserve the original symmetry). Obviously, the simplest case that can be considered is that of the linear limit, in which perturbations have vanishingly small amplitudes, much as in the case of small oscillations for a system of point masses in classical mechanics.

In this chapter, general remarks are made about this common framework, and some interesting consequences for the internal structure of collisionless stellar systems are noted. By means of a nontrivial example from plasma physics and an elementary example from classical mechanics, some key concepts are demonstrated and a few important issues are raised that will be found later to be a common thread in the study of the dynamics of normal galaxies. In Chapter 10, the main results and some technical aspects of the process of studying equilibrium and stability for one relatively simple class of self-gravitating fluids, that is, the classical ellipsoids, are summarized.

Within the above-mentioned framework it is implicitly assumed that in many observed systems, some simple symmetries are approximately realized, which is intuitively true. However, strictly speaking, real galaxies are time-evolving systems in which, even for the case where a simple basic state is reasonably well justified as a model, deviations from it are characterized by finite amplitudes. In addition, for large stellar systems, we know (see Chapter 7) that standard collisional relaxation is insufficient to drive the system efficiently toward a state of thermodynamical equilibrium. Thus the paradigm mentioned earlier is only meant to provide useful tools to understand the dynamics of galaxies and should not be taken literally to describe how objects have evolved to reach their current state. In particular, there is no claim that the current state of a galaxy should be that of an exact equilibrium configuration; furthermore, if we relate an observed feature, such as spiral arms, to the presence of an unstable mode on an axisymmetric disk, we are not implying that the galaxy disk originally was axisymmetric and that, because of instability, it later developed spiral structure. The general perception will be better clarified by analysis of the motivations that underlie the framework of equilibrium-stability analyses (see Section 9.3).

Even for systems that appear rather smooth, almost featureless, and regular, such as is the case of some elliptical galaxies, the search for equilibrium models as representative of the observed objects should be carried out with caution. Simple guiding arguments, such as the Jeans theorem (see Section 9.1), may indeed lead to useful models and explain the character of some observed inhomogeneities (e.g., the vertical density profile in a galaxy disk; see Chapter 14; or the luminosity profile of elliptical galaxies; see Chapter 22). However, if the same arguments leave some puzzles unresolved (e.g., the difficulty in constructing triaxial equilibrium models with realistic density profiles; see Chapter 22), it may just be that the actual stellar systems we wish to model simply do not conform to the mathematical assumptions made when the Jeans theorem is formulated. It might even be that a strict equilibrium is not available, and slow evolution is inevitable. Therefore, a deep analysis of the Jeans theorem, beyond the naive formulation that is required for the simplest applications, is not pursued here.

In practice, when applied to normal galaxies, the paradigm of separating the study of equilibrium and stability has a rather different impact depending on whether we refer to spiral or elliptical galaxies. The reason is that the two classes of objects raise different dynamical issues (see Section 9.4).

9.1   The Jeans Theorem

We have seen that the collisionless Boltzmann equation states that the one-particle distribution function f is conserved. Therefore, for an assumed potential $\Phi(\mathbf{x})$, the problem of finding an equilibrium solution $f(\mathbf{x}, \mathbf{v})$ to describe a collisionless stellar system reduces to that of finding an appropriate integral of the motion. This statement may be loosely called the *Jeans theorem*. The standard thermodynamical limit is the limit at which the relevant integral is the exponential of a dimensionless energy so that the distribution function is a Maxwellian. When there is no reason to argue that thermodynamical equilibrium is ensured, in principle, any positive-definite integral of the motion f can be associated with an equilibrium solution of the collisionless Boltzmann equation.

If we are interested in investigating the mathematical problem of finding all possible solutions for an assumed potential $\Phi(\mathbf{x})$, we thus must give a clear definition of what we mean by integral of the motion, discuss whether all integrals have the same value in the construction of the distribution function, and explore what happens when $\Phi(\mathbf{x})$ is associated with a significant number of irregular orbits. All this would soon bring us to very subtle issues at the forefront of theoretical mechanics. Some of these questions are briefly discussed in Part IV (especially in Chapters 21 and 22). For the following, a simpler approach is taken, and important consequences that derive by an inductive use of the Jeans theorem are pointed out. When well-known integrals, such as energy and angular momentum, are referred to, many interesting points can be made.

One important lesson is that, in general, inhomogeneities, anisotropies, and flows for the equilibrium cannot be specified in advance in an arbitrary manner because they depend in a complex, nonlocal manner on the way the integrals of the motion enter the distribution function. In turn, the natural choice $f = f(E)$, where $E = m(v^2/2 + \Phi)$ is the one-particle energy, which seems to be the simplest generalization of a Maxwellian, forces us to systems characterized by an isotropic pressure tensor (because the kinetic energy is a scalar) and no internal streaming (because E is an even function of $\mathbf{v}$). Furthermore, if the form of $f(E)$ is specified in advance, the associated density profile results from the Poisson equation, for which it makes a big difference whether Φ is fully or only partly determined by the mass density associated with f. At the other extreme, if f describes a dust solution of test particles in an external potential Φ, a given form of $f(E)$ implies a well-defined density profile. The inversion from density $\rho(\mathbf{x})$ to $f(E)$ may be well posed (e.g., for a spherically symmetric system) but may lead to unacceptable functions f (in particular, if f turns out to be negative for some values of E). Thus, constructing a self-consistent equilibrium configuration with some desired macroscopic properties in a collisionless environment may be a subtle and sometimes very difficult game.

Physical arguments may constrain the form of the distribution function f (e.g., that it should be close to a Maxwellian or that it should show the effects of partial relaxation in a certain way), thus leading to specific macroscopic profiles. The physical arguments are usually derived from formation scenarios. As a result, observed macroscopic properties, such as the $R^{1/4}$ law for elliptical galaxies, may provide important clues to the relevant formation processes.

9.1.1 A Self-Consistent Collisionless Plasma Configuration with a Current Sheet

To introduce the strategy involved in the construction of collisionless equilibrium configurations separately from the specific issues that pertain to the dynamics of spiral or elliptical galaxies, it may be instructive to recall a well-known example from the theory of high-temperature plasmas. Let us consider a plane-parallel plasma configuration[1] made of ions and electrons, both in the collisionless regime. We look for stationary equilibrium solutions for which the system is homogeneous along the y and z directions and inhomogeneous in x, with the magnetic field along z, and a vanishing electric field. We may thus refer to a vector potential $\mathbf{A} = A(x)\mathbf{e}_y$ such that the magnetic field is given by $\mathbf{B} = (dA/dx)\mathbf{e}_z$. For each species we have three natural integrals of the motion: the energy $E = mv^2/2$ and the two canonical momenta $p_y = mv_y + (q/c)A(x)$ and $p_z = mv_z$. Here q and m are the charge and the mass, respectively, and c is the speed of light. For simplicity, we take ions with charge $q = +e$, with e the magnitude of the electron charge. The collisionless Boltzmann equation for each species is automatically satisfied if we take distribution functions of the form

$$f^{(j)} = f^{(j)}(E^{(j)}, p_y^{(j)}, p_z^{(j)}) . \tag{9.1}$$

Here $j = e, i$ identifies electron and ion quantities, and the distributions are meant to be normalized to the number density. For distributions that are even functions of p_z, a self-consistent equilibrium is found if the following equations are satisfied:

$$0 = -4\pi \sum q \int f \, d^3v, \tag{9.2}$$

$$\frac{d^2A}{dx^2} = -\frac{4\pi}{c} \sum q \int f v_y \, d^3v. \tag{9.3}$$

Equation (9.2) is the neutrality condition consistent with the assumption of vanishing electric field; Eq. (9.3) indicates that the magnetic field is generated self-consistently by the currents associated with the two-component plasma.

The simplest expression for the distribution functions is that of a displaced Maxwellian, that is,

$$f^{(j)} = n_0 \left[\frac{m^{(j)}}{2\pi kT} \right]^{3/2} \exp\left\{ -\frac{1}{kT} \left[E^{(j)} - p_y^{(j)} u^{(j)} + \frac{m^{(j)} u^{(j)2}}{2} \right] \right\}, \tag{9.4}$$

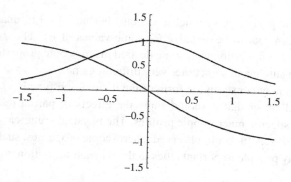

Fig. 9.1. A current sheet in a plasma slab geometry; the plasma density and the magnetic field are plotted as functions of the coordinate perpendicular to the sheet.

where $u^{(j)}$ are constant velocities. We may now take $u^{(e)} = -u^{(i)} = -u$ so that neutrality condition (9.2) is ensured, and we are left with a single ordinary differential equation for A:

$$\frac{d^2A}{dx^2} = -\frac{8\pi n_0 eu}{c} \exp\left(\frac{eu}{kTc}A\right),$$ (9.5)

where a current sheet is produced by the term in the distribution function that is odd in v_y.

If we take the natural boundary conditions for a configuration that is symmetric with respect to the $x = 0$ plane (i.e., $A = 0$ and $B = dA/dx = 0$ at $x = 0$), we find the exact solution (Fig. 9.1):

$$A = -\frac{2kTc}{eu} \log \cosh\left(\frac{ux}{c\lambda_D}\right),$$ (9.6)

$$B = -\sqrt{16\pi n_0 kT} \tanh\left(\frac{ux}{c\lambda_D}\right),$$ (9.7)

where $\lambda_D = (kT/4\pi n_0 e^2)^{1/2}$ is the relevant Debye length. If we insert the expression for A in the distribution functions, we find, for the number density of the plasma,

$$n^{(e)} = n^{(i)} = \frac{n_0}{\cosh^2(ux/c\lambda_D)}.$$ (9.8)

We can easily extend this isotropic configuration (with scalar pressure) to the anisotropic case by adding terms proportional to p_y^2 and p_z^2 in the exponential of Eq. (9.4). This solution anticipates the properties of the self-gravitating plane-parallel slab that will be discussed in Chapter 14. The full analogy will be better appreciated after introduction of the Jeans length in Chapter 12.

In the process of constructing a collisionless equilibrium configuration within a given symmetry, many options are available that are consistent with the Jeans theorem. This example illustrates that if we focus on physically plausible distribution functions, the macroscopic profiles have shapes with a well-defined character that depend in a nonlocal manner on the way the integrals of the motion enter the distribution function. In the corresponding search of fluid models, the physical arguments to be invoked would set the relevant equation of state.

9.2   Symmetry Breaking

In the construction of equilibrium configurations under given assumptions, we often find not just one equilibrium but entire families of solutions, parameterized by some dimensionless quantities (such as the parameters u/c and $u/\sqrt{kT/m}$ for the current sheet discussed earlier or the concentration parameter C for the King models to be discussed in Chapter 22). In some cases we find that less symmetric equilibrium configurations become available in certain domains of the parameter space, which is signaled, in a study of perturbations over the basic state, by the fact that there exist neutral modes, that is, modes that correspond to a time-independent displacement from a more symmetric family to a less symmetric family. These considerations have a wide range of applications and may be suitably introduced by an example involving the dynamics of a single point-mass.

9.2.1   An Elementary Example from Classical Mechanics

An elementary but instructive example of the mechanism of symmetry breaking is offered by one type of spherical pendulum.[2] (Another problem with many analogies with the case presented here will be described in Chapter 13 and refers to a charge that is taken to move close to the top of a hill in the presence of a vertical magnetic field.) We consider a spherical bowl, made to rotate at an angular velocity Ω about the vertical axis, and ask whether there are equilibrium configurations for a ball constrained to move on such a bowl in the presence of the gravity acceleration g. Thus formulated, with Ω as the controlling parameter,[3] we determine the equilibrium configuration by extremizing the mechanical energy

$$E_{\text{mec}} = W - \frac{1}{2}\Omega^2 I_z = -mgL\cos\theta - \frac{1}{2}m\Omega^2 L^2 \sin^2\theta, \tag{9.9}$$

where m is the mass of the ball, L is the radius of the bowl, and θ is the poloidal angle with respect to the vertical axis ($\theta = 0$ denotes the position at the bottom of the bowl). A natural dimensionless parameter $\xi = L\Omega^2/g = \Omega^2/\omega_p^2$ controls the shape of the mechanical energy as a function of θ. For a slowly rotating bowl, $\xi < 1$, E_{mec} is characterized by a unique minimum at $\theta = 0$. By raising the value of Ω, we find that at $\xi = 1$, the minimum is transformed into a maximum, whereas a minimum develops at $\theta = \arccos(1/\xi)$ for $\xi > 1$ (Fig. 9.2). Thus a new equilibrium configuration occurs with $\theta \neq 0$, that is, off the symmetry axis, if the bowl rotates sufficiently fast. The breaking of the symmetry around the z axis produced by a given realization of the latter equilibrium is not a true symmetry break, in the sense that the rotation invariance around z guarantees the existence of an infinity of solutions, each identified by a different toroidal angle.

Naively, we might expect that for $\xi > 1$, the maximum of the mechanical energy corresponds to a case of dynamical instability; that is, that a small perturbation on a ball placed at the bottom of the bowl would make it leave the $\theta = 0$ position exponentially fast (at least in the linear limit). The situation is not so trivial because we find that instability occurs only in the presence of friction between the ball and the bowl. In fact, if we consider the full equations of the motion in the vicinity of $\theta = 0$, including a friction term proportional to the velocity of the ball with

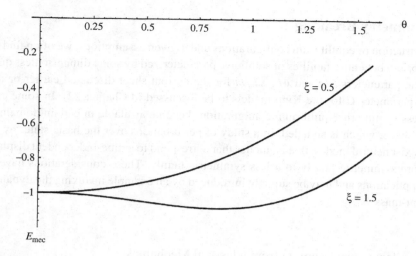

Fig. 9.2. Symmetry breaking, exemplified by the rotating bowl.

respect to the rotating bowl, in Cartesian coordinates we have

$$\ddot{x} - 2\Omega\dot{y} - \Omega^2 x = -\mu\dot{x} - \omega_p^2 x, \tag{9.10}$$

$$\ddot{y} + 2\Omega\dot{x} - \Omega^2 y = -\mu\dot{y} - \omega_p^2 y. \tag{9.11}$$

This system admits nontrivial normal-mode solutions of the form $(x, y) = (\tilde{x}, \tilde{y}) \exp(i\omega t)$, provided that the determinant associated with the linear system vanishes:

$$D(\omega) \equiv (\omega^2 + \Omega^2 - \omega_p^2 - i\mu\omega)^2 - 4\omega^2\Omega^2 = 0. \tag{9.12}$$

In the limit $\mu \to 0$, we find the approximate relation

$$D(\omega) \sim D_0(\omega) - 2i\mu\omega(\omega^2 + \Omega^2 - \omega_p^2), \tag{9.13}$$

where

$$D_0(\omega) \equiv (\omega^2 + \Omega^2 - \omega_p^2)^2 - 4\omega^2\Omega^2. \tag{9.14}$$

From here it is straightforward to identify the following eigenvalues for the small-amplitude displacements from the $\theta = 0$ equilibrium point:

$$\omega \sim \pm\omega_p(1 - \xi^{1/2}) + \frac{i}{2}\mu(1 - \xi^{1/2}). \tag{9.15}$$

(Two other solutions have a positive imaginary part for any value of ξ.) Thus, by including the effects of a small amount of friction, we have proved that for $\xi > 1$, the equilibrium point at $\theta = 0$ is unstable. The instability just discussed requires the presence of dissipation and thus is called *secular instability*. The onset of secular instability occurs precisely when, in the absence of viscosity, the relevant mode becomes neutral ($\omega = 0$), that is, at $\xi = 1$, where the new nontrivial equilibrium sequence becomes available. Similarly, one could prove that for $\xi > 1$, the nontrivial equilibrium point at $\theta = \arccos(1/\xi)$ is stable.

We will find an interesting analogue of this process of symmetry breaking in Chapter 10, for the case of classical ellipsoids, in which the stability analysis on the axisymmetric equilibrium sequence, in the vicinity of the point where the family of triaxial ellipsoids bifurcates, matches in detail the behavior just described.

9.2.2 Classification of Modes

For continuous media, described by either fluid or kinetic equations (see Chapter 8), a systematic approach can be developed for the study of small-amplitude perturbations over a given equilibrium, based on the symmetry group characterizing the undisturbed configuration. In general, if the equilibrium is invariant under a certain transformation associated with an ignorable coordinate x, we can Fourier analyze a generic perturbation with respect to such a coordinate, and in the linear limit, each elementary wave component $\exp(ik_x x)$ decouples, subject to independent dynamics. Indeed, the modal analysis consists of separating the elementary-wave components $\exp(i\omega t)$ as allowed by the time invariance associated with the basic state. In contrast, if symmetry does not exist with respect to a coordinate y, the various k_y elementary waves that may be considered to interact with each other, and the associated Fourier analysis is less useful. Other types of symmetry, such as parity (e.g., the reflection transformation $x \to -x$ in the plane-parallel slab constructed in the preceding section), make it possible to study separately perturbations that are either odd or even with respect to a symmetry plane (Fig. 9.3). In Part III we will see that these two types of perturbations correspond to either density or bending waves over the disk, which have interesting astrophysical counterparts in the dynamics of spiral galaxies.

Therefore, perturbations are usually classified in a way that depends on the symmetry of the basic state, much like eigenstates in quantum mechanics.[4] These ideas, which are best clarified in the framework of dispersive waves (see Chapter 11), are not elaborated on further here; it is

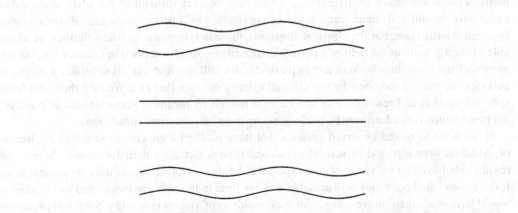

Fig. 9.3. Even (or density; *top*) and odd (or bending; *bottom*) disturbances of a sheet (*middle*) in a slab geometry (after Bertin, G., Lin, C. C. 1996. *Spiral Structure in Galaxies: A Density Wave Theory*, MIT Press, Cambridge, MA, p. 73; ©1996 Massachusetts Institute of Technology).

mentioned only that this general approach will have important applications in the dynamics of galaxies because it provides a powerful quantitative tool of investigation.

9.3   Motivation for Perturbation Analyses

From a physical point of view, different motivations are at the basis of equilibrium-stability analyses. These are often tacitly assumed, but it may be useful to list them in general before the detailed discussion of Parts III and IV, in order to bring out the main physical scopes of this type of investigation. In relation to currently observed systems, the basic issues addressed by an equilibrium-stability analysis are appropriate modeling, formation processes, response to external driving forces, and long-term evolution.

One broad goal of perturbation analysis is the modeling of specific observed structures. For example, in Part III we will approach the description of spiral arms or a bar in terms of perturbations over an otherwise axisymmetric disk. Here generally two alternative perceptions are available. In one point of view, the structures are argued to be evolving on the dynamical time scale. Observations then would provide a picture at a given instant of a time-dependent process; by means of a perturbation analysis over a given basic state, we may try to reproduce the relevant evolutionary process associated with the observed structures. In the second point of view, which has semiempirical justification in the case of large-scale spiral structure in galaxies (see Chapter 18) and may be appropriate in many other physical contexts, we consider that the structures to be modeled are quasi-stationary; that is, we are dealing with an approximate equilibrium configuration that is less symmetric with respect to the case in which the structures have vanishingly small amplitudes. In this case, the perturbation analysis is primarily a device to gather information on this less symmetrical basic state and on the conditions for its realization; there is little emphasis here on the precise evolutionary process that has led to the observed state. In this latter perception, the observed structure is argued to correspond to a state of lower free energy and may be approximated in terms of moderately unstable modes, subject to a nonlinear equilibration, that break the underlying symmetry.

Symmetric states are easier to handle. Thus self-consistent models corresponding to less symmetrical quasi-stationary configurations, which may be very difficult to calculate or available under only exceptional circumstances, can be approached by a perturbation analysis over a more symmetric basic state. For the classical ellipsoids, the less symmetric triaxial solutions are available explicitly without the help of a perturbation analysis (as is described in Chapter 10), but for more realistic mass distributions, and in particular, for collisionless models suitable for elliptical galaxies, the triaxial case may be too difficult to carry out (see Part IV). We may then start from spherical models and reach the less symmetrical models by means of perturbations. In practice, the perturbation procedure may be usefully set up even by numerical simulations.

If we want to model observed systems that have reached their current structure by means of evolution over a period of several dynamical time scales after their formation, there is no reason to believe that by virtue of some special conditions, violent instabilities are present now. If the systems had been violently unstable at some time in the past, the associated fast evolution would have made them move away from the conditions of violent instability. Such fast evolution would stop once a state of marginal or moderate instability is reached. This argument suggests that in our search for realistic models, we may discard all equilibrium configurations that we recognize to be in a condition of violent instability. This simple idea is one important motivation for stability analyses, which thus help to ensure that models suitable for the observed systems are at most subject to slow evolution. At the interpretation level, we may in some cases imagine that the current state results from instabilities developed in the distant past, for example, at the time of formation, as in interpretation of the Jeans instability that is described in Chapter 12.

Knowledge of the intrinsic modes of a dynamical system is an important prerequisite for study of the response to external driving. In particular, tidal effects on spiral galaxies depend on the internal properties of the disk, much as the tides realized on our planet depend not only on the properties of the Sun-Moon configuration but significantly on the intrinsic free modes of oscillation of the driven system. Thus modal analyses are important also in view of the study of driven processes.

Another goal sometimes considered is that of a derivation of the relevant transport processes. Long-term evolution of dynamical systems is also due to the presence of collective modes. This may occur either because of the presence of specific coherent structures (e.g., large-scale spiral structure is associated with torques that redistribute the angular momentum of the disk) or because of the action of less-organized waves, small-scale instabilities, and turbulent activity (e.g., it has been argued that the thickening of galaxy disks results mainly from the slow cumulative action of collective effects). It is well known that such anomalous transport may be significantly more efficient than the standard collisional transport that can be derived from studies such as those indicated in Chapter 7. In the theory of electromagnetic plasmas, this has become evident for both laboratory and space plasmas[5] (in the behavior of toroidal magnetic confinement configurations, in the solar wind, or in the physics of accretion disks). So far, many useful ideas have been gathered in an effort to relate collective mechanisms of instability to momentum and energy-transport processes, but a quantitative deductive framework to explain given situations is not yet available.

A final point of interest of equilibrium-stability analyses is obviously just the study of dynamical mechanisms. The physical processes involved may be interesting in principle, even when they are not applied or applicable to specific astrophysical issues. In practice, many research papers focus on discussion of the dynamical mechanisms, leaving the imagined physical situation only sketched or for the reader to fill in.

9.4 Spiral Galaxies versus Elliptical Galaxies

In relation to the discussion given in this chapter, the study of spiral galaxies is very different from that of elliptical galaxies. A few points are summarized that emerge from Part I and that represent the focus for the rest of this book.

The intrinsic symmetry of spiral galaxies is reasonably well established. Although bars, oval distortions, some triaxiality in the bulge-halo, large-scale spiral structure, and warps are known to occur to some extent, it is appropriate to consider a basic state that is generally well represented by an axisymmetric disk embedded in a spheroidal bulge-halo. The orbital structure in the disk for such a basic state is rather simple, with most orbits confined to relatively thin annuli. The disk is a dynamically cool, rotation-supported system. Probably the central theme for dynamical studies is how deviations from such a basic state develop. From the clues gathered from these studies, in general, we would like to construct a dynamical framework for interpretation of the correlations that define the morphological classifications established empirically. Eventually we would like to understand the long-term evolution of these galaxies. Much of what can be done about these issues in terms of dynamical modeling depends in a sensitive way on the actual amount and spatial distribution of dark matter and on the amount and distribution of cold gas in molecular form. The models have to take into account the uncertainty of these factors that remains despite the many observational constraints available.

In contrast, for elliptical galaxies, the main issue involves what is the applicable basic symmetry. A significant fraction of ellipticals even may be well represented by a spherical basic state. For many objects, axisymmetry is a reasonable starting point. Still the scarcity of a straightforward regular tracer of the field, such as the cold HI disk in spiral galaxies, leaves the general issue and the role of intrinsic triaxiality unresolved. In any case, the very existence of collisionless triaxial stellar systems poses severe challenges and subtle dynamical questions. Bright ellipticals are dynamically hot, pressure-supported systems. The importance of systematic motions (i.e., of ordered kinetic energy) often increases in the outer parts of the galaxy, but here the kinematic data are more difficult to obtain. In any case, there is no doubt that the orbital structure is complex, with many orbits crossing the whole stellar system. This makes equilibrium-stability analyses nontrivial even in the simplest case of spherical geometry. The contribution of dark matter to elliptical galaxies and whether its distribution differs significantly from that of luminous matter are less easy to determine here than in the case of spiral galaxies. Long-term evolution is often addressed in the context of cluster dynamics or in relation to the observed properties of galaxies at relatively high redshifts, which correspond to significantly large look-back times.

The preceding comments identify the main issues that we can explore by looking through the dynamical window. Dynamical studies are not meant to deduce mechanisms but rather to help clarify what the observational windows leave only ambiguously determined. Unfortunately, we still must resort to significant amounts of unseen material to obtain a satisfactory interpretation of the data. At this stage, it is plausible that for both spiral and elliptical galaxies, dark halos are rather diffuse.

Notes

1. This solution was found by Harris, E. G. 1962. *Il Nuovo Cimento*, **23**, 115. See also Attico, N., Pegoraro, F. 1999. *Phys. Plasmas*, **6**, 767.
2. The following example of secular instability has been credited to Lamb in the monograph by Jeans, J. H. 1929. *Astronomy and Cosmogony*, Cambridge University Press, Cambridge (reprinted in 1961 by Dover, New York).
3. A different problem would be to ask what are the circular orbits with assigned angular momentum J_z for a particle constrained to move on the spherical surface.
4. An interesting discussion of symmetry breaking in quantum mechanics is given by Peierls, R. 1991. *J. Phys. A*, **24**, 5273.
5. Coppi, B. 1980. *Comments Plasma Phys. Cont. Fusion*, **5**, 261.

10   Classical Ellipsoids

The themes introduced in Chapter 10 are demonstrated in one class of dynamical systems, which, because of its simplicity, serves as a bridge from the case of single-particle dynamics to the more complex collective behavior of continuous media. The study of incompressible self-gravitating fluids with ellipsoidal shape focuses on an environment with a limited number of degrees of freedom. Still it opens the way to several subtle phenomena that have attracted the interest of many eminent mathematicians and physicists, starting with Newton.[1]

The classical ellipsoids were originally studied as models, as a starting point, to interpret the properties of rotating celestial bodies, such as planets and stars.[2] Curiously, similar concepts have recently found applications in completely different fields, such as the physics of atomic nuclei[3] and black holes.[4] Tools related to the simple properties of ellipsoidal mass distributions have found applications in the study of the formation of structures by collapse ("pancakes" or protogalaxies) in the cosmological context.[5]

In galactic dynamics, classical ellipsoids might be thought of as the natural representation of elliptical galaxies.[6] Unfortunately, spectroscopic observations in the mid-1970s proved that bright ellipticals are not flattened by rotation and thus do not fit this simple picture (see Subsection 4.2.2 and Part IV). Nonetheless, classical ellipsoids have been very useful because of the way the potential theory can be handled analytically, with application to the construction of rotation curves[7] and various stellar dynamical models[8] (see also Section 14.3 and Part IV). In the problem of spiral structure, one of the first explanations proposed for the generation of spiral arms was in terms of the orbit instability that is known to occur at the edge of a sufficiently flat spheroid.[9] Revived interest grew in the early 1970s when the bar instability was found in self-gravitating disks (morphologically similar to the bar instability that can occur in flat spheroids); this was argued as making a case for the need for dark halos.[10]

One interesting aspect of the sequences of ellipsoidal figures of equilibrium is the mechanism of bifurcation. Theoretically, this bears important relations to the general behavior of nonlinear systems.[11] Because the symmetry break along the axisymmetric Maclaurin sequence is found to occur at sufficiently high values of eccentricity when the internal energy contributes less to the equilibrium of the configuration, from the thermodynamical point of view we may be brought to interpret the bifurcation to the triaxial Jacobi ellipsoids in terms of a second-order phase transition.[12]

10.1 Ellipsoidal Figures of Equilibrium

10.1.1 Some Properties of Homogeneous Ellipsoids

In Cartesian coordinates, an ellipsoid is defined by the quadratic relation

$$S(x_1, x_2, x_3) = \frac{x_1^2}{a_1^2} + \frac{x_2^2}{a_2^2} + \frac{x_3^2}{a_3^2} = 1. \tag{10.1}$$

We assume that $a_1 \geq a_2 \geq a_3$. The volume is given by $V = (4/3)\pi a_1 a_2 a_3 = (4/3)\pi a_1^3 \sqrt{1 - e^2}\sqrt{1 - \eta^2}$. Thus the ellipsoid is identified by a scale length a_1 and by two dimensionless form factors, the eccentricities[13] $e = \sqrt{1 - a_3^2/a_1^2}$ and $\eta = \sqrt{1 - a_2^2/a_1^2}$, which define its shape. The axisymmetric oblate case is given by $\eta = 0$; the third axis is the symmetry axis. The axisymmetric prolate case is given by $\eta = e$; the first axis is the symmetry axis.

For a constant density ρ, the mass of the ellipsoid is given by $M = \rho V$. The three moments $I_i = \int x_i^2 \rho dV$ are easily calculated to be $I_1 = Ma_1^2/5$, $I_2 = (1 - \eta^2)I_1$, and $I_3 = (1 - e^2)I_1$.

The potential inside a homogeneous ellipsoid generalizes the potential inside a sphere, with

$$\Phi^{(in)} = -\pi G\rho \sum A_i(a_i^2 - x_i^2), \tag{10.2}$$

where

$$A_i = a_1 a_2 a_3 \int_0^\infty \frac{du}{\Delta(a_i^2 + u)}, \tag{10.3}$$

with

$$\Delta = \sqrt{(a_1^2 + u)(a_2^2 + u)(a_3^2 + u)}. \tag{10.4}$$

A simple identity easily follows:

$$\sum A_i = 2a_1 a_2 a_3 \int_0^\infty \frac{1}{\Delta^2}\frac{d\Delta}{du}\, du = 2. \tag{10.5}$$

The quantities A_i can be expressed as a function of the two eccentricities:

$$A_1 = \frac{2\sqrt{(1 - e^2)(1 - \eta^2)}}{e\eta^2}[F(e, \eta) - E(e, \eta)], \tag{10.6}$$

$$A_2 = \frac{2e\sqrt{(1 - e^2)(1 - \eta^2)}}{\eta^2(e^2 - \eta^2)}\left[E(e, \eta) - F(e, \eta)\frac{e^2 - \eta^2}{e^2} - \frac{\sqrt{1 - e^2}}{\sqrt{1 - \eta^2}}\frac{\eta^2}{e}\right], \tag{10.7}$$

$$A_3 = \frac{2\sqrt{(1 - e^2)(1 - \eta^2)}}{e(e^2 - \eta^2)}\left[\frac{e\sqrt{1 - \eta^2}}{\sqrt{1 - e^2}} - E(e, \eta)\right]. \tag{10.8}$$

Here the two incomplete elliptical integrals are defined as

$$E(e, \eta) = \int_0^{\arcsin e}\left(1 - \frac{\eta^2}{e^2}\sin^2 x\right)^{1/2} dx, \tag{10.9}$$

$$F(e, \eta) = \int_0^{\arcsin e}\left(1 - \frac{\eta^2}{e^2}\sin^2 x\right)^{-1/2} dx. \tag{10.10}$$

For the oblate case,

$$A_1 = A_2 = \frac{\sqrt{1-e^2}}{e^3} \arcsin e - \frac{1-e^2}{e^2}, \tag{10.11}$$

$$A_3 = \frac{2}{e^2} - \frac{2\sqrt{1-e^2}}{e^3} \arcsin e. \tag{10.12}$$

For the prolate case,

$$A_1 = \frac{1-e^2}{e^3} \ln \frac{1+e}{1-e} - \frac{2(1-e^2)}{e^2}, \tag{10.13}$$

$$A_2 = A_3 = \frac{1}{e^2} - \frac{1-e^2}{2e^3} \ln \frac{1+e}{1-e}. \tag{10.14}$$

From the preceding expressions it is possible to derive the value of the total gravitational energy:

$$W = \frac{1}{2} \int \rho \Phi \, dV = \frac{1}{2} \int_{S \leq 1} \rho \Phi^{(\text{in})} \, dV = -2\pi G \rho \sum A_i I_i \tag{10.15}$$

$$= -\frac{4\pi}{5} \left(\frac{3}{4\pi}\right)^{2/3} \frac{GM^2}{V^{1/3}} g(e, \eta), \tag{10.16}$$

with

$$g(e, \eta) = \frac{(1-e^2)^{1/6}(1-\eta^2)^{1/6}}{e} F(e, \eta). \tag{10.17}$$

The function F is the same elliptical integral defined in Eq. (10.10).

Similarly, if we consider an ellipsoid that is rigidly rotating around the third axis at angular velocity Ω, the kinetic energy K can be written as

$$K = \frac{1}{2}(I_1 + I_2)\Omega^2 = \frac{J^2}{2(I_1 + I_2)} = \frac{5}{4} \left(\frac{4\pi}{3}\right)^{2/3} \frac{J^2}{MV^{2/3}} f(e, \eta), \tag{10.18}$$

with

$$f(e, \eta) = \frac{2(1-e^2)^{1/3}(1-\eta^2)^{1/3}}{2 - \eta^2}. \tag{10.19}$$

An interesting measure of the amount of rotational support to the structure is the parameter $t = K/|W|$, which, for equilibrium configurations, is bound by the virial theorem to be in the range $0 \leq t < 1/2$. By combining the preceding relations, we find

$$t(e, \eta) = \frac{25}{12} \left(\frac{4\pi}{3}\right)^{1/3} \frac{J^2}{GM^3 V^{1/3}} \frac{f(e, \eta)}{g(e, \eta)}. \tag{10.20}$$

For completeness, we record the expression[14] of the potential external to a homogeneous oblate spheroid in the equatorial plane ($x_3 = 0$):

$$\Phi^{(\text{ext})}(x_1, x_2) = \pi G \rho a_1^2 \frac{\sqrt{1-e^2}}{e} \left[\left(\frac{R^2}{e^2} - 2\right) \arcsin\left(\frac{e}{R}\right) - \frac{R}{e}\sqrt{1 - \frac{e^2}{R^2}} \right], \tag{10.21}$$

where the dimensionless distance from the symmetry axis is defined as $R \equiv \sqrt{x_1^2 + x_2^2}/a_1$. The expression is applicable for $R \geq 1$, but it is defined with a different choice of constant (so that it vanishes for $R \to \infty$) with respect to the one used in Eq. (10.2) for $\Phi^{(\text{in})}$.

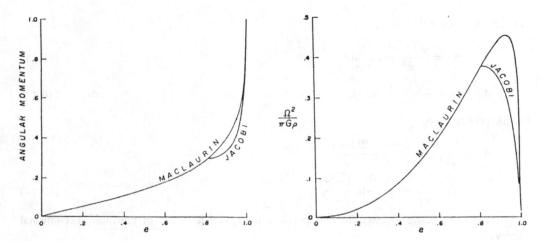

Fig. 10.1. Maclaurin and Jacobi sequences for the classical ellipsoids (from Chandrasekhar, S. 1969. *Ellipsoidal Figures of Equilibrium*, Yale University Press, New Haven, CT., p. 79; ©1969 by Yale University).

10.1.2 Classification of the Equilibrium Configurations

The study of self-gravitating equilibrium configurations for rotating fluids leads to several classes of solutions depending on the adopted constraints. We may consider as a starting point (1) a homogeneous fluid (of constant density ρ) with (2) ellipsoidal shape and (3) a stationary linear flow field. The last assumption refers, in the formulation of Dirichlet and Riemann, to the case where the Lagrangian coordinates of the fluid elements in an inertial frame of reference are related by a linear transformation to the initial coordinates. There are, in general, two natural frames of reference: the inertial frame and the frame moving with the ellipsoidal figure. It can be shown that the most general linear flow consistent with assumption (3) can be seen as the superposition of a uniform rotation with vector angular velocity Ω and a uniform internal streaming defined by a vector vorticity ζ (with respect to the frame of reference rotating at Ω, in which the ellipsoidal figure is seen as stationary). All the properties of the equilibrium configurations that may be found under these assumptions can be worked out on the basis of the second-order tensor virial relations[15] (see also Subsection 10.2.4).

The most natural solutions are those characterized by solid-body rotation, that is, by $\zeta = 0$ (Fig. 10.1). There is an axisymmetrical sequence of such equilibria, with the axis of symmetry provided by the direction of Ω. This is called the *Maclaurin sequence*. Curiously, for a given value of angular velocity, with $\Omega^2 < 0.449331\pi G\rho$, there are two possible values for the eccentricity e of the oblate spheroid for which equilibrium can be ensured; no solutions are available at higher values of the angular rotation. In turn, if we focus on the total angular momentum in a suitable dimensionless form, the connection between flattening and rotation is one to one. At high values of angular momentum and $e > e_c \approx 0.81267$, a sequence of rigidly rotating triaxial configurations known as *Jacobi ellipsoids* becomes available. This is quite surprising, and the bifurcation from the axisymmetrical sequence raises a number of interesting dynamical questions.

A theorem that is due to Dedekind states that any stationary linear flow equilibrium configuration is associated with an adjoint solution in which the roles of rotation Ω and vorticity

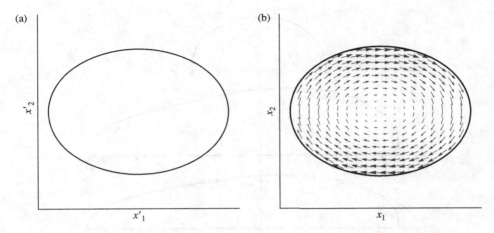

Fig. 10.2. Illustration of the Dedekind theorem. (*a*) Polar view of an equatorial section of a triaxial Jacobi ellipsoid in solid-body rotation (no motions are observed in the frame of reference corotating with the ellipsoid). (*b*) Polar view of the corresponding Dedekind ellipsoid, seen in the inertial frame of reference, where the triaxial shape is stationary; the internal flow field indicates the presence of vorticity (adapted from Tazzari, M. 2011. Thesis, Università degli Studi di Milano, Milano, Italy).

ζ are interchanged. Therefore, the bifurcation to the Jacobi ellipsoids is accompanied by a bifurcation to a sequence of triaxial ellipsoids that are stationary in the inertial frame of reference and are fully supported by internal streaming (Fig. 10.2). These are called *Dedekind ellipsoids*.

Riemann has shown that other families of solutions exist. The Jacobi ($f = 0$) and Dedekind ($f = \infty$) solutions are extreme limits of a family characterized by a parameter $f = \zeta / \Omega$ in the case in which the axes of the rotation and the vorticity vectors coincide and are parallel to one of the principal axes of the ellipsoid. There are three other classes of solutions that can be defined when the two vectors lie on a plane that coincides with one of the principal planes of the ellipsoid.

It was also realized that for a homogeneous rotating fluid, high values of angular momentum may open the way to nonellipsoidal figures of equilibrium (the Poincaré "pears" bifurcate at $e \approx 0.9386$, $\eta \approx 0.9018$; Fig. 10.3). This led to speculations on the possible origin of planets and binary stars by a process of fission at very high values of angular momentum. In the context of binary stars, a different but related problem is that in which the ellipsoidal shape in a rotating fluid is induced by the tidal forces created by a primary body. The *Roche ellipsoids* are constructed with the assumption that the primary is spherical. General configurations of this type are extreme idealizations of very interesting and modern astrophysical environments, such as those characterizing X-ray binaries and the associated accretion disks, and the coalescence processes that are thought to lead to some types of supernovae.[16]

Some of the properties of the main equilibrium sequences hold unchanged if assumption (1) is replaced by the less restrictive condition (1a) that the fluid is heterogeneous, with density stratified on similar concentric ellipsoids.[17] Although the homogeneous limit corresponds to the case

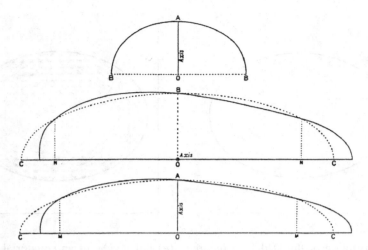

Fig. 10.3. Illustration of the transition from the Jacobi sequence to pear-shaped rotating fluids (Darwin, G. H. "On the Pear-Shaped Figure of Equilibrium of a Rotating Mass of Liquid," 1902. *Philos. Trans. Roy. Soc.*, **198**, 301; © 1902 by the Royal Society).

of an incompressible fluid, it remains to be proved that for significant departures from spherical symmetry, the heterogeneous case thus formulated is consistent with a physically sensible equation of state. Another more physical approach to the study of self-gravitating rotating fluids, of direct interest to the physics of rotating stars, is to also relax ellipsoidal assumption (2) and impose a physically justified equation of state, for example, a barotropic $p = p(\rho)$ or a polytropic $p = K\rho^{1+1/n}$. At this stage, we are naturally led to consider the detailed properties of stellar interiors and the possibility of differential rotation. Here the results are generally based on extensive numerical studies.[18]

Collisionless stellar systems (such as elliptical galaxies) go well beyond these categorizations because the concept of equation of state is not well defined,[19] and the interchange among internal streaming, random motions, and figure rotation can in principle occur in very complicated manners. Even though some collisionless counterparts to the classical ellipsoids can be constructed, we are far from having any simple results on rotating nonspherical collisionless equilibria, as will be discussed in Part IV.

10.2 Rigidly Rotating Equilibrium Ellipsoids

Inside a rigidly rotating ellipsoid, equilibrium is ensured by the balance among pressure, gravitational, and centrifugal forces. We can find the ellipsoidal equilibrium sequences by imposing that the surface of the ellipsoid be an equipotential surface with respect to the sum of the gravitational and centrifugal potentials:

$$\Phi^{(in)} - \frac{1}{2}\Omega^2(x_1^2 + x_2^2) = \text{constant} \qquad \text{at } S(\mathbf{x}) = 1. \tag{10.22}$$

This provides the two conditions

$$2A_3(1 - e^2) = 2A_2(1 - \eta^2) - \frac{\Omega^2}{\pi G\rho}(1 - \eta^2) = 2A_1 - \frac{\Omega^2}{\pi G\rho}. \tag{10.23}$$

One condition can be used to define the shape of the ellipsoid, $\eta = \eta(e)$, and the other fixes the amount of rotation Ω compatible with such shape.

10.2.1 Axisymmetrical Maclaurin Spheroids

The axisymmetrical Maclaurin sequence is identified by setting $\eta = 0$ so that one condition is satisfied trivially by symmetry, whereas the other reduces to

$$\frac{\Omega^2}{\pi G \rho} = 2A_1 - 2A_3(1 - e^2) = 2 - A_3 - 2A_3(1 - e^2) \tag{10.24}$$

$$= 2\left[\frac{(3 - 2e^2)\sqrt{1 - e^2}}{e^3}\arcsin e - \frac{3(1 - e^2)}{e^2}\right]. \tag{10.25}$$

Note that the equilibrium angular rotation vanishes in both limits $e \to 0$ and $e \to 1$. In contrast, the dimensionless angular momentum,

$$\frac{J^2}{GM^3 V^{1/3}} = \frac{3}{25}\left(\frac{3}{4\pi}\right)^{1/3}\left(\frac{a_3}{a_1}\right)^{-4/3}\frac{\Omega^2}{\pi G \rho} = \frac{12}{25}\left(\frac{3}{4\pi}\right)^{1/3} j(e), \tag{10.26}$$

with

$$j(e) = \frac{(1 - e^2)^{-2/3}}{2}\left[\frac{(3 - 2e^2)\sqrt{1 - e^2}}{e^3}\arcsin e - \frac{3(1 - e^2)}{e^2}\right], \tag{10.27}$$

is a monotonic function of the eccentricity e. If we now recall Eq. (10.20), we find that for the rotation parameter on the Maclaurin equilibrium sequence,

$$t = t(e) = j(e)\frac{f(e, \eta = 0)}{g(e, \eta = 0)} = \frac{3 - 2e^2}{2e^2} - \frac{3\sqrt{1 - e^2}}{2e\arcsin e}. \tag{10.28}$$

This function is also monotonic with e. (It is essentially on the basis of this expression that the observed amount of rotation relative to the central velocity dispersion, i.e., $\sim [2t(e)/(1 - 2t(e))]^{1/2}$, is proved to be too small in bright elliptical galaxies, so the observed flattening has to be argued to be produced by pressure anisotropy;[20] see also Chapter 22.)

10.2.2 Triaxial Jacobi Ellipsoids

For $e > e_c \approx 0.81267$, the equilibrium conditions can be satisfied with a nonvanishing value of η. The resulting equilibrium sequence of triaxial ellipsoids is called the *Jacobi sequence*. For a given sufficiently high value of dimensionless angular momentum, there are unique values for e and $\eta \neq 0$ that ensure equilibrium. Thus the Jacobi sequence is characterized by a function $\eta = \eta(e)$ and bifurcates at $e = e_c$ from the Maclaurin sequence (which remains available at higher values of e). Along the Jacobi sequence, the value of $\Omega^2/\pi G \rho$ monotonically decreases with e, and the dimensionless angular momentum increases. The rotation parameter $t = t(e, \eta(e))$ increases with e more slowly than the function $t(e)$ on the Maclaurin sequence [see Eq. (10.28)].

10.2.3   Extremizing the Mechanical Energy

Another approach[21] to finding the relevant equilibrium sequences consists of extremizing the mechanical energy

$$E_{mec} = k \frac{J^2}{MV^{2/3}} f(e, \eta) - w \frac{GM^2}{V^{1/3}} g(e, \eta) \tag{10.29}$$

at fixed values of the volume and angular momentum [the constants w and k are given in Eqs. (10.16) and (10.18), respectively]. Thus we should have for the Maclaurin sequence

$$k \frac{J^2}{MV^{2/3}} \left(\frac{\partial f}{\partial e} \right)_{\eta=0} = w \frac{GM^2}{V^{1/3}} \left(\frac{\partial g}{\partial e} \right)_{\eta=0}. \tag{10.30}$$

Indeed, it can be checked that $w/k = (12/25)(3/4\pi)^{1/3}$ and that

$$\left(\frac{\partial g}{\partial e} \right)_{\eta=0} = j(e) \left(\frac{\partial f}{\partial e} \right)_{\eta=0}, \tag{10.31}$$

consistent with Eq. (10.26). The general conditions corresponding to Eq. (10.23) are recovered when the condition of stationary E_{mec} is imposed with respect to both form factors e and η.

10.2.4   Tensor Virial Equations

We can also derive the relevant equilibrium configurations by imposing the second-order virial relations (which provide a sufficient set of constraints in the problem of ellipsoidal figures of equilibrium). The time-dependent equations can be written as

$$\frac{1}{2} \frac{d^2 I_{ij}}{dt^2} = 2K_{ij} + W_{ij} + \Pi \delta_{ij}, \tag{10.32}$$

where the inertia tensor I_{ij}, the kinetic-energy tensor K_{ij}, and the gravitational-energy tensor W_{ij} are given by

$$I_{ij} = \int \rho x_i x_j dV, \tag{10.33}$$

$$K_{ij} = \frac{1}{2} \int \rho u_i u_j \, dV, \tag{10.34}$$

$$W_{ij} = \int \rho x_i \frac{\partial}{\partial x_j} \left(G \int \frac{\rho(\mathbf{x}')}{|\mathbf{x} - \mathbf{x}'|} dV' \right) dV, \tag{10.35}$$

and the pressure contribution is given by

$$\Pi = \int p \, dV. \tag{10.36}$$

We obtain the standard scalar virial constraint $W + 2K + 3\Pi = 0$ for a stationary system by taking the trace of the preceding tensor relations in the time-independent case. We obtain the two conditions given in Eq. (10.23) by eliminating Π from the three conditions required by the diagonalized virial tensor relations. We recall that the virial equations are derived directly by taking moments of the Euler equations. In particular, we obtain the second-order virial relations by integrating over dV the jth component of the Euler equations multiplied by x_i. Under the ellipsoidal constraint, the second-order moments provided earlier contain all the dynamics.

10.3   **The Mechanism of Bifurcation**

At a bifurcation point we naturally expect the existence of neutral modes that describe the dynamics of the symmetry break. These neutral modes are zero-frequency displacements in the frame of reference where the equilibrium configuration is at rest, marking the access to the new equilibrium sequence. From a different point of view, if two different configurations are available at given values of the global properties (volume, angular momentum, and internal energy), one basic state should be more favored in terms of free energy than the other; it should correspond to the existence of an instability in a modal analysis. However, if only one equilibrium configuration is available (e.g., for rigidly rotating ellipsoids at $e < e_c$), presumably that equilibrium is also a minimum energy state. These considerations apply to many dynamical systems and suggest that either a linear modal analysis or a more refined thermodynamical argument should be able to clarify further the bifurcation mechanism. The transition from the Maclaurin to the Jacobi sequence offers a good example from which these concepts can be quantified in detail.

10.3.1   Linear Modal Analysis

For a spherical basic state, the natural decomposition of a linear (small-amplitude) modal analysis is in terms of spherical harmonics (see also Chapter 23). For an incompressible sphere, the first significant modes are those characterized by $l = 2$ because $l = 0$ would be inconsistent with the assumption of incompressibility, whereas $l = 1$ would violate momentum conservation. The five $l = 2$ modes ($m = -2, -1, 0, 1,$ and 2) correspond to the f modes or Kelvin modes of the incompressible sphere.[22] For compressible spherical models applicable to the description of stars, there are also p modes with relatively high frequency, mostly acoustic and radial, and g modes with relatively low frequency, mostly tangential and dominated by gravity; the characterization of these modes is best illustrated by a homogeneous compressible model.[23]

It is convenient to look at the linearized (small-amplitude) perturbations, characterized by a vector Lagrangian displacement $\boldsymbol{\xi}(\mathbf{x}) \exp(i\hat{\omega}t)$, in the frame of reference rotating with the Maclaurin spheroid. Linear displacements in the spatial coordinates, $\xi_i = \sum \Xi_{ik} x_k$ with constant Ξ_{ik}, are perturbations that preserve the ellipsoidal shape and leave the ellipsoid within the linear-flow hypothesis considered by Dirichlet and Riemann. The study of these modes, which leave the shape ellipsoidal, is naturally approached by means of the second-order virial equations introduced earlier.[24] The modes have an $l = 2$ character. Given the symmetry of the basic state, we consider elementary solutions with spatial structure $\exp(im\phi)$ with respect to the toroidal angle. These perturbations appear with frequency $\omega = \hat{\omega} + m\Omega$ in the inertial frame of reference or, in general, with frequency $\omega' = \hat{\omega} + m(\Omega - \Omega')$ in a frame rotating at Ω'. The dynamical equations provide a relation between the frequency $\hat{\omega}$ in the corotating frame and the eccentricity e that defines the basic state in which the modes are considered to take place; we call this the *dispersion relation* for reasons that will be best appreciated from Chapter 11. In the absence of viscosity or other forms of dissipation, it can be proved[25] that the frequency of the modes enters the dispersion relation as $\hat{\omega}^2$, so if $\hat{\omega} = d(e)$ is a solution, $\hat{\omega} = -d(e)$ is also a solution. We can visualize the spatial structure of the perturbation by considering the component of the Lagrangian displacement of the fluid normal to the bounding surface [i.e., $X = (\boldsymbol{\xi}\nabla)S(\mathbf{x})$].

The modes that are odd with respect to x_3 (with an $m = 1$ character) are called *transverse-shear modes*. These correspond to tilting or bending perturbations and are somehow analogous to some modes that will be discussed in Chapter 19. They distort the surface of the spheroid by a normal displacement

$$X_t \propto x_3 (x_1^2 + x_2^2)^{1/2} \exp(i\hat{\omega}_t t + i\phi). \tag{10.37}$$

Here ϕ is the azimuthal toroidal angle in the rotating frame. It can be shown that the relevant eigenfrequencies for these modes are all real; that is, tilting modes are pure oscillations for any value of eccentricity e.

One mode (even with respect to x_3, with an $m = 0$ character) does not change the symmetry of the basic state:

$$X_p \propto \left(\frac{x_1^2 + x_2^2}{a_1^2} - 2 \frac{x_3^2}{a_3^2} \right) \exp(i\hat{\omega}_p t); \tag{10.38}$$

such a pulsation mode is also found to be stable, that is, to correspond to an oscillatory solution.

Finally, there are modes (even with respect to x_3, with an $m = 2$ character) that break the symmetry of the spheroid:

$$X_{\text{bar}} \propto (x_1^2 + x_2^2) \exp(i\hat{\omega}_{\text{bar}} t + 2i\phi); \tag{10.39}$$

such toroidal modes have a natural connection to a triaxial ellipsoid and thus may be called *bar modes*. They resemble some of the bar modes that will be discussed in Part III (see Section 17.5). The associated frequency satisfies the dispersion relation

$$\hat{\omega}_{\text{bar}}^2 = \Omega^2 \left[1 \pm \left(\frac{4\pi G\rho B_{11}}{\Omega^2} - 1 \right)^{1/2} \right]^2, \tag{10.40}$$

with [see Eq. (10.3)]

$$B_{11} = a_1 a_2 a_3 \int_0^\infty \frac{u \, du}{\Delta (a_1^2 + u)^2} = \frac{A_1}{2} + \frac{(1 - e^2)}{4e} \frac{dA_1}{de}. \tag{10.41}$$

The latter expression for B_{11}, with A_1 given in Eq. (10.11), holds for the axisymmetric case that we are considering. Therefore, along the Maclaurin sequence we find that $\Omega^2/\pi G\rho < 2B_{11}$ for $e < e_c$, $2B_{11} < \Omega^2/\pi G\rho < 4B_{11}$ for $e_c < e < e_d$, and $\Omega^2/\pi G\rho > 4B_{11}$ for $e > e_d$. As a result, one of the eigenfrequencies $\hat{\omega}_{\text{bar}}$ associated with these modes does vanish at $e = e_c \approx 0.81267$ (i.e., $a_3/a_1 \approx 0.5827$). Such a root remains stable at $e > e_c$ until instability takes place at much higher values of the flattening, at $e > e_d \approx 0.952887$ (i.e., $a_3/a_1 \approx 0.3033$), where one of the roots corresponds to an exponentially growing solution. It is worth noting that one of the roots at $e = e_c$ has $\hat{\omega}_{\text{bar}} = -2\Omega$, that is, $\omega = 0$, which thus marks the bifurcation to the Dedekind ellipsoids (with a nonrotating triaxial figure, adjoint to the Jacobi ellipsoids).

With similar tools we can study the bifurcations among the various ellipsoidal figures of equilibrium; transitions from Maclaurin spheroids to some Riemann ellipsoids are realized by modes that appear as neutral in a suitable rotating frame because, we recall, $\omega' = \hat{\omega} + m(\Omega - \Omega')$. With the help of higher harmonics for the fluid displacements and higher-order tensor virial equations, we can also study transitions to the Poincaré pears.

10.3.2 Dynamical and Secular Stability

The outline of the linear stability analysis given earlier remains incomplete. How should we interpret the situation in the eccentricity range $e_c < e < e_d$, in which the Maclaurin spheroids appear to be stable, whereas triaxial configurations are available? The theory of classical ellipsoids offers here another example of important concepts that are often met in many different contexts, not only in hydrodynamics but also in plasma physics. This is the possibility that small dissipative effects, such as those induced by viscosity in a fluid or by resistivity in a plasma, under certain circumstances may help to release some free energy that otherwise would remain unavailable because of the presence of constraints posed by the ideal equations of dissipationless dynamics. A very simple related mechanical example was mentioned in Subsection 9.2.1; an additional example will be presented in Subsection 13.4.1.

If we carry out the linear stability analysis of the modes described in Subsection 10.3.1, including the effects of viscous dissipation, we find that the bar modes in the range $e_c < e < e_d$ are indeed unstable. This type of instability is often called *secular instability* to distinguish it from the more common dynamical instability, which acts on the dynamical time scale (e.g., the instability of the bar modes occurring at $e > e_d$). For secular instability of the bar modes on Maclaurin spheroids, the instability mechanism is regular, and the growth rate is inversely proportional to the viscous time scale. In different problems there are singular dissipative instabilities for which the relevant growth rate is proportional to fractional powers of the dissipative time scale.[26]

10.3.3 Thermodynamical Analogue of Phase Transitions

Another way of understanding the mechanism of bifurcation is based on an analogy with second-order phase transitions in thermodynamics.[27] Second-order phase transitions[28] are common phenomena in crystals (or in alloys such as CuZn)[29] that occur when the invariance group of the crystal suddenly reduces to one of its subgroups as the temperature decreases.[30] In the phase with lower symmetry, a new observable, which vanishes in the symmetric phase, is necessary to describe the state of the system (e.g., in the phase with lower symmetry, a crystal may acquire a magnetic dipole or an electrical quadrupole moment). In contrast to first-order phase transitions, in second-order phase transitions, the state functions change continuously, with a discontinuity in the specific heat. If we expand the relevant Gibbs free energy in the neighborhood of the transition point, we find that while the temperature decreases, a minimum of free energy corresponding to the symmetric phase is transformed into a maximum while a new minimum appears that identifies the less symmetric phase.[31]

We thus conclude that the transition from axisymmetric to triaxial ellipsoids falls within the preceding framework. In particular, the thermodynamical analogue is further strengthened by the fact that the amount of support that is due to internal energy, as measured by the parameter $1/2 - t$ [see Eq. (10.20)], monotonically decreases along the Maclaurin sequence so that indeed the less symmetric triaxial phase is realized for cold systems.[32] We recall that for the function in Eq. (10.28), $t(e_c) \approx 0.13752$ and $t(e_d) \approx 0.27383$. For rigidly rotating ellipsoidal equilibria, the t parameter is monotonic and increases more slowly with e along the Jacobi sequence with respect to the way it increases along the Maclaurin sequence.

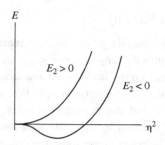

Fig. 10.4. Transition from the Maclaurin to the Jacobi sequence.

For the bifurcation from the Maclaurin to the Jacobi sequence, we thus may adopt the Landau approach and consider as order parameter the quantity $\varphi = \eta^2$. The total energy E, taken as a function of the state variables Π, volume V, and angular momentum J, in the neighborhood of $\varphi = 0$ (i.e., in the neighborhood of the symmetric phase) is then expanded as

$$E = E_0(\Pi, V, J) + \varphi E_1(\Pi, V, J) + \varphi^2 E_2(\Pi, V, J) + \cdots. \tag{10.42}$$

We calculate the energy coefficients E_n by considering expression (10.17) for the gravitational energy and by noting that the quantity $\partial e/\partial \varphi$ at fixed values of Π, V, and J can be determined by imposing the scalar virial constraint, which also gives $dE = dW/2$. Then it can be shown[33] that $E_1 = 0$, and

$$E_2 = -W_0 \frac{(1 - e^2)^{1/6}}{64 e^5} \psi(e), \tag{10.43}$$

with

$$\psi(e) = (-3 - 8e^2 + 8e^4) \arcsin e + e(3 + 10e^2)(1 - e^2)^{1/2}. \tag{10.44}$$

Here we have introduced $W_0 = -wGM^2 V^{-1/3}$, following Subsection 10.2.3. We then find that E_2 changes from positive to negative (thus marking the change from minimum to maximum free energy; Fig. 10.4) precisely at $e = e_c$, with $\psi(e_c) = 0$.

Notes

1. A thorough treatise on the subject is given by Chandrasekhar, S. 1969. *Ellipsoidal Figures of Equilibrium*, Yale University Press, New Haven, CT.
2. Currently, the flattest star known is Achernar (α Eridani), for which an apparent axial ratio of 1.56 ± 0.05 has been measured with interferometric observations at VLT. The measured value of the apparent equatorial velocity for the same star is ≈ 220 km s^{-1}, well below the theoretically expected speed for such a flattening, may indicate that the star is in a state of differential rotation; Domiciano de Souza, A., Kervella, P., et al. 2003. *Astron. Astrophys.*, **407**, L47. Other well-known and well-studied rapidly rotating stars are Altair (α Aquilae) and Regulus (α Leonis).
3. Troudet, T., Arvieu, R. 1981. *Ann. Phys.*, **134**, 1.
4. Davies, P. C. W. 1978. *Rep. Progr. Phys.*, **41**, 1313.
5. Zel'dovich, Ya. B. 1970. *Astrofizika*, **6**, 319; *Astron. Astrophys.*, **5**, 84; Lynden-Bell, D. 1964. *Astrophys. J.*, **139**, 1195; Lin, C. C., Mestel, L., Shu, F. H. 1965. *Astrophys. J.*, **142**, 1431.
6. See Freeman, K. C. 1975. In *Galaxies and the Universe*, eds. A. Sandage, M. Sandage, J. Kristian. University of Chicago Press, Chicago, p. 409.

7. See Burbidge, E. M., Burbidge, G. R. 1975. In *Galaxies and the Universe*, eds. A. Sandage, M. Sandage, J. Kristian. University of Chicago Press, Chicago, p. 81.

8. For example, the homogeneous rotating ellipsoids and the inhomogeneous, uniformly rotating elliptical disks constructed by Freeman, K. C. 1966. *Mon. Not. Roy. Astron. Soc.*, **134**, 1 and 15. See also Bisnovatyi-Kogan, G. S., Zel'dovich, Ya. B. 1970. *Astrofizika*, **6**, 387; Polyachenko, V. L. 1976. *Sov. Phys. Dokl.*, **21**, 417; Lynden-Bell, D. 1962. *Mon. Not. Roy. Astron. Soc.*, **123**, 447; Hunter, C. 1974. *Mon. Not. Roy. Astron. Soc.*, **166**, 633. The difficulties in the construction of stellar dynamical analogues of the homogeneous classical fluid ellipsoids are reviewed by Hunter, C. 1975. In *IAU Symposium*, Vol. 69, ed. A. Hayli. Reidel, Dordrecht, The Netherlands, p. 195. An interesting approximate study of the three-dimensional case of ellipsoidal stellar dynamical polytropes is given by Vandervoort, P. O. 1980. *Astrophys. J.*, **240**, 478; Vandervoort, P. O., Welty, D. E. 1982. *Astrophys. J.*, **263**, 654.

9. These investigations were made by B. Lindblad; see Chandrasekhar, S. 1942. *Principles of Stellar Dynamics*, University of Chicago Press, Chicago. For a discussion of the potential external to a classical ellipsoid, see MacMillan, W. D. 1958. *Theoretical Mechanics: The Theory of the Potential*, Academic, New York.

10. Ostriker, J. P., Peebles, P. J. E. 1973. *Astrophys. J.*, **186**, 467; see also Part III.

11. For example, see Poston, T., Stewart, I. 1978. *Catastrophe Theory and Its Applications*, Pitman, London.

12. Bertin, G., Radicati, L. A. 1976. *Astrophys. J.*, **206**, 815.

13. Another option is to introduce the related ellipticities; the polar ellipticity is defined as $\varepsilon = 1 - a_3/a_1 = 1 - \sqrt{1 - e^2}$.

14. Chandrasekhar, S. 1942. *op. cit.*, sect. 4.5.

15. Chandrasekhar, S. 1969. *op. cit.*

16. For example, see Frank, J., King, A., Raine, D. 2002. *Accretion Power in Astrophysics*, 3rd ed., Cambridge University Press, Cambridge, UK.

17. See also Roberts, P. H. 1962. *Astrophys. J.*, **136**, 1108; Bertin, G., Radicati, L. A. 1976. *op. cit.*

18. See Tassoul, J.-L. 1978. *Theory of Rotating Stars*, Princeton University Press, Princeton, NJ.

19. See appendix A in Bertin, G., Pegoraro, F., Rubini, F., Vesperini, E. 1994. *Astrophys. J.*, **434**, 94.

20. Binney, J. J. 1978. *Comments Astrophys.*, **8**, 27; *Mon. Not. Roy. Astron. Soc.*, **183**, 501; 2005. *Mon. Not. Roy. Astron. Soc.*, **363**, 937. See also Cappellari, M., Emsellem, E., et al. 2007. *Mon. Not. Roy. Astron. Soc.*, **379**, 418.

21. See Zel'dovich, Ya. B., Novikov, I. D. 1971. *Relativistic Astrophysics, 1: Stars and Relativity*, University of Chicago Press, Chicago; Dyson, F. J. 1971. *Neutron Stars and Pulsars*, Fermi Lectures 1970, Accademia Nazionale dei Lincei, Rome.

22. Thomson, W. 1863. *Phil. Trans. Roy Soc. London*, **153**, 603.

23. Cowling, T. G. 1941. *Mon. Not. Roy. Astron. Soc.*, **101**, 367; Cox, J. P. 1980. *Theory of Stellar Pulsation*, Princeton University Press, Princeton, NJ.

24. All the relevant derivations can be found in the monograph by Chandrasekhar, S. 1969. *op. cit.*

25. This is related to the self-adjointness of the relevant operators.

26. For example, in the problem of magnetic reconnection; e.g., see Biskamp, D. 1993. *Nonlinear Magnetohydrodynamics*, Cambridge University Press, Cambridge, UK. An interesting discussion of the secular instability, connecting the Maclaurin to the Jacobi sequence, in relation to the Kelvin circulation theorem, has been given by Lynden-Bell, D. 1965. *Astrophys. J.*, **142**, 1648.

27. This approach, suggested by Bertin, G., Radicati, L. A. 1976. *Astrophys. J.*, *op. cit.*, has recently revived interest with new applications to the dynamics of galaxies; see Christodoulou, D. M., Kazanas, D., Shlosman, I., Tohline, J. E. 1995. *Astrophys. J.*, **446**, 472; 485; 500; 510. A general mathematical study of the possible symmetry breakings from the group theory point of view has been given by Constantinescu, D., Michel, L., Radicati, L. A. 1979. *J. Phys.*, **40**, 147.

28. Landau, L., Lifchitz, E. 1967. *Physique Statistique*, MIR, Moscow.

29. Other examples of second-order phase transitions are the transition from ferromagnetic to paramagnetic phases or the transition from liquid to superfluid helium; the transition point is often called the *Curie point*.

30. The less symmetric phase is usually realized at lower temperatures, but there are exceptions to this behavior.

31. This is the same qualitative description that is invoked in the discussion of inflation in cosmology; see Linde, A. D. 1974. *JETP Lett.*, **19**, 183; Guth, A. 1981. *Phys. Rev. D*, **23**, 347; for a general discussion of the ideas developed in this scenario, see Linde, A. D. 1990. *Particle Physics and Inflationary Cosmology*, Harwood, New York; Peebles, P. J. E. 1993. *Principles of Physical Cosmology*, Princeton University Press, Princeton, NJ.

32. Ostriker, J. P., Peebles, P. J. E. 1973. *op. cit.*, indeed argued that as a general property of self-gravitating rotating systems, a bar instability takes place when the system is sufficiently cold.

33. The calculations for E_2 are nontrivial; see Bertin, G., Radicati, L. A. 1976. *op. cit.*

11 Introduction to Dispersive Waves

This short introduction to dispersive waves[1] provides an important prerequisite for moving from the dynamics of individual particle orbits to the dynamics of collective effects. As such, the ideas summarized here have a large impact on this book, especially on Part III, in which they are seen at work in the description of density and bending waves. The basic concepts (waves, wave packets, wave trains, group propagation, etc.) define a framework that has found wide and successful applications in hydrodynamics, geophysical fluid dynamics, and plasma physics.

In contrast with the case of hyperbolic waves, for which the description is centered on the properties of one class of partial differential equations, the study of dispersive waves focuses on one characteristic property of the solutions of many different types of dynamical equations that is called the *dispersion relation*. This is a relation, very often of a rather trivial algebraic form, between space and time modulation of elementary components of the wave process. The relation incorporates the constraints set by the dynamical equations.

It is sometimes believed that the main goal of the dynamicist should be to derive the dispersion relation for a given process and for a given model; the derived expression then would mark the end of the investigation. In a sense, this is true, to the extent that the dispersion relation does summarize properly the relevant underlying dynamical equations. In practice, the most interesting part of the work begins after the derivation. This is to interpret the dispersion relation, that is, to extract from it the explanation for the variety of collective phenomena that can occur in the same model. For this reason, in Part III, lower priority is given to the derivation aspect of the various dispersion relations, and instead, the assumptions made, the key features of the models adopted, and the impact on the interpretation of many astrophysical effects are emphasized.

Several conclusions can be drawn by a simple use of algebraic dispersion relations (see Chapters 15, 16, and 19). However, for the purpose of describing the dynamics of wave-wave interaction and of wave-particle interaction (these processes may occur where wave branches meet or where the dispersion relation formally becomes singular; see Chapter 17) or to set the stage for the description of nonlinear evolution, the properties of certain partial differential equations must be addressed. Curiously, sometimes the algebraic dispersion relation is derived and used first, and the relevant partial differential equation is constructed later (basically by taking a step backward, $k \rightarrow -i\partial/\partial x$) under special physical motivations.

The most important concepts at the basis of collective effects in dispersive media can be illustrated without resorting to exotic environments. For this reason, it may be useful to have here a short digression on the relatively well-known case of water waves that offers a number of familiar examples within which the richness of dispersive waves can be easily appreciated without distractions.

11.1   **Hyperbolic Waves**

Waves are one of the most intuitive and yet most elusive concepts in physics. It is convenient to avoid the issue of providing the most general definition of a wave and to keep in mind the simple guiding rule that any recognizable propagating signal in a continuous medium should be associated with a wave. Two wide classes can then be introduced, the class of hyperbolic waves and the class of dispersive waves.

Hyperbolic waves are the solutions of a class of partial differential equations called *hyperbolic equations*. The prototype of these equations is

$$\frac{\partial^2 u}{\partial t^2} - c^2 \nabla^2 u = 0. \tag{11.1}$$

The most elementary, one-dimensional case of a hyperbolic equation is actually given by

$$\frac{\partial u}{\partial t} + c \frac{\partial u}{\partial x} = 0. \tag{11.2}$$

Equations (11.1) and (11.2) are generally introduced in the linear context, so c is taken to be a constant, which is then identified with the relevant wave velocity. The basic properties of the theory of hyperbolic waves are generally presented in courses of classical electrodynamics.[2]

Many problems in dynamics and, in particular, the study of the collisionless Boltzmann equation (e.g., see Subsection 15.1.1), have solutions based on the method of integration of hyperbolic equations along their characteristics, which reduces the study of hyperbolic equations to the more manageable domain of ordinary differential equations. This opens the way to interesting nonlinear phenomena. Without entering the vast field of nonlinear hyperbolic waves, we recall here that the simplest quasi-linear generalization of Eq. (11.2) is that in which c is allowed to be a function of u:

$$\frac{\partial u}{\partial t} + c(u) \frac{\partial u}{\partial x} = 0, \tag{11.3}$$

which can be solved as

$$\frac{du}{dt} = 0 \tag{11.4}$$

on the characteristics defined by

$$\frac{dx}{dt} = c(u). \tag{11.5}$$

If we refer to the initial-value problem under the condition $u(t=0, x_0) = f(x_0)$, with $f(x_0)$ assigned on the real axis $-\infty < x_0 < +\infty$, Eq. (11.5) for the characteristics has the solution

$$x = x_0 + c(u)t = x_0 + c(f(x_0))t. \tag{11.6}$$

Equation (11.6) should be read as an implicit equation that gives $x_0 = x_0(x, t)$. In other words, it traces the space coordinate x at time t to the space coordinate x_0 at time $t = 0$. Then the desired

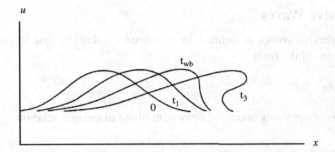

Fig. 11.1. Wave breaking: successive profiles at times t_1, t_{wb}, and t_3 of a signal that originates at $t = 0$.

solution to the proposed initial-value problem is

$$u(x,t) = f(x_0(x,t)). \tag{11.7}$$

Usually presented first in elementary courses in fluid dynamics, the preceding analysis leads to the interesting process of *wave breaking* (Fig. 11.1), which occurs when c depends on the amplitude of the solution u; in this case, the characteristics may force the solution to become multivalued beyond a certain time t_{wb}. For some physical problems (e.g., in the dynamics of ray tracing; see Subsection 11.2.3), there is no harm in this. In other contexts, for example, in gas dynamics, in which the function u may be associated with a physical variable, such as density or pressure that is inherently single-valued, wave breaking marks the actual breakdown of the validity of the equations that we are dealing with. In particular, the gradients that formally diverge at t_{wb} may make physical effects that are usually ignored become important, thus regularizing the relevant differential equation and associated solution. One example of this mechanism is the formation of shock waves, in which dissipative effects associated with a small amount of viscosity or with finite thermal conductivity are involved.

In this respect, one interesting benchmark is given by the classical case of the Burgers equation:[3]

$$\frac{\partial u}{\partial t} + u \frac{\partial u}{\partial x} = \epsilon \frac{\partial^2 u}{\partial x^2}. \tag{11.8}$$

For this equation we may derive the asymptotic shock-wave solutions available in the limit $\epsilon \to 0$ by matching an outer solution (outside the shock), for which ϵ effectively vanishes, to an inner solution (across the shock), for which the diffusive term proportional to ϵ becomes important because the gradients involved are large. In addition, the Burgers equation admits exact solutions because it can be mapped into the classical heat equation by means of the Cole-Hopf transformation.[4] The asymptotic analysis of shock formation thus can be tested here in full detail. Otherwise, the range of applicability of asymptotic solutions is generally tested by means of numerical integrations. Some key ideas characterizing the role of asymptotic analysis, so often used in the dynamics of galaxies (e.g., in the study of inhomogeneous disks), are well illustrated by this classical example of applied mathematics.

11.2   Dispersive Waves

The class of dispersive waves is defined by elementary solutions that in the simplest one-dimensional case are of the form

$$u = a\cos(kx - \omega t), \tag{11.9}$$

where a, k, and ω are constants under the constraint of the dispersion relation

$$\omega = \omega(k), \tag{11.10}$$

provided that

$$\omega'' = \frac{d^2\omega}{dk^2} \neq 0. \tag{11.11}$$

Many partial differential equations admit solutions of the form of Eq. (11.9). Hyperbolic Eq. (11.1), which may represent sound waves, does also, but it implies that $\omega^2 = c^2 k^2$, so the associated hyperbolic waves are not dispersive [condition (11.11) is violated]. In contrast, other equations, such as the Klein-Gordon equation[5]

$$\frac{\partial^2 u}{\partial t^2} - c^2 \nabla^2 u + \omega_0^2 u = 0, \tag{11.12}$$

are hyperbolic and, at the same time, lead to dispersive waves; note that the dispersion relation associated with Eq. (11.12) is $\omega^2 = c^2 k^2 + \omega_0^2$, so condition (11.11) is satisfied. The class of dispersive waves encompasses a very wide variety of problems. In many cases the boundary conditions set the wave character of the solution even when, at first sight, the underlying differential equation would have little to do with the more traditional wave equation. For example, we will see that water waves descend from the study of the equations of an incompressible fluid, described by $\nabla^2 \psi = 0$.

11.2.1   Fourier Analysis

The preceding discussion has been given in terms of elementary solutions. For the study of linear problems, we consider arbitrary superpositions of elementary solutions and thus can construct more general wave perturbations called *wave packets*.

The most familiar procedure is that of Fourier analysis, for which the general solution is constructed as

$$u = \int_{-\infty}^{\infty} F(k)\exp(ikx - i\omega t)\, dk, \tag{11.13}$$

supplemented by the dispersion relation $\omega = \omega(k)$. The choice of $F(k)$ can be made so as to satisfy certain initial conditions. (The way to specify a consistent set of initial conditions depends on the number of wave branches allowed by the dispersion relation.) Here the focus is on the *wave-number space*. Note that the wave number k is a dummy variable and thus does not describe the spatial modulation of the solution $u(x,t)$. The evolution properties of the wave packet in physical space are all contained implicitly in its spectral representation $F(k)$.

In this context, the notion of group velocity c_g is very familiar and can be introduced by reference to the asymptotic behavior of the wave packet [Eq. (11.13)] at large values of t. In particular, let us determine the structure of the wave packet at $t \to \infty$ for a fixed ratio $x/t = c$. That is, we want to study the integral

$$u(x,t) = \int_{-\infty}^{\infty} F(k) \exp(-i\chi t) \, dk \tag{11.14}$$

as $t \to \infty$, with the definition

$$\chi = \chi(k) = \omega(k) - kc \tag{11.15}$$

and c taken to be a given constant. The asymptotic behavior of the integral is estimated by means of the method of stationary phase.[6] The generic leading behavior of the integral is derived by the expansion of

$$\chi \sim \chi(k_0) + \frac{1}{2}\chi''(k_0)(k-k_0)^2 + \cdots \tag{11.16}$$

$$= \chi(k_0) + \frac{1}{2}\omega''(k_0)(k-k_0)^2 + \cdots \tag{11.17}$$

around the stationary points k_0 for which $\chi'(k_0) = 0$. Note that the definition of χ implies the relation

$$\left(\frac{d\omega}{dk}\right)_{k_0} = c = \frac{x}{t}, \tag{11.18}$$

which implicitly defines the function $k_0 = k_0(x,t)$. Each stationary point thus contributes to the resulting generic expression for the wave packet by a term

$$u(x,t) \sim a(x,t) \exp[i\theta(x,t)] \tag{11.19}$$

$$\sim F(k_0) \exp[-i\chi(k_0)t] \int_{-\infty}^{\infty} \exp\left[-\frac{i}{2}(k-k_0)^2 \chi''(k_0)t\right] dk \tag{11.20}$$

$$\sim F(k_0) \sqrt{\frac{2\pi}{t|\omega''(k_0)|}} \exp\left[-i\chi(k_0)t - \frac{i\pi}{4}\,\text{sgn}(\omega'')\right]. \tag{11.21}$$

The integration in k has been performed following the steepest descent after a suitable rotation in the complex plane.

Equations (11.19) through (11.21) show that sufficiently far from the initial conditions the wave packet has an oscillating behavior dominated by a wave number k_0 that is exactly the one for which the group velocity $c_g(k_0) = \omega'(k_0)$ equals x/t. Note that the phase $\theta(x,t)$ can be written as

$$\theta(x,t) = -\chi(k_0)t, \tag{11.22}$$

so

$$\frac{\partial\theta}{\partial x} = k_0 + [x - \omega'(k_0)t]\frac{\partial k_0}{\partial x} = k_0, \tag{11.23}$$

$$\frac{\partial\theta}{\partial t} = -\omega(k_0) + [x - \omega'(k_0)t]\frac{\partial k_0}{\partial t} = -\omega(k_0), \tag{11.24}$$

with $k_0 = k_0(x,t)$ defined by Eq. (11.18). Finally, the behavior of the amplitude

$$a = F(k_0) \sqrt{\frac{2\pi}{t|\omega''(k_0)|}} \exp\left[-\frac{i\pi}{4}\, \mathrm{sgn}(\omega'')\right] = a(x,t) \tag{11.25}$$

can be interpreted in terms of the transport of energy and wave action.

The asymptotic analysis also gives the higher-order terms in expansion (11.20), together with the conditions on $F(k)$ for applicability of the preceding estimate. One obvious situation in which the preceding analysis breaks down is at the turning points, for which $\chi''(k_0) = \omega''(k_0) = c_g'(k_0) = 0$. The behavior of wave packets then has the form

$$u(x,t) \sim F(k_0) \exp[-i\chi(k_0)t] \int_{-\infty}^{\infty} \exp\left[-\frac{i}{6}(k-k_0)^3 \chi'''(k_0)t\right] dk \tag{11.26}$$

and thus involves Airy functions.[7] We will find an example of this in the analysis of simple turning points for density waves in Chapter 17.

11.2.2 Modulation Theory

Modulation theory starts from the asymptotic behavior of a wave packet as given in approximation (11.19) and focuses directly on physical space (not on the wave-number space). This approach to dispersive waves is very intuitive and also has been found to give a more convenient representation for the study of nonlinear waves. In addition, it is naturally suited for a generalization to descriptions of inhomogeneous and time-dependent media. In fact, the WKBJ analysis[8] can be seen as one particular case of modulation theory.

The idea is to refer to general solutions of the form (11.19), that is, to assume as an ansatz

$$u = a\cos\theta \tag{11.27}$$

and to define the generalized wave number k and generalized frequency ω as [see Eqs. (11.23) and (11.24)]

$$k = \frac{\partial\theta}{\partial x}, \tag{11.28}$$

$$\omega = -\frac{\partial\theta}{\partial t}. \tag{11.29}$$

A uniform wave train, with $a = $ constant and $\theta = kx - \omega t$, would then be reduced to Eq. (11.9). However, in the general context, the wave number and the frequency are now meant to be local concepts. We then assume that the dynamical equations impose the constraint

$$\frac{\partial\theta}{\partial t} = -\omega\left(\frac{\partial\theta}{\partial x}\right), \tag{11.30}$$

which is called the *dispersion relation*. The study of the relevant dynamical equations will determine whether and under what conditions Eq. (11.27), together with constraint (11.30), is an exact solution or only an approximate representation of the solutions involved.

The phase velocity c_{ph} is defined from the condition $\theta(x,t) = $ constant, that is,

$$\frac{\partial\theta}{\partial x}\frac{dx}{dt} + \frac{\partial\theta}{\partial t} = 0, \tag{11.31}$$

which gives $c_{ph} = \omega/k$. The group velocity now has a very simple interpretation in terms of transport of the wave number. From the definitions of Eqs. (11.28) and (11.29), we have

$$0 = \frac{\partial k}{\partial t} + \frac{\partial \omega}{\partial x} = \frac{\partial k}{\partial t} + c_g(k)\frac{\partial k}{\partial x}; \tag{11.32}$$

that is, the wave number propagates at the group velocity c_g. Curiously, this is the same as Eq. (11.3), which returns us to a nonlinear hyperbolic problem. This is not surprising because it is well known that the dynamics of wave packets closely resembles the dynamics of particles.

11.2.3 Dynamics of Ray Tracing in Inhomogeneous Media

In the modulation approach, dispersive waves in inhomogeneous media are described by the same set of equations [(11.27) through (11.29)], but now the dispersion relation has the form

$$\omega = \omega(k; x, t) \tag{11.33}$$

and thus depends explicitly on x and t. Several examples will be given in the context of density waves, but we may refer to more familiar examples, for example, the case of water waves in a basin with changing depth.

From the condition $\partial k/\partial t + \partial \omega/\partial x = 0$, the transport equation for the wave number now becomes

$$\frac{\partial k}{\partial t} + \left(\frac{\partial \omega}{\partial k}\right)_x \frac{\partial k}{\partial x} = -\left(\frac{\partial \omega}{\partial x}\right)_k. \tag{11.34}$$

Therefore, we can write

$$\frac{dk}{dt} = -\left(\frac{\partial \omega}{\partial x}\right)_k \tag{11.35}$$

on the characteristics

$$\frac{dx}{dt} = \left(\frac{\partial \omega}{\partial k}\right)_x. \tag{11.36}$$

The characteristics are no longer straight lines. The equations for the dynamics of ray tracing are identical to those of particle dynamics, provided that we interpret the function $\omega(k; x, t)$ as the relevant Hamiltonian. In fact, we can check that, as for Hamiltonian systems,

$$\frac{d\omega}{dt} = -\left(\frac{\partial \omega}{\partial t}\right)_{x,k}; \tag{11.37}$$

furthermore, the relation between θ and ω is just the eikonal relation[9] between the action and the Hamiltonian:

$$\frac{\partial \theta}{\partial t} + \omega\left(\frac{\partial \theta}{\partial x}; x, t\right) = 0. \tag{11.38}$$

11.3 **Water Waves**

The subject of water waves[10] defines a very broad research area, for which most phenomena associated with small-amplitude perturbations can be explained starting from the dispersion relation for linear perturbations:

$$\omega^2 = gk \tanh(hk) \left(1 + \frac{T}{\rho g} k^2 \right). \tag{11.39}$$

Here g is the gravitational acceleration, T represents the surface tension associated with capillary forces, ρ is the mass density, and h is the depth of the water reservoir.

11.3.1   Derivation of the Dispersion Relation

Consider an incompressible, nonviscous fluid in a constant-gravity field in the simple case for which $T = 0$. Let $\mathbf{u}$ describe the flow field and p represent the pressure. Then the relevant equations are

$$\nabla \mathbf{u} = 0, \tag{11.40}$$

$$\frac{\partial \mathbf{u}}{\partial t} + (\mathbf{u}\nabla)\mathbf{u} = -\frac{1}{\rho}\nabla p - \mathbf{g}. \tag{11.41}$$

The Euler equation can be curled to yield the equation for the vorticity $\nabla \times \mathbf{u}$:

$$\frac{\partial(\nabla \times \mathbf{u})}{\partial t} + (\mathbf{u}\nabla)(\nabla \times \mathbf{u}) = [(\nabla \times \mathbf{u})\nabla]\mathbf{u}, \tag{11.42}$$

which shows that an initially irrotational flow remains so at later times. We then focus on an irrotational flow ($\nabla \times \mathbf{u} = 0$) and thus introduce a potential ψ for the flow such that $\mathbf{u} = \nabla \psi$. Continuity Eq. (11.40) becomes the Laplace equation

$$\nabla^2 \psi = 0, \tag{11.43}$$

whereas the Euler equation can be integrated to give

$$\frac{p - p_0}{\rho} = -\frac{\partial \psi}{\partial t} - \frac{1}{2}(\nabla \psi)^2 - gz, \tag{11.44}$$

where z represents the vertical coordinate and p_0 represents the pressure of the air at the upper boundary of the fluid.

  We now ignore the horizontal y coordinate and choose the undisturbed water level to be at $z = 0$. Let the function $-h(x)$ represent the depth of the water reservoir and describe the upper-boundary surface of the fluid by the function $\eta(x,t)$. There are three natural boundary conditions for this problem. One is for the fixed bottom of the reservoir, where we impose $\mathbf{u}_\perp = 0$:

$$\frac{\partial \psi}{\partial z} + \frac{dh}{dx}\frac{\partial \psi}{\partial x} = 0 \qquad \text{at } z = -h(x). \tag{11.45}$$

The free surface requires the other two boundary conditions; one is a geometric condition related to $\mathbf{u}_\perp$:

$$\frac{\partial \eta}{\partial t} + \frac{\partial \eta}{\partial x}\frac{\partial \psi}{\partial x} = \frac{\partial \psi}{\partial z} \qquad \text{at } z = \eta(x,t), \tag{11.46}$$

and the other sets the equilibrium condition with the air:

$$\frac{\partial \psi}{\partial t} + \frac{1}{2}\left(\frac{\partial \psi}{\partial x}\right)^2 + \frac{1}{2}\left(\frac{\partial \psi}{\partial z}\right)^2 + gz = 0 \qquad \text{at } z = \eta(x,t). \tag{11.47}$$

In the linearized limit of vanishingly small-amplitude perturbations with respect to a quiet water reservoir characterized by $\mathbf{u} = 0$, $\eta = 0$, we have the following equation and boundary conditions:

$$\frac{\partial^2 \psi}{\partial x^2} + \frac{\partial^2 \psi}{\partial z^2} = 0, \tag{11.48}$$

$$\frac{\partial \psi}{\partial z} + \frac{dh}{dx}\frac{\partial \psi}{\partial x} = 0 \qquad \text{at } z = -h. \tag{11.49}$$

$$\frac{\partial \eta}{\partial t} = \frac{\partial \psi}{\partial z} \qquad \text{at } z = 0, \tag{11.50}$$

$$\frac{\partial \psi}{\partial t} + g\eta = 0 \qquad \text{at } z = 0. \tag{11.51}$$

The two conditions at $z = 0$ can be combined into

$$\frac{\partial^2 \psi}{\partial t^2} + g\frac{\partial \psi}{\partial z} = 0 \qquad \text{at } z = 0. \tag{11.52}$$

If, for simplicity, we consider the case in which $h = $ constant, we can proceed to a Fourier analysis and study elementary waves of the form $\psi = \hat{\psi}(z)\exp[i(kx - \omega t)]$, so the relevant equation and boundary conditions are

$$\frac{d^2 \hat{\psi}}{dz^2} - k^2 \hat{\psi} = 0, \tag{11.53}$$

$$\frac{d\hat{\psi}}{dz} = 0 \qquad \text{at } z = -h. \tag{11.54}$$

$$g\frac{d\hat{\psi}}{dz} - \omega^2 \hat{\psi} = 0 \qquad \text{at } z = 0. \tag{11.55}$$

This homogeneous linear problem is solved by the function

$$\hat{\psi} = A \cosh[k(z+h)], \tag{11.56}$$

provided that

$$\omega^2 = gk \tanh(hk). \tag{11.57}$$

Equation (11.57) is the desired dispersion relation [see Eq. (11.39)].

11.3.2   Regimes

Several different regimes can be considered. Shallow water waves are approximately hyperbolic, in the sense that, for $h|k| \ll 1$,

$$\omega^2 \sim (gh)k^2 \left[1 + \left(\frac{T}{\rho gh^2} - \frac{1}{3}\right)h^2 k^2 + \cdots\right]. \tag{11.58}$$

The term in brackets $\propto k^2$ disappears for $h \approx 0.48$ cm, which gives a quantitative estimate of the range of capillary forces in water. The structure of the dispersion relation for shallow-water theory suggests two interesting differential equations (taking one step backward, $k \to -i\partial/\partial x$), the linear Korteweg-de Vries equation[11]

$$\frac{\partial \psi}{\partial t} + c\frac{\partial \psi}{\partial x} + v\frac{\partial^3 \psi}{\partial x^3} = 0, \tag{11.59}$$

associated with the dispersion relation $\omega = ck - vk^3$, and the Boussinesq equation[12]

$$\frac{\partial^2 \psi}{\partial t^2} - \alpha^2\frac{\partial^2 \psi}{\partial x^2} - \beta^2\frac{\partial^4 \psi}{\partial x^2 \partial t^2} = 0, \tag{11.60}$$

associated with the dispersion relation $\omega^2 = \alpha^2 k^2/(1 + \beta^2 k^2)$.

The opposite limit of deep water waves is fully dispersive because, for $h|k| \gg 1$, we get

$$\omega^2 \sim g|k|. \tag{11.61}$$

For short waves, capillarity effects become important, and we have

$$\omega^2 \sim g|k| + \frac{T}{\rho}|k|^3. \tag{11.62}$$

11.3.3 Examples

A number of examples based on water waves illustrate important aspects of the theory of density waves for galaxy disks. Some analogies lie deep at the conceptual level but should not be taken to imply that the specific dynamical mechanisms involved are literally the same in the two environments.[13]

One interesting example shows the interplay of different wave branches, with different propagation properties, in the dynamics of short waves. If we refer to the dispersion relation for short waves [approximation (11.62)], we can easily show that the phase velocity c_{ph} has a minimum $c_{phm} \approx 23.2$ cm s^{-1} at $\lambda_{min} = 2\pi/k_{min} \approx 1.73$ cm, where it intersects the curve for the group velocity c_g; the latter curve has a minimum $c_{gm} \approx 17.9$ cm s^{-1} at $\lambda_m = 2\pi/k_m \approx 4.39$ cm. The capillary wave branch occurs for $\lambda = 2\pi/k < \lambda_{min}$ and is characterized by the property $c_g > c_{ph}$. The gravity wave branch, at higher wavelengths, has the opposite behavior, $c_g < c_{ph}$. In general, for a given choice of the group velocity above c_{gm}, there are two possible waves characterized by that group velocity.

The preceding properties of short waves give an interesting way of explaining how we can create standing-wave patterns by putting an obstacle across a stream characterized by speed U. The generation of a steady crest pattern requires the equality $c_{ph}(k) = U$ for the wavelengths involved. This already shows that for the steady pattern to appear, the stream must be sufficiently fast ($U > c_{phm}$). However, the obstacle must be the source of the pattern generated (causality condition or radiation boundary condition), and therefore, group velocity must allow the signal to propagate from the obstacle to where the standing wave is observed. Because $c_g > c_{ph} = U$ for capillary waves and $c_g < c_{ph} = U$ for gravity waves, we conclude that the standing pattern is created by capillary waves upstream and by longer gravity waves downstream. From a picture of the pattern, we are thus able to infer the direction in which the stream is moving.

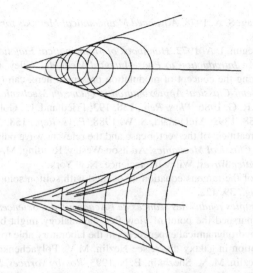

Fig. 11.2. The formation of a ship wave pattern (from Whitham, G. B. 1974. *Linear and Non-linear Waves*, Wiley, New York, pp. 410, 413; ©1974 by John Wiley & Sons, Inc.; reprinted by permission of John Wiley & Sons, Inc.).

Another very interesting example is that of the formation of ship waves in deep water[14] (Fig. 11.2). Here we are considering the deep-water limit of approximation (11.61), for which $c_{ph} = 2c_g$. This relation between phase and group velocity explains the opening angle of the wake behind a ship, which is given by $2 \arcsin(1/3) \approx 39°$ and is independent of the speed of the ship. The explanation is best provided in graphical form. Let $\mathbf{U}$ be the speed of the ship. Then water-wave signals emitted by the ship with wave number $\mathbf{k}$ can appear to be stationary, provided that $\mathbf{Uk} = \omega$ (this is analogous to the phase-velocity condition for the obstacle on a stream described earlier). The angle α between the point reached by the ship at a time t after the signal was emitted and the point reached by a wave packet moving at speed c_g in the direction of $\mathbf{k}$ (with respect to the line drawn by the ship in its motion) must satisfy the relation

$$\sin \alpha = \frac{(1/4)Ut}{(3/4)Ut} = \frac{1}{3}, \tag{11.63}$$

which is fixed and independent of t, U, and the relative angle between $\mathbf{U}$ and $\mathbf{k}$. This proves that the cone of water waves in deep water is geometrically similar to the Čerenkov cone[15] produced by supersonic motion in a gas by a projectile traveling at Mach 3.

Notes

1. This chapter basically follows Whitham, G. B. 1974. *Linear and Nonlinear Waves*, Wiley, New York. One excellent introduction to the theory of partial differential equations is given by Sommerfeld, A. 1949. *Partial Differential Equations in Physics*, Academic, New York.
2. For example, see Jackson, J. D. 1962. *Classical Electrodynamics*, Wiley, New York.
3. Burgers, J. M. 1948. *Adv. Appl. Mech.*, **1**, 171.
4. Hopf, E. 1950. *Comm. Pure Appl. Math.*, **3**, 201; Cole, J. D. 1951. *Q. Appl. Math.*, **9**, 225.
5. For example, see Messiah, A. 1962. *Quantum Mechanics*, Wiley, New York.

6. See Bender, C. M., Orszag, S. A. 1978. *Advanced Mathematical Methods for Scientists and Engineers*, McGraw-Hill, New York.

7. See Abramowitz, M., Stegun, I. A. 1972. *Handbook of Mathematical Functions*, Dover, New York.

8. See Heading, J. 1962. *An Introduction to Phase-Integral Methods*, Wiley, New York. More modern developments, generalizing the concept of modulation presented here, can be found in Maslov, V. P., Fedoriuk, M. V. 1981. *Semi-Classical Approximation in Quantum Mechanics*, Reidel, Dordrecht, The Netherlands; Littlejohn, R. G. 1986. *Phys. Rep.*, **138**, 193; Friedland, L., Goldner, G., Kaufman, A. N. 1987. *Phys. Rev. Lett.*, **58**, 1392; McDonald, S. W. 1988. *Phys. Rep.*, **158**, 337. In plasma physics, a major issue is a proper treatment of the vector case and the relevant wave polarizations.

9. See Goldstein, H. 1950. *Classical Mechanics*, Addison-Wesley, Reading, MA.

10. See Stoker, J. J. 1957. *Water Waves*, Wiley Interscience, New York.

11. This is a linear version of the famous equation associated with soliton solutions; Korteweg, D. J., de Vries, G. 1895. *Phil. Mag.*, **39**, 422.

12. Boussinesq, J. 1871. *Comptes rendus des séances de l'Academie des Sciences*, **72**, 755.

13. Some researchers have pursued the point of view that the analogy might be made in such stringent terms as to even devise hydrodynamical experiments in the laboratory able to simulate the mechanisms of spiral structure formation in galaxy disks; see Nezlin, M.V., Polyachenko, V. L., et al. 1986. *Sov. Astron. Lett.*, **12**, 213; Nezlin, M. V., Snezhkin, E. N. 1993. *Rossby Vortices, Spiral Structure, Solitons*, Springer-Verlag, Heidelberg. We do not share their hope because in our view different mechanisms operate in the excitation and maintenance of spiral structure in galaxies, as explained in Part III.

14. The original result is due to Kelvin and was later reformulated concisely by Lighthill; see Whitham, G. B. 1974. *op. cit.*, from which most of this chapter has been taken. But see the interesting article by Rabaud, M., Moisy, F. 2013. *Phys. Rev. Lett.*, **110**, id.214503.

15. See Jackson, J. D. 1962. *op. cit.*

12   Jeans Instability

In this chapter we derive the basic equations for Jeans instability[1] in the simplest formulation. This is meant to be the first example in the astrophysical context of the concepts introduced in Chapter 11. In reality, the study of stability brings us one step beyond the concept of dispersive waves. Most of what has been described in Chapter 11 is applicable in cases in which the frequency in the relevant dispersion relation is allowed to be complex, $\omega = \omega_R + i\omega_I$, provided that $|\omega_I| \ll |\omega_R|$. But we should be aware that in the opposite limit we are no longer talking about waves in the usual sense.

In the simplest case, gravity and pressure forces are the main ingredients in the Jeans instability mechanism. Quite different results are derived from the same ingredients, depending on the model considered. In particular, although the fluid model is characterized by a dispersion relation for which neutral waves ($\omega_I = 0$) are allowed, a kinetic analysis leads to a dispersion relation in which stable perturbations are generally damped. At the end of the chapter, we briefly examine the linearized equations for perturbations of a plane-parallel slab, demonstrating how, in general, the collisionless Boltzmann equation is solved by integration along the unperturbed characteristics.

The analysis also points out some analogies to the plasma context and thus justifies some of the points made in Chapter 6. We do not deal here with the variety of applications that the Jeans instability has in astrophysics because this is widely available in the literature. The more specific applications to the large-scale dynamics of galaxy disks are postponed until Part III.

12.1   Homogeneous Fluid Model

We consider a formal equilibrium configuration of a homogeneous self-gravitating fluid initially at rest. This basic state is taken as the limit of more realistic inhomogeneous basic states for which all the scale lengths characterizing the density and pressure distributions are much longer than the scales of the perturbations that are considered. This qualification thus should be taken into account at the interpretation level of the derived dispersion relation.

Let $\mathbf{u}_1$, ρ_1, Φ_1, and p_1 represent perturbations of the fluid velocity, density, gravitational potential, and pressure, respectively. Then the linearized equations for momentum and mass conservation, the equation of state, and the Poisson equation become

$$\rho_0 \frac{\partial \mathbf{u}_1}{\partial t} = -\nabla p_1 - \rho_0 \nabla \Phi_1, \tag{12.1}$$

$$\frac{\partial \rho_1}{\partial t} + \rho_0 \nabla \mathbf{u}_1 = 0, \tag{12.2}$$

$$p_1 = c_s^2 \rho_1, \tag{12.3}$$

$$\nabla^2 \Phi_1 = 4\pi G \rho_1. \tag{12.4}$$

Here ρ_0 and c_s^2 are constants characterizing the homogeneous basic state. Taking the divergence of Eq. (12.1) and eliminating $\mathbf{u}_1$, Φ_1, and p_1 by means of Eqs. (12.2) through (12.4), we are left with a single partial differential equation for ρ_1 that resembles Eq. (11.12):

$$\frac{\partial^2 \rho_1}{\partial t^2} - c_s^2 \nabla^2 \rho_1 - \omega_J^2 \rho_1 = 0, \tag{12.5}$$

with

$$\omega_J^2 = 4\pi G \rho_0. \tag{12.6}$$

The corresponding dispersion relation is thus

$$\omega^2 = c_s^2 k^2 - 4\pi G \rho_0. \tag{12.7}$$

In the limit $G \to 0$, we recover sound waves, whereas in the cold case $c_s^2 \to 0$, we find the free-fall kinematic limit. In general, the frequency of the perturbation is determined by a balance between the stabilizing pressure contribution and the destabilizing self-gravity term. For sufficiently long waves, $|k| < k_J$, with

$$k_J = \frac{\omega_J}{c_s}, \tag{12.8}$$

Jeans instability occurs.

12.1.1   Inhomogeneous Systems and Relation to the Virial Theorem

The closest realistic basic state where we might try to apply the preceding concepts could be a small region of a nonrotating spherical system of finite mass M, total kinetic energy K, and gravitational energy W. In this case, we may check what would be the relevant parameter regime for the Jeans dispersion relation, given the fact that the virial theorem for such a system imposes a condition between K and W. An average thermal speed c_s may be defined from the relation $2K = Mc_s^2$, and a scale length R for the size of the system may be introduced by means of the relation $W = -GM^2/R$; a typical density thus would be $\bar{\rho} = 3M/(4\pi R^3)$. Therefore, the virial constraint would require that

$$c_s^2 = \frac{GM}{R} = \frac{1}{3}(4\pi G \bar{\rho})R^2, \tag{12.9}$$

so that

$$R = \sqrt{3}\frac{c_s}{\bar{\omega}_J}. \tag{12.10}$$

The quantity $\bar{\omega}_J$ is constructed as in Eq. (12.6) in terms of $\bar{\rho}$. We have thus proved that the system is not large enough to accommodate Jeans unstable perturbations. In other words, it appears that a finite spherical, nonrotating, self-gravitating system has its size determined in such a way that the Jeans instability is suppressed. This is, of course, an indication of a dimensional-analysis argument only. It suggests that the actual stability of a finite-mass spherical system should be checked by a different analysis for which the inhomogeneity of the system is fully recognized.

The study of the stability of a fluid self-gravitating system contained in a finite volume identifies interesting phenomena that are related to the so-called gravothermal catastrophe. In Part V we will devote one section (Section 25.1) to a brief summary of the main results obtained in that context and to some issues that still remain open.

Some physical ingredients, not present in the simplest Jeans stability analysis just given, may open the way to other types of instability (e.g., see Chapter 23).

12.1.2 Electrostatic Plasma Waves

In the plasma context, an analysis similar to that leading to Eq. (12.5) can be carried out for electrostatic waves in an ionized gas.[2] Here a homogeneous basic state is more natural because we can start from a neutrality condition ensured by the presence of ions and electrons. In the simplest derivation, we imagine that the ions are fixed and only electrons are mobile. (A kinetic analysis, in line with what will be given in Section 12.2, is required for defining the limitations and the merits of this simple fluid model;[3] in particular, the fluid model used here would be justified only for sufficiently high-frequency waves.) The different behavior of electrostatic forces with respect to charges of the same sign produces a difference in the Poisson equation, which now becomes

$$\nabla^2 \Phi_1 = -4\pi n_1 q, \tag{12.11}$$

where n_1 is the number-density perturbation for the electron component, and $q = -e$ is the electron charge. The final dispersion relation for electrostatic plasma waves[4] is

$$\omega^2 = c_s^2 k^2 + \omega_{pe}^2, \tag{12.12}$$

with

$$\omega_{pe}^2 = \frac{4\pi n e^2}{m_e}. \tag{12.13}$$

As a result, plasma waves are stable because Coulomb forces help to prevent the development of charge clumps. The formal analogue to the Jeans wave number is

$$k_D = \frac{\omega_{pe}}{c_s}. \tag{12.14}$$

The associated length scale, called the *Debye length* (expressed in terms of temperature T and particle number density n, the standard definition is $\lambda_D = \sqrt{kT/4\pi n e^2}$, where k is here the Boltzmann constant), sets the size beyond which low-frequency electrostatic forces are screened collectively by the plasma. If a point charge is imposed, charges rearrange themselves inside a Debye sphere to neutralize the field outside the sphere.[5]

12.1.3 Simple Generalizations

One may easily generalize the Jeans stability analysis to other simple basic configurations,[6] for example, to the case in which the system is initially in solid-body rotation with angular velocity $\mathbf{\Omega}$. In the rotating frame of reference, Eq. (12.1) is replaced with

$$\rho_0 \frac{\partial \mathbf{u}_1}{\partial t} = -\nabla p_1 - \rho_0 \nabla \Phi_1 + 2\rho_0 \mathbf{u}_1 \times \mathbf{\Omega}, \tag{12.15}$$

with the appearance of a Coriolis correction. If we introduce the angle θ defined by the two vectors $\mathbf{k}$ and $\mathbf{\Omega}$, the relevant dispersion relation is found to be

$$\omega^4 - (\omega_1^2 + \omega_2^2)\omega^2 + \omega_1^2\omega_2^2 = 0, \tag{12.16}$$

with

$$\omega_1^2 + \omega_2^2 = 4\Omega^2 + c_s^2 k^2 - 4\pi G\rho_0, \tag{12.17}$$

$$\omega_1^2\omega_2^2 = 4\Omega^2(c_s^2 k^2 - 4\pi G\rho_0)\cos^2\theta. \tag{12.18}$$

Therefore, for $\cos\theta \neq 0$, the Jeans criterion for stability, $|k| > k_J$, holds unmodified, whereas for $\mathbf{k}$ perpendicular to the rotation axis, the dispersion relation is

$$\omega^2 = 4\Omega^2 + c_s^2 k^2 - 4\pi G\rho_0. \tag{12.19}$$

In the presence of differential rotation (see Chapters 13 and 15), Eq. (12.19) has $4\Omega^2$ replaced with κ^2 (κ is the epicyclic frequency). Note how rotation, by means of Coriolis forces, is able to suppress Jeans instability (in principle, if $4\Omega^2 > 4\pi G\rho_0$) on the plane perpendicular to the rotation axis.

If we refer to typical values for ρ_0 and c_s applicable to the solar neighborhood, we find a Jeans length close to 1 kpc. This indeed compares well with the thickness of the disk; along the vertical direction the rotation of our Galaxy has no effect (see also Chapter 14).

If we are interested in generalizing the analysis to a thin-sheet geometry (such as a homogeneous slab of the type discussed in Section 12.3, in the limit in which the thickness becomes vanishingly small), the Poisson equation for even density perturbations $\rho_1 = \sigma_1\delta(z)$ [to replace Eq. (12.4)] becomes

$$\frac{d^2\Phi_1}{dz^2} - k^2\Phi_1 = 4\pi G\sigma_1\delta(z), \tag{12.20}$$

where k refers to a Fourier analysis in the plane of the sheet. This is solved under the condition of evanescence at $|z| \to \infty$, leading to the Gauss jump condition

$$-|k|\Phi_1 = 2\pi G\sigma_1. \tag{12.21}$$

The resulting replacement of ρ_0 with $|k|\sigma_0/2$ in the final dispersion relation is a signature of surface waves, which we have already met in the context of water waves in Chapter 11. These simple generalizations are the starting point for the study of density waves in galaxy disks, as will be discussed in Chapter 15.

12.2 Homogeneous Kinetic Model and Landau Damping

We now consider Jeans instability analysis in a homogeneous kinetic model.[7] We refer to an isotropic equilibrium described by a Maxwellian distribution

$$f_0 = \rho_0(\sqrt{2\pi}c)^{-3}\exp(-v^2/2c^2). \tag{12.22}$$

Equations (12.1) through (12.3) are replaced with the linearized collisionless Boltzmann equation for f_1

$$\frac{Df_1}{Dt} = \frac{\partial f_1}{\partial t} + \mathbf{v}\nabla_\mathbf{x} f_1 = (\nabla_\mathbf{x}\Phi_1)(\nabla_\mathbf{v} f_0), \tag{12.23}$$

which is a simple linear hyperbolic equation (see Section 11.1). Thus, in this linearized analysis, f_1 is taken to evolve along the unperturbed orbits, that is, along the characteristics given by straight lines at velocity $\mathbf{v}$. The Poisson equation in terms of f_1 is

$$\nabla^2 \Phi_1 = 4\pi G \int f_1 \, d^3 v. \tag{12.24}$$

In this simple case, we Fourier analyze in space and time; in particular, we take $\Phi_1 = \hat{\Phi}_1 \exp(-i\omega t + i\mathbf{kx})$. The preceding equations are then reduced to

$$(\omega - \mathbf{kv})f_1 = \frac{\mathbf{kv}}{c^2} f_0 \Phi_1, \tag{12.25}$$

$$-k^2 \Phi_1 = \omega_J^2 \frac{1}{\rho_0} \int f_1 \, d^3 v. \tag{12.26}$$

The formal dispersion relation obtained when Φ_1 is eliminated is

$$\frac{k^2}{k_J^2} = \frac{1}{\sqrt{\pi}} \int_{-\infty}^{\infty} \frac{x}{x - \zeta} \exp(-x^2) \, dx, \tag{12.27}$$

with

$$k_J = \frac{\omega_J}{c}, \tag{12.28}$$

$$\zeta = \frac{\omega}{\sqrt{2}|k|c}. \tag{12.29}$$

The integral in Eq. (12.27) is singular at $x = \zeta$, and therefore the result demands a physical interpretation. The Landau prescription[8] consists of enforcing a causality requirement by defining the integral by analytical continuation starting from the physically justified case of growing modes. Thus the dispersion relation is

$$\frac{k^2}{k_J^2} = 1 + \zeta Z(\zeta), \tag{12.30}$$

where Z is the plasma dispersion function.[9]

12.2.1 Plasma Dispersion Function

The plasma dispersion function (Fig. 12.1) can be written as[10]

$$Z(\zeta) = 2i \exp(-\zeta^2) \int_{-\infty}^{i\zeta} \exp(-t^2) \, dt = i\sqrt{\pi} \exp(-\zeta^2)[1 + \mathrm{erf}(i\zeta)]. \tag{12.31}$$

It satisfies the differential equation

$$Z' + 2\zeta Z + 2 = 0, \quad Z(0) = i\sqrt{\pi}, \tag{12.32}$$

or

$$Z'' + 2\zeta Z' + 2Z = 0. \tag{12.33}$$

Sometimes we refer to the associated function [suggested by Eq. (12.30)]

$$W = -[1 + \zeta Z(\zeta)] = \frac{Z'}{2}. \tag{12.34}$$

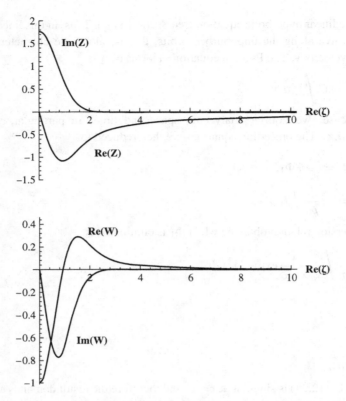

Fig. 12.1. The plasma dispersion function for $\text{Im}(\zeta) = 0$.

The plasma dispersion function has two interesting asymptotic limits, one for $|\zeta| \ll 1$,

$$Z(\zeta) = i\sqrt{\pi}\exp(-\zeta^2) - 2\zeta + \frac{4\zeta^3}{3} + O(\zeta^5), \tag{12.35}$$

and one regular limit for $|\zeta| \gg 1$,

$$Z(\zeta) = -\frac{1}{\zeta} - \frac{1}{2\zeta^3} - \frac{3}{4\zeta^5} + O(\zeta^{-7}). \tag{12.36}$$

12.2.2 Marginal Stability and the Fluid Limit

We derive the marginal stability condition by considering the dispersion relation in the limit of the asymptotic expansion given by Eq. (12.35). Thus we find that

$$\frac{k^2}{k_J^2} \sim 1 + i\sqrt{\pi}\zeta, \tag{12.37}$$

so $k^2/k_J^2 < 1$ gives $\zeta_I > 0$ (growing in time) and $k^2/k_J^2 > 1$ gives $\zeta_I < 0$ (damped in time). This behavior is markedly different from that of the fluid model [see Eq. (12.7)], where at the threshold of Jeans instability we move from instability to neutral oscillations. This property, characteristic of the collisionless Boltzmann equation, is called *Landau damping*. In the stellar dynamics of galaxy disks (see Sections 16.4 and 19.2) we will find interesting phenomena associated with it.

We obtain the fluid limit of the dispersion relation by considering the opposite regime, in which $|\zeta| \gg 1$, because, under this condition, the details of the distribution functions are lost, and the Maxwellian is seen as a fluid beam.[11] Then the relevant dispersion relation becomes

$$\frac{k^2}{k_J^2} \sim -\frac{1}{2\zeta^2} - \frac{3}{4\zeta^4}, \tag{12.38}$$

and hence

$$\omega^2 \sim -\omega_J^2 \left(1 - \frac{3k^2c^2}{\omega_J^2}\right), \tag{12.39}$$

which should be compared with the result of the fluid model given in Eq. (12.7).

12.3   Kinetic Analysis of a Plane-Parallel Slab

We now consider an equilibrium configuration made of a plane-parallel slab, homogeneous in the horizontal directions and inhomogeneous in z, so that the relevant integrals of the motion are $\mathbf{v}_\perp$ and $E_z = (1/2)v_z^2 + \Phi_0(z)$. Thus we can take $f_0 = f_0(\mathbf{v}_\perp, E_z)$. For the following discussion, no more is written about the equilibrium; we just assume that the vertical orbits are bound and periodic, with bounce frequency $\omega_b = \omega_b(E_z)$. A detailed study of a fully self-consistent basic state of this type will be given in Section 14.1. Only some key steps of the related kinetic stability analysis are outlined here.[12]

The linearized collisionless Boltzmann equation for a perturbation f_1 in terms of the perturbed potential Φ_1 is [see Eq. (12.23)]

$$\frac{Df_1}{Dt} = \frac{\partial f_1}{\partial t} + \mathbf{v}_\perp \nabla_{\mathbf{x}_\perp} f_1 + v_z \frac{\partial f_1}{\partial z} - \frac{\partial \Phi_0}{\partial z} \frac{\partial f_1}{\partial v_z} = (\nabla_{\mathbf{x}} \Phi_1)(\nabla_{\mathbf{v}} f_0). \tag{12.40}$$

We can rearrange the various terms and Fourier analyze in time and space, except for the vertical coordinate, so that

$$\frac{Df_1}{Dt} = (\nabla_{\mathbf{x}_\perp} \Phi_1)(\nabla_{\mathbf{v}_\perp} f_0) + v_z \frac{\partial \Phi_1}{\partial z} \frac{\partial f_0}{\partial E_z} \tag{12.41}$$

$$= i\mathbf{k}_\perp (\nabla_{\mathbf{v}_\perp} f_0)\Phi_1 + \left(\frac{D\Phi_1}{Dt} - \frac{\partial \Phi_1}{\partial t} - \mathbf{v}_\perp \nabla_{\mathbf{x}_\perp} \Phi_1\right) \frac{\partial f_0}{\partial E_z} \tag{12.42}$$

$$= \frac{\partial f_0}{\partial E_z} \frac{D\Phi_1}{Dt} + i\left(\hat{\omega} \frac{\partial f_0}{\partial E_z} + \mathbf{k}_\perp \nabla_{\mathbf{v}_\perp} f_0\right) \Phi_1. \tag{12.43}$$

In Eq. (12.41) we have made use of the identity $\partial f_0/\partial v_z = v_z \partial f_0/\partial E_z$. In Eq. (12.43) we have introduced the natural Doppler-shifted frequency $\hat{\omega} = \omega - \mathbf{k}_\perp \mathbf{v}_\perp$. This equation can be integrated along the unperturbed characteristics (see Section 11.1) to obtain

$$f_1 = \frac{\partial f_0}{\partial E_z} \Phi_1 + i\left(\hat{\omega} \frac{\partial f_0}{\partial E_z} + \mathbf{k}_\perp \nabla_{\mathbf{v}_\perp} f_0\right) \int_{-\infty}^{t} dt' \, \Phi_1(\mathbf{x}'(t'), t'). \tag{12.44}$$

The first term on the right-hand side of Eq. (12.44) is traditionally called the *adiabatic response*, although it is only a part of the actual response in the adiabatic zero-frequency limit (see discussion of this issue in Subsection 22.2.2). The integral appearing in the second term is meant to be calculated with integrals of the motion fixed at $\mathbf{v}_\perp$ and E_z. The structure of the response f_1 to the perturbed potential Φ_1 identified here is very general and will be met again in Part III (see especially Subsections 15.1.1 and 16.4.1) and in Part IV (see Chapter 23, especially Section 23.1). This chapter concludes with the following short digression on the origin of resonances in stellar dynamics.

12.3.1   Bounce Orbit Expansion

The integration along the unperturbed characteristics is trivial in the horizontal variables because we just have to take [see also Eq. (11.6)]

$$\mathbf{x}'_\perp = \mathbf{x}_\perp + \mathbf{v}_\perp (t' - t). \tag{12.45}$$

As a result, the relevant integral becomes

$$\int_{-\infty}^{t} dt' \, \Phi_1(\mathbf{x}'(t'), t') = \exp(-i\omega t + i\mathbf{k}_\perp \mathbf{x}_\perp) \int_{-\infty}^{0} d\tau \, \hat{\Phi}_1(z') \exp(-i\hat{\omega}\tau). \tag{12.46}$$

Because the z motion is periodic, with bounce frequency $\omega_b = \omega_b(E_z)$, $\hat{\Phi}_1(z')$ is periodic too, and we can expand it as

$$\hat{\Phi}_1(z') = \sum_{n=-\infty}^{+\infty} \hat{\Phi}_n \exp(in\omega_b \tau), \tag{12.47}$$

so the integral becomes

$$\int_{-\infty}^{t} dt' \, \Phi_1(\mathbf{x}'(t'), t') = \exp(-i\omega t + i\mathbf{k}_\perp \mathbf{x}_\perp) \sum_{n=-\infty}^{+\infty} \hat{\Phi}_n \int_{-\infty}^{0} d\tau \exp[i(n\omega_b - \hat{\omega})\tau]), \tag{12.48}$$

thus giving rise to an infinite number of resonances. These should be dealt with following the Landau prescription introduced in Section 12.2.

Notes

1. Jeans, J. H. 1902. *Phil. Trans. Roy. Soc. London*, **199**, 1; 1929. *Astronomy and Cosmogony*, Cambridge University Press, Cambridge, UK; see also chapter 13 of the monograph by Chandrasekhar, S. 1961. *Hydrodynamic and Hydromagnetic Stability*, Oxford University Press, Oxford, UK.
2. These are often referred to as *Langmuir waves*; see Stix, T. H. 1962. *The Theory of Plasma Waves*, McGraw-Hill, New York; Krall, N. A., Trivelpiece, A. W. 1973. *Principles of Plasma Physics*, McGraw-Hill, New York.
3. See chapters 4 and 8 of Krall, N. A., Trivelpiece, A. W. 1973. *op. cit.*
4. Tonks, L., Langmuir, I. 1929. *Phys. Rev.*, **33**, 195; Bohm, D., Gross, E. P. 1949. *Phys. Rev.*, **75**, 1851.
5. In reality, the shielding mechanism is generally shared by both electrons and ions because in the zero-frequency limit both components necessarily participate in the process.

6. For the generalization to a differentially rotating fluid, see Schatzman, E., Bel, N. 1955. *Comptes rendus des séances de l'Academie des Sciences*, **241**, 20; for the generalization to the magnetohydrodynamic context, see Chandrasekhar, S., Fermi, E. 1953. *Astrophys. J.*, **118**, 116.

7. See the excellent discussion by Lynden-Bell, D. 1967. In Vol. 9 of *Lectures in Applied Mathematics* Series, American Mathematical Society, Providence, RI, p. 131.

8. Landau, L. D. 1946. *J. Phys. USSR*, **10**, 25.

9. Fried, B. F., Conte, S. 1961. *The Plasma Dispersion Function*, Academic, New York.

10. See Abramowitz, M., Stegun, I. 1970. *Handbook of Mathematical Functions*, Dover, New York.

11. See Krall, N. A., Trivelpiece, A. W. 1973. *op. cit.*

12. A thorough analysis has been given by Mark, J. W.-K. 1971. *Astrophys. J.*, **169**, 455.

Part III
Spiral Galaxies

13 Orbits

The theory of orbits is important in the study of the dynamics of galaxies. Because the relaxation times for the relevant processes are very long (see Chapter 7), much of the interest lies in the study of orbits in a mean field. What is learned from these studies applies beyond the case of single particles, for the evolution operator in a continuum description (either in a fluid or in a stellar dynamical description; see Chapter 8) can be essentially identified with the Hamiltonian that governs single-particle orbits.

Galaxy disks are relatively cool systems; that is, most star (or gas cloud) orbits are very close to being circular, so a typical star (or gas cloud) velocity at a given location is very close to the average velocity of rotation of the disk. A proper description of a collection of orbiting particles should be given in terms of a distribution function (see Chapter 14). However, the physically intuitive properties of such cool disks can be easily traced back to the characteristics of quasi-circular orbits of individual particles, which are outlined in this chapter. The deviations from circular orbits are called *epicycles*. To lowest order, they correspond to a harmonic oscillator in the radial direction characterized by the epicyclic frequency κ. In cool disks, the typical radial velocity of a star c is much smaller than the local average velocity u. Thus the epicycle is small because the typical radial excursion c/κ is much smaller than r. Note that the typical restoring force associated with the radial oscillations (which is basically determined by the conservation of angular momentum) is $c\kappa$, which turns out to be of the order of the vertical force due to the self-gravity of the disk $2\pi G\sigma$. This point will be discussed further in relation to the problem of density waves in Chapter 15.

The epicyclic motion is analogous to the Larmor oscillation of a charged particle in a magnetic field (Fig. 13.1). This analogy is the basis for the development of common tools of investigation, as described at the end of this chapter. These studies separate the orbit into two parts, the orbit of a guiding center, which reacts to perturbing forces perpendicular to the gyration axis in terms of drifts rather than accelerations, and the rapid oscillations (Larmor or epicyclic) around it. One useful concept of classical dynamics that has found wide application in the description of the motion of charged particles in plasma physics is that of the adiabatic invariant.[1] A well-known example is that of the ratio of the energy associated with the Larmor oscillation and the cyclotron frequency. Under appropriate conditions, this quantity is approximately conserved when the particle moves in an inhomogeneous and/or time-dependent field, which is the key to a simple interpretation of a number of interesting phenomena, such as the trapping of particles in a magnetic mirror. This has a simple counterpart in the study of star orbits in a cool disk.

From the empirical point of view, the best chance we have for a direct measurement of the properties of individual stellar orbits in a galaxy disk is provided by the solar neighborhood.

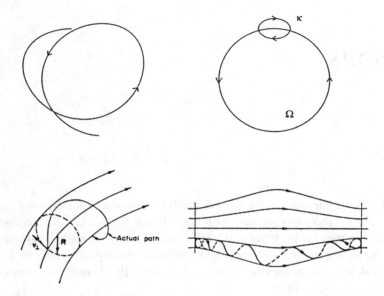

Fig. 13.1. Decomposition of a quasi-circular orbit into circular and epicyclic motion (*upper figures*; after Bertin, G., Lin, C. C. 1996. *Spiral Structure in Galaxies: A Density Wave Theory*, MIT Press, Cambridge, MA, p. 70; ©1996 Massachusetts Institute of Technology). Sketch of a Larmor gyration of a charge moving in a magnetic field and the phenomenon of magnetic trapping in a mirror field (*lower figures*; these figures are taken from Schmidt, G. 1979. *Physics of High Temperature Plasmas*, 2nd ed., Academic, New York, pp. 19, 22; ©1979 Academic Press, Inc.).

However, the systematic motions occurring in the disk of our Galaxy are not easily derived from the observations, simply because the Sun is part of such general flow. A classical study by Oort[2] provided the empirical demonstration that our Galaxy is in a state of differential rotation, as proposed earlier by Bertil Lindblad. The basis for that study was the idea of expanding the average velocity field (flow field) in Taylor series around the location of the Sun, from which we could easily express the properties of the average motion relative to the Sun expected in the various directions in the sky, along the galactic plane, and at various distances from the Sun. In formulas, we have

$$U_R \sim R[K + C\cos(2l) + A\sin(2l)], \tag{13.1}$$

$$U_l \sim R[B + A\cos(2l) - C\sin(2l)], \tag{13.2}$$

where R and l denote distance from the Sun's location and the galactic longitude. For a generic flow field **u**, the four constants that result from the Taylor expansion can be written as

$$A = \frac{r_\odot}{2}\left[\frac{\partial(u_\theta/r)}{\partial r} + \frac{1}{r^2}\frac{\partial u_r}{\partial \theta}\right]_\odot, \tag{13.3}$$

$$B = \frac{1}{2r_\odot}\left[\frac{\partial(ru_\theta)}{\partial r} - \frac{\partial u_r}{\partial \theta}\right]_\odot, \tag{13.4}$$

$$C = \frac{r_\odot}{2} \left[\frac{\partial(u_r/r)}{\partial r} - \frac{1}{r^2} \frac{\partial u_\theta}{\partial \theta} \right]_\odot, \tag{13.5}$$

$$K = \frac{1}{2r_\odot} \left[\frac{\partial(ru_r)}{\partial r} + \frac{\partial u_\theta}{\partial \theta} \right]_\odot. \tag{13.6}$$

Because the disk is cool, by observing the motion of groups of stars in various directions, we can easily check that the flow is basically a pure differential rotation; that is, we find $K \approx 0$ and $C \approx 0$. Furthermore, the parameters A and B, called the *Oort constants*, obtained by fitting the data, give the local properties of the galactic rotation curve. Note that for a pure differential rotation, $u_r = 0$ and $u_\theta \sim r\Omega(r)$, so that

$$A = \left(\frac{r}{2} \frac{d\Omega}{dr} \right)_\odot, \qquad B = \left[\frac{1}{2r} \frac{d(r^2\Omega)}{dr} \right]_\odot. \tag{13.7}$$

In general, the sign of A is taken to be positive and that of B to be negative because galactic longitudes are defined in such a way that the rotation velocity is negative.[3]

These and other kinematic studies provide a rather detailed picture of the orbits of stars in the solar neighborhood,[4] where it is found that $\Omega < \kappa < \Omega_z$ ($\Omega \approx 26$ km s^{-1} kpc^{-1} so that the rotation period is ≈ 240 Myr; based on the Oort limit for the local mass density, described in Chapter 14, the vertical period of oscillation is ≈ 70 Myr). The vertical frequency of oscillation about the galactic plane is thus the highest.

13.1 Action and Angle Variables

For the dynamics of disk galaxies, the basic situation to be considered is when a mean-field potential Φ is stationary and axisymmetric around the z axis. Note that such a potential is very different from the Keplerian potential generated by a point mass located at $r = 0$. In the equatorial plane defined by $z = 0$, the calculation of orbits is reduced to a one-dimensional problem by the introduction of an effective potential

$$\Phi_{\text{eff}}(r, J) = \Phi(r) + \frac{J^2}{2r^2}, \tag{13.8}$$

so the energy integral can be written as

$$E = \frac{1}{2}p_r^2 + \Phi_{\text{eff}}. \tag{13.9}$$

Thus the radial momentum (in our case this is identified with the radial velocity) can be expressed as a function of r and of the integrals of the motion E and J, where J is the specific angular momentum. For a large class of potentials, the function Φ_{eff} exhibits one minimum at $r = r_0$ (Fig. 13.2), which identifies the radius of circular orbits with angular momentum J. If we take $J > 0$ and define

$$\Omega^2(r) = \frac{1}{r} \frac{d\Phi}{dr}, \tag{13.10}$$

the guiding center radius is related to the specific angular momentum by

$$r_0^2 \Omega(r_0) = J, \tag{13.11}$$

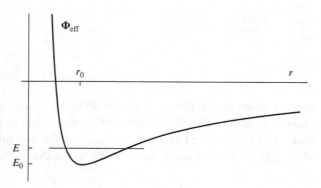

Fig. 13.2. Sketch of the effective potential for equatorial orbits in an axisymmetric field.

which is generally a one-to-one relationship; for Φ_{eff} to exhibit a minimum at r_0, the function J^2, with $J = J(r_0)$ defined by Eq. (13.11), must be monotonically increasing. Typically, for a given value of J, bound orbits are associated with energies in the range $E_0 < E < 0$, with

$$E_0 = \frac{1}{2} r_0^2 \Omega^2(r_0) + \Phi(r_0) \tag{13.12}$$

the minimum energy, which corresponds to the circular orbit. In the radial coordinate, the motion is periodic and takes place between two turning points $r_{\text{in}}(E, J) < r_0 < r_{\text{out}}(E, J)$. A radial action variable thus can be set, that is,

$$J_r = \oint p_r(r, E, J) \, dr, \tag{13.13}$$

with the property

$$dE = \Omega_\theta dJ + \Omega_r dJ_r. \tag{13.14}$$

Here the radial frequency is defined as $\Omega_r = 2\pi / \tau_r$, with the bounce time given by

$$\tau_r = \oint \frac{dr}{p_r(r, E, J)}. \tag{13.15}$$

In turn, the angular frequency is defined by

$$\Omega_\theta = \left\langle \frac{J}{r^2} \right\rangle = \frac{1}{\tau_r} \oint \frac{J}{r^2} \frac{dr}{p_r(r, E, J)}. \tag{13.16}$$

Orbits are closed (in the inertial frame of reference) if the ratio between the two frequencies is rational.

13.2   Epicyclic Orbits

In the limit of quasi-circular orbits, which can be quantified as

$$\left| \frac{E - E_0}{E_0} \right| \ll 1, \tag{13.17}$$

we have $\Omega_r(E,J) \to \kappa(r_0)$, $\Omega_\theta(E,J) \to \Omega(r_0)$, and the radial periodic motion can be approximated by a harmonic oscillator with the epicyclic frequency defined by

$$\kappa^2 = 4\Omega^2 \left(1 + \frac{1}{2}\frac{d\ln\Omega}{d\ln r}\right) = \frac{1}{r^3}\frac{d}{dr}[r^4\Omega^2(r)]. \tag{13.18}$$

{We also have $J_r \sim [E - E_0(r_0)]/\kappa(r_0)$, which is analogous to $v_\perp^2/B$ for a gyrating charged particle in a magnetic field.} Thus, if we separate out the motion of the guiding center by writing $r(t) = r_0 + r_1(t)$ and $\dot{\theta}(t) = \Omega(r_0) + \dot{\theta}_1(t)$, we find that the linearized equation for the conservation of angular momentum leads to

$$r_0\dot{\theta}_1(t) = -2\Omega(r_0)r_1(t); \tag{13.19}$$

thus the epicycles are ellipses characterized by aspect ratio $2\Omega/\kappa$ (therefore, they are usually elongated in the direction of the motion), with the star running in the opposite direction with respect to the guiding center (i.e., the motion in the epicycle is clockwise if the motion on the circular orbit at r_0 is counterclockwise).

Note that from Eq. (13.18) the condition for the stability of circular orbits ($\kappa^2 > 0$) formally coincides with the classical Rayleigh's criterion for the stability of a rotating fluid.[5] A few important special cases should be noted. A pure harmonic potential (i.e., the mean-field potential associated with a homogeneous sphere) implies solid-body rotation, in the sense that $\Omega^2 = 4\pi G\rho/3 = \text{constant}$; in this case, we have $\kappa = 2\Omega$, and orbits are closed in the form of ellipses centered at $r = 0$. A point mass generates a Keplerian potential; from the third law of planetary motion, we see that in this case $\kappa = \Omega$, and thus orbits are closed in the form of ellipses with one focus at $r = 0$. For galaxy disks, because they are often characterized by a flat rotation curve, the typical relation should be $\kappa \approx \sqrt{2}\Omega$, and orbits are generally not closed. Some simple cases of orbits with Ω and κ in a rational ratio are shown in Fig. 13.3.

In Chapter 14 it is shown that the velocity distribution for a relatively cool disk, because of the epicyclic constraints, has an anisotropic pressure tensor for which the radial pressure exceeds the tangential pressure by the ratio $4\Omega^2/\kappa^2$.

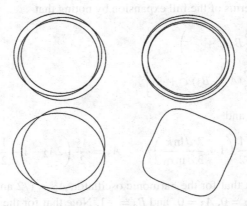

Fig. 13.3. Quasi-circular orbits when the ratio of angular to radial frequency is rational (3/2, *upper left*; 2/3, *lower left*; 4, *upper right*; 1/4, *lower right*). (In a frame rotating with angular velocity Ω_p, the relevant angular frequency is the frequency in the inertial frame reduced by the value of Ω_p.)

For some purposes (e.g., for some detailed stellar dynamical studies of density waves in which an integration along the unperturbed orbits is performed), it is of interest to have a full description of the epicyclic expansion, beyond the lowest-order harmonic oscillator obtained by approximating the potential Φ_{eff} with a parabola in r_0. We obtain such a systematic expansion[6] by introducing an appropriate phase variable λ. To do this, we first consider the transformation $(E, J) \rightarrow (a, r_0)$, where the dimensionless epicyclic energy a is given by

$$a^2 = \frac{2[E - E_0(r_0)]}{r_0^2 \kappa^2(r_0)}. \tag{13.20}$$

Thus the radial momentum can be expressed as a function:

$$p_r(r, a, r_0) = \left\{ a^2 \kappa^2(r_0) r_0^2 + 2[\Phi_{\text{eff}}(r_0, J) - \Phi_{\text{eff}}(r, J)] \right\}^{1/2}. \tag{13.21}$$

Now we introduce the phase variable λ by replacing the radial-velocity coordinate p_r with

$$p_r(r, a, r_0) = r_0 \kappa(r_0) a \sin \lambda. \tag{13.22}$$

The complete epicyclic expansion is thus obtained by Taylor expansion of Eq. (13.21) around $r = r_0$, which, when inserted in Eq. (13.22), gives

$$\frac{r_0}{r} = 1 + \sum_{n=1}^{\infty} A_n(r_0) a^n \cos^n \lambda. \tag{13.23}$$

From here we obtain the expression for $dr/d\lambda$ as well. Then, from

$$\frac{dr}{dt} = a r_0 \kappa(r_0) \sin \lambda = \frac{dr}{d\lambda} \frac{d\lambda}{dt}, \tag{13.24}$$

$$\frac{d\theta}{dt} = \frac{r_0^2 \Omega(r_0)}{r^2} = \frac{d\theta}{d\lambda} \frac{d\lambda}{dt} \tag{13.25}$$

we obtain the desired expressions for $\lambda = \lambda(t)$ and $\theta = \theta(t)$, which completes the derivation. We can summarize the first terms of the full expansion by noting that

$$\Omega_r = \kappa(r_0) \left(1 + \frac{3}{2} B_3 a^2 + \cdots \right), \tag{13.26}$$

$$\Omega_\theta = \Omega(r_0) \left[1 + \frac{3}{2} (A_3 + B_3) a^2 + \cdots \right], \tag{13.27}$$

with $B_3 = -A_3 + 2A_2 - 1$ and

$$A_1 = 1, \qquad A_2 = \frac{1}{2} \left(1 + \frac{2}{3} \frac{d \ln \kappa}{d \ln r_0} \right), \qquad A_3 = \frac{A_2}{2} \left(2A_2 - 1 - \frac{1}{2} \frac{d \ln A_2}{d \ln r_0} \right). \tag{13.28}$$

Here we can easily check that for the harmonic oscillator, $A_2 = 1/2$ and $A_3 = B_3 = 0$, whereas for the Keplerian case, $A_2 = 0$, $A_3 = 0$, and $B_3 = -1$. Note that for the isochrone potential (see Chapter 21), $A_1 = 1$, and all the other A_n vanish.

Many of these results find application in the study of the dynamics of galaxies. They are also of interest in some simple problems of celestial mechanics, in which the potential Φ is often close to Keplerian. For example, the potential of the Earth in space $(r > r_T)$, because of its flattening at the poles, is approximately given by

$$\Phi(r, \vartheta) \sim -\frac{GM_T}{r} \left[1 - J_{20} \frac{r_T^2}{r^2} P_{20}(\cos \vartheta) \right], \tag{13.29}$$

where we have retained only the quadrupole term in the general solution to the Laplace equation (here at $r = r_T$ the quantity $\pi/2 - \vartheta$ represents the geographic latitude); for the Earth, we have $J_{20} \approx 10^{-3}$. We recall the expression for the Legendre polynomial $P_{20}(x) = (3x^2 - 1)/2$. The epicyclic theory easily allows us to study the precession of the perigee of a satellite on the equatorial plane, where $\vartheta = \pi/2$. The precession rate is proportional to the difference between κ and Ω.

13.3 Rotating Frame

It should be emphasized that an orbit can appear open in one frame of reference and closed in another. In fact, suppose that we move to a rotating frame for which the polar coordinates are (r, ϕ), with $\dot{\phi} = \dot{\theta} - \Omega_p$; here Ω_p is the angular velocity of the rotating frame. Then orbits are described by the new Hamiltonian (Jacobi integral)

$$H = \frac{1}{2} \left(p_r^2 + \frac{p_\phi^2}{r^2} \right) + \Phi(r) - p_\phi \Omega_p, \tag{13.30}$$

with $p_\phi = J$, so $H = E - J\Omega_p$. In the rotating frame, the important ratio $2\Omega/\kappa$ becomes $2(\Omega - \Omega_p)/\kappa$, which then may be rational or not depending on our choice of Ω_p. In the dynamics of galaxies, there are sometimes physical reasons that identify a specific value of the angular velocity of the rotating frame. The three important possible conditions of $2(\Omega_p - \Omega)/\kappa = -1, 0$, and $+1$ are often called *conditions of inner Lindblad resonance, corotation*, and *outer Lindblad resonance*, respectively (Fig. 13.4). Thus, in the rotating frame, at the Lindblad resonances, orbits appear closed into ellipses centered at $r = 0$. This feature and the fact that $\Omega - \kappa/2$ can be approximately constant on a wide radial range led Bertil Lindblad to conjecture that a two-armed spiral structure could persist as a kinematic wave in a differentially rotating disk (Fig. 13.5).

The shear flow pattern associated with the differential rotation in an axisymmetric disk, with the flow reversal at the corotation circle, is somewhat reminiscent of certain magnetic surface configurations noted in magnetically confined toroidal plasmas, in which a suitable projection of the magnetic field changes sign on a neutral surface.

13.4 Trapping at the Lagrangian Points

We now consider a weakly nonaxisymmetric potential of a form that is seen as stationary in a suitable rotating frame (rotating at angular velocity Ω_p):

$$\Phi = \Phi_0(r) + \Phi_s(r, \phi). \tag{13.31}$$

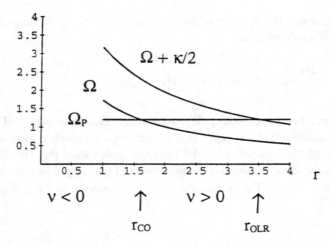

Fig. 13.4. Location of corotation and outer Lindblad resonance for a differentially rotating disk with respect to an assigned frequency Ω_p. The corotation circle divides the disk into two regions, one rotating faster ($\nu < 0$) and the other rotating slower ($\nu > 0$) than Ω_p (after Bertin, G., Lin, C. C. 1996. *Spiral Structure in Galaxies: A Density Wave Theory*, MIT Press, Cambridge, MA, p. 79; ©1996 Massachusetts Institute of Technology).

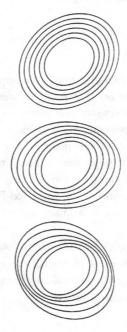

Fig. 13.5. Illustration of Lindblad's kinematic density waves (after Bertin, G., Lin, C. C. 1996. *Spiral Structure in Galaxies: A Density Wave Theory*, MIT Press, Cambridge, MA, p. 72; ©1996 Massachusetts Institute of Technology).

For simplicity (and in view of a number of simple applications), we take

$$\Phi_s(r,\phi) = A(r)\cos[-m\phi + \psi(r)],$$ (13.32)

where $A, m > 0$; m is an integer. If the amplitude $A(r)$ varies slowly and the phase $\psi(r)$ is instead varying rapidly, the potential Φ_s may describe an m-armed spiral pattern with local pitch angle i determined by

$$\tan i = \frac{m}{rk},$$ (13.33)

where we have set $k = d\psi/dr$. With respect to the azimuthal coordinate, at a given radial location, the perturbing potential presents m maxima and m minima.

The perturbing potential Φ_s is taken to be small; that is, we consider the case in which the nonaxisymmetric forces are small:

$$\epsilon_A = \frac{mA}{r^2\Omega_0^2} \ll 1.$$ (13.34)

Here Ω_0 is defined in terms of Eq. (13.10) based on Φ_0. We choose our frame of reference so that $\psi(r_{co}) = 0$, where r_{co} is the corotation radius.

Given the stationarity of Φ_s, the Jacobi integral is conserved, whereas in the inertial frame, the energy E and the angular momentum J vary along the orbits, with $\dot{E} = \Omega_p\dot{J}$.

In general, the orbits are very complicated. However, in the rotating frame, we may easily identify $2m$ points that are stationary, that is, points where a particle released at rest can, in principle, stay in equilibrium. These are a generalization of the Lagrangian points that are known in the context of celestial mechanics. The equations of the motion show that these points occur at the stationary points of Φ_s with respect to the azimuthal coordinate $(m\phi - \psi) = b\pi$, with $b = 0, 1, 2, \ldots, (2m-1)$, so that the tangential force vanishes. The vanishing of the radial force requires that the following relation be satisfied:

$$r(\Omega_p^2 - \Omega_0^2) - (-1)^b \frac{dA}{dr} = 0,$$ (13.35)

where b is even or odd depending on whether Φ_s is a maximum or a minimum (with respect to the azimuthal coordinate). Thus we see that the Lagrangian points occur in the vicinity of the corotation circle.

A curious behavior is found by checking the stability of these equilibrium points. For simplicity, we focus on a case for which $A = $ constant so that the stationary points are at $r = r_{co}$ and $\phi = b\pi/m$. Then, if we make the linearized stability analysis for the motion in the neighborhood of $r = r_{co}$, $\phi = 0$ (which under the preceding assumptions corresponds to a maximum of the perturbing potential along the corotation circle), we find the following relation for the eigenfrequencies ω of the normal modes of oscillations:

$$\hat{\omega}^4 + \left(-\frac{\kappa^2}{\Omega_p^2} + \epsilon_A \frac{|rk|^2}{m} + m\epsilon_A\right)\hat{\omega}^2 + 2m\epsilon_A\left(-\frac{d\ln\Omega_0}{d\ln r}\right) = 0,$$ (13.36)

where $\hat{\omega} = \omega/\Omega_p$, and all the r-dependent quantities are evaluated at the corotation circle. Thus we see that under the normal condition of monotonically decreasing $\Omega(r)$, the point of maximum of the perturbing potential is stable. For small values of ϵ_A, one of the two frequencies derived from Eq. (13.36) corresponds to a modification of the (fast) radial epicyclic oscillation, whereas the new solution, which corresponds to symmetry breaking and modifies the neutral displacements along the corotation circle at $\omega = 0$, is a slow libration frequency given by

$$\omega_L^2 \sim 2m\epsilon_A \left(-\frac{d\ln\Omega_0}{d\ln r} \right) \left(\frac{\Omega_p^2}{\kappa^2} \right) \Omega_p^2. \tag{13.37}$$

Note the scaling $\hat{\omega}_L \sim \epsilon_A^{1/2}$. In contrast, a similar analysis carried out at the perturbing potential minima at corotation would show that they are generally unstable.

The preceding study of the stationary points for a nonaxisymmetric rigidly rotating potential is analogous to the classical restricted three-body problem in celestial mechanics.[7] A two-star or a star-planet (in circular orbit) configuration gives rise to five Lagrangian points. Two points are at $60°$ with respect to the two masses on opposite sides on the plane of the orbit. The other three points are on the axis passing through the two masses. If the mass ratio ϵ is small (such as for the Sun-Jupiter case, for which the mass ratio is $\approx 10^{-3}$), then all five points are close to the corotation circle, that is, the circle of the orbit of the lighter mass. Then two of the three axial Lagrangian points are at a distance $O(\epsilon^{1/3})r_{co}$ from the lighter mass, whereas the third is on the opposite side with respect to the heavier mass. If ϵ is sufficiently small (below a threshold value, i.e., ≈ 0.04), then the two Lagrangian points off axis are stable, as is well known from the trapping of two separate families of asteroids in the solar system. The *libration* period also scales as $\epsilon^{-1/2}$ in this case.

Another application is found in the study of geostationary satellites, many of which are used for telecommunication purposes. In this case, the modification to the Earth's basically Keplerian potential occurs via a quadrupole term that fits in the description of Eq. (13.32) with $\psi = 0$ and $A(r) \sim 1/r^3$ (as can also be checked from the potential theory of the classical ellipsoids; see Chapter 10). This is so because the mass distribution of the Earth is not perfectly axisymmetric. The rotating frame for which such nonaxisymmetric perturbation is stationary is obviously the one that makes one turn in 24 h. The weakness of the nonaxisymmetric field is quantified by the fact that the libration period about the two stable potential maxima is of ≈ 200 days.

The trapping of orbits close to a resonant circle is a rather general process that would be best described by use of action and angle variables.[8] For example, at a Lindblad resonance for an $m = 2$ perturbation, a trapping phenomenon occurs that refers to the major axis of the elliptical orbits that are closed in the axisymmetric case but that are not closed because of the presence of Φ_s. Thus, for an $m = 2$ stationary perturbation, the major axis slowly librates and is trapped around an appropriate direction, whereas it precesses away from a direction at $90°$ from that.

13.4.1 Trapping of a Charged Particle on the Top of a Hill

The preceding discussion should be a reminder that the presence of rotation may bring in some counterintuitive features in the properties of orbits. A simple problem in mechanics, where the presence of a magnetic field replaces the role of rotation, presents some analogies with the preceding case. It should also be stressed that the concept of stability investigated here refers to cases where no dissipation is present. From these examples it could also be appreciated why, in some circumstances, dissipation may have a destabilizing role[9] (see also Chapter 10).

Consider a charged particle (charge q and mass m) constrained to move on a hill described by an axisymmetric surface $z = z(r)$ in the presence of a constant vertical magnetic field $\mathbf{B} = B_0 \mathbf{e}_z$ and constant gravity $\mathbf{g} = -g\mathbf{e}_z$. Then, in polar cylindrical coordinates, the energy is

$$E = \frac{m}{2}[\dot{r}^2 + r^2\dot{\theta}^2 + (z')^2\dot{r}^2] + mgz, \tag{13.38}$$

which, because of the conservation of canonical momentum associated with the axial symmetry,

$$p_\theta = mr^2\dot{\theta} + \frac{q}{2c}B_0 r^2, \tag{13.39}$$

can be written as

$$E = \frac{m}{2}[1 + (z')^2]\dot{r}^2 + \Phi_{eff}, \tag{13.40}$$

with

$$\Phi_{\mathrm{eff}}(r, p_\theta) = \frac{1}{2mr^2}\left(p_\theta - \frac{q}{2c}B_0 r^2\right)^2 + mgz(r). \tag{13.41}$$

Suppose that

$$z(r) = -Cr^\alpha, \tag{13.42}$$

with $1 < \alpha < 2$ and $C > 0$. Then the orbit of the particle stays close to the top of the hill, which behaves as a stable equilibrium point even though it corresponds to the maximum of the potential energy.

13.5 Equations for the Guiding Centers

13.5.1 Larmor Oscillations and Drifts

We briefly recall the basic idea about the guiding-center description of particle orbits[10] (Fig. 13.6). Consider a charged particle (charge q and mass m) moving in a constant magnetic field $\mathbf{B}$ in the presence of a constant force $\mathbf{F}$:

$$m\frac{d\mathbf{v}}{dt} = \frac{q}{c}(\mathbf{v} \times \mathbf{B}) + \mathbf{F}. \tag{13.43}$$

We can separate the equations in the parallel and perpendicular directions (with respect to the magnetic field) to obtain

$$m\frac{d\mathbf{v}_\parallel}{dt} = \mathbf{F}_\parallel, \tag{13.44}$$

$$m\frac{d\mathbf{v}_\perp}{dt} = \frac{q}{c}(\mathbf{v}_\perp \times \mathbf{B}) + \mathbf{F}_\perp; \tag{13.45}$$

then we separate the motion of the guiding center $\mathbf{v}_0$ from the Larmor oscillation by means of

$$\mathbf{v}_\perp = \mathbf{v}_0 + \mathbf{w} \tag{13.46}$$

so that

$$m\frac{d\mathbf{w}}{dt} = \frac{q}{c}(\mathbf{w} \times \mathbf{B}). \tag{13.47}$$

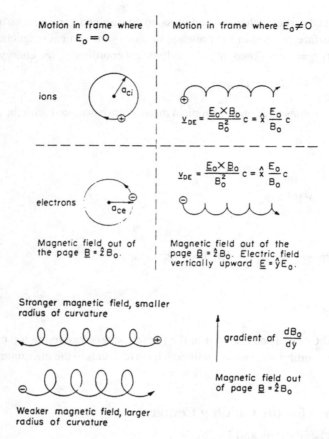

Fig. 13.6. Qualitative representation of the $\mathbf{E} \times \mathbf{B}$ drift (*above*) and magnetic gradient drift (*below*) for charged particles (from Krall, N. A., Trivelpiece, A. W. 1973. *Principles of Plasma Physics*, McGraw-Hill, New York, pp. 612, 625; with kind permission from the authors).

The resulting guiding-center motion is given by the relation

$$\mathbf{v}_0 = \frac{c}{q} \frac{\mathbf{F}_\perp \times \mathbf{B}}{B^2}, \tag{13.48}$$

from which it is evident that a perpendicular force produces a velocity and not an acceleration.

The more general inhomogeneous and/or time-dependent case in which $\mathbf{B}$ and $\mathbf{F}$ are not constant can still be worked out with a similar strategy if the resulting drifts turn out to be slow and the basic gyration frequency is sufficiently high. For example, a gradient in $\mathbf{B}$ can be absorbed in $\mathbf{F}_\perp$ in Eq. (13.45). Some effects, such as those that are due to the curvature of the field, may require some detailed analysis, which is not discussed here.[11] When the curvature is negligible, the parallel motion is basically reduced to the solution of the one-dimensional problem

$$E = \frac{m}{2} v_\parallel^2 + \mu B + U = \text{constant}, \tag{13.49}$$

where U is the potential associated with $\mathbf{F}_\parallel$, and $\mu = m v_\perp^2 / 2B$ is the adiabatic invariant associated with the Larmor gyration. Essentially the basic condition for this type of description is that the variations in the magnetic field encountered by the particle on its orbit must be slow with respect to the relevant cyclotron frequency.

13.5.2 Star Drifts and Stellar Hydrodynamics

As in the case of Larmor gyrations of charged particles in a magnetic field, the epicyclic theory can be extended to the case of weakly nonaxisymmetric and weakly time-dependent fields. This tool allows us to gain a physical perception of the general properties of orbits without resorting to numerical surveys and to move well beyond the simple small oscillations that are considered in the standard stability analysis of the Lagrangian points, as was briefly given in the preceding section. The specific theory in the context of stellar dynamics,[12] formally developed for cool disks in the vicinity of the relevant corotation radius, is based on the generalization of the definitions of angular and epicyclic frequency as

$$\Omega^2 = \frac{1}{r}\frac{\partial \Phi}{\partial r}, \qquad \kappa^2 = 4\Omega^2\left(1 + \frac{1}{2}\frac{\partial \ln \Omega}{\partial \ln r}\right), \tag{13.50}$$

where now Φ is the sum of an axisymmetrical potential and a weakly nonaxisymmetric and weakly time-dependent perturbation. Thus the guiding center orbits $[r_0(t), \phi_0(t)]$ can be derived from an effective Jacobi integral

$$H_0 = E_0(r_0, \phi_0, t) - r_0^2\Omega(r_0, \phi_0, t)\Omega_p, \tag{13.51}$$

where E_0 generalizes the concept of energy associated with the circular orbits [Eq. (13.12)]:

$$E_0 = \frac{1}{2}r_0^2\Omega^2(r_0, \phi_0, t) + \Phi(r_0, \phi_0, t). \tag{13.52}$$

The relevant equations of the motion are

$$\dot{r}_0 = -\frac{2\Omega}{r_0\kappa^2}\frac{\partial H_0}{\partial \phi_0} - \frac{r_0}{\kappa^2}\frac{\partial \Omega^2}{\partial t}, \tag{13.53}$$

$$\dot{\phi}_0 = \frac{2\Omega}{r_0\kappa^2}\frac{\partial H_0}{\partial r_0}. \tag{13.54}$$

The second term on the right-hand side of Eq. (13.53) corresponds to the well-known polarization drift of plasma physics. It should be stressed that in the presence of rotation, the perturbation forces induce drift velocities in the motion of the guiding centers and not accelerations.

We then obtain the complete properties of the orbits by combining the information on the motion of the guiding centers with the fact that the adiabatic invariant μ is essentially constant:

$$\mu = \frac{H - H_0(r_0, \phi_0, t)}{\kappa(r_0, \phi_0, t)}, \tag{13.55}$$

where H is the star Jacobi integral, which depends on the physical coordinates (r, ϕ) and the conjugate momenta.

When the perturbing potential is independent of time, the guiding-center orbits are then simply obtained as contours of the function $H_0(r_0, \phi_0)$. The Lagrangian points of the star orbit analysis described earlier are then recovered as stationary points for H_0; their stability properties are

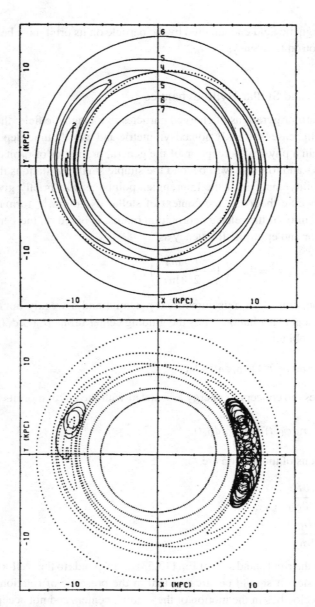

Fig. 13.7. Vorticity islands marked, by the guiding center orbits (in the corotating frame) in an axisymmetrical disk in the presence of a two-armed rigidly rotating spiral field (from Berman, R. H., Mark, J. W.-K. "Stellar dynamics in thin disk galaxies. II: Stellar hydrodynamic equations," 1979. *Astrophys. J.*, **231**, 388; reproduced by permission of the AAS). The structures created are reminiscent of magnetic islands that may result from reconnection in sheared magnetic configurations or cat's eyes that may originate from a shear flow. The lower frame shows full orbits, which can be seen as a result of the superposition of the slow libration of the guiding center and the rapid epicyclic oscillation (see also Contopoulos, G. 1973. *Astrophys. J.*, **181**, 657).

easily reconstructed by inspection of the relevant Hessian. If we take as a perturbation a two-armed spiral potential of the form given in Eq. (13.32), we find that the contours of constant H_0 identify two islands at corotation connected by a separatrix passing through the unstable potential minima (Fig. 13.7). Outside the islands, moving away from corotation either inside or outside the corotation circle, the contours form distorted circles. This new shear-flow configuration should be compared with the unperturbed shear flow. The trapped orbits at the stable Lagrangian points define a kind of cat's eyes.[13] There is a clear analogy with the structure of magnetic islands that originate in plasma configurations by means of magnetic reconnection.[14]

The concepts introduced in this section find many applications in the context of magnetically confined plasmas, in particular, in the description of trapped and circulating particles in toroidal plasma configurations.[15] These sources of analogies are not discussed any further here. In closing, another development, which is conceptually very interesting, is briefly touched on. For a collisionless system, it is possible to construct fluid equations from the moments of the collisionless Boltzmann equation (see Chapter 8). In the absence of collisions, a well-known problem is how to close the fluid equations into a finite set or, in more physical terms, how to define an appropriate equation of state. This question has found a solution in plasma physics in terms of the double adiabatic theory,[16] which makes use of the conservation of the adiabatic invariant to set a constraint equivalent to that of an equation of state. This leads to the justification of magnetohydrodynamic (MHD) – like equations for a collisionless plasma, with the peculiarity that pressure is considered anisotropic. It should be stressed that the closure is obtained under a set of assumptions that make the double adiabatic theory applicable to only a rather limited class of perturbations. Still the procedure is very interesting, especially from the physical point of view. A similar theory has been developed for the context of the stellar dynamics of galaxy disks.[17]

13.6 **Exact Orbits and the Problem of Self-Consistency**

Many interesting aspects of the theory of orbits relevant to the dynamics of spiral galaxies are not treated in this book. In this section at least some key issues are mentioned, and then the subtle step of moving from the one-particle description to collective behavior is discussed.

13.6.1 Three-Dimensional Orbits and Polar-Ring Galaxies

One important study addresses the properties of orbits outside the galactic plane. Here the basic reference case is that of an axisymmetrical potential, with the equatorial plane defined by the galaxy disk. The field is in general significantly nonspherical because of the presence of the disk but also as a result of the flattening of the bulge-halo component. Thus, even for the quasi-planar case, we expect out-of-disk orbits to be subject to a wobble, that is, a precession. In general, a description of fully three-dimensional orbits requires the use of numerical integrations. Some well-known methods of classical mechanics, such as inspection of the relevant surfaces of section, help to point out the integrability properties of the assumed potential.

These investigations have found at least two important applications. In one study, which addresses mainly quasi-planar orbits, the relation between vertical and equatorial frequencies opens the way to a discussion of bending waves (see Chapter 19), much as the case of epicyclic oscillations properly introduces the analysis of density waves in the disk. Here one point should be stressed because it defines a major difference between bending and density waves. The vertical restoring force on a star just off the galactic plane is due to two contributions. One is associated with the inhomogeneity of the disk and with the spheroidal bulge-halo component; even for a spherical field, at a radial location r on the disk, the equatorial force is given by $\Omega^2 r$, which translates, by projection, into a restoring force just off the plane $F_z^{(1)} \approx \Omega^2 z$. However, the local disk self-gravity contributes (even in the absence of such a term; see also Chapter 14) by $F_z^{(2)} \approx 4\pi G\rho z$. The fact that Ω_z in the solar neighborhood exceeds Ω by a factor of approximately $\sqrt{10}$ shows that at least in the middle of the optical disk, the local self-gravity term $F_z^{(2)}$ generally should dominate. However, this strong restoring force is not the one relevant for the problem of bending waves because these waves displace the disk as a whole, so the local self-gravity of a patch of the disk does not act on itself when the patch is displaced from the equatorial plane. The subtlety involved in this point can be rephrased in the following way: With respect to the limit of zero thickness, a bending wave is intrinsically nonlinear. In any case, it may be misleading to assign directly a term Ω_z^2 as the kinematic ingredient to the dispersion relation of bending waves. In contrast, there is no doubt that κ^2 does play such a role for density waves, as the pioneering work of Bertil Lindblad first pointed out. These points will be discussed further in Chapter 19.

In a second application, the observation of kinematic properties of either individual objects (such as globular clusters) or gaseous rings, significantly off the galactic plane, can be used to put a constraint on the underlying potential well of the galaxy.[18] Indeed, there is an interesting class of galaxies with prominent rings (the so-called polar-ring galaxies;[19] Fig. 13.8) for which we have a chance of additional dynamical constraints beyond those provided by the standard rotation curve associated with the motion of the material in the disk. A general working hypothesis in many models of gaseous rings is that they are to be identified with closed orbits (for single particles) associated with the underlying potential. In reality, the dynamics of an observed ring is related to the settling of the gas into such an orbit,[20] which may involve some nontrivial collective processes. Thus the related problem of diagnostics, that is, the problem of reconstructing the underlying field from observed orbits, is generally marked by severe ambiguities. The astrophysical problem has an additional source of uncertainty, which is the fact that we generally have incomplete empirical knowledge of the orbit under consideration. Within this general context, a significant effort is currently under way with the goal to extract information on the structure and evolution of the three-dimensional mass distribution of our Galaxy from a detailed study of orbits of stars, star streams, globular clusters, and satellites belonging to the galactic halo.[21] A particularly interesting problem is raised by the possibility[22] and discovery[23] of hypervelocity stars for which the origin is under current debate. Neutron stars also may be characterized by very high velocities[24] but are generally not included in the category of hypervelocity stars because their origin is likely to be related to completely different mechanisms.

13.6.2   Exact Orbits in Nonaxisymmetrical Disks

Within the context of orbits in the galactic plane, the axisymmetric case is trivial; the case of nonaxisymmetric potentials opens instead a number of interesting possibilities even in the case in which the perturbation is stationary in a suitable rotating frame. (A nonaxisymmetric potential stationary in the inertial frame is probably of secondary interest from the physical point of view.) In general, the orbits that violate the conditions for the guiding-center description outlined earlier

Fig. 13.8. The famous polar-ring galaxy NGC 4650A [credit: The Hubble Heritage Team (AURA/STScI/NASA)]. Optical images of S0 galaxies with polar rings (in particular, of UGC 7576, A 0136-0801, AM 2020-5050, ESO 603-G21, NGC 2685, ESO 199-IG12, ESO 474-G26, and II Zw 73) can be found in the article by Schweizer, F., Whitmore, B. C., Rubin, V. C. 1983. *Astron. J.*, **88**, 909. There are galaxies such as NGC 3998 (see Knapp, G. R., van Driel, W., van Woerden, H. 1985. *Astron. Astrophys.*, **142**, 1) where the polar ring is observed only in HI.

must be calculated numerically. Indeed, large surveys of orbits have been performed both in the presence of barred or ovally distorted potentials and in the presence of spiral fields.[25] The orbits thus have been classified systematically.[26]

13.6.3 Orbital Response and Self-Consistency

The simple interpretation of spiral arms as due to a collective crowding of orbits (as Bertil Lindblad initially suggested) has the major merit of focusing the discussion on the problem of quasi-stationary structures as being due to waves,[27] with remarkable attention being paid to two-armed structures. However, such kinematic waves have the obvious shortcomings that they do not take into account the velocity dispersion of stars and, more important, that, when the spiral arms are set up, the gravitational field is no longer axisymmetric. We will study in Chapter 15 how these difficulties can be overcome. This point naturally introduces the problem of moving from a single-particle description to the relevant collective behavior.

In practice, a quite severe problem arises immediately when we address the issue of nonaxisymmetric fields. The study of orbits, when thought of in terms of a relatively large collection of orbits, is essentially an indication of the response of a specific component (such as a set of stars) to an imposed field. The trapping of stars at the Lagrangian points associated with the potential maxima is rather disconcerting because potential theory tells us that the potential maxima should roughly coincide with the density minima. Then the indication is that the orbits would act opposite to (out of phase with) the assumed spiral field; in other words, there is little chance that these nonaxisymmetric fields would be naturally supported. We can partially understand this puzzling behavior by noting the general property that stars with guiding centers close to the separatrix (i.e., those that are barely trapped) spend a long time away from the trapping point, staying indeed close to the region where their presence would be needed to support the assumed field. This is just a suggestion, which shows that the problem of demonstrating how a field can be supported by the orbits that it imposes is nontrivial. Nevertheless, as is recognized in a full analysis of density waves, the physical conditions for the establishment of density waves involve a number of issues (especially for nonbarred spiral modes, in particular, the dissipation in the gas) that go well beyond the single-particle orbits.

In any case, a systematic program to see whether stationary nonaxisymmetric fields can receive adequate orbital support has been undertaken, and there are a number of interesting results.[28] To some extent, the issue addressed by these investigations is the same as the issue of finding triaxial self-consistent models by direct superposition of orbits, which has led to some notable results in the context of elliptical galaxies.[29]

Notes

1. See Kulsrud, R. M. 1957. *Phys. Rev.*, **106**, 205; Kruskal, M. 1962. *J. Math. Phys.*, **3**, 806; Northrop, T. G. 1963. *The Adiabatic Motion of Charged Particles*, Wiley Interscience, New York.
2. Oort, J. H. 1927. *Bull. Astron. Inst. Neth.*, **3** (120), 275.
3. For example, see Chandrasekhar, S. 1960. *Principles of Stellar Dynamics*, Dover, New York (originally published in 1942).
4. The precise numbers are subject to several sources of uncertainties; see Kerr, F. J., Lynden-Bell, D. 1986. *Mon. Not. Roy. Astron. Soc.*, **221**, 1023.
5. See, for example, Chandrasekhar, S. 1961. *Hydrodynamic and Hydromagnetic Stability*, Oxford University Press, Oxford, UK; reprinted in 1981 by Dover, New York.
6. Shu, F. H. 1969. *Astrophys. J.*, **158**, 505; Mark, J. W.-K. 1976. *Astrophys. J.*, **203**, 81. Mark's analysis removes an undesired secular term present in the original derivation.
7. Contopoulos, G. 1973. *Astrophys. J.*, **181**, 657.
8. Lynden-Bell, D. 1973. In *Dynamical Structure and Evolution of Stellar Systems*, eds. L. Martinet, M. Mayor. Geneva Observatory, Sauverny, Switzerland.
9. M. D. Kruskal, as quoted in Coppi, B. 1966. In *Non-Equilibrium Thermodynamics, Variational Techniques, and Stability*, eds. R. J. Donnelly, R. Herman, I. Prigogine. Chicago University Press, Chicago, p. 259.

10. For example, see Schmidt, G. 1979. *Physics of High Temperature Plasmas*, 2nd ed., Academic, New York.

11. See the book by Schmidt mentioned earlier; see also Krall, N. A., Trivelpiece, A. W. 1973. *Principles of Plasma Physics*, McGraw-Hill, New York.

12. Berman, R. H. 1975. Ph.D. dissertation, Massachusetts Institute of Technology, Cambridge, MA; Berman, R. H., Mark, J. W.-K. 1977. *Astrophys. J.*, **216**, 257; Berman, R. H., Mark, J. W.-K. 1979. *Astrophys. J.*, **231**, 388.

13. Thomson, W. 1880. *Nature (London)*, **23**, 45.

14. See, e.g., White, R. 1983. *Handbook of Plasma Physics*, Vol. 1, eds. A. A. Galeev, R. N. Sudan. North-Holland, Dordrecht, The Netherlands.

15. Bertin, G., Coppi, B., Taroni, A. 1977. *Astrophys. J.*, **218**, 92.

16. Chew, G. F., Goldberger, M. L., Low, F. E. 1956. *Proc. Roy. Soc. London A*, **236**, 112.

17. See Berman, R. H., Mark, J. W.-K. 1977 and 1979. *op. cit.*.

18. See, e.g., Sackett, P. D., Rix, H.-W., et al. 1994. *Astrophys. J.*, **436**, 629.

19. See Moiseev, A. V., Smirnova, K. I., et al. 2011. *Mon. Not. Roy. Astron. Soc.*, **418**, 244, and references therein.

20. See, e.g., Christodoulou, D. M., Katz, et al. 1992. *Astrophys. J.*, **395**, 113.

21. For example, see Lynden-Bell, D., Lynden-Bell, R. M. 1995. *Mon. Not. Roy. Astron. Soc.*, **275**, 429; Helmi, A. 2004. *Mon. Not. Roy. Astron. Soc.*, **351**, 643; Belokurov, V., Zucker, D. B., et al. 2006. *Astrophys. J.*, **642**, L137; Koposov, S. E., Rix, H.-W., Hogg, D. W. 2010. *Astrophys. J.*, **712**, 260.

22. Hills, J. G. 1988. *Nature (London)*, **331**, 687.

23. Brown, W. R., Geller, M. J., et al. 2005. *Astrophys. J. Lett.*, **622**, L33.

24. Hui, C. Y., Becker, W. 2006. *Astron. Astrophys.*, **457**, L33; Winkler, P. F., Petre, R. 2007. *Astrophys. J.*, **670**, 635.

25. See, e.g., Contopoulos, G. 1973. *Astrophys. J.*, **181**, 657; Contopoulos, G., Papayannopoulos, T. 1980. *Astron. Astrophys.*, **92**, 33.

26. See Athanassoula, E., Bienaymé, O., et al. 1983. *Astron. Astrophys.*, **127**, 349.

27. A more detailed description of this important point can be found in Bertin, G., Lin, C. C. 1996. *Spiral Structure in Galaxies: A Density Wave Theory*, MIT Press, Cambridge, MA.

28. See Patsis, P. A., Hiotelis, N., Contopoulos, G., Grosbøl, P. 1994. *Astron. Astrophys.*, **286**, 46, and references therein; see also Lynden-Bell, D. 1979. *Mon. Not. Roy. Astron. Soc.*, **187**, 101.

29. Schwarzschild, M. 1979. *Astrophys. J.*, **232**, 236; Schwarzschild, M. 1982. *Astrophys. J.*, **263**, 599.

14 The Basic State: Vertical and Horizontal Equilibrium of the Disk

As described in Chapter 9, the concept of the basic state is a key step in the modeling process. As in other contexts, for galaxy disks, identification of an appropriate basic state is done for several purposes. One interesting aspect is that by studying the internal structure of the basic state, we can clarify the overall constraints imposed by self-consistency; in this respect, the two examples that follow, dealing with the vertical and horizontal equilibrium of a rotation-supported disk, are rather simple (in collisionless stellar dynamics, self-consistency imposes much stronger and less intuitive constraints in pressure-supported systems) but very instructive. The models that are set up can then be used to fit the data and thus derive information on the structure of galaxies. Furthermore, a broad choice of equilibrium models can form the basis for systematic stability analyses. In this case, the basic state is meant to be a tool to study evolution, but it should not be assumed to represent necessarily the actual conditions of the galaxy in the distant past.

From the point of view of stellar dynamics, the case of axisymmetric galaxy disks is relatively simple, because for realistic models most orbits are quasi-circular and the disk is rather thin. The related description can then be decoupled at two different levels: (1) the vertical dynamics can be treated separately from the horizontal dynamics; and (2) for the horizontal structure, because the angular momentum essentially identifies the radial coordinate, the problem of self-consistency can be worked out in two separate steps: (2.1) choose a density-potential pair, and (2.2) find a (quasi-Maxwellian) distribution function that supports the assumed density. In contrast, this procedure of going from ρ to f would generally lead to nonphysical models in the spherical case.

The disk is generally embedded in a bulge-halo component that tends to act dynamically as a fixed external field. The cool, thin disk can easily carry waves and instabilities to which the hotter bulge-halo hardly responds. Still, there are interesting effects related to the interaction between the bulge-halo and the disk that should be taken into account, some of which will be described in Chapter 19. However, even as a frozen external force field, the presence of the bulge-halo changes the stability characteristics of the disk considerably and thus should be properly included in the galaxy models. Therefore, for realistic disk models, the density-potential pairs are not to be regarded as fully self-consistent; that is, the gravitational field is partly generated by nondisk material.

As described in Part I, galaxy disks are made of several star components (stellar populations with different kinematic characteristics) and several dissipative components (clouds of cold atomic hydrogen and molecular gas, with a wide range of sizes and distributions, and warmer, diffuse gas). Typically, the outer disk ($r \geq 2h_\star$) comprises a region for which gas is dynamically important. Furthermore, the central part of the disk generally behaves more like a spheroidal system than a flat layer simply because its finite thickness becomes appreciable in relation to the radial distance. Actually, for galaxies that possess significant bulges, such as M81, the observations are often unable to distinguish whether the disk component continues all the way in to the center or is replaced there with the bulge altogether. This situation is physically very interesting and complex and poses difficult challenges in the modeling process (see also Chapter 8). The main consequence of these empirical remarks is that some elegant models that are constructed and can be studied in great detail (e.g., a stellar dynamical model of a self-gravitating disk with flat rotation curve and analytical distribution function) generally lack some important physical ingredients.

14.1 Vertical Equilibrium

14.1.1 The Isothermal Self-Gravitating Slab

Consider an ideal disk model[1] that is homogeneous in the horizontal directions (x and y coordinates) and inhomogeneous in the vertical direction z. Imagine that the density distribution ρ is even with respect to the equatorial ($z = 0$) plane so that $\rho(z) = \rho(-z)$ with a maximum $\rho_0 = \rho(0)$ and a monotonically decreasing profile away from the equatorial plane. We wish to find a self-consistent equilibrium solution, that is, a pair $[\rho(z), \Phi(z)]$, under the assumption of isothermality. The equilibrium that we have in mind is kinetic (i.e., a solution of the Vlasov-Poisson system), but it is sufficient to start with the hydrostatic condition

$$\frac{1}{\rho}\frac{dp}{dz} + \frac{d\Phi}{dz} = 0 \tag{14.1}$$

supplemented by the self-consistency relation

$$\frac{d^2\Phi}{dz^2} = 4\pi G\rho \tag{14.2}$$

and by the equation of state

$$p = c^2\rho, \tag{14.3}$$

where c^2 is a constant. Using dimensionless quantities based on the natural scale height defined by $z_0^2 = c^2/2\pi G\rho_0$, we can combine the preceding relations to give

$$\frac{d^2\psi}{d\zeta^2} = -2\hat{\rho} = -2\exp(\psi), \tag{14.4}$$

where $\hat{\rho} = \rho/\rho_0$, $\psi = -\Phi/c^2$, and $\zeta = z/z_0$.

Note that we have taken $\psi(0) = 0$, consistent with the condition $\hat{\rho}(0) = 1$. This nonlinear equation for ψ is easily integrated once to give

$$\frac{1}{2}\left(\frac{d\psi}{d\zeta}\right)^2 = 2[1 - \exp(\psi)], \tag{14.5}$$

where the energy integration constant has been set to 2 to satisfy the zero-field condition on the equatorial plane $(d\psi/d\zeta)_0 = 0$ required by symmetry. This first-order equation then leads to the well-known $\hat{\rho} = (1/\cosh\zeta)^2$ (see also Subsection 9.1.1). Explicitly, returning to the dimensional variables, we have

$$\rho = \rho_0 \frac{4}{[\exp(-z/z_0) + \exp(+z/z_0)]^2} = \rho_0 \exp(-\Phi/c^2). \tag{14.6}$$

With this choice of gravitational potential, the self-consistent distribution function for the isothermal slab is

$$f = f_0 \exp(-E/c^2), \tag{14.7}$$

where $f_0 = \rho_0/(\sqrt{2\pi}\,c)$ and $E = v_z^2/2 + \Phi(z)$.

Note that the constant projected disk density σ is related to the scales of the slab as

$$\sigma = \int_{-\infty}^{\infty} \rho\, dz = 2\rho_0 z_0 \int_0^{\infty} \hat{\rho}\, d\zeta = -\rho_0 z_0 \left[\frac{d\psi}{d\zeta}\right]_0^{\infty} = 2\rho_0 z_0. \tag{14.8}$$

In the application of Eqs. (14.4) and (14.5) to evaluate the integral, we recognize the Gauss theorem. Note also the two interesting limits

$$\Phi \sim \frac{2c^2}{z_0}|z| = 2\pi G\sigma|z|, \qquad z \gg z_0 \tag{14.9}$$

(consistent with the notion of a constant field outside a plane capacitor), and

$$\Phi \sim \frac{1}{2}\left(\frac{d^2\Phi}{dz^2}\right)_0 z^2 = \frac{1}{2}\omega_b^2 z^2, \qquad z \ll z_0, \tag{14.10}$$

where $\omega_b = \sqrt{4\pi G\rho_0}$ is the natural bounce frequency provided by gravity.

Note that in a two-component slab model (of stars and gas) in which each component is vertically isothermal, the eigensolution for the potential Φ is no longer given by Eq. (14.6). Nonetheless, an interesting global constraint is provided by the Gauss theorem:

$$(1+\alpha)^2 = \frac{c_\star^2}{\pi G\sigma_\star z_\star} + \frac{c_g^2}{\pi G\sigma_g z_g}\alpha^2, \tag{14.11}$$

where the subscripts refer to the gas and star components, and $\alpha = \sigma_g/\sigma_\star$ is the relative density ratio, following the notation that will be adopted in Chapter 16. Here the thicknesses are defined in such a way that the projected column densities are related to the peak (equatorial) densities by $\sigma_\star = 2\rho_\star(0)z_\star$ and $\sigma_g = 2\rho_g(0)z_g$. This relation thus generalizes Eq. (14.8), which can be seen as the limit where $\alpha = 0$. Another trivial limit occurs when the two components are identical to each other, with $\sigma_g = \sigma_\star = \sigma/2$ and $\alpha = 1$.

14.1.2 The Inhomogeneous Disk and Study of the Solar Neighborhood

The discussion of the simple isothermal self-gravitating slab suggests that by measuring the length scale of the vertical density distribution and the vertical velocity dispersion in the solar neighborhood, we should be able to obtain a dynamical determination[2] of the local mass density $\rho_\odot$. Of course, our disk is inhomogeneous in the equatorial plane and is made of many components, so the treatment performed earlier is not directly applicable. However, the basic principle involved can be extended to more realistic models.

The first difficulty, the inhomogeneity of the disk, can be looked at in terms of successive approximations. Basically, for a thin disk, the gradients in the vertical direction are stronger than those in the radial direction by the geometric factor r/z_0. In this sense, the homogeneous slab is just the lowest-order description of a two-scale analysis, which can be carried through to higher levels of approximations if so desired. For example, for an inhomogeneous axisymmetric basic state, the Poisson equation is

$$\frac{\partial^2 \Phi}{\partial z^2} + \frac{1}{r}\frac{\partial}{\partial r}\left(r\frac{\partial \Phi}{\partial r}\right) = 4\pi G\rho, \qquad (14.12)$$

which includes a second-order correction to Eq. (14.2) in the thickness geometric expansion. Curiously, the correction vanishes at $z = 0$ for a flat rotation curve. We can best express the thickness parameter in terms of other frequently used parameters of the *zero-thickness* fluid model by noting that formally $z_0/r = \epsilon_0 Q^2$, where $\epsilon_0 = \pi G\sigma/r\kappa^2$ is the self-gravity parameter and $Q = c\kappa/\pi G\sigma$.

In the remaining part of this subsection, we find it instructive to describe briefly the arguments of a long-term controversy about the possible existence of a discrepancy between dynamically measured and observed density of the disk in the solar neighborhood. The controversy went on during a period of about seventy years and was finally resolved after the acquisition of accurate astrometric data by *Hipparcos* (see Chapter 2). Such a discrepancy would have important implications as to the distribution of dark matter in our Galaxy (see also Chapter 20).

One analysis[3] uses Eq. (14.1), with the first term determined by the observations of a well-defined (and, it is hoped, dynamically well-mixed, i.e., steady-state) class of stars, such as K giants, to determine the vertical-force term and especially its gradient in the vertical direction. Thus we have an estimate of the left-hand side of Eq. (14.12), which leads to a determination of the total volume mass density on the equatorial plane. The class of stars used for the vertical-force determination is best chosen so that their distance is accurately measured and their vertical distribution is sufficiently spread out to give a good sampling in z. In this latter respect, K giants have a velocity dispersion (≈ 20 km s^{-1}) that is approximately twice that of A0 stars, thus extending approximately four times more in terms of vertical height. The vertical range that is best sampled turns out to be between 300 and 1,500 pc. Oort noted a serious discrepancy between observed material and dynamically estimated mass and considered three options: (1) that the unseen material, possibly in the form of molecular clouds, is distributed in a thin layer as the cold gas, (2) that the unseen material is distributed as the disk stars (i.e., as the K giants he was using), and (3) that the unseen matter is like Population II stars (with high vertical velocity dispersion, of ≈ 50 km s^{-1}). Based on a number of considerations, his best solution is the one that follows option (2), with total mass density $\approx 10^{-23}$ g cm$^{-3} \approx 0.15\ M_\odot$ pc^{-3} and local mass-to-light ratio $M/L \approx 2$, for which ≈ 60 percent is in the form of observed gas (20 percent) and visible stars (40 percent), whereas ≈ 40 percent is dark matter. He also noted that the number for the total density at $z = 0$ thus determined is the same when A0 stars are used instead of K giants.

The project has been revived more recently by Bahcall and colleagues.[4] A new feature in the discussion is that of allowing for the presence of a halo density ρ_h, which is taken to be constant in z. The local density ratio $\epsilon_\rho = \rho_h/\rho_{\text{disk}}(0)$ in the solar neighborhood is estimated to be in the range 0.03 to 0.3. In this simple model, the two-component self-consistent problem of an isothermal slab in the halo background can be solved. For small values of ϵ_ρ, the one-component isothermal solution (14.6) is modified as

$$\rho_{\text{disk}} \approx \rho_0 \frac{4}{[\exp(-z/z_0) + \exp(+z/z_0)]^2} \exp[-\epsilon_\rho (z/z_0)^2],$$ (14.13)

with

$$\sigma_{\text{disk}} \approx 2\rho_0 z_0 (1 - \epsilon_\rho \pi^2/12).$$ (14.14)

The analysis can be extended to include several disk components with different densities and scale heights, as well as cases in which there are significant departures from vertical isothermality. Thus a global self-consistent fit, with an objective statistical analysis, can be performed simultaneously for all the observed components (based on detailed spatial and kinematic data for one specific tracer). This has basically led to a confirmation of the Oort mass discrepancy: It has confirmed the need for the presence of significant amounts of unseen material in the disk.[5] The method appears to constrain the total volume density at $z=0$ ($0.20 \pm 0.04\, M_\odot\, \text{pc}^{-3}$) better; wider ranges of options are allowed for the column disk density.

In a different approach,[6] attention is given to a sample of stars (512 K dwarfs) very far out (at ≈ 3 kpc from the equatorial plane) with a method that appears to place a strong constraint on the total column disk density within 1.1 kpc from the galactic plane. This method intentionally avoids the assumption of isothermality of the tracer distribution function by applying an Abel inversion (from ρ to f) technique that uses the relation

$$f_t(E_t) = -\frac{1}{\pi} \int_{E_t}^{\infty} \frac{d\rho_t}{d\Phi} \frac{d\Phi}{\sqrt{2(\Phi - E_t)}},$$ (14.15)

where $E_t = \Phi + v_z^2/2$ is the energy of the tracer star. The distribution function f_t depends on the volume density ρ_t of the tracer (by means of the Abel inversion) only at points where the potential is high. Thus it is possible to derive the potential at large distances from the plane from high-z data alone, without having to model the potential close to the equatorial plane. The integral requires good knowledge of the various profiles in their tails. However, instead of leaving the possibility of a potential freely determined from the self-consistency requirement, this approach parameterizes it as

$$\Phi = A_d \left(\sqrt{z^2 + z_d^2} - z_d \right) + A_h z^2,$$ (14.16)

or, even more simply, as

$$\Phi = B_d |z| + B_h z^2,$$ (14.17)

where the constants A_d and A_h (or B_d and B_h) determine the relative weight of disk and halo contributions to the field, and the scale height z_d sets the thickness of the total disk mass [for comparison, see also the two limits of the isothermal slab model, relations (14.9) and (14.10)]. The numbers that come out of this different study give $71 \pm 6\, M_\odot\, \text{pc}^{-2}$ for the disk plus halo column density within 1.1 kpc from the galactic plane, and the disk mass that is found is $48 \pm 9\, M_\odot\, \text{pc}^{-2}$, which thus appears to be compatible with the observed disk material (characterized by $48 \pm 8\, M_\odot\, \text{pc}^{-2}$). Thus apparently the local disk mass discrepancy is avoidable, so the dark matter in our Galaxy is not in the disk, but in a diffuse halo.

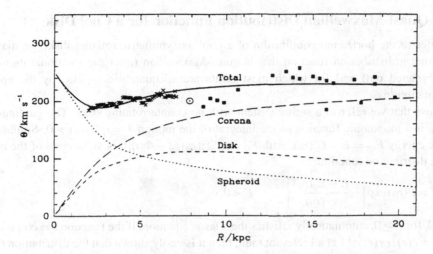

Fig. 14.1. A disk-bulge-halo decomposition for the rotation curve (HI data) of the Milky Way (from Merrifield, M. R. "The rotation curve of the Milky Way to 2.5 R_0 from the thickness of the HI layer," 1992. *Astron. J.*, **103**, 1552; reproduced by permission of the AAS). (The bulge contribution to the "Total" rotation curve is denoted by "Spheroid" and the halo contribution by "Corona.")

A similar conclusion, in favor of a light disk, also was reached by other studies that tried to reassess the dynamical determination of the density in the solar neighborhood, either by combining local and global constraints on detailed parametric models[7] or by reconsidering critically the statistical analysis of star counts for F dwarfs and K giants as density tracers made earlier by Oort and Bahcall.[8]

A proper resolution of the controversy would have important implications for the overall picture of the structure of our Galaxy. In fact, some of the modeling procedures for the solar neighborhood already use as input parameters quantities suggested by global models of the Galaxy,[9] for which admittedly there exist important sources of uncertainty, such as our incomplete knowledge of the bulge parameters and the value of the horizontal scale of the disk. Clearly, the disk suggested by the studies mentioned earlier (even in the interpretation of Bahcall and colleagues) is definitely light, and the contribution of the dark halo to the rotation curve at the location of the Sun is very important.[10] In this sense, the disk of our Galaxy (Fig. 14.1) does not appear to conform to the maximum-disk hypothesis (see Chapter 20).

Fortunately, resolution of the controversy was made possible as soon as new data from *Hipparcos* became available. In particular, accurate astrometric measurements (distances and proper motions) for $\approx 2,500$ bright A-F dwarf stars within 125 pc of the Sun showed that indeed the lighter disk interpretation is the correct one.[11] This is one very important point in favor of the picture that the dark halos required by the study of rotation curves are likely to be round and not disk-shaped. In turn, this is one important point in favor of the picture of collisionless dark matter (see discussion in Chapter 20). In addition, the case for a light disk has important implications for the problem of spiral structure (see Chapter 18). Eventually, the problem of the mass density of the solar neighborhood turned out to be a very instructive example of a controversy that has led to the development of improved modeling procedures and was resolved by the acquisition of better data.

14.2   Quasi-Maxwellian Distribution Function for a Cool Disk

If we refer to the horizontal equilibrium of a cool axisymmetric rotating disk, we may wish to construct a distribution function that is quasi-Maxwellian (after the rotational motion has been subtracted out) and for which most orbits are adequately described by the epicyclic approximation.[12]

Suppose that we refer to a stellar system with no counterrotating stars. The guiding-center radius r_0 is a monotonic function of the integral of the motion $J = r_0^2 \Omega(r_0) > 0$. Similarly, the epicyclic energy $E_{ep} = E - E_0(r_0)$, with $E_0 = r_0^2 \Omega^2(r_0)/2 + \Phi(r_0)$, is an integral of the motion. Thus the distribution function

$$f = P(r_0)\exp\left[-\frac{E - E_0(r_0)}{c^2(r_0)}\right], \qquad E < 0, \tag{14.18}$$

and $f = 0$, for $E \geq 0$, automatically satisfies the Vlasov equation. If the function $c = c(r)$ is chosen so that $\epsilon = c(r)/r\kappa(r) \ll 1$ at all relevant radii, then it is easily shown that the distribution (14.18) is an anisotropic quasi-Maxwellian applicable to cool stellar disks of essentially any desired density and temperature profiles. In fact, if we subtract the circular velocity $r\Omega(r)$, making a change of variables from (v_r, v_θ) to (w_r, w_θ), with $v_r = w_r$ and $v_\theta = w_\theta + r\Omega(r)$, and we recall that the lowest-order epicyclic expansion requires

$$w_\theta = \frac{\kappa^2(r)}{2\Omega(r)}(r_0 - r), \tag{14.19}$$

then we find the approximate expression for the distribution function associated with Eq. (14.18):

$$f = P(r)\exp\left[-\frac{w_r^2 + (\frac{2\Omega}{\kappa})^2 w_\theta^2}{2c^2(r)}\right][1 + O(\epsilon)]. \tag{14.20}$$

Note that given the fact that the epicyclic excursion is small, the dependence on r_0 determines the lowest-order dependence on the radial coordinate r. The function c is subject to only the epicyclic constraint (which may become a nontrivial constraint only if we insist on finding a solution down to $r \to 0$), whereas the normalization function P can be taken to be

$$P(r) = \left(\frac{2\Omega}{\kappa}\right)\frac{\sigma}{2\pi c^2} \tag{14.21}$$

to generate the desired density distribution $\sigma(r)$. This solution allows us to choose a fully self-consistent pair (σ, Φ) or, as is often the case in the description of disk galaxies, a pair for which the potential Φ is partly supported by the disk and partly by a different component (bulge-halo).

One important result of this derivation is that for a cool stellar disk, the pressure tensor is naturally anisotropic, with

$$\frac{c_r^2}{c_\theta^2} = \frac{4\Omega^2}{\kappa^2} \geq 1, \tag{14.22}$$

that is, the velocity ellipsoid is at $90°$ with respect to the orbit epicycle. The vertex of stellar motions should point exactly toward the galactic center.[13]

Another interesting feature of the quasi-Maxwellian solution constructed earlier is that it is associated with a drift velocity v_D, which may become important for relatively warm disks. In

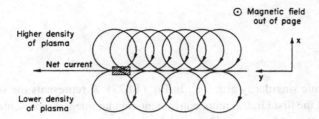

Fig. 14.2. Illustration of the diamagnetic current induced by pressure gradients in a plasma (from Krall, N. A., Trivelpiece, A. W. 1973. *Principles of Plasma Physics*, McGraw-Hill, New York, p. 99; with kind permission from the authors).

the context of stellar dynamics, this is usually called the *asymmetric drift*.[14] We might calculate the value of the drift directly from the definition

$$u_\theta = \langle v_\theta \rangle = r\Omega(r) + v_D, \tag{14.23}$$

with the average based on distribution (14.18). In practice, the drift velocity is more readily evaluated from the relevant momentum-balance condition (8.11) derived in Chapter 8, which implies that

$$v_D = \frac{c^2}{2\Omega r} \frac{d\ln}{d\ln r} \left(\frac{\sigma c^2}{\Omega^{1/2}} \right) = O(\epsilon^2), \tag{14.24}$$

where $c = c_r$. In general, at location r, the stellar disk rotates slower than a single star in circular orbit because the gravitational field is partly balanced by the pressure gradient (complicated by the existence of anisotropic pressure).

The analogous drift in the case of magnetized plasmas is called *diamagnetic velocity*; if a plasma species (charge q) in a magnetic field **B** is treated as a fluid (of particle density n and pressure p), the relevant diamagnetic velocity is given by (here c is the speed of light)

$$\mathbf{v_D} = -c\frac{\nabla p \times \mathbf{B}}{qnB^2}, \tag{14.25}$$

which can be visualized by means of a simple geometric model[15] (Fig. 14.2).

14.3 **Exact Models**

14.3.1 Density-Potential Pairs

The potential $\Phi_D(r)$ generated by an axisymmetric disk of zero thickness and mass density σ on the equatorial plane $z = 0$ is given by

$$\Phi_D(r) = -2\pi G \int_0^\infty K^{(0)}(r, r')\sigma(r')\,dr', \tag{14.26}$$

where the well-known kernel

$$K^{(0)}(r, r') = \frac{r'}{r+r'} \frac{2}{\pi} \int_0^{\pi/2} \frac{1}{\sqrt{1 - \zeta \cos^2 x}} dx = \frac{1}{\pi}\sqrt{\frac{r'\zeta}{r}} K(\zeta), \tag{14.27}$$

with

$$\zeta = \frac{4rr'}{(r+r')^2},\tag{14.28}$$

exhibits a logarithmic singularity at $r = r'$. In Eq. (14.27), K represents the standard complete elliptical integral of the first kind. A more convenient, regular integral representation can be given in terms of Fourier-Bessel transforms[16]

$$\Phi_D(r) = -2\pi G \int_0^\infty dk J_0(kr) \int_0^\infty dr' r' \sigma(r') J_0(kr'),\tag{14.29}$$

where J_0 indicates the standard Bessel function. Then the associated force field is given by

$$\Omega_D^2(r)r = \frac{d\Phi_D}{dr} = 2\pi G \int_0^\infty dk k J_1(kr) \int_0^\infty dr' r' \sigma(r') J_0(kr').\tag{14.30}$$

From these expressions we can derive the properties of some disk-density distributions and discern some interesting (σ, Φ_D) pairs. In particular, an exponential disk[17]

$$\sigma = \sigma_0 \exp(-r/h_\star)\tag{14.31}$$

generates a rotation curve

$$V_D^2 = \Omega_D^2(r)r^2 = (\pi G h_\star \sigma_0)[I_0(R/2)K_0(R/2) - I_1(R/2)K_1(R/2)]R^2,\tag{14.32}$$

where $R \equiv r/h_\star$, and I and K denote standard modified Bessel functions; the peak of the rotation curve $[(V_D^2)_{max} \approx 0.78 \, \pi G \sigma_0 h_\star]$ is reached at $r \approx 2.2 h_\star$. These expressions find frequent application given the general exponential character of stellar disks (see Chapter 4).

There are other elegant density-potential pairs for disks that are significantly far from the empirically suggested exponential profile. Among others is the Maclaurin disk, that is, the zero-thickness limit of the classical spheroids[18] (see Chapter 10), with

$$\sigma = \sigma_0 \left[1 - \left(\frac{r}{R}\right)^2\right]^{1/2}, \qquad r < R,\tag{14.33}$$

$$\Omega_D^2 = \frac{\pi^2 G \sigma_0}{2R} = \frac{3\pi GM}{4R^3}, \qquad r < R.\tag{14.34}$$

Here R is the finite radius of the disk and M is its total mass; for $r > R$, $\sigma = 0$, and the angular velocity Ω_D is no longer constant. Then the self-similar density distribution[19] (which is singular at the origin and generates a disk of infinite mass)

$$\sigma = \sigma_0 \frac{r_s}{r},\tag{14.35}$$

where σ_0 and r_s are arbitrary constants, leads to a perfectly flat rotation curve with

$$V_D^2 = 2\pi G \sigma_0 r_s = V_0^2.\tag{14.36}$$

A variation on the self-similar model, which can cure its singularity at the origin,[20] is given by the pair

$$\sigma = \sigma_0 \frac{r_s}{\sqrt{r^2 + r_s^2}}, \tag{14.37}$$

$$V_D^2 = 2\pi G \sigma_0 r_s \left(1 - \frac{r_s}{\sqrt{r^2 + r_s^2}}\right). \tag{14.38}$$

14.3.2   Distribution Function for the Isothermal Self-Similar Disk

Some density-potential pairs have been extended into a fully self-consistent collisionless model. One interesting case is the self-similar stellar dynamical model considered by Zang.[21] The relevant distribution function is taken to be

$$f_S = CJ^a \exp\left(-E/c^2\right) \tag{14.39}$$

for positive values of the angular momentum J and $f_S = 0$ for $J < 0$, where c and C are constants and

$$a = \frac{V_0^2}{c^2} - 1. \tag{14.40}$$

If $E = v^2/2 + V_0^2 \ln(r/r_s)$, integration over velocity space shows that the associated disk density is the self-similar one, with

$$\sigma = \sqrt{\pi} C\Gamma\left[\frac{(a+1)}{2}\right](2c)^{(a+2)} r_s^a \frac{r_s}{r}. \tag{14.41}$$

Here Γ is the gamma function. Thus, for a fully self-consistent solution, we obtain the value of C by combining this relation with Eqs. (14.35) and (14.36). Note that this solution is isothermal, with the radial velocity dispersion given by $c_r = c$, whereas in the azimuthal direction,

$$u_\theta = \sqrt{2}c \frac{\Gamma[(a+2)/2]}{\Gamma[(a+1)/2]}, \tag{14.42}$$

$$c_\theta = \left\{(a+1) - 2\frac{\Gamma^2[(a+2)/2]}{\Gamma^2[(a+1)/2]}\right\}^{1/2} c. \tag{14.43}$$

The cold-disk limit ($a \to 0$) for c_r/c_θ is consistent with the epicyclic condition [Eq. (14.22)]. A variation on this self-similar model is the cut-out model of index N defined by

$$f_S^{(\mathrm{cut})} = f_S \frac{J^N}{J^N + (r_s V_0)^N}. \tag{14.44}$$

In the cold limit, the reduction factor applied to f_S, being dependent on only J, has a simple interpretation as a reduction factor in the density profile [see the comment following Eq. (14.20)]. It should be stressed that although part of the mass is thus carved out from the central region, which may be taken to mimic the effect of the presence of a bulge, the underlying potential is assumed to remain exactly logarithmic, and the rotation curve remains perfectly flat. Thus part of the potential is considered to be provided by an external immobile component not associated with $f_S^{(\mathrm{cut})}$.

14.4   A Reference Basic State in View of the Problem of Spiral Structure

So far in this chapter it has been shown that we can describe in great detail axisymmetric self-consistent models of galaxy disks, and the basic ingredients relevant for the construction of such models in relation to their vertical and horizontal dynamics have been discussed. It is clear that, in principle, by means of these tools we can produce models consistent with the basic observational constraints.

However, for cases such as that of our own Galaxy, for which these constraints refer to several components, the models are bound to be very complicated, with many separate collisionless and fluid components needed to describe just the disk material. In addition, a proper model for the bulge and the dark halo also should be incorporated, and this has yet to be addressed. The set of equations required for describing such a full model would be incredibly large, but as we saw in the case of the solar neighborhood, in which the data at our disposal are plentiful and accurate, severe ambiguities would remain as to identification of the relevant model for the system under investigation. Thus it is a fact that the goal of constructing *the* basic model for galaxy disks, as a prerequisite for studying the dynamics and the evolution of spiral galaxies, is simply hopeless.

In the face of such enormous difficulties, one attitude could be that of limiting dynamical studies, of the type that will be covered in the following chapters, to the investigation of simple toy models, for which the analysis can be carried through with exact results on the evolutionary process (see Chapter 8). This attitude has the serious shortcoming of widening the gap between theory and the observed systems. In fact, formulations that are easily amenable to exact analytical solutions often lack key physical ingredients that are known to be present in the systems we would like to study. A more constructive approach in this very difficult modeling process is to adopt an iterative procedure in which dynamical and empirical discoveries help each other to identify better and better dynamical models. The models that are thus set up are realistic, but they should be continually revised while progress is made in either observational or theoretical terms.

In obvious contrast with the expectations of a deductive approach, the important features that a satisfactory dynamical model must include depend on the dynamical issues and questions for which the model is set up. For example, for study of the stability of galaxy disks, the three-dimensional character of the distribution of matter, even if the disk is thin, plays a crucial role. This will be discussed in Chapter 15. In addition, the dissipative character of the cold-gas component plays a major role, especially for tightly wound waves, even if the total amount of gas in a spiral galaxy is typically a small percentage of the total mass. This will be demonstrated in Chapter 16. In contrast, for the purpose of producing disk-bulge-halo decompositions in the study of rotation curves of external galaxies, some of these ingredients have a lesser impact. It should be reiterated that these conclusions, although not a surprise to many who have been working in the field, have arisen in clear terms only recently, mostly on the basis of observational and theoretical progress of the 1980s, and were as such not at all anticipated in the early studies of density waves in the 1960s.

Thus this chapter proceeds with a description of the basic properties of a family of simple one-component, zero-thickness fluid models of the disk that has been the basis of an extensive linear modal stability analysis (see Chapter 17). A thorough discussion of the physical justification of these models is given elsewhere.[22] These models and the following modal survey have their natural application in the study of global spiral structures in galaxies (see Chapter 18).

An axisymmetric one-component model of a galaxy disk is specified by three functions: the rotation curve $V(r)$, the *active* density profile $\sigma(r)$, and the equivalent acoustic speed $c(r)$. The rotation curve gives the radial force field in the equatorial plane of the disk by means of the definition $V^2(r) = r d\Phi/dr$ and is the quantity that is best constrained by the data (see Chapter 4). The profiles for $V(r)$ chosen later are very simple and do not cover all types of observed rotation curves, but it is easy to change them whenever desired. In any case, it should be emphasized that the exact properties of the adopted rotation curve, for a study of large-scale spiral structure, matter mainly in the radial range in which spiral structure is observed; for example, it is generally of little importance how the rotation curve behaves beyond the bright optical disk (except for the interesting phenomenon of prominent gaseous spiral arms often observed beyond the bright optical disk, which will be described in Chapter 18). The specification of the word *active* is meant to emphasize the fact that the disk density profile $\sigma(r)$ is only partly related to the disk density as determined, for example, in mass models of galaxies. In practice, the profile has to be tailored by a suitable reduction factor [see Eq. (14.48)] to include the two important previously mentioned effects: the three-dimensional character of the disk and the presence of the cooler, dissipative gas component. A proper justification of the relevant prescription can be given only after the dynamics of waves has been analyzed (see Section 16.2). In practice, these motivations lead to a cut-out prescription that is similar to the one that has been used in the context of the isothermal self-similar stellar disk [see Eq. (14.44)]. Here the idea is to parameterize these effects in a simple and flexible way. Studies of individual galaxies require a detailed investigation of this modeling aspect, which may lead to significant deviations from the choice of $\sigma(r)$ given later. Finally, the equivalent acoustic speed is specified directly in terms of the parameter $Q(r) = c\kappa/\pi G\sigma$. Again, even if the data play a major role in the discussion of a realistic choice for this profile, $c(r)$ should *not* be meant to be any of the possible choices provided by the observations (i.e., the velocity-dispersion profile for K giant stars or the velocity-dispersion profile for cold atomic hydrogen), but rather it is meant to be a representative profile that, within our simple one-component model, best includes the effects of the three-dimensional structure of the disk and especially the presence of the dissipative gas on the basis of what is learned from various physical discussions. In particular, the Q-profile given later, with the flat behavior in the outer disk, is taken to represent the effect of self-regulation. This dynamical "thermostat" will be discussed in Chapter 16.

Thus the following choice of profiles attempts to restrict our attention to a set of physically realistic models out of the infinite number of possible mathematical models that may be imagined. The idea is to consider the fluid model not as a toy model but to supplement it as much as possible with information derived from the data, independent stellar-dynamical and multi-component studies, and all plausible physical considerations that are deemed to be relevant. In summary, these are the profiles that are considered:

$$V(r) = V_\infty \frac{r}{r_\Omega}\left[1 + \left(\frac{r}{r_\Omega}\right)^2\right]^{-1/2}, \tag{14.45}$$

$$\sigma(r) = \sigma_0 \exp(-r/h)f + \sigma_g, \tag{14.46}$$

$$Q(r) = Q_{OD}\{1 + q\exp[-(r/r_Q)^2]\}. \tag{14.47}$$

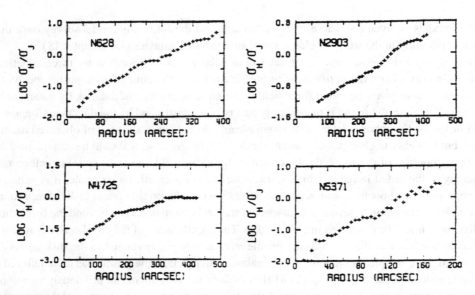

Fig. 14.3. Gas-to-star density ratio in four galaxy disks of Sc galaxies, as illustrated by radio (HI) and optical (J emulsion) observations (after Wevers, B. M. R. H. 1984. Ph.D. thesis, University of Groningen, The Netherlands).

Here V_∞, σ_0, Q_{OD}, and q and the scale lengths r_Ω, h, and r_Q are constants that set the general properties of the profiles. For many purposes, given the fact that the gas distribution often has a longer radial scale than that of the stars, the term σ_g can be taken as a constant (Fig. 14.3). Finally, the *mass-reduction factor f*, introduced to incorporate the physical effects mentioned earlier in this section, can be given the following simple form:

$$f = 1 - f_0(r/r_{cut}) + \frac{1}{6}f_0(r/r_{cut})\exp(r/h), \tag{14.48}$$

with

$$f_0(x) = (1+4x)(1-x)^4, \quad x \le 1, \tag{14.49}$$

$$f_0(x) = 0, \quad x > 1, \tag{14.50}$$

which thus introduces the new (constant) scale length r_{cut} for the mass reduction.

Note that the active disk density profile is given independently of the rotation curve. Implicitly, it is thus assumed that part of the force field is provided by a mass component, with the possible contribution of a bulge and a halo, which does not participate actively in the dynamics of the disk.

The preceding functions are meant to be varied in a rather arbitrary manner in a survey of models. However, to describe cool self-gravitating disks, the self-gravity parameter

$$\epsilon_0 = \frac{\pi G \sigma}{r \kappa^2} \tag{14.51}$$

and the epicyclic parameter

$$\epsilon = \frac{c}{r \kappa} \tag{14.52}$$

should be small throughout the disk.

14.5   A Reference Basic State in View of the Problem of Dark Matter

The problem of dark-matter halos in spiral galaxies will be treated separately in Chapter 20. Here we present a simple reference basic state that can be used as a starting point for the discussion of the disk-halo decomposition of rotation curves.[23] For simplicity, we refer to pure stellar-disk galaxies, that is, disk galaxies for which the contributions of a visible bulge and a gaseous disk are negligible. (In the presence of a bulge and a gaseous disk, the analysis is more complicated but can be performed in a straightforward manner because the situation remains conceptually similar.)

As briefly introduced in Subsection 4.2.1, the contribution of a dark halo to the rotation curve is usually expressed in quadrature [see Eq. (4.8)]. The often-observed trend of an asymptotically flat rotation curve (at large radii, $V \sim V_\infty$) is reminiscent of the properties of the isothermal sphere (see Chapter 22), and stellar disks are often characterized by an exponential luminosity profile. Therefore, parametric decompositions sometimes consider

$$V^2 = V_\star^2 + V_h^2,$$
(14.53)

where, for example, the halo contribution is that associated with the density distribution of a pseudoisothermal sphere so that

$$V_h^2 = V_\infty^2 \left[1 - \sqrt{\frac{2}{\alpha} \frac{1}{R}} \arctan \left(\sqrt{\frac{\alpha}{2}} R \right) \right],$$
(14.54)

and the disk contribution is given by

$$V_\star^2 = V_D^2 \, ;$$
(14.55)

the function V_D^2 is defined by Eq. (14.32). Here $R = r/h_\star$ is the dimensionless radial coordinate in units of the exponential scale, and $\alpha \equiv 8\pi G \rho_h(0) h_\star^2 / V_\infty^2$ is a dimensionless measure of the central halo density. A dimensionless measure of the central disk mass can be defined as $\beta \equiv 8\pi G \sigma_0 h_\star / V_\infty^2$. Thus a parametric decomposition of this kind is carried out by looking for the best pair (α, β) of dimensionless parameters or, equivalently, for the best pair of dimensional quantities $[\rho_h(0), M/L]$ able to fit the observed rotation curve V. As will be described in Chapter 20, this leads to a definition of the maximum-disk decomposition (i.e., the decomposition characterized by a maximum value of the disk mass-to-light ratio M/L) and to a serious problem of disk-halo degeneracy because the fit can be equally well performed by means of lighter disks and heavier halos. The situation is exemplified in Fig. 14.4 for the case of a fit to the rotation curve of the galaxy NGC 3198. Similar parametric decompositions can be carried out based on other popular forms of the assumed spherical halo density distribution.[24]

The preceding approach is clearly not self-consistent because it ignores the effect of the disk mass distribution on the assumed spherically distributed density distribution. Intuitively, we expect the halo density distribution (and its shape) to be affected more by relatively heavier disks. Under the hypothesis of an isothermal halo, a self-consistent approach starts from the distribution function

$$f_h = A \exp(-aE),$$
(14.56)

where a and A are constants, and $E = v^2/2 + \Phi$ is the specific energy of the halo particles. Here the total potential $\Phi = \Phi_h + \Phi_\star$ is not spherically symmetric, but only axisymmetric.

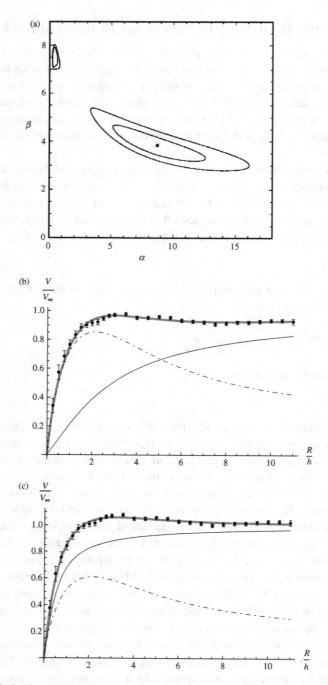

Fig. 14.4. Disk-halo degeneracy for a parametric nonconsistent decomposition of the rotation curve of NGC 3198. (*a*) Contours of 68 and 95 percent confidence regions in the (α, β) parameter plane. (*b*) Parametric disk-halo decomposition corresponding to the local maximum of the likelihood with higher disk importance. (*c*) Parametric disk-halo decomposition corresponding to the local maximum of the likelihood with lower disk importance. (From Amorisco, N. C., Bertin, G. 2010. *Astron. Astrophys.*, **519**, A47).

In dimensionless cylindrical polar coordinates ($\zeta \equiv z/h_\star$), the equation that determines the dimensionless halo potential is

$$\left(\frac{1}{R}\frac{\partial}{\partial R}R\frac{\partial}{\partial R} + \frac{\partial^2}{\partial \zeta^2}\right)\psi_h = -\alpha \exp[\psi_h(R,\zeta) + \psi_\star(R,\zeta)], \qquad (14.57)$$

which is a nonlinear equation for ψ_h (a dimensionless function proportional to $-\Phi_h$) to be solved for assigned $\psi_\star$. (For a zero-thickness disk, even in the nonexponential case, the potential $\psi_\star$ can be calculated over the entire space with the techniques referred to in Subsection 14.3.1; study of the case of finite thickness opens up new issues and will not be addressed here.) The solution is constructed with the boundary conditions that the total potential vanishes at the origin and ψ_h reduces asymptotically to the corresponding dimensionless potential of the isothermal sphere at large radii (see Chapter 22). It is constructed by iteration, starting from the isothermal sphere solution for ψ_h inserted as a seed on the right-hand side of the equation.[25]

The role of self-consistency in the decomposition of rotation curves is exemplified in Fig. 14.5. Together with the results of a fit (not shown here) to a fiducial rotation curve $V = V_\infty[1 - \exp(-R)]$, representative of a sample of thirty-five pure disk galaxies and consistent with the empirical relation described in Subsection 4.3.2, the results presented in Fig. 14.5 and in this subsection suggest that self-consistency removes the disk-halo degeneracy and points in the direction of significantly submaximal disks.[26] The role of the disk-halo decomposition in the problem of the morphology of spiral galaxies will be described in Chapter 18.

The flattening of the halo in the presence of an embedded disk is differential;[27] it is illustrated in Fig. 14.6. This result is of interest for studies of galaxy warps (see Chapter 19), for studies of three-dimensional orbits and polar-ring galaxies (see Subsection 13.6.1), for the modeling of spiral galaxies as gravitational lenses,[28] and for a proper modeling of the phenomenon to be described in the next section.

14.6 The Problem of Extraplanar Gas

In recent years, a new structural problem relevant to spiral galaxies has been brought to our attention by deep radio observations of the distribution of atomic hydrogen around galaxy disks. Together with the observation of diffuse warm and hot gas, this is commonly referred to as the *problem of extraplanar gas*.[29] Some galaxies, among them NGC 891,[30] NGC 2403,[31] UGC 7321,[32] and NGC 4559,[33] show the existence of cold atomic hydrogen, characterized by slow rotation, well outside the thin-disk region. Given the fact that deep radio observations and careful data analysis are needed to detect such a thick, slowly rotating HI component, it may well be that the cases that have been collected and studied in detail so far are representative of a situation quite common in spiral galaxies.

These results have suggested different alternative interpretations, either based on the possibility that ionized gas, swept up by stellar winds and supernova explosions, rises above the disk, cools down, and falls back to the plane[34] or on the accretion of intergalactic primordial gas.[35]

Independent of the astrophysical scenario underlying the origin of the extraplanar gas, the phenomenon is likely to correspond to a quasi-stationary structural feature because of its overall regularity and symmetry and its presumed frequency of occurrence. It is thus natural to ask what could be the dynamical basis for such an unusual quasi-equilibrium state. One surprising

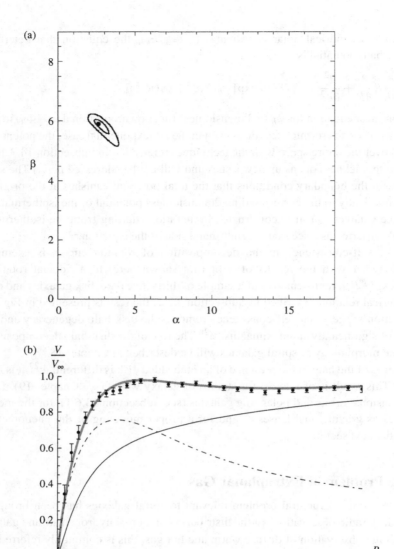

Fig. 14.5. Removal of the disk-halo degeneracy by a self-consistent decomposition of NGC 3198. (*a*) Contours of 68 and 95 percent confidence regions in the (α, β) parameter plane. (*b*) Best-fit disk-halo decomposition. (From Amorisco, N. C., Bertin, G. 2010. *Astron. Astrophys.*, **519**, A47).

aspect of it is the violation of cylindrical rotation, expected for a barotropic fluid,[36] given the fact that a barotropic equation of state is a commonly used starting point of many astrophysical models.

In the following, we adopt polar cylindrical coordinates, denoted by (r, z, ϕ). For a one-component axisymmetric fluid under the influence of the total gravitational potential $\Phi(r, z)$, with a velocity field characterized by only rotation around the symmetry axis, the dynamical

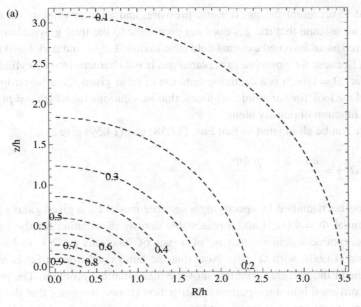

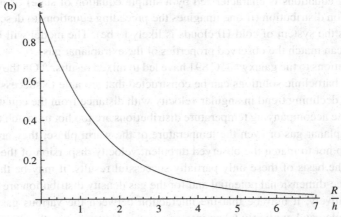

Fig. 14.6. Differential flattening of the isothermal halo embedding a pure disk. (*a*) Contours of the dark-halo density distribution (normalized to its central value) for the best-fit isothermal halo model fitting a fiducial galaxy rotation curve consistent with the scaling relation (4.12); note that the contours coincide with those of the total gravitational potential. (*b*) Ellipticity profile of the dark-matter halo illustrated in panel (*a*). (From Amorisco, N. C., Bertin, G. 2010. *Astron. Astrophys.*, **519**, A47).

equilibrium conditions can be written as[37]

$$\frac{1}{\rho}\frac{\partial p}{\partial z} = -\frac{\partial \Phi}{\partial z}, \tag{14.58}$$

$$\frac{1}{\rho}\frac{\partial p}{\partial r} = -\frac{\partial \Phi}{\partial r} + r\Omega^2. \tag{14.59}$$

Here ρ, p, and Ω represent the gas density, pressure, and angular velocity. In the discussion reported here, we assume that the gas does not contribute to the total gravitational field, which is thus considered as an assigned external field (due to disk, bulge, and dark halo) to be modeled independently. Because the observed extraplanar gas is not characterized by cylindrical rotation, and in particular, $\Omega = \Omega(r, z)$ is a declining function of $|z|$ at given r, the barotropic assumption is dropped, and we look for baroclinic solutions, that is, solutions for which $p = p(r, z)$ cannot be expressed as a function of density alone.

The pressure can be eliminated so that Eqs. (14.58) and (14.59) give

$$\frac{\partial}{\partial z} \left(\rho r \Omega^2 \right) = \frac{\partial \rho}{\partial z} \frac{\partial \Phi}{\partial r} - \frac{\partial \rho}{\partial r} \frac{\partial \Phi}{\partial z}. \tag{14.60}$$

The problem has been studied by specifying a sensible model for a given galaxy [i.e., by specifying the potential $\Phi = \Phi(r, z)$] and a reasonable density distribution for the gas $\rho = \rho(r, z)$, in search of acceptable solutions, that is, of a pair of functions (p, Ω^2), to be derived from Eqs. (14.58) and (14.60), with $\Omega^2 > 0$. Note that the latter condition on Ω^2 is not guaranteed by the equations in the present formulation of the mathematical problem. The pressure p thus found can be converted into a temperature distribution (if one imagines that the gas described by the preceding equations is characterized by a simple equation of state) or into an effective velocity-dispersion distribution (if one imagines the preceding equations to describe a turbulent medium, such as the system of cold HI clouds is likely to be). The model will be judged to be satisfactory if it can match the observed properties of the extraplanar gas.

So far, applications to the galaxy NGC 891 have led to mixed results.[38] On the one hand, it has been shown that baroclinic solutions can be constructed that are able to successfully match the observed overall declining trend in angular velocity with distance from the equatorial plane. On the other hand, the accompanying temperature distributions are too hot to match the temperature of the cold extraplanar gas or even the temperature of the warm phase; they are also found to be effectively too hot to match the observed turbulent velocity dispersion of the extraplanar HI component. At the basis of these only partially successful results, it may be that the assumed forms for the three-dimensional potential and for the gas density distribution are not sufficiently realistic. Alternatively, it may be that the interaction between the various gas phases should be properly modeled and explicitly taken into account.[39] However, another natural explanation could be that the model itself, in which the fluid motions are restricted to be in the form of pure rotation around the symmetry axis and a priori exclude the possibility of radial and vertical components in the velocity field, is bound to fail because it lacks some important qualitative ingredients (in particular, the presence of meridional circulation[40]).

In any case, this is one interesting general problem that stimulates the search for complex quasi-stationary three-dimensional basic states. A successful solution would lead to a better understanding of the interactions between disk and halo and between galaxy and intergalactic medium and would be highly relevant to the general issue of galaxy evolution.

Notes

1. Spitzer, L., Jr. 1942. *Astrophys. J.*, **95**, 329.
2. Kapteyn, J. C. 1922. *Astrophys. J.*, **55**, 302; Jeans, H. 1922. *Mon. Not. Roy. Astron. Soc.*, **82**, 122.

3. Oort, J. H. 1932. *Bull. Astron. Inst. Neth.*, **6** (238), 249; 1960. *Bull. Astron. Inst. Neth.*, **15** (494), 45; 1965. In *Galactic Structure*, eds. A. Blaauw, M. Schmidt. University of Chicago Press, Chicago, p. 455. The data from which the analysis is made include data from K giants by Hill, E. R. 1960. *Bull. Astron. Inst. Neth.*, **15**, 1; and from A0 stars by Jones, D. H. P. 1962. *Roy. Obs. Bull.*, **52**.

4. Bahcall, J. N. 1984. *Astrophys. J.*, **276**, 156 and 169. The first study focused on the data for a sample of F stars by Hill, G., Hilditch, R. W., Barnes, J. V. 1979. *Mon. Not. Roy. Astron. Soc.*, **186**, 813, giving good coverage out to $z \approx 200$ pc. Bahcall, J. N. 1984. *Astrophys. J.*, **287**, 926, extends the study to new data available for a sample of K giants. Bahcall, J. N., Flynn, C., Gould, A. 1992. *Astrophys. J.*, **389**, 234, again discuss the basis of the method adopted and include the study of new extensive data for a sample of K giants by Flynn, C., Freeman, K. C. 1993. *Astron. Astrophys. Suppl.*, **97**, 835, providing good coverage out to 500 pc; they focus on a homogeneous subsample of 125 stars extracted from the original of 560 K giants by using metallicity filters to eliminate young stars and luminosity filters to minimize distance error determinations.

5. The first study by Bahcall, J. N. 1984. *Astrophys. J.*, **276**, 169, gives as a best solution that with total mass density at $z = 0$ of $0.185 \, M_\odot \, \mathrm{pc}^{-3}$, $(M/L)_V = 2.9$ for the disk material, total disk density of $67 \, M_\odot \, \mathrm{pc}^{-2}$, more than 50 percent of unobserved dark matter in the disk, $\epsilon_\rho = 0.1$. The more recent analysis of Bahcall, J. N., Flynn, C., Gould, A. 1992. *Astrophys. J.*, **389**, 234, recognizes that although the best models point to even more dark matter in the disk, the confidence levels of the statistical analysis are relatively weak, with very wide $1 \, \sigma$ uncertainty ranges.

6. Kuijken, K. H., Gilmore, G. 1989. *Mon. Not. Roy. Astron. Soc.*, **239**, 571, 605, and 651; 1991. *Astrophys. J. Lett.*, **367**, L9.

7. Bienaymé, O., Robin, A. C., Crézé, M. 1987. *Astron. Astrophys.*, **180**, 94, starting from a synthetic model of galactic stellar populations formulated earlier by Robin, A., Crézé, M. 1986. *Astron. Astrophys.*, **157**, 71.

8. Crézé, M., Robin, A. C., Bienaymé, O. 1989. *Astron. Astrophys.*, **211**, 1.

9. Bahcall, J. N., Soneira, R. M. 1980. *Astrophys. J. Suppl.*, **44**, 73; Bahcall, J. N., Schmidt, M., Soneira, R. M. 1982. *Astrophys. J. Lett.*, **258**, L23; 1983. *Astrophys. J.*, **265**, 730; Caldwell, J. A. R., Ostriker, J. P. 1981. *Astrophys. J.*, **251**, 61; Ostriker, J. P., Caldwell, J. A. R. 1983. In *Kinematics, Dynamics, and Structure of the Milky Way*, ed. W. H. Shuter. Reidel, Dordrecht, The Netherlands, p. 249.

10. In contrast, traditional models for our Galaxy, before the issue of flat rotation curves made dark halos a natural ingredient of mass models, had a very heavy disk and a relatively low escape velocity; the model by Schmidt, M. 1965. In *Galactic Structure*, eds. A. Blaauw, M. Schmidt. University of Chicago Press, Chicago, p. 513, was consistent with the Oort limit for the volume density, but in the solar neighborhood had a disk density of $114 \, M_\odot \, \mathrm{pc}^{-2}$ with an escape velocity of only 380 km s^{-1} and a Keplerian decline of the rotation curve in the outer disk. This should be compared with more recent mass models of the Galaxy; see Merrifield, M. R. 1992. *Astron. J.*, **103**, 1552.

11. Crézé, M., et al. 1998. *Astron. Astrophys.*, **329**, 920. This result was basically confirmed by a thorough analysis from different authors involved in the controversy; Holmberg, J., Flynn, C. 2000. *Mon. Not. Roy. Astron. Soc.*, **313**, 209; Holmberg, J., Flynn, C. 2004. *Mon. Not. Roy. Astron. Soc.*, **352**, 440.

12. Here the analysis follows that of Shu, F. H. 1969. *Astrophys. J.*, **158**, 505. The anisotropy of the distribution of stellar motions had been noted by Kapteyn, J. C. 1905. *Rept. Brit. Assoc. 1905*, 257, and the concept of velocity ellipsoid was proposed by Schwarzschild, K. 1907. *Göttingen Nachr. 1907*, 614; 1908. *Göttingen Nachr. 1908*, 191. The constraints imposed by the collisionless Boltzmann equation on a quasi-Maxwellian distribution function are extensively discussed by Chandrasekhar, S. 1960. *Principles of Stellar Dynamics*, Dover, New York (originally published in 1942). In a later paper, Shu, F. H. 1978. *Astrophys. J.*, **225**, 83, remarks that a singularity noted in the function defined in Eq. (14.18) can be removed by truncating the distribution function much as in the manner of the King models (see Subsection 22.3.1) without any significant change in the asymptotic discussion presented in this section.

13. Especially in the early studies of stellar dynamics, much work has been devoted to studying and interpreting the deviation of the vertex in the solar neighborhood, which is particularly evident for stars of early spectral types; e.g., see Blaauw, A., Schmidt, M., eds. 1965. *Galactic Structure*, University of Chicago Press, Chicago.

14. Only partly related to the phenomenon of the asymmetric drift, a classical problem in stellar dynamics is the strong asymmetry in the tail of the distribution of peculiar velocities (**w**), with a sharp decrease of stars observed to move in the direction of the overall rotation with $w_\theta > 63$ km s^{-1}, as noted by Oort, J. H. 1926. *Kapteyn Astron. Lab. Groningen Publ.*, **40**. Initially, the phenomenon of the so-called high-velocity stars was interpreted as a kind of loss cone, that is, as being due to the escape of stars from the Galaxy, whereas it is now recognized that the escape velocity at the location of the Sun is far higher than that required by this interpretation; more simply, the asymmetries are now thought to reflect the properties of a system made of two components, a cool, rapidly rotating disk and a hot, slowly rotating or nonrotating halo. A discussion of the early studies is given by Chandrasekhar, S. 1960. *op. cit.*

15. See, for example, the third chapter of Chen, F. F. 1984. *Introduction to Plasma Physics and Controlled Fusion*, Vol. 1: *Plasma Physics*, 2nd ed., Plenum, New York.

16. Toomre, A. 1963. *Astrophys. J.*, **138**, 385.

17. Freeman, K. C. 1970. *Astrophys. J.*, **160**, 811.

18. Hunter, C. 1963. *Mon. Not. Roy. Astron. Soc.*, **126**, 299, studied the stability of cold disks represented by this density-potential pair.

19. Mestel, L. 1963. *Mon. Not. Roy. Astron. Soc.*, **126**, 553.

20. Rybicki, G. B. 1974 (cited as private communication by T. A. Zang 1976).

21. Zang, T. A. 1976. Ph.D. dissertation, Massachusetts Institute of Technology, Cambridge, MA; another interesting case is that of the collisionless analogues of the uniformly rotating Maclaurin disks; see Freeman, K. C. 1966. *Mon. Not. Roy. Astron. Soc.*, **134**, 15; and Kalnajs, A. J. 1972. *Astrophys. J.*, **175**, 63.

22. Bertin, G., Lin, C. C. 1996. *Spiral Structure in Galaxies: A Density Wave Theory*, MIT Press, Cambridge, MA; many other technical details are given by Bertin, G., Lin, C. C., Lowe, S. A., Thurstans, R. P. 1989. *Astrophys. J.*, **338**, 78.

23. This section is based on the article Amorisco, N. C., Bertin, G. 2010. *Astron. Astrophys.*, **519**, A47.

24. For example, models of the halo density distribution suggested by cosmological simulations: Navarro, J. F., Frenk, C. S., White, S. D. M. 1996. *Astrophys. J.*, **462**, 563; Einasto, J. 1969. *Astron. Nachr.*, **291**, 97. See Chemin, L., de Blok, W. J. G., Mamon, G. A. 2011. *Astron. J.*, **142**, 109.

25. The method extends the iteration scheme devised by Prendergast, K. H., Tomer, E. 1970. *Astrophys. J.*, **75**, 674, for the construction of nonspherical models of elliptical galaxies to the present case of nonzero outer boundary conditions. It makes use of a standard multipole expansion in Legendre polynomials and is described in detail in appendix A of the paper by Amorisco, N. C., Bertin, G. 2010. *op. cit.* A similar iteration procedure also can be used to extend the spherical King models to the triaxial case, as required by imposing the presence of an external tidal potential (see Chapter 22).

26. To some extent this conclusion anticipates the results of a long-term project called *The Disk Mass Project* designed to determine whether disks are maximal or lighter; see Verheijen, M. A. W., Bershady, M. A., et al. 2004. *Astron. Nachr.*, **325**, 151; Westfall, K. B., Bershady, M. A., et al. 2011. *Astrophys. J.*, **742**, 18; Bershady, M. A., Martinsson, T. P. K., et al. 2011. *Astrophys. J. Lett.*, **739**, L47; basically, the project extends the dynamical study of the solar neighborhood, described in Subsection 14.1.2, to the case of external galaxies.

27. The flattening of a bulge by a massive disk was studied by Monet, D. G., Richstone, D. O., Schechter, P. L. 1981. *Astrophys. J.*, **245**, 454.

28. Suyu, S. H., Hensel, S. W., et al. 2012. *Astrophys. J.*, **750**, id.10.

29. See Sancisi, R., Fraternali, F., et al. 2008. *Astron. Astrophys. Reviews*, **15**, 189.

30. Swaters, R. A., Sancisi, R., van der Hulst, J. M. 1997. *Astrophys. J.*, **491**, 140; Oosterloo, T., Fraternali, F., Sancisi, R. 2007. *Astron. J.*, **134**, 1019.

31. Fraternali, F., van Moorsel, G., et al. 2002. *Astron. J.*, **123**, 3124.

32. Matthews, L. D., Wood, K. 2003. *Astron. J.*, **593**, 721.

33. Barbieri, C. V., Fraternali, F., et al. 2005. *Astron. Astrophys.*, **439**, 947.

34. This is commonly called the *galactic fountain model*; Shapiro, P. R., Field, G. B. 1976. *Astrophys. J.*, **205**, 762; Bregman, J. N. 1980. *Astrophys. J.*, **236**, 577. For a study of NGC 891, see Bregman, J. N., Miller, E. D., et al. 2013. *Astrophys. J.*, **766**, id.57.

35. As proposed for the high-velocity clouds in our Galaxy by Oort, J. H. 1970. *Astron. Astrophys.*, **7**, 381.

36. The Poincaré-Wavre theorem; see Lebovitz, N. R. 1967. *Annu. Rev. Astron. Astrophys.*, **5**, 465; Tassoul, J.-L. 1980. *Theory of Rotating Stars*, Princeton University Press, Princeton, NJ.

37. Here we briefly summarize the analysis given by Barnabè, M., Ciotti, L., et al. 2006. *Astron. Astrophys.*, **446**, 61; and Marinacci, F., Fraternali, F., et al. 2010. *Mon. Not. Roy. Astron. Soc.*, **401**, 2451.

38. Barnabè, M., Ciotti, L., et al. 2006. *op. cit.*; Marinacci, F., Fraternali, F., et al. 2010. *op. cit.*

39. See Marinacci, F., Fraternali, F., et al. 2011. *Mon. Not. Roy. Astron. Soc.*, **415**, 1534, and references therein.

40. Waxman, A. M. 1978. *Astrophys. J.*, **222**, 61.

15 Density Waves

Density waves are thought to be at the basis of the explanation of spiral structures in galaxies, especially of the so-called grand-design structure whose extent is on the global scale. A physical discussion of the problem of spiral structure in galaxies will be given in Chapter 18. In this and Chapters 16 and 17 we focus instead on some relevant dynamical mechanisms. As anticipated earlier (Chapter 9), density waves are one of the two main classes of natural perturbations that are expected in a disk (the other class, bending waves, will be addressed in Chapter 19). They leave the equatorial symmetry of the galaxy disk unchanged and are associated with density enhancements and rarefactions that usually break the axisymmetry of the basic state. A special class of density waves ($m = 0$) leaves the disk axisymmetric.

The concept of density waves is a general one because the phenomenon simply reflects the oscillatory character of the disk that has been described in terms of single-star orbits. In practice, the detailed properties of density waves depend on the model considered. In early investigations, the galaxy disk was thought of mainly as a *stellar* disk, and thus the studies focused on the equations of stellar dynamics. On the one hand, attention was immediately drawn to a semiempirical approach, whereby an attempt was made at interpreting large-scale grand-design spiral structures in terms of a quasi-stationary pattern;[1] if this is approximated by a single wave (with a well-defined pattern frequency), for a given basic state, the requirements of self-consistency impose a specific spatial structure of the density and the velocity-field perturbation that depends on the pattern speed, which is considered a free parameter to be determined from observations[2] (Fig. 15.1). On the other hand, the stability of galaxy disks, that is, the excitation of density waves, was the primary concern.[3] Actually, the separation between the study of the properties of a self-sustained wave pattern and the study of the excitation mechanisms that may (or may not) lead to the eventual establishment of such a quasi-stationary structure is not so sharp in the early papers, becoming better defined later on in the course of development of the theory.

It soon became clear that a fluid model is able to capture important aspects of the physics of density waves and, because of its simplicity, can lead the way more easily than the rather intricate set of stellar dynamical equations.[4]

A major stumbling block to systematic studies of density waves, even in the simplest one-component, zero-thickness models, is imposed by the potential theory: The perturbed gravitational field at one location depends on the density perturbation at every other location of the disk by means of an integral relation similar to Eq. (14.26). This difficulty is sharpened by the fact that a realistic basic state is inhomogeneous (in the radial direction). From the methodological point of view, these obstacles have been approached in two different ways. One possibility is to consider special (sometimes singular) models for which the analysis can be carried through,

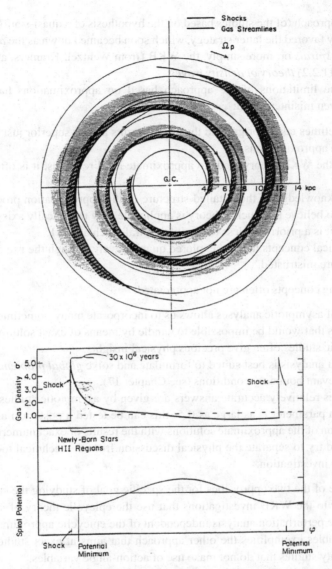

Fig. 15.1. Illustration of the shock scenario under the hypothesis of quasi-stationary spiral structure (from Roberts, W. W. "Large-scale shock formation in spiral galaxies and its implications on star formation," 1969. *Astrophys. J.*, **158**, 123; reproduced by permission of the AAS) for a two-armed trailing structure inside the corotation circle (gas streamlines are arrowed). The inner side of the bright optical arms (as traced by newly formed stars and HII regions) should be marked by a sharp gas peak and narrow dust lanes, locating the pattern of a large-scale shock.

with exact results, in a reasonably convenient manner. The other approach, quite common in hydrodynamics and plasma physics, is that of considering large classes of models and perturbations for which an approximate analysis can be carried out. The initial thrust in the direction of a

semiempirical approach (of the studies based on the hypothesis of a quasi-stationary spiral structure) immediately favored the latter strategy, which soon became known as the *asymptotic theory
of tightly wound spirals* or, more simply, the WKB (from Wentzel, Kramers, and Brillouin; see
also Subsection 11.2.2) *theory of density waves.*

Sometimes the limitations of the approach based on approximations have been unduly
emphasized or even misinterpreted:

1. It is sometimes taken for granted that exact results are far superior just because they are
   free from approximations.
2. Because the WKB theory leads to approximate local relations, it is often called a *local
   theory*.
3. Without knowledge of the detailed structure of the approximation process, we may be
   brought to believe that such a theory is applicable only to basically axisymmetric waves,
   and thus it is a priori hopeless to use it for spirals such as M81.
4. The physical concepts that are explored quantitatively through the use of these approximations are mistrusted.

Thus the following concepts often are not appreciated:

1. The use of asymptotic analyses allows us to incorporate many, sometimes crucial, physical effects that would be impossible to handle by means of exact solutions; furthermore,
   asymptotic studies often give precious physical insights.
2. The WKB analysis is best suited to formulate and solve *global problems*, with inclusion
   of the relevant boundary conditions (see Chapter 17).
3. Sometimes relatively accurate answers are given by asymptotic studies, even when the
   expansion parameter is as large as 0.5; this can be tested a posteriori and quantified by
   comparison of the approximate solutions with the results of exact numerical integrations.
4. We should try to separate the physical discussion from the technical tools that are used
   in a given investigation.

As an example of the two approaches for the specific goal of studying the stability of a disk
of stars, we refer to the WKB investigations that use the epicyclic theory of stellar orbits.[5] A
formulation of the perturbation analysis independent of the epicyclic approximation in terms of
action-angle variables exemplifies the other approach that aims at exact studies, but there are
some exact stability studies that do not make use of action-angle variables.[6]

A very popular model often considered for stability investigations is that of the infinite homogeneous shearing sheet.[7] This is basically a slab model of the corotation region in which the only
quantity assumed to vary in the basic state is the velocity profile (in a linear fashion, i.e., with a
constant shear rate) so as to focus on the effects of the underlying shear flow and the excitation
mechanisms that take place at corotation. The geometry considered is usually Cartesian, sometimes with doubly periodic boundary conditions. This model has given us very important insights
into the local processes occurring in a small patch of a differentially rotating disk.[8] The main
problem, to our knowledge still unresolved, is that of a proper matching with the inhomogeneous
disk.[9] Until this issue is resolved, the shearing sheet is of little help for global stability analyses.
Indeed, because of its inherent homogeneity, it has often generated misleading conclusions.

The study of an infinite homogeneous sheet can be conveniently carried out in terms of a
time-dependent wave-number representation.[10] This representation emphasizes the possibility of

transient, fast-evolving structures. Thus it is often believed that these studies prove that density-wave patterns must be transient and fast evolving. In reality, there are two separate issues: the adopted mathematical representation of the perturbations[11] and the viability of the hypothesis of a quasi-stationary spiral structure for grand-design galaxies (see Chapter 18). Resolution of the latter issue has its dynamical foundation in the existence of large-scale global modes, which are supported by the inhomogeneous disk through the action of the relevant boundary conditions. The time-dependent wave-number representation is thus best suited to describe small-scale transient spiral activity. There are well-known analogues in hydrodynamics[12] and plasma physics.[13]

In this chapter we develop the dispersion-relation approach based on suitable asymptotic expansions, which is best suited to describe the dynamics of an inhomogeneous disk and large-scale global modes. The general outlook is that of dispersive waves in inhomogeneous media, in line with what was introduced in Chapter 11.

15.1 Dispersion Relations

The detailed properties of linear density waves are contained in the form of the local dispersion relation, which we may write generically as

$$D(\omega, m, k; r) = 0. \tag{15.1}$$

Here the perturbations are taken to have an elementary dependence on time and azimuth, so the associated potential can be written as

$$\Phi_1 = \tilde{\Phi} \exp[i(\omega t - m\theta)], \tag{15.2}$$

where $\tilde{\Phi}$ is independent of t and θ, and the radial wave number is formally introduced as

$$k \equiv -\frac{i}{\tilde{\Phi}} \frac{\partial \tilde{\Phi}}{\partial r}. \tag{15.3}$$

Thus k is a local wave number. The dispersion relation depends on r through the properties of the various quantities, such as $\sigma(r)$ and $\Omega(r)$, that characterize the axisymmetric equilibrium basic state (see Chapter 14).

Given the fact that the basic state is in differential rotation, we anticipate that the ω dependence in D occurs by means of a Doppler-shifted frequency, which can be written in dimensionless form as

$$\nu = \frac{\omega - m\Omega}{\kappa}. \tag{15.4}$$

The detailed form of the dispersion relation and thus the properties of the associated density waves depend on which model is chosen. A few significant cases are described here. In all these cases, the dispersion relation is invariant under the transformations $\nu \to -\nu$ and $k \to -k$.

15.1.1 Dispersion Relation for Tightly Wound Waves in a Disk of Stars

The dispersion relation for tightly wound density waves on a zero-thickness collisionless disk of stars[14] is derived starting from the linearized collisionless Boltzmann equation for the perturbed distribution function f_1:

$$\frac{\partial f_1}{\partial t} + \{f_1, H_0\} = \{H_1, f_0\} = \{\Phi_1, f_0\}, \tag{15.5}$$

where H_0 is the unperturbed Hamiltonian, and $H_1 = \Phi_1$ is the perturbing spiral potential. Because the equilibrium is supposed to be stationary and axisymmetric, the unperturbed function f_0 is taken to be a function of the specific energy E and angular momentum J (see Chapter 14). We Fourier analyze the perturbing potential Φ_1 and the response distribution function f_1 in time and azimuth: $f_1 = \tilde{f} \exp[i(\omega t - m\theta)]$, $\Phi_1 = \tilde{\Phi} \exp[i(\omega t - m\theta)]$. Integrating Eq. (15.5) along the unperturbed orbits, we obtain

$$\tilde{f} = \frac{\partial f_0}{\partial E}\tilde{\Phi} + \hat{f}, \tag{15.6}$$

where

$$\hat{f} = -\frac{\omega \partial f_0/\partial E + m\partial f_0/\partial J}{2\sin(\omega\tau_e - m\theta_e)} \int_{-\tau_e}^{\tau_e} d\tau \, \tilde{\Phi}(r_\star(\tau)) \exp\{i[\omega\tau - m\theta_\star(\tau)]\}. \tag{15.7}$$

Here the quantities $2\tau_e = \tau_r$ and $2\theta_e$ are functions of E and J and denote the radial period of oscillation of the stars in the equilibrium potential and the azimuthal angle traversed in such a period (see Chapter 13). The functions $r_\star$ and $\theta_\star$ indicate the unperturbed orbits, as parameterized by the time variable τ, with the conditions $r_\star(\pm\tau_e) = r$, $\theta_\star(\pm\tau_e) = \pm\theta_e$. The full derivations of Eqs. (15.6) and (15.7) are not given here because the basic ideas are best appreciated in the simpler context of a homogeneous basic state; these were given in Chapter 12.

Then we consider the ordering

$$\epsilon = \frac{c}{r\kappa} \sim \frac{1}{|rk(r)|} \ll 1, \tag{15.8}$$

where for the epicyclic parameter we follow the notation of Chapter 14. The ordering of the radial wave number k [defined in Eq. (15.3)] is a kind of WKB condition and will be discussed more thoroughly in Chapter 17.

The preceding ordering characterizes tightly wound waves on a cool stellar disk and is found to be appropriate for relatively light active disks. (The term *light* basically refers to the parameter $\epsilon_0 = \pi G\sigma/r\kappa^2$ in a sense that will be clarified in Section 15.3 below. For heavier disks, a more general treatment is required, as will be outlined in Subsection 15.1.3.) The ordering allows us to derive an approximate local relation between perturbed potential and perturbed density (see Chapter 17). We then obtain the dispersion relation by evaluating $\tilde{f}$ under the same ordering, based on a quasi-Maxwellian distribution function f_0 (see Chapter 14), and by equating the density response associated with $\tilde{f}$ to the density calculated from the potential theory. To leading order, the self-consistency equation gives

$$\frac{k_0}{|k|} = -2\frac{e^{-x}}{x}\sum_{n=1}^{\infty}\frac{I_n(x)}{(\frac{\nu}{n})^2 - 1} = \frac{\mathcal{F}_\nu(x)}{1 - \nu^2}, \tag{15.9}$$

where ν is as defined in Eq. (15.4), $k_0 = \kappa^2/2\pi G\sigma$, and $x = k^2c^2/\kappa^2$. Here c is the radial velocity dispersion. In Eq. (15.9), two equivalent forms of the dispersion relation are given, one in terms

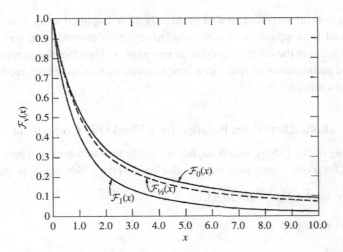

Fig. 15.2. The stellar reduction factor (from Lin, C. C., Yuan, C., Shu, F. H. "On the spiral structure of disk galaxies. III: Comparison with observations," 1969. *Astrophys. J.*, **155**, 721; reproduced by permission of the AAS).

of the stellar reduction factor (Fig. 15.2)

$$\mathcal{F}_\nu(x) = \frac{1-\nu^2}{x}\left\{1 - \frac{\nu\pi}{\sin\nu\pi}\frac{1}{2\pi}\int_{-\pi}^{\pi}ds\cos(\nu s)\exp[-x(1+\cos s)]\right\}, \tag{15.10}$$

following the traditional notation of the density-wave theory and the other in terms of standard Bessel functions, which brings out the analogy between this problem of stellar dynamics and a similar problem of waves in a magnetized plasma.[15]

The dispersion relation breaks down at resonances, that is, whenever ν takes on integer values, including the corotation case ($\nu = 0$). At these locations, a WKB analysis can be worked out that incorporates the important Landau damping effects[16] (for a simple description of Landau damping, see Chapter 12). In the linear theory, energy and angular-momentum exchange between the wave and the basic state occurs only at these resonant locations.

In the early developments of the theory, most of the efforts were devoted to finding observational support for the density-wave concept by using the preceding dispersion relation as a tool to fit observed spiral patterns. These were assumed to be characterized by one value of the pattern speed, $\Omega_p = \mathrm{Re}(\omega)/m$, and the short-wave branch of the dispersion relation (which was judged to be the safest to use from the point of view of the approximations made) was applied to derive the spatial structure of the pattern in terms of the function $k(r)$ (see also the discussion of propagation diagrams given later). In this way, the dynamical problem of explaining what sets the corotation circle (i.e., what determines the pattern frequency) was given lower priority, and Ω_p was considered as a parameter to be determined from the observations. In this context, when the dispersion relation was applied, it was limited to its principal range (i.e., for $-1 < \nu < 1$), and it was immediately realized that the interstellar gas is bound to play an important role.[17]

Resolution of the dynamical problem of what sets the corotation circle requires global analysis with a careful consideration of the relevant boundary conditions. Even though such a global analysis in the stellar dynamical context was carried out, it was soon realized that the simpler fluid model could be a faithful and efficient tool of investigation if applied away from the resonances; such a model must be supplemented with detailed results borrowed from stellar dynamics in order to properly include the effects expected at resonances. Thus, for the purpose of discussing the excitation and maintenance of spiral structure in galaxies, it is more convenient to refer to the more intuitive fluid model.

15.1.2 The Quadratic Dispersion Relation for a Fluid One-Component Model

Under the ordering of Eq. (15.8), which applies to tightly wound waves in a cool disk, the linear stability analysis for a one-component fluid model leads to the following dispersion relation:

$$(\omega - m\Omega)^2 = \kappa^2 + k^2 c^2 - 2\pi G\sigma |k|, \tag{15.11}$$

which also can be written as

$$D(\nu, |\hat{k}|) = 0, \tag{15.12}$$

where

$$D = \nu^2 - 1 - \frac{1}{4}Q^2\hat{k}^2 + |\hat{k}|, \tag{15.13}$$

and $\hat{k} = 2\epsilon_0 rk$. We recall from Chapter 14 that $Q = c\kappa/\pi G\sigma$ and that the self-gravity parameter is given by $\epsilon_0 = \pi G\sigma/r\kappa^2$. In this dispersion relation, c is the equivalent acoustic speed associated with the fluid model. A simple heuristic derivation is available elsewhere[18] and will not be repeated here because the full derivation will be given in Chapter 17. Note that in order to follow by analogy the notation of Eq. (15.9), the relevant fluid reduction factor could be written as $\mathcal{F}_\nu^{(g)} = (1 - \nu^2)/(1 - \nu^2 + x)$, where $x = c^2 k^2/\kappa^2$. This reduction factor tends to unity in the cold fluid limit $(c \to 0)$.

15.1.3 The Cubic Dispersion Relation for the Fluid Model

In the dispersion relations given in previous subsections, the azimuthal wave number m/r does not enter explicitly, except for its presence in the Doppler-shifted frequency ν. This fact was immediately noted as an unsatisfactory feature of the asymptotic analysis available in the early development of the theory.[19] However, with much of the attention focused on bisymmetric $(m = 2)$ spiral structure, the ordering of Eq. (15.8) appeared to be the only regime in which the potential theory could be treated in terms of approximate local relations.

This impasse is solved by a systematic asymptotic analysis[20] based on the following ordering:

$$\epsilon_0^2 \ll 1, \qquad K^2 = O(1), \tag{15.14}$$

where the total dimensionless wave number is defined as

$$K^2 \equiv 4\epsilon_0^2(k^2 r^2 + m^2). \tag{15.15}$$

Thus even very long waves with $k \approx 0$ can be described by means of a local treatment of the potential theory, provided that the quantity m^2 is taken to be formally large. This is a different type of WKB approximation with respect to that described earlier. Sometimes asymptotic schemes can be stretched to unfavorable regimes with accurate results. This turned out to be the case for this study because numerical surveys later proved[21] that excellent insights can indeed be gathered with this asymptotic analysis, even for the case of $m = 2$.

Based on the preceding ordering, the one-component fluid model[22] leads to the following dispersion relation:

$$\frac{Q^2}{4} = \frac{1}{K} - \frac{1 - \nu^2}{K^2 + \frac{J^2}{(1 - \nu^2)}}. \tag{15.16}$$

Here a new important parameter, independent of c, is introduced:

$$J = 2m\epsilon_0 \left(\frac{2\Omega}{\kappa}\right) \left|\frac{d \ln \Omega}{d \ln r}\right|^{1/2}. \tag{15.17}$$

Note that for $m = 0$ or for the case of solid-body rotation, the new parameter vanishes. For $J \to 0$, the dispersion relation reduces to the quadratic relation of Eq. (15.11). When $J \neq 0$, the relation is a cubic in the dimensionless total wave number K.

Much of the analysis of the stellar dynamical equations within the more general asymptotic ordering of Eq. (15.14), which covers the case of open waves, has been worked out;[23] for simplicity, as stated earlier, most of the following quantitative discussions are based on the more convenient fluid model.

15.1.4 Finite-Thickness Effects

Under the ordering of tightly wound waves of Eq. (15.8), the effects associated with the finite thickness of the disk, which dilutes the gravity field, are incorporated into the quadratic dispersion relation of the fluid model when the self-gravity term is reduced by a factor[24] inversely proportional to $(1 + |kz_0|)$:

$$D = \nu^2 - 1 - \frac{1}{4}Q^2\hat{k}^2 + \frac{|\hat{k}|}{1 + |\hat{k}|\hat{z}_0}, \tag{15.18}$$

where

$$\hat{z}_0 = \frac{1}{2\epsilon_0}\frac{z_0}{r}. \tag{15.19}$$

Here the disk density entering the definition of Q and of the dimensionless wave number $\hat{k}$ is the projected disk density (see Chapter 14). Note that with respect to the estimate given after Eq. (14.12), here we prefer to leave the value z_0/r unspecified because for an anisotropic disk the vertical velocity dispersion can be significantly different from that in the radial direction; indeed, for the stellar disk in the solar neighborhood, the vertical dispersion is smaller by a factor of 2. In principle, the simple fluid model also can be applied to describe the properties of an anisotropic disk.

15.2 **Marginal Stability**

One way of looking at the dispersion relation is by asking whether there are any instabilities, that is, values of the wave number for which $\text{Im}(\omega) < 0$; in this case, given the choice of phase

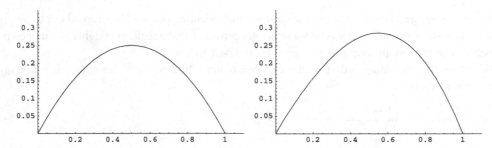

Fig. 15.3. *Left:* The marginal-stability curve for the zero-thickness fluid model [quadratic dispersion relation; see Eq. (15.20)]. *Right:* The marginal stability curve in stellar dynamics, corresponding to Eq. (15.27) of the text; the original diagram was given by Toomre, A. 1964. *Astrophys. J.*, **139**, 1217. In both frames, the variable on the horizontal axis is the dimensionless radial wavelength k_0/k, and the quantity on the vertical axis is proportional to Q^2; note that $(1.069)^2/4 = 0.2857$, that is, the value of the peak of the curve plotted in the right frame.

made in Eq. (15.2), the perturbation would grow exponentially in time. Because the dispersion relations depend on ω through the term ν^2, it is clear that if a transition occurs from the stable to an unstable case (while the relevant parameters are varied), this would correspond to a transition from positive to negative values of ν^2. Thus it is customary to inspect the properties of the dispersion relation in the vicinity of marginal stability defined by $\nu^2 = 0$. The discussion is most obvious if we consider axisymmetric disturbances, with $m = 0$, because in this case $\nu^2 = \omega^2/\kappa^2$.

15.2.1   For the Quadratic Dispersion Relation of the Fluid Model

If we focus on the quadratic dispersion relation of the fluid model, the analysis is very simple. In fact, by setting $\nu^2 = 0$ in Eq. (15.13), in terms of the dimensionless wavelength defined as $\hat{\lambda} \equiv 1/|\hat{k}|$, the dispersion relation becomes

$$\frac{Q^2}{4} = \hat{\lambda}(1 - \hat{\lambda}), \tag{15.20}$$

which is a parabola in the $(\hat{\lambda}, Q^2)$ plane (see the left frame of Fig. 15.3). For points of the plane below such a curve, the dispersion relation requires that $\nu^2 < 0$ (i.e., the disk is unstable). Thus, for $Q > 1$, the disk is stable for any value of $\hat{\lambda}$. In this sense, the condition of $Q = 1$ is often referred to as that of *marginal stability*. Note that the most dangerous perturbations are characterized by $\hat{\lambda} = 1/2$; at short wavelengths the Jeans instability is stabilized by pressure, whereas at long wavelengths it is stabilized by rotation.[25]

15.2.2   For the Quadratic Dispersion Relation of the Fluid Model with Finite Thickness

If we take instead the dispersion relation defined by Eq. (15.18), we find the marginal-stability condition

$$\frac{Q^2}{4} = \hat{\lambda}^2 \left(\frac{1}{\hat{\lambda} + \hat{z}_0} - 1 \right). \tag{15.21}$$

For a given value of $\hat{z}_0$, this can be drawn in the $(\hat{\lambda}, Q^2)$ plane as a curve that has its maximum,

$$Q^2_{\max} = \left(\frac{1 + \sqrt{1 + 8\hat{z}_0}}{2} - 2\hat{z}_0 \right)^2 \frac{3 - \sqrt{1 + 8\hat{z}_0}}{1 + \sqrt{1 + 8\hat{z}_0}}, \tag{15.22}$$

at the most dangerous wavelength given by

$$\hat{\lambda}_{\max} = \frac{1 + \sqrt{1 + 8\hat{z}_0}}{4} - \hat{z}_0. \tag{15.23}$$

In these expressions we should keep in mind that both Q^2 and $\hat{\lambda}$ are meant to be positive real quantities. Thus we readily see that a disk with $\hat{z}_0 > 1$ is hostile to tightly wound waves. The effects of finite thickness are stabilizing because the marginal stability condition becomes

$$Q = Q_{\max}(\hat{z}_0) < 1. \tag{15.24}$$

Thus a cooler disk is allowed by marginal stability for $\hat{z}_0 > 0$.

15.2.3 For the Quadratic Dispersion Relation of a Fluid Fully Self-Gravitating Disk with Finite Thickness

If we wish to address the marginal-stability condition for a fluid disk (not a fluid model), for which the velocity dispersion is isotropic, in the simple fully self-gravitating isothermal slab model, we should combine Eq. (15.24) with the relation deriving from the condition of vertical hydrostatic equilibrium [see the comment after Eq. (15.19) and the estimate given after Eq. (14.12)]:

$$Q^2 = 2\hat{z}_0. \tag{15.25}$$

This identifies the pair of values $(Q, \hat{z}_0)$ appropriate for a marginally stable thick fluid disk (with isotropic pressure), as illustrated in Fig. 15.4.

15.2.4 For the Zero-Thickness Case in Stellar Dynamics

If we now go back to the zero-thickness case but consider the stellar dynamical dispersion relation, we may set $\nu^2 = 0$ in Eq. (15.9) and find

$$\frac{k_0}{|k|} = 2 \frac{e^{-x}}{x} \sum_{n=1}^{\infty} I_n(x) = \frac{1 - e^{-x} I_0(x)}{x}, \tag{15.26}$$

where we have used a standard addition property of Bessel functions. If we keep the definition of $Q = c\kappa/\pi G\sigma$ that is convenient for the fluid model, we can rewrite the marginal stability condition as (see the right frame of Fig. 15.3)

$$Q = 2 \frac{1 - e^{-x} I_0(x)}{\sqrt{x}}. \tag{15.27}$$

The maximum of this curve,

$$Q^{(*)}_{\max} \approx 1.069, \tag{15.28}$$

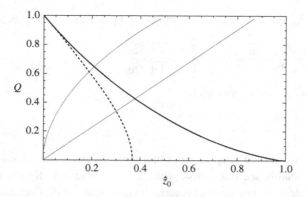

Fig. 15.4. The thick solid declining line represents the marginal stability condition $Q = Q_{\max}(\hat{z}_0)$ of Eq. (15.24) and the dashed declining curve the corresponding condition for the case in which the dilution of the gravity field (associated with the finite thickness of the disk) in the relevant dispersion relation is described by an exponential factor [instead of the rational factor used in Eq. (15.18)]. The thin rising curves represent the function $Q = Q(\hat{z}_0)$ for the two cases of a fully self-gravitating slab model [upper curve; see Eq. (15.25)] and for a non-self-gravitating layer (lower curve). (From Bertin, G., Amorisco, N. C. 2010. *Astron. Astrophys.*, **512**, A17.)

occurs at

$$x_{\max} \approx 0.9481. \tag{15.29}$$

This result[26] is usually given by the statement that the marginal stability condition for a stellar disk is the same as that for a fluid disk, provided that in the definition of Q the constant π is replaced with $\pi \times Q_{\max}^{(*)} \approx 3.3583$.

15.2.5   For the Cubic Dispersion Relation of the Fluid Model

From the preceding examples it has been shown that for the purpose of a simple summary of the stability analysis, we may refer to an effective stability parameter introduced as

$$Q_{\text{eff}} \equiv \frac{Q}{Q_{\max}}, \tag{15.30}$$

where $Q_{\max}$ is identified by the discussion of the relevant dispersion relation to retain the condition of marginal stability at $Q_{\text{eff}} = 1$. It is clear that such a simplification makes sense only to the extent that the dispersion relation involved is qualitatively similar to the quadratic equation for which the discussion of marginal stability leads to considering the parabola of Eq. (15.20). Otherwise, in the presence of significant differences, we should refer directly to the relevant parameter space (e.g., see the discussion of stability in Chapter 16 in the case in which the star and the gas components are dynamically decoupled). For these reasons, the discussion of the cubic dispersion relation was initially made in the regime of small J; the destabilizing effects of tangential forces were described in terms of the effective stability parameter defined by

$$\frac{1}{Q_{\text{eff}}^2} = \frac{1}{Q^2} + \frac{1}{4}J^2 Q^4. \tag{15.31}$$

When J is not small, the qualitative structure of the dispersion relation changes (the cubic admits one real root only; see following sections), and the most dangerous perturbations are those with $k = 0$ [i.e., $K^2 = 4m^2\epsilon_0^2 \equiv J^2/\chi_0(s)$]. The quantity $\chi_0(s)$ defined here depends on only the shear rate $s \equiv -d\ln\Omega/d\ln r$ and is written as

$$\chi_0(s) = \frac{s}{1 - s/2} = -\frac{4\Omega^2}{\kappa^2}\frac{d\ln\Omega}{d\ln r}. \tag{15.32}$$

Thus the marginal stability condition in this new regime can be drawn, for various values of s, in a (J, Q) diagram as curves defined by

$$\frac{Q^2}{4} = \frac{\sqrt{\chi_0(s)}}{J} - \frac{\chi_0(s)}{J^2[1 + \chi_0(s)]}. \tag{15.33}$$

These curves are bounded by two lines:

$$Q^2 < 1 + \chi_0(s), \tag{15.34}$$

$$J^2 > \frac{\sqrt{\chi_0(s)}}{1 + \chi_0(s)}. \tag{15.35}$$

Each of them should be considered as a necessary condition of instability in this new regime. The latter relation corresponds to the condition for the efficiency of the so-called swing amplifier.[27]

15.3 Wave Branches

A different way of looking at the dispersion relation is to take a given (usually real) value of the frequency ω (or the pattern frequency ω/m) and a fixed azimuthal number (or number of arms; usually taken to be positive) m and to ask which are the allowed real solutions for the radial wave number k. (Note that for given values of ω and m, the Doppler-shifted frequency ν is a function of r specified by the rotation curve in the basic state.) This leads to identification of the relevant wave branches of the dispersion relation. Thus the quadratic dispersion relation generally admits four separate wave branches (usually called *short trailing*, *long trailing*, *short leading*, and *long leading*). Recall that the dispersion relations described earlier all depend on k^2; that is, every trailing wave branch ($k < 0$) has its leading ($k > 0$) counterpart. The expressions for the various wave branch solutions are trivial for the quadratic dispersion relation (Fig. 15.5). Interest in the simple fluid model is largely based on the fact that in its principal range ($-1 < \nu < 1$), the more complicated Eq. (15.9) derived from stellar dynamics admits a set of four wave branches with similar properties (Fig. 15.6). There is one important difference to be emphasized: the trend at $|\nu| \to 1$, at which the short-wave branch tends to a finite wave number ($|\hat{k}| \to 4/Q^2$) in the quadratic dispersion relation, whereas the stellar dynamic wave number diverges. However, both dispersion relations fail at the Lindblad resonances, for which a separate analysis has to be carried out.

For a given value of ν, when Q is too large, there are no real solutions for k; that is, the four wave branches are lost. We can best appreciate the general situation by taking the reference case $Q = 1$ and by studying the properties of the various wave branches while Q is varied from its reference value.

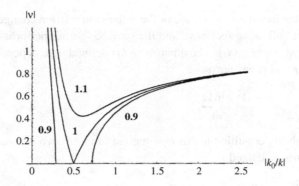

Fig. 15.5. Wave branches for the fluid model (quadratic dispersion relation) plotted for $Q = 0.9$, $Q = 1$, and $Q = 1.1$.

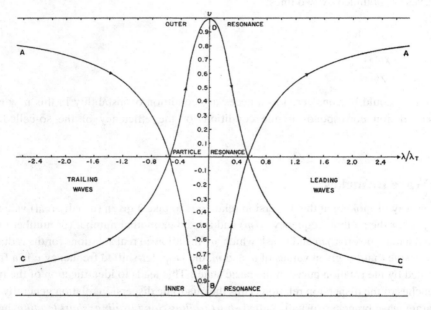

Fig. 15.6. Wave branches for the dispersion relation for tightly wound density waves in stellar dynamics (from Shu, F. H. "On the density-wave theory of galactic spirals. II: The propagation of the density of wave action," 1970. *Astrophys. J.*, **160**, 99; reproduced by permission of the AAS), for the marginally stable case (corresponding to the $Q = 1$ case of Fig. 15.5). Here λ/λ_T is the dimensionless radial wavelength k_0/k. Arrows indicate direction of group propagation.

The situation is qualitatively different for the cubic dispersion relation when J is not too small. The reason is that a cubic in general admits either one or three real roots. Thus we may have up to six wave branches. The condition for the transition from one to three real roots is given by

$$JQ^3 = \frac{16\sqrt{2}}{27}. \tag{15.36}$$

For small J, there are three roots; thus, besides short and long waves, a new (longer) wave branch may be available. This longer wave is not available if the real solution for K violates the geometric condition $K \geq 2m\epsilon_0$. In any case, for larger values of J (as specified by the preceding condition), the two long-wave branches are lost, and the dispersion relation admits as a possibility only the short-wave branch (of either the leading or trailing form), which actually may be quite open. Also here (as for the quadratic dispersion relation), if the disk is too hot, even the short-wave branch may be lost.

This simple discussion readily shows that the waves available in a disk have their properties set by the parameter regime of the basic state, which is specified by the values of Q and J. The parameter J is proportional to the disk density, and thus we see that light and heavy disks (in terms of ϵ_0) are expected to be characterized by qualitatively different dynamics.

15.3.1 Group Propagation and Wave Action

Following the theory of dispersive waves, each wave branch identified in a given dispersion relation $D(\omega, m, k; r) = 0$ is associated with well-defined propagation properties and energy content. These can be easily calculated from the dispersion relation using standard expressions. In particular, the radial group velocity is given by

$$c_g = -\frac{\partial \omega}{\partial k} = \frac{\partial D/\partial k}{\partial D/\partial \omega}, \tag{15.37}$$

and the wave-action density[28] for waves of amplitude a can be written as

$$A = \frac{\partial D}{\partial \omega} a^2, \tag{15.38}$$

so the angular-momentum density and energy density of the wave are given by

$$\mathcal{G} = m\mathcal{A}, \tag{15.39}$$

$$\mathcal{E} = \omega\mathcal{A}. \tag{15.40}$$

From these densities, the relevant fluxes are defined by multiplication by the group velocity. For example, for the action flux, we have

$$\mathcal{F} = c_g \mathcal{A}. \tag{15.41}$$

For many purposes, the most important piece of information that can be derived from the preceding expressions is the sign of each quantity; for example, we may wish to know whether wave propagation is inward or outward or in which direction angular momentum is carried by the wave. In the limit of tightly wound spiral structure [from either Eq. (15.9) or Eq. (15.11)], this is summarized by the following rules:

$$\mathrm{sgn}(c_g) = s_k s_\nu s_b, \tag{15.42}$$

$$\mathrm{sgn}(\mathcal{A}) = s_\nu, \tag{15.43}$$

where $s_\nu = \mathrm{sgn}(\nu)$, $s_k = \mathrm{sgn}(k)$, and $s_b = 1$ for long waves ($s_b = -1$ for short waves). Thus short trailing waves propagate away from the corotation circle.

Obviously, a linear theory has nothing to provide about the amplitude a of the wave. In contrast, it may provide important information on the amplitude profile of the wave, for example, by considering the constraint of conservation of wave action (where applicable) based on the preceding expressions.

The issue of identification of the energy density of the wave, especially in relation to the possibility of negative energy density (inside the corotation circle), is not at all trivial in a system governed by long-range forces.[29] The proper identification provides a powerful tool in the discussion of dynamical mechanisms and has led to the discovery of the mechanism of overreflection (see Subsection 15.5.3).

15.3.2 Propagation Diagrams

For a given basic state, the information on the availability of wave branches is well summarized by the so-called propagation diagrams. These are plots of the real solutions for the radial wave number

$$k = k(r; \omega, m) \tag{15.44}$$

that are obtained from the dispersion relation $D(\omega, m, k; r) = 0$ at fixed values of ω and m. The wave branches are thus followed as the radial coordinate is varied. Because of the inhomogeneity of the basic state, in some radial ranges, propagation may be forbidden if the dispersion relation does not admit real solutions for k. The complex solutions usually describe regions where the waves become evanescent (Fig. 15.7). Because, for a given basic state, ν is generally a monotonic function of r, and given the simple geometric interpretation of the quantity $\mu = rk/m$

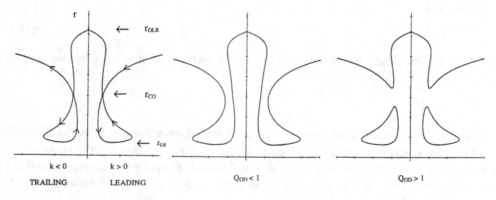

Fig. 15.7. Propagation diagrams based on the quadratic dispersion relation of the fluid model for a basic state that might support a tightly wound spiral mode. The Q profile is taken from Eq. (14.47) for three different values of Q_{OD} ($Q_{OD} = 1$ in the left frame). The radial coordinate is on the vertical axis. Outer Lindblad resonance (OLR) and corotation (CO) are marked. For $r < r_{ce}$, propagation is not allowed because the value of Q is too high (from Bertin, G., Lin, C. C. 1996. *Spiral Structure in Galaxies: A Density Wave Theory*, MIT Press, Cambridge, MA, p. 212; ©1996 Massachusetts Institute of Technology).

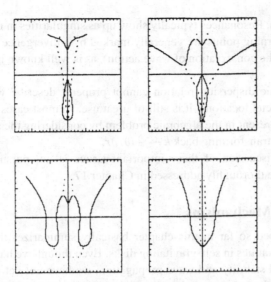

Fig. 15.8. Propagation diagrams based on the cubic dispersion relation for four basic states that are subject to four different types of global modes, prototypes of observed morphologies (from Bertin, G., Lin, C. C., Lowe, S. A., Thurstans, R. P. 1989. *Astrophys. J.*, **338**, 104). In each frame, the dimensionless frequency ν is plotted on the vertical axis, which runs from -1 (*bottom*) to $+1$ (*top*), and the quantity $(rk/2)$ is plotted on the horizontal axis, running from -15 to $+15$. Solid curves represent the wave branches, and dotted curves represent evanescent solutions. According to the cubic dispersion relation, the very long waves cannot reach Lindblad resonances; the wave branches are drawn up to $\nu = 0.75$.

[Eq. (13.33)], the propagation diagrams are often plotted as (Fig. 15.8)

$$\mu = \mu(\nu; \omega, m). \tag{15.45}$$

The latter representation more readily shows the behavior of the various wave branches in the vicinity of the relevant resonances ($\nu = 0, \pm 1$).

The propagation diagrams usually incorporate, by means of appropriate arrows, the information on the direction of group propagation given by Eq. (15.37). Thus we can easily follow how an imagined wave packet propagates in the inhomogeneous disk.

15.4 Turning Points and Resonances

It has often been mentioned that simple algebraic dispersion relations break down at resonances [e.g., see comments after Eq. (15.10)], at which waves can be emitted or absorbed. A similar failure occurs at turning points, that is, at the locations where different wave branches meet and interact with each other (linear wave interaction). The term *turning point* refers to the fact that at these locations, a wave packet, much like a particle against a potential barrier, can be turned back (or refracted back, changing from one wave branch to another), which is a frequent simple case.[30] Other, more complex behavior is also possible; in particular, besides the resonant effects due to corotating material, waves reaching the corotation circle display a very interesting turning-point behavior responsible for the subtle dynamical mechanism of overreflection (see Subsection 15.5.3). Resonances typically show up as singularities in the relevant dispersion relation; in contrast, turning points are generally marked by a divergence in wave amplitude (if we try to make use of the conservation of wave action), as is well known in the context of WKB analyses.

Even if the algebraic dispersion relation cannot properly describe wave propagation and absorption at these special locations, it is still of great use. In most cases we easily resolve the apparent singularities present in the algebraic problem by considering the associated differential problem[31] obtained by transforming back $k \to -i\partial/\partial r$.

At this stage the discussion of these important issues is not pursued because the most important points will be thoroughly addressed in Chapter 17.

15.5 Dynamical Mechanisms

What has been described so far in this chapter basically summarizes the main tools for the description of wave dynamics in self-gravitating disks. Even though we have discussed key concepts, such as marginal stability and wave propagation, we have not yet faced the issue of the potential physical situations that may describe the dynamics of spiral galaxies or any specific dynamical mechanisms that may affect the evolution of wave packets or the establishment of wave trains and global modes in models of galaxy disks. In this section, the key mechanisms that play an important role in the dynamics of density waves in a disk, mostly in view of a global modal analysis, are briefly outlined. A physical description of the main processes, with a focus on the astrophysical problem of spiral structure in galaxies, is amply provided in a separate monograph.[32] A full description of the associated mathematical theory is best appreciated by studying the regime of tightly wound spiral structure and will be the focus of Chapter 17. A summary of the astrophysical problem will be given in Chapter 18.

15.5.1 Parameter Regimes

From simple inspection of the propagation diagrams in a few cases, especially with reference to the cubic dispersion relation, it is clear that the dynamics of waves in a disk can be very different, depending on the parameter regime that characterizes the basic state. Various physical effects are governed by specific parameters. In this chapter, Q and J have already been introduced, which basically measure how hot and how heavy the disk is. We have also shown that thickness is naturally measured by $\hat{z}_0$. In Chapter 16 we will study the effects of the presence of a separate cold (gas) component that is characterized by relative temperature β and relative density α. All these important physical ingredients have to be taken into account in discussion of the astrophysical problem, and, indeed, it appears that we are able to relate the various morphological categories that are observed to specific domains in parameter space, thus setting up a dynamical framework for the classification of spiral galaxies (see Chapter 18).

One important aspect of the problem of wave dynamics is that the disk is intrinsically inhomogeneous. Thus the parameters mentioned earlier should not be thought of as constant quantities in a given galaxy model because they are functions of the radial coordinate. By inspecting the parameter profiles, we may judge where wave propagation is likely to occur (e.g., the bulge region is then found to be hostile to waves) and where certain dynamical mechanisms may develop (e.g., the outer disk in normal spiral galaxies often turns out to be gas-rich and thus may be considered to be cool from the dynamical point of view, with the possibility of wave excitation). In particular, it is the parameter regimes attained at small and large radii that set up the physical boundaries over which discrete global spiral modes can develop; this important physical aspect translates into the appropriate boundary conditions for the mathematical model that is aimed at describing realistic galaxy disks.

Marginal stability, as introduced earlier in terms of local dispersion relations, identifies a very important condition for the basic state. In fact, if the system is far from it on the side of instability, then it is expected to be subject to rapidly growing perturbations, which are bound to change the properties of the basic state in a short time (the dynamical time scale). For astrophysical applications, we often make this point by saying that violently unstable models are just the wrong choice of basic state. Observed systems are generally well beyond such a transient dynamical state, or, in other words, such a rapidly evolving dynamical state, even if present in a few systems at a certain time, would be very hard to catch by the observer. On the other side of marginal stability, if the disk is well within a (locally) stable regime (e.g., if the disk is too thick and/or too hot), then not only would the local instabilities be absent, but, unfortunately, wave propagation also would be inhibited altogether. Therefore, especially now that from the empirical point of view we have strong evidence for density waves, the relevant regimes in many cases *must* be close to marginal stability. It is up to the dynamicist to provide a plausible physical mechanism that is able to explain how a disk can be maintained close to such a marginal state of stability. In Chapter 16 we will discuss how, for the purpose, in the subtle case of light disks, self-regulation can be provided by cold dissipative gas.

In the simplest description of a one-component zero-thickness fluid disk we have seen that the (cubic) dispersion relation is characterized by two parameters. Thus marginal stability is best described in terms of the so-called (J, Q) diagram (Fig. 15.9), usually drawn in the $(\ln J, \ln Q^2)$

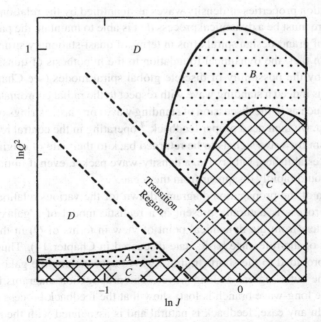

Fig. 15.9. The (J, Q) diagram (from Bertin, G., Lin, C. C., Lowe, S. A., Thurstans, R. P. 1989. *Astrophys. J.*, **338**, 104). The dotted region, from regime A to regime B, represents conditions of moderate instability. The disk self-gravity increases from left to right.

plane. The strip of moderate growth presents a clear discontinuity around a value of J that depends on the shear rate, typically ≈ 0.6 [see inequality (15.35)]. Light disks (low J) fall in the regime in which marginal stability is given by $Q \approx 1$, and wave propagation and stability are then very sensitive to the value of Q. Heavier disks may allow for considerably larger values of Q {in principle, up to $[1 + \chi_0(s)]^{1/2}$; see Eq. (15.34)}, and especially close to the transition region, they are not very sensitive to the precise value of Q. At low values of J, up to six wave branches may be available, whereas at higher values of J, wave propagation occurs in the form of either leading or trailing waves (these latter waves formally belong to the short branch but actually may be characterized by small values of k).

15.5.2   Feedback and Maintenance

Density waves were initially conceived as the natural way to resolve the winding dilemma, that is, the difficulty of explaining spiral arms in grand-design galaxies as quasi-stationary structures in the presence of differential rotation. The key interesting factor is obviously that the structure in a wave phenomenon is produced by collective effects, so the motion of the single particles that make a wave can be significantly decoupled from that of the wave. The wave interpretation of spiral structure, as mentioned at the beginning of this chapter, received a number of empirical confirmations by studies in which the spiral pattern is approximated by a single, quasi-stationary wave, with its pattern frequency as a free parameter to be determined from the observational fit; the approximation was made in terms of the short trailing branch identified by the dispersion relation for tightly wound spiral waves.

However, such a simplified, straightforward description has a conceptual difficulty, which was clear from the very beginning, and this is related to the phenomenon of group propagation. Given the propagation properties of density waves as quantified by the relations obtained earlier in this chapter, there must be a dynamical process that is able to maintain the pattern if we follow the interpretation of grand-design spiral arms in terms of quasi-stationary structure produced by density waves.[33] In fact, the dynamical foundation to the hypothesis of quasi-stationary spiral structure is given by the possibility of discrete global spiral modes (see Chapters 17 and 18). Each global mode is indeed a standing wave with respect to the radial coordinate, rotating rigidly around the disk. Much as in the discussion of standing waves on elastic strings or membranes, for a mode to be set up, we then need proper feedback[34] operating in the central regions of the disk, that is, a mechanism for wave signals to be returned back to their site of origin so that a proper wave cycle can be established. Otherwise, a density-wave packet, even if initially configured in the desired way, would rapidly disappear from the scene.

From inspection of the propagation diagrams drawn for the various relations, it is clear that such a feedback process is naturally present in a realistic model of a galaxy disk. The disk is bound to be hotter (from the dynamical point of view in terms of Q) in the central regions (see the properties of the reference basic state discussed in Chapter 14). Thus, for low values of J, feedback is provided to the short (trailing) waves approaching the galaxy center that are returned back to the outer regions as long waves. The propagation diagrams for the regime of higher J (where the long-wave branch is lost) show that the feedback process occurs by means of leading waves. In any case, feedback is natural and is associated with the intrinsic inhomogeneity of the disk, for which the center is generally protected by a kind of refractive wave barrier.

One important point to keep in mind in this context is the radial location of the inner Lindblad resonance (ILR) with respect to the central wave barrier. (Note that propagation diagrams drawn on the basis of the fluid-dispersion relations may be easily misinterpreted in this respect because they do not reveal immediately the presence of the Lindblad resonances.) If the ILR is exposed (i.e., if a short trailing wave can reach it before being returned back toward the outer regions), then the feedback process is inhibited because it can be shown that waves are completely absorbed at such a resonance in a stellar disk. In a fluid disk there is partial transmission across the resonance (see Chapter 16), so partial feedback can be ensured. However, because the central regions of the disk are star-dominated, we conclude that in general the necessary feedback cannot be provided for mode maintenance when the ILR is exposed. This interesting dynamical mechanism is responsible for one important feature in the wave dynamics of galaxy disks: The models that support a discrete spectrum of spiral modes usually have their growing modes limited to a set of two or three. This is why modes with large m and modes with too low a pattern frequency cannot be easily supported by a galaxy disk, which has a natural counterpart in the observed phenomenology (see Chapter 18).

15.5.3 Overreflection and Excitation

If a wave signal coming from the central regions of the disk impinges on the corotation zone, it is subject to a turning-point effect with consequent evolution into a pair of waves, a transmitted wave (moving farther out toward the outer disk) and a reflected wave (sent back toward the galaxy center). Simply because of conservation of wave action (i.e., without considering possible resonant effects that are due to corotating material), the fact that wave-action density is negative inside the corotation circle and positive outside it requires that the transfer of angular momentum toward the outer regions of the disk be accompanied by an amplification of the incoming wave signal, which is thus returned toward the center as a stronger reflected wave. This dynamical process is called *overreflection*[35] (Fig. 15.10).

In its original formulation, in the early 1970s, the process of overreflection was studied in the regime of only low-J dynamics, in which wave cycles involve short and long waves. Thus the picture given was that a long trailing wave propagating outward, while reaching the corotation circle, gets transmitted as a short trailing wave to the outer parts of the disk, with a stronger (reflected) short trailing wave being returned in the direction of the galaxy center. However, wave dynamics depends on the parameter regime considered, and we have seen that there are regimes in which the long-wave branch is lost. Almost ten years later it was found[36] that overreflection can occur in two different forms, one in the regime of low J with the properties of long-wave–short-wave interaction described earlier and the other, at higher J, in which overreflection operates by conversion of a leading wave into a pair of trailing waves. In other words, if the parameter regime does not allow for long waves, overreflection still can take place, but it must involve leading waves. This latter form of overreflection (of a leading into a stronger trailing wave) is precisely the counterpart of the so-called swing mechanism that was found (and is usually presented) in terms of the transient evolution of wave packets in the homogeneous shearing sheet.[37] This point finds further simple proof in that the criterion on disk density for onset of the amplification mechanism is indeed the same in both the description of overreflection given here and the time-dependent studies of the homogeneous shearing sheet [see inequality (15.35) and the following comment].

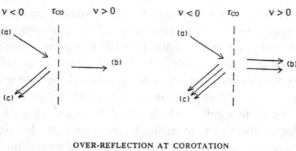

Fig. 15.10. Wave overreflection at corotation in the two different regimes of low *J* (type *A*) and of higher *J* (type *B*). The two upper frames show the effect of lowering the level of local stability at the corotation circle. The case on the left represents the marginal case, and the case on the right corresponds to local instability (after Bertin, G., Lin, C. C. 1996. *Spiral Structure in Galaxies: A Density Wave Theory*, MIT Press, Cambridge, MA, pp. 107, 223; ©1996 Massachusetts Institute of Technology). Arrows refer to the direction of group propagation; the number of arrows represents the intensity of the wave.

The amplification (or excitation) process just described has two important properties. First of all, it is distinct from the local (Jeans-type) instability addressed earlier in this chapter. In fact, if we consider the regime of very low *J*, significant amplification occurs even if the system is marginally stable, with $Q = 1$ (for a quantitative analysis of the amplification factor, see Chapter 17). The second important aspect is that the amplification process depends on the boundary conditions of the problem that is studied. Amplification occurs if the overall process transfers angular momentum outward. In this respect, trailing and leading waves have a different dynamical behavior, and the symmetry $k \rightarrow -k$ of the dispersion relation is broken. Therefore, the mechanism is very appealing in view of the excitation of large-scale structure with application to observed grand-design spirals (which are found to be generally trailing).

15.5.4 Wave Cycles for Self-Excited Modes

The mechanisms of feedback and overreflection described earlier can be combined in the form of a wave cycle operating as a resonant cavity responsible for unstable global spiral modes.[38] Thus, overreflection is sometimes called *waser*. The global modes grow by transferring angular momentum outward. If a wave cycle is set up between an inner wave barrier (providing feedback) and the corotation zone (providing amplification by means of overreflection), a standing-wave structure can be set up that grows exponentially in time. The growth rate corresponds to the amplification factor gained at the rate determined by the group propagation time required for a wave packet to travel along the wave cycle, following the relevant propagation diagram. The global modes are discrete because the wave packet imagined in the time-dependent description just given has to return to the corotation circle always with the right phase so as to imply, in a steady-state situation, positive angular momentum transport to the outer regions.

As can be appreciated by looking at the propagation diagrams drawn on the basis of the cubic dispersion relation, several options are available depending on the existing wave channels for a given galaxy model. Two reference cases have been identified. One, corresponding to the regime of light galaxy disks (low J), in which the relevant wave cycle can be all trailing, based on the long- and the short-wave branches, gives rise to self-excited normal (unbarred) spiral modes. The other, corresponding to the regime of heavy disks (high J), in which the relevant wave cycle is based on a leading and a trailing wave, generates self-excited barred spiral modes (which are characterized by a two-blob structure inside the corotation circle owing to the superposition of the leading wave on the trailing wave). Although the study of global modes in the general case requires the use of numerical integration, for the two regimes just mentioned, a full analytical solution can be carried out. This will be given for the regime of low J in Chapter 17. (For the other regime, it can be worked out by complete analogy.)

15.6 Equations of the Homogeneous Shearing Sheet

For completeness, at the end of this chapter the basic equations for density waves in the homogeneous shearing sheet are given; a detailed comparison between the properties of this homogeneous model and those that can be derived directly from a description that incorporates the important role of inhomogeneity (in particular, from the cubic dispersion relation) is given in the article that is the basis for this short section.[39]

We focus on a reference point in a fluid axisymmetric disk, at radius r_c with $\theta = 0$, and assume that all the quantities that characterize the basic state are constant, except for the angular velocity, which is taken to vary in a linear fashion (constant shear rate s), that is,

$$\Omega(r) = \Omega_c[1 - s(r - r_c)/r_c], \tag{15.46}$$

so the local value of the epicyclic frequency is given by $\kappa_c^2 = 4\Omega_c^2(1 - s/2)$. The standard polar cylindrical coordinate system (r, θ) may be replaced with (x, y), where $x = r_c \ln(r/r_c)$ and $y = r_c\theta$, which thus defines a local Cartesian system of coordinates centered on our reference point. We then take a Fourier representation for any physical quantity associated with a linear perturbation Ψ on such a model

$$\Psi(x, y, t) = \sum_{m=-\infty}^{\infty} \int_{-\infty}^{\infty} \Psi_m(\xi, t) \exp[i(\xi x - \eta y)] \, d\xi, \tag{15.47}$$

where $\eta = m/r_c$ is the azimuthal wave number. With respect to the radial coordinate, this corresponds to a decomposition in logarithmic spirals.

The equations of the motion and continuity in the plane of the disk are then studied, supplemented by an approximate local relation between perturbed density and perturbed potential, which is the same as the local WKB relation adopted in the derivation of the cubic dispersion relation described earlier in this chapter (see also Chapter 17 for a quantitative analysis). The general solution for Ψ in such a model can be written as

$$\Psi_m(\xi,t) = \sum_{i=1}^{2} f_i^{(m)}(\xi + s\Omega_c\eta t)\hat{\Psi}_i^{(m)}(\xi),$$ (15.48)

where $f_i^{(m)}$ are two arbitrary functions deriving from the conservation equation for the vorticity, and $\hat{\Psi}_i^{(m)}(\xi)$, representing the waves, are two linearly independent solutions of the time-independent oscillator equation in the pitch-angle variable $\mu \equiv \xi/\eta$,

$$\frac{d^2\hat{\Psi}^{(m)}}{d\mu^2} + \frac{s\chi_0(s)}{4}\left[1 - K + \frac{1}{4}Q^2K^2 - \frac{J^2}{K^2} + \frac{3}{4}\frac{J^4}{K^4}\left(1 - \frac{s}{2}\right)\right]\hat{\Psi}^{(m)} = 0.$$ (15.49)

Here J, Q, and $\chi_0(s)$ are the parameters defined earlier that are taken to be constant (i.e., calculated at $r = r_c$), and [see Eq. (15.15)]

$$K^2 = (2m\epsilon_0)^2(\mu^2 + 1).$$ (15.50)

By a proper choice of the two arbitrary functions $f_i^{(m)}$, we can describe[40] either time-dependent swinging wave packets, corresponding to the solutions obtained by separation of variables in a comoving coordinate system,[41] or steady wave trains, which are more natural in a Eulerian description.

Notes

1. Lindblad, B. 1963. *Stockholm Observ. Ann.*, **22**, 3.
2. The beginning of the density-wave theory is generally associated with the two articles: Lin, C. C., Shu, F. H. 1964. *Astrophys. J.*, **140**, 646; 1966. *Proc. Natl. Acad. Sci. USA*, **55**, 229. The second paper includes a quantitative discussion of the contribution of interstellar gas.
3. Starting especially with Toomre, A. 1964. *Astrophys. J.*, **139**, 1217; Kalnajs, A. J. 1965. Ph.D. dissertation, Harvard University, Cambridge, MA.
4. The idea of using the fluid model as a quick tool of investigation is already apparent, for example, in the above-mentioned paper by Lin and Shu, 1964, and in the study by Goldreich, P., Lynden-Bell, D. 1965. *Mon. Not. Roy. Astron. Soc.*, **130**, 125. A more systematic use of the fluid-model equations for density waves starts later; see Feldman, S. I., Lin, C. C. 1973. *Stud. Appl. Math.*, **52**, 1. A separate line of research has aimed at a description of a collisionless stellar disk by means of fluid equations, that is, has tried to develop the model of stellar hydrodynamics; see Hunter, C. 1970. *Stud. Appl. Math.*, **49**, 59; 1979. *Astrophys. J.*, **227**, 73; Berman, R. H., Mark, J. W.-K. 1977. *Astrophys. J.*, **216**, 257.
5. After the pioneering articles mentioned earlier, we should mention especially the work by Shu, F. H. 1968. Ph.D. dissertation, Harvard University, Cambridge, MA; 1969. *Astrophys. J.*, **158**, 505; 1970. *Astrophys. J.*, **160**, 89 and 99; and the long series of papers by Mark, J. W.-K. 1971. *Proc. Natl. Acad. Sci. USA*, **68**, 2095; 1974. *Astrophys. J.*, **193**, 539; 1976. *Astrophys. J.*, **203**, 81; 1976. *Astrophys. J.*, **205**, 363; 1976. *Astrophys. J.*, **206**, 418; 1977. *Astrophys. J.*, **212**, 645; and Bertin, G., Lau, Y. Y., Lin, C. C., Mark, J. W.-K., Sugiyama, L. 1977. *Proc. Natl. Acad. Sci. USA*, **74**, 4726; Bertin, G., Mark, J. W.-K. 1978. *Astron. Astrophys.*, **64**, 389; Bertin, G. 1980. *Phys. Rep.*, **61**, 1; Bertin, G., Haass, J. 1982. *Astron. Astrophys.*, **108**, 265; Bertin, G. 1983. *Astron. Astrophys.*, **127**, 145.
6. Kalnajs, A. J. 1971. *Astrophys. J.*, **166**, 275; 1976. *Astrophys. J.*, **205**, 745 and 751; 1977. *Astrophys. J.*, **212**, 637. See also Lynden-Bell, D., Kalnajs, A. J. 1972. *Mon. Not. Roy. Astron. Soc.*, **157**, 1, in which an action-angle variable study is reduced to the epicyclic limit. Some exact analyses take advantage of the special properties of the assumed basic state: Kalnajs, A. J. 1972. *Astrophys. J.*, **175**,

63, studies the stability of the collisionless Maclaurin disk; and Zang, T. A. 1976. Ph.D. dissertation, Massachusetts Institute of Technology, Cambridge, MA, investigates the stability of the collisionless self-similar disk.

7. Goldreich, P., Lynden-Bell, D. 1965. *Mon. Not. Roy. Astron. Soc.*, **130**, 125.

8. Goldreich, P., Lynden-Bell, D. 1965. *op. cit.*; Julian, W. H., Toomre, A. 1966. *Astrophys. J.*, **146**, 810. Goldreich, P., Tremaine, S. 1978. *Astrophys. J.*, **222**, 850; Toomre, A. 1981. In *The Structure and Evolution of Normal Galaxies*, eds. S. M. Fall, D. Lynden-Bell. Cambridge University Press, Cambridge, UK, p. 111; Toomre, A., Kalnajs, A. J. 1991. In *Dynamics of Disk Galaxies*, ed. B. Sundelius. Göteborg University, Göteborg, Sweden, p. 341. The demonstration of the group propagation of density-wave packets by Toomre, A. 1969. *Astrophys. J.*, **158**, 899, is also carried out in the shearing sheet.

9. See also Hunter, C. 1983. In *Fluid Dynamics in Astrophysics and Geophysics*, ed. N. R. Lebovitz, American Mathematical Society, Providence, RI, p. 179. For the problem of magnetic reconnection in electromagnetic plasmas, the matching of the the *inner* solution for the slab model (with constant magnetic shear) with the *outer* solutions is indeed the main focus of attention; see Furth, H. P., Killeen, J., Rosenbluth, M. N. 1963. *Phys. Fluids*, **6**, 459.

10. Goldreich, P., Lynden-Bell, D. 1965. *op. cit.*

11. Drury, L. O. C. 1980. *Mon. Not. Roy. Astron. Soc.*, **193**, 337; Lin, C. C., Thurstans, R. P. 1984. In *Plasma Astrophysics*, eds. J. Hunt, T. D. Guyenne. ESA SP-207, European Space Agency Publications Division, Noordwijk, The Netherlands, p. 121.

12. Orr, W. McF. 1907. *Proc. Roy. Irish Acad. Sect. A*, **27**, 9 and 69; Marcus, P., Press, W. H. 1977. *J. Fluid Mech.*, **79**, 525.

13. For example, in the study of ballooning modes. Here great care has been taken with the issue of a proper matching with the outer equations; see Pegoraro, F., Schep, T. J. 1986. *Plasma Phys. Cont. Fusion*, **28**, 647.

14. A detailed derivation can be found in the original papers, especially Shu, F. H. 1970. *Astrophys. J.*, **160**, 89 and 99; and in the Brandeis Lectures Series by Lin, C. C., Shu, F. H. 1968. In *Astrophysics and General Relativity*, Vol. 2, eds. M. Chrétien, S. Deser, J. Goldstein. Gordon & Breach, New York, p. 239.

15. Bernstein, I. B. 1958. *Phys. Rev.*, **109**, 10.

16. These studies have been carried out systematically by Mark, starting with 1971. *Proc. Natl. Acad. Sci. USA*, **68**, 2095; at the Lindblad resonances ($\nu = \pm 1$) the solution that formally, from Eq. (15.9), would have an infinite wave number is regularized in one with finite but *complex* wave number, describing the processes of absorption by means of Landau damping. A synthetic physical description of the various processes of energy emission and absorption at the resonances is given by Lynden-Bell, D., Kalnajs, A. J. 1972. *op. cit.*

17. For example, see Lin, C. C., Yuan, C., Shu, F. H. 1969. *Astrophys. J.*, **155**, 721; Roberts, W. W., Roberts, M., Shu, F. H. 1975. *Astrophys. J.*, **196**, 381; Visser, H. C. D. 1978. Ph.D. thesis, Groningen University, The Netherlands.

18. For example, see Bertin, G., Lin, C. C. 1996. *Spiral Structure in Galaxies: A Density Wave Theory*, MIT Press, Cambridge, MA.

19. One major concern was related to the very existence of the long-wave branch, associated with Eq. (15.9) or Eq. (15.11), because its limit ($k \to 0$) is obviously incompatible with the WKB ordering of Eq. (15.8), and it was even conjectured that the long-wave branch might disappear altogether. Attempts were thus made to retain some m-dependent terms that were naturally expected to be present just by simple geometric arguments; e.g., see Lynden-Bell, D., Kalnajs, A. J. 1972. *op. cit.*; but these early attempts lacked a clear asymptotic justification.

20. The derivation of the dispersion relation for the fluid model based on this new ordering was given by Lau, Y. Y., Bertin, G. 1978. *Astrophys. J.*, **226**, 508; the dispersion relation was first announced by Bertin, G. 1978. In *Structure and Properties of Nearby Galaxies*, eds. E. M. Berkhuijsen, R. Wielebinski. Reidel, Dordrecht, The Netherlands, p. 128.

21. Bertin, G., Lin, C. C., Lowe, S. A., Thurstans, R. P. 1989. *Astrophys. J.*, **338**, 104.

22. The full derivation, which starts from the same set of equations as those given in Chapter 17, and a discussion of the limits of the cubic dispersion relation can be found in the paper by Lau, Y. Y., Bertin, G. 1978. *op. cit.*; and in appendix A of the paper by Bertin, G., Lin, C. C., Lowe, S. A., Thurstans, R. P. 1989. *op. cit.*

23. The modes shown by Bertin, G., Lau, Y. Y., Lin, C. C., Mark, J. W.-K., Sugiyama, L. 1977. *Proc. Natl. Acad. Sci. USA*, **74**, 4726, are based on stellar dynamical equations that include the effects of finite inclination of spiral arms, as derived by Bertin, G., Mark, J. W.-K. 1978. *Astron. Astrophys.*, **64**, 389. In general, the equations of stellar dynamics are found to be more sensitive to J. Furthermore, we should keep in mind that for very open waves, the relevant velocity dispersion (the one that enters the stability and propagation discussion) is the tangential dispersion, which in the anisotropic stellar disk is reduced by the factor $\kappa/2\Omega$.

24. To lowest order, this correction is identical to the exponential correction mentioned by Toomre, A. 1964. *Astrophys. J.*, **139**, 1217; and followed in the Brandeis Lectures Series by Lin and Shu. 1968. *op. cit.*; the polynomial correction adopted here appears to be preferable in view of the analysis of Vandervoort, P. O. 1970. *Astrophys. J.*, **161**, 87; and of Yue, Z. Y. 1982. *Geophys. Astrophys. Fluid Dyn.*, **20**, 1.

25. See Safronov, V. S. 1960. *Ann. Astrophys.*, **23**, 979, who states that a qualitatively similar result had been obtained earlier by Gurevich, L. E., Lebedinsky, A. I. 1950. *Izvestia Academii Nauk USSR*, **14**, 765. The Jeans instability in the presence of differential rotation was examined in cylindrical geometry by Bel, N., Schatzman, E. 1958. *Rev. Mod. Phys.*, **30**, 1015.

26. Toomre, A. 1964. *Astrophys. J.*, **139**, 1217.

27. Toomre, A. 1981. In *The Structure and Evolution of Normal Galaxies*, eds. S. M. Fall, D. Lynden-Bell. Cambridge University Press, Cambridge, UK, p. 111.

28. See also Whitham, G. B. 1965. *J. Fluid Mech.*, **22**, 273; Bretherton, F. P., Garrett, C. J. R. 1968. *Proc. Roy. Soc. London, A*, **302**, 529.

29. See also Synge, J. L. 1974. *Hermathena*, **117**. The conservation of wave action is the natural interpretation of a second-order analysis of WKB studies. For tightly wound density waves within stellar dynamics, the concept of density of wave action became established in the late 1960s; see Toomre, A. 1969. *Astrophys. J.*, **158**, 899; Shu, F. H. 1970. *Astrophys. J.*, **160**, 99, and their reference to the work of Kalnajs.

30. This situation is well studied in many different fields of research; see Budden, K. G. 1985. *The Propagation of Radio Waves*, Cambridge University Press, Cambridge, UK; Cairns, R. A. 1991. *Radiofrequency Heating of Plasmas*, Hilger, Bristol, UK.

31. Or, for resonant absorption, by carefully studying the situation in which k is allowed to take on a complex value. A systematic analysis of these issues has been given by J. W.-K. Mark in his papers (quoted earlier in this chapter).

32. Bertin, G., Lin, C. C. 1996. *op. cit.*

33. This point was stressed as a serious concern especially by Toomre, A. 1969. *Astrophys. J.*, **158**, 899.

34. This is well described by Lin, C. C. 1970. In *IAU Symposium*, Vol. 38, eds. W. Becker, G. Contopoulos. Reidel, Dordrecht, The Netherlands, p. 377.

35. This important dynamical mechanism was discovered by Mark, J. W.-K. 1974. In *IAU Symposium*, Vol. 58, ed. J. R. Shakeshaft. Reidel, Dordrecht, The Netherlands, p. 417 (see his fig. 3 and the related turning-point equation); 1976. *Astrophys. J.*, **205**, 363. In the context of plasma physics, it is related to processes involving negative energy waves; see Sturrock, P. A. 1960. *J. Appl. Phys.*, **31**, 2052; Kadomtsev, B. B., Mikhailovsky, A. B., Timofeyev, A. V. 1964. *Zh. Eksper. Teor. Fiz.*, **47**, 2266.

36. Bertin, G. 1983. In *IAU Symposium*, Vol. 100, ed. E. Athanassoula. Reidel, Dordrecht, The Netherlands, p. 119; a complete description is given in the papers by Bertin, G., Lin, C. C., Lowe, S. A. 1984. In *Plasma Astrophysics*, eds. J. Hunt, T. D. Guyenne. ESA SP-207, European Space Agency Publications Division, Noordwijk, The Netherlands, p. 115; and Bertin, G., Lin, C. C., Lowe, S. A., Thurstans, R. P. 1989. *Astrophys. J.*, **338**, 104.

37. Goldreich, P., Lynden-Bell, D. 1965. *op. cit.*; Julian, W. H., Toomre, A. 1966. *op. cit.*; Toomre, A. 1981. *op. cit.*

38. With a clear analogy to the laser process. See Mark, J. W.-K. 1976. *Astrophys. J.*, **205**, 363; 1977. *Astrophys. J.*, **212**, 645.

39. Here we follow the discussion by Lin, C. C., Thurstans, R. P. 1984. In *Plasma Astrophysics*, eds. J. Hunt, T. D. Guyenne. ESA SP-207, European Space Agency Publications Division, Noordwijk, The Netherlands, p. 121.

40. This was shown explicitly by Lin and Thurstans, 1984, in the article on which this section is based.

41. As was done by Kelvin, 1877, quoted by Marcus, P., Press, W. H. 1977. *J. Fluid Mech.*, **79**, 525; see also Orr, W. McF. 1907. *Proc. Roy. Irish Acad. Sect. A*, **27**, 9 and 69.

16   Roles of Gas

In Part I we saw that galaxy disks can be thought of as basically comprising two components, Population I and Population II. One component is dominated by cold gas, in atomic or molecular form, but contains significant amounts of stars recently born in the interstellar medium. This component is in a thin layer (at least within the bright optical disk) and is characterized by very low velocity dispersion (often below 10 km s^{-1}) with respect to the circular motions associated with the differential rotation. The other component, Population II, is dominated by relatively old stars, in a thicker layer, and is characterized by higher velocity dispersions; that is, it is warmer from the dynamical point of view. Such separation is only a simplifying tool for dynamical investigations, whereas, in reality, continuous changes of dynamical properties are associated with the many components of a galaxy disk. (The properties of the extraplanar gas were briefly summarized at the end of Chapter 14.)

The gas-dominated component is characterized by small epicycles. Thus, from the dynamical point of view, it is naturally responsible for small-scale features in the disk, which are probably rapidly evolving. (The large-scale spiral structure that will be described in the following chapters then must draw its main support from the stars.) Furthermore, being cold, the gaseous disk is expected to provide an important contribution to the Jeans instability of the disk[1] and, in this sense, to be an important source of excitation also for large-scale spiral modes.

Another important property of the Population I subsystem is that it is characterized by significant dissipation, which acts on a short time scale by means of cloud-cloud collisions. Being cold and dissipative, it is subject to shocks,[2] which can provide a saturation mechanism at finite amplitudes for the growing global spiral modes associated with grand-design spiral structure.[3] Being dissipative, the gas component can provide a dynamical thermostat for the two-component disk by means of a process of self-regulation, so the conditions for the occurrence of large-scale normal spiral structure can be maintained for a relatively long time.[4] *Self-regulation* is a term used in a variety of contexts (e.g., in the case of the chemical reactions that participate in the combustion of flames) to indicate the presence of competing mechanisms, usually characterized by multiple time scales, that manage to bring and keep a system in a quasi-steady state. Thus a system under self-regulation is expected to evolve slowly and to be associated with a fairly small region of the relevant parameter space.[5] Self-regulation processes within the gas may determine some properties of the interstellar medium.[6] In this chapter the discussion is restricted to a self-regulation process that is expected to have an impact on the large-scale spiral structure in galaxies.

In many galaxies, the gas distribution is rather diffuse. As a result, its dynamical impact is concentrated mainly in the outer regions, beyond two exponential scale lengths of the stellar disk. Thus the processes mentioned earlier, which are very important for relatively light disks, tend to operate mainly in the outer disk; this explains why corotation, for normal (nonbarred) spiral structure, is generally expected in the outer disk.

The Population I subsystem behaves mainly as a fluid. As such, it is less vulnerable to Lindblad resonance absorption, in contrast to the stars, for which strong absorption occurs by means of Landau damping.

It is the purpose of this chapter to illustrate some key dynamical aspects of a two-component system in which the two components will often be called simply *gas* and *stars*. The topics addressed here are relevant to the physical foundation of the global modal analysis presented in Chapter 17 and to the discussion of the problem of spiral structure in galaxies given in Chapter 18. It should be emphasized that the roles of gas mentioned earlier must be properly taken into account as part of a nontrivial modeling process if we want to apply our dynamical studies to the observed spiral structure in galaxies. This is especially important for relatively light disks and for normal (nonbarred) spiral structure. In relation to large-scale open and barred spiral structure, the gas is expected to play mainly a passive role (particularly in the form of large-scale shocks).

16.1 Waves and Effective Stability in a Two-Component Disk

We consider a zero-thickness disk made of two components and follow the assumptions that lead to the quadratic dispersion relation [Eq. (15.11)] for linear perturbations. The two components are labeled by a subscript i, which takes on the values g (for gas) and $\star$ (for stars). Both components are treated as a fluid; the gaseous disk is cooler $(c_g^2 < c_\star^2)$ and generally lighter $(\sigma_g < \sigma_\star)$ (but we will keep open the possibility that $\sigma_g > \sigma_\star$).

16.1.1 Two-Fluid, Zero-Thickness Dispersion Relation

The density response to a perturbed potential Φ_1, as calculated for each component from the equations of continuity and motion, is[7]

$$\sigma_{1i} = -\frac{\sigma_i}{c_i^2(1-v^2)} x^{(i)} \mathcal{F}_v^{(i)} \Phi_1,$$ (16.1)

where $\mathcal{F}_v^{(i)}$ is the fluid reduction factor introduced after Eq. (15.13) referred to the component i and $x^{(i)} = k^2 c_i^2/\kappa^2$. To leading order for tightly wound waves, the Poisson equation requires that

$$-|k|\Phi_1 = 2\pi G(\sigma_{1g} + \sigma_{1\star}).$$ (16.2)

By referring all the quantities to the stellar component, that is, by setting

$$Q_\star \equiv \frac{c_\star \kappa}{\pi G \sigma_\star}, \qquad \hat{k} \equiv 2\frac{\pi G \sigma_\star}{r\kappa^2}rk, \qquad \alpha \equiv \frac{\sigma_g}{\sigma_\star}, \qquad \beta \equiv \frac{c_g^2}{c_\star^2},$$ (16.3)

we can write the dispersion relation obtained by eliminating Φ_1 as

$$D(v, \hat{k}; Q_\star, \alpha, \beta) = 0,$$ (16.4)

where

$$D = (1-v^2)^2 + (1-v^2)|\hat{k}|[(Q_\star^2/4)(1+\beta)|\hat{k}| - (1+\alpha)]$$
$$+ (Q_\star^2/4)|\hat{k}|^3[(Q_\star^2/4)\beta|\hat{k}| - (\alpha+\beta)]. \tag{16.5}$$

This is a quartic in $|k|$.

Note that the one-component limit obtained by setting $\alpha \to 0$ gives

$$D \to [1 - v^2 + \beta(Q_\star^2/4)\hat{k}^2][1 - v^2 + (Q_\star^2/4)\hat{k}^2 - |\hat{k}|], \tag{16.6}$$

which still retains the possibility of sound waves in the gas tracer in addition to the density waves in the stellar component [see Eq. (15.13)]. We obtain another one-component limit by taking $\beta \to 1$ and noting that the density $\sigma_\star(1+\alpha)$ can be replaced with a total mass density σ. In this case, by rescaling the definitions of $Q_\star$ and $\hat{k}$, the dispersion function is found to tend to

$$D \to [1 - v^2 + (Q^2/4)\hat{k}^2 - |\hat{k}|]^2, \tag{16.7}$$

as desired.

In the principal range defined by $v^2 < 1$ (see the discussion following the stellar dynamic dispersion relation for tightly wound waves in Chapter 15), the waves are even, in the sense that much as in one mode of oscillation for two coupled pendula, the two components are perturbed in phase. This is apparent from the relation

$$\frac{\sigma_{1g}}{\sigma_{1\star}} = \alpha \frac{1 - v^2 + x^{(\star)}}{1 - v^2 + x^{(g)}} = \alpha \frac{1 - v^2 + (Q_\star^2/4)\hat{k}^2}{1 - v^2 + \beta(Q_\star^2/4)\hat{k}^2}, \tag{16.8}$$

which gives the relative response of the two components. This simple relation also shows how the gas response can indeed be large at short wavelengths.

16.1.2   Marginal Stability and Decoupling

We study marginal stability by setting $v^2 = 0$ in the dispersion relation Eq. (16.5). By referring to the dimensionless wavelength $\hat{\lambda} \equiv 1/|\hat{k}|$, we have

$$\frac{Q_\star^2}{4} = \left(\frac{\hat{\lambda}}{2\beta}\right)\left[(\alpha+\beta) - \hat{\lambda}(1+\beta) + \sqrt{\hat{\lambda}^2(1-\beta)^2 - 2\hat{\lambda}(1-\beta)(\alpha-\beta) + (\alpha+\beta)^2}\right]. \tag{16.9}$$

This is to be considered in the domain $0 < \hat{\lambda} < 1 + \alpha$. The two limits $\alpha \to 0$ and $\beta \to 1$ can be easily checked to yield the one-component limit, that is, the parabola of Eq. (15.20).

A simple numerical study gives the maximum of the marginal stability curve in the $(\hat{\lambda}, Q_\star^2)$ plane for given values of the relative density and relative temperature, that is, the function (Fig. 16.1)

$$Q_{\max}^2 = Q_{\max}^2(\alpha, \beta) > 1. \tag{16.10}$$

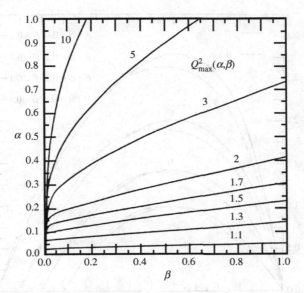

Fig. 16.1. Contours of $(Q_{max})^2$ in the (β, α) plane (from Bertin, G., Romeo, A. B. 1988. *Astron. Astrophys.*, **195**, 105).

The maximum is reached at a wavelength $\hat{\lambda}_{max}(\alpha, \beta)$ ($\rightarrow 1/2$ in the one-component limit). We can thus proceed to define an effective stability parameter

$$Q_{eff} = \frac{Q_\star}{Q_{max}(\alpha, \beta)} \tag{16.11}$$

following the discussion leading to Eq. (15.30). If the gas is not too cold ($\beta > 0.03$), the shape of the marginal-stability curve is basically the same as that for the one-component quadratic dispersion relation (Fig. 16.2); in addition, the curves drawn for different values of α and β resemble the curves drawn by means of a more complicated model in which the stars are treated by the equations of stellar dynamics [for this purpose, it is sufficient to replace the reduction factor $\mathcal{F}_\nu^{(\star)}$ in Eq. (16.1) with its appropriate stellar dynamical counterpart,[8] Eq. (15.10)]. Thus in this regime the use of the effective stability parameter Q_{eff} is sufficient to capture the effects of the presence of gas on the Jeans stability of the disk. Here, for small values of α, we have $Q_{max}^2 \approx 1 + 4\alpha$.

A qualitatively different behavior is found when the gas component is sufficiently cold (Fig. 16.3) and can be summarized by a phase diagram drawn in the (β, α) plane (Fig. 16.4). Here a two-phase region can be identified roughly with a triangle with its vertex at a triple point defined by $(\beta_0, \alpha_0) = (17 - 12\sqrt{2}, 3 - 2\sqrt{2}) \approx (0.03, 0.17)$. The other two points that define this triangular region are the origin and the point on the vertical axis at $\alpha = 1/8$. For regimes inside such a two-phase region of the plane, the marginal-stability curve presents two peaks. In particular, below the transition curve defined by

$$\alpha^2 = \beta, \tag{16.12}$$

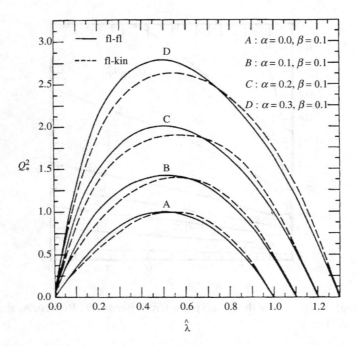

Fig. 16.2. Marginal stability curves showing the destabilizing role of gas in a two-fluid analysis (from Bertin, G., Romeo, A. B. 1988. *Astron. Astrophys.*, **195**, 105) compared with the fluid-kinetic curves (readapted from Lin, C. C., Shu, F. H. 1966. *Proc. Natl. Acad. Sci. USA*, **55**, 229).

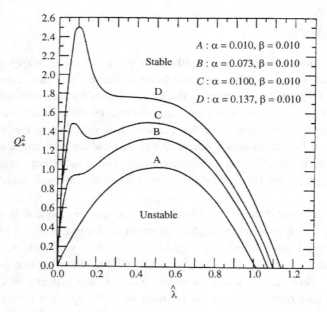

Fig. 16.3. Marginal-stability curves showing the decoupling of the gas component from the stellar component observed when the gas is cold (from Bertin, G., Romeo, A. B. 1988. *Astron. Astrophys.*, **195**, 105).

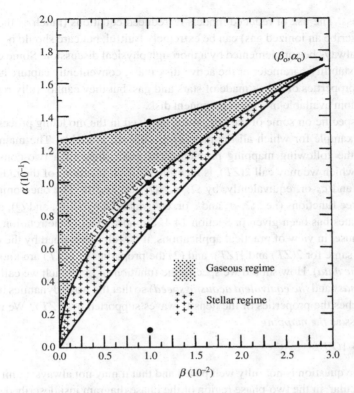

Fig. 16.4. Phase diagaram for a two-fluid system representing a disk made of stars and gas (from Bertin, G., Romeo, A. B. 1988. *Astron. Astrophys.*, **195**, 105). The four black dots mark the locations of cases *A, B, C, D* of Fig. 16.3.

the stellar peak (i.e., the one at larger wavelengths) is higher; above the transition curve, the gaseous peak (which develops at small wavelengths) dominates. At the triple point, the marginal-stability curve in the $(\hat{\lambda}, Q_\star^2)$ plane exhibits only one flat maximum with $Q_{max}^2(\alpha_0, \beta_0) = 2$ at $\hat{\lambda}_{max}(\alpha_0, \beta_0) = 1 - \sqrt{2}/2 \approx 0.29$. Note that on the transition curve defined by $\alpha^2 = \beta$ we have $Q_g \equiv c_g \kappa / \pi G \sigma_g = Q_\star$.

These results show that even a relatively small amount of cold gas can have a significant impact on the stability and wave dynamics of a disk. In the presence of gas, a stellar disk can be relatively warm ($Q_\star > 1$), and still the combined disk of stars and gas can be Jeans-unstable. The effective stability parameter Q_{eff} conveniently summarizes the destabilizing role of gas. However, in the two-phase region of the parameter space, Q_{eff} should be used cautiously because there the disk has a qualitatively different behavior with respect to the one-component case. In fact, the presence of the two peaks proves that the two components are, to a large extent, decoupled from the dynamical point of view. Thus some of the related wave dynamics escapes any one-component description. This point is further clarified in the following discussion.

16.2 Modeling Process for a Disk of Stars and Gas

The importance of gas and the complexity of two-component systems make it clear that if we plan to apply what has been learned from Chapter 15 to spiral galaxies, we have to face a very subtle modeling process. The use of reduced models or reduced equations (much as the use of MHD equations to describe an ionized gas) can be extremely fruitful, but care should be taken, and the models should always be supplemented by a thorough physical discussion. Some concepts, such as the effective stability parameter or the active disk mass, conveniently capture and summarize many physical properties of a disk made of stars and gas, but they cannot fully represent all the degrees of freedom available in a two-component disk.

To be more specific on some of the key issues involved in the modeling process, we address here a simple example for which all the necessary tools are available. The main question can be reduced to the following mapping problem. The basic state of a two-component, zero-thickness disk, which we may call $2(ZT)$, is identified by five functions of the radial coordinate: Ω, $\sigma_\star$, $c_\star$, σ_g, and c_g or, equivalently, by Ω, $\sigma_\star$, $Q_\star$, α, and β. The one-component disk is identified by three functions (i.e., Ω, σ, and c or, equivalently, by Ω, σ, and Q); a reference set of such basic states has been given in Section 14.4. Suppose that (1) the rotation curve is fixed (especially because, in view of practical applications, it is well determined by the observations), that is, Ω is the same for $2(ZT)$ and $1(ZT)$, and (2) the properties of $2(ZT)$ are known (they may thus be called *the data*). How should we specify the functions σ, c (which we call *the model*, or *the active disk mass* and *the equivalent acoustic speed*) so that the wave dynamics in the resulting $1(ZT)$ best matches the properties of the density waves supported by $2(ZT)$? We refer to such a modeling process as *the mapping*

$$2(ZT) \rightarrow 1(ZT). \tag{16.13}$$

It is clear that the question is not fully well posed and that it may not always admit a satisfactory answer. In particular, in the two-phase region of the phase diagram just described, $1(ZT)$ simply lacks the wave channels that are available in $2(ZT)$, and the mapping is then bound to fail. Thus, in order to proceed, we have to be more specific in reference to the goals and limits of the modeling that we have in mind.

Suppose that we are interested in the study of global modes based on tightly wound density waves, and let us assume that the two components are dynamically coupled [in particular, $2(ZT)$ is outside the two-phase region]. Then we could define the mapping so that the tightly wound waves supporting the modes are as close as possible to each other in the two systems. If this is the goal (see Chapter 15), the relevant propagation diagrams, drawn for various choices of the corotation circle, should be as close as possible to each other (Fig. 16.5). In principle, we might try to find the appropriate σ, c by setting up such an optimization process. In practice, we easily reach this by noting that outside the two-phase region of parameter space, the marginal stability curves are qualitatively similar in the two models and that the propagation diagrams follow the properties of the waves in the vicinity of the peak of the marginal-stability curve (i.e., there are long and short waves with a relatively small difference in wavelength). In other words, the natural assumption is that we should map the height and location of the peak of the marginal-stability curve in physical units in the two systems. Thus, for the height, we have

$$Q = \frac{Q_\star}{Q_{\max}(\alpha, \beta)}, \tag{16.14}$$

which is the Q_{eff} prescription introduced earlier in a less formal description, and

$$\frac{[1/(2\epsilon_0 r|k|)]}{(1/2)} = \frac{\hat{\lambda}}{\hat{\lambda}_{\max}(\alpha, \beta)} \tag{16.15}$$

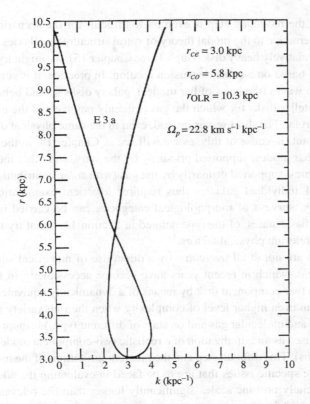

E 3 a

$r_{ce} = 3.0$ kpc

$r_{co} = 5.8$ kpc

$r_{OLR} = 10.3$ kpc

$\Omega_p = 22.8$ km s^{-1} kpc^{-1}

Fig. 16.5. A propagation diagram (from Bertin, G., Romeo, A. B. 1988. *Astron. Astrophys.*, **195**, 105) for a two-component fluid model turns out essentially to coincide with that for the one-component model constructed following the mapping $2(ZT) \rightarrow 1(ZT)$. The success of such a simple mapping depends on the physical conditions considered.

for the location, which implies that

$$\sigma = 2\sigma_\star \hat{\lambda}_{max}(\alpha, \beta). \tag{16.16}$$

In general, Eq. (16.16) is different from the intuitive prescription[9] given by $\sigma = \sigma_\star(1 + \alpha)$. The quantities that define the $1(ZT)$ model appear on the left-hand side of Eqs. (16.14) through (16.16), whereas on the right-hand side we find the functions (the data) that define $2(ZT)$. The use of these prescriptions in the case of decoupled wave dynamics is discouraged. When decoupling occurs, we should consider the full two-component disk; for many purposes, we can obtain some useful indications, in this case, by studying the waves in the gaseous disk and the stellar disk separately.

When finite-thickness effects are included, the mapping of a two-component disk $2(FT)$ (which is identified by seven functions) into a one-component, zero-thickness effective disk $1(ZT)$ can be made similarly,[10] even though the discussion becomes more complicated and more conditions are to be met for the mapping to be meaningful. In any case, this mapping, or modeling procedure, requires specifying the kind of wave dynamics that is under investigation. Thus the procedure is not yet available for time-dependent or nonlinear perturbations. Therefore, it should be emphasized that the modeling step is even more difficult for the scenario of fast evolution considered as an alternative to the modal theory of spiral structure in galaxies.

In the regime of relatively heavy disks (high J; see Chapter 15), we might try to develop a two-component analysis based on a cubic dispersion relation. In practice, it is sufficient to note that with respect to open waves (the basis of bar modes), galaxy disks should behave approximately as one-component stellar disks for which the gas is mainly passive and the effects of thickness should be less important. For short waves instead, even in the same physical disk, gas and thickness may be important. Because of this, as we will see in Chapter 18, in the same galaxy disk there may be open bar modes, supported primarily by the stars, and other more tightly wound, normal spiral structures, supported primarily by the gas (with some contribution from the stars).

The modeling of individual galaxies thus requires a patient examination of all possible regimes. Fortunately, surveys of morphological categories can be carried out by means of a reference family of basic states, of the type defined in Section 14.4, that try to incorporate in a flexible manner the relevant physical factors.[11]

These difficulties are not at all overcome by a direct use of numerical simulations of disks made of stars and gas, which in recent years have become accessible.[12] In fact, the obstacles found in modeling a two-component disk by means of a dynamically equivalent one-component disk are present to an even higher level of complexity when the rich variety of available astro-nomical data (on HI and molecular gas and on stars of different types) is mapped into the choice of parameters to be used as an initialization of a realistic two-component model (to be studied by numerical simulations). This difficult step and the inherent limitations of the modeling procedure add up to other more specific issues that are to be faced in evaluating the adequacy of N-body simulations,[13] especially on time scales significantly longer than the relevant dynamical time (see also Subsection 18.1.1).

16.3   Self-Regulation for a Disk of Stars and Gas

In the two-fluid model 2(ZT) described earlier, consider the possibility that the basic state evolves in time as a result of heating and cooling mechanisms.[14] We assume that the density distributions and the rotation curve (i.e., the quantities $\Omega, \sigma_\star,$ and σ_g) remain unchanged, so the only evolving profiles are those of $c_\star$ and c_g. For simplicity, we focus attention on a given location of the disk (which may be a typical location in the outer disk at $r \approx 3h_\star$) and introduce a heuristic description of the process by means of the first-order equations

$$\frac{d\ln Q_\star}{dt} = f_\star, \tag{16.17}$$

$$\frac{d\ln Q_g}{dt} = -g + f_g, \tag{16.18}$$

where $Q_g \equiv (\sqrt{\beta}/\alpha)Q_\star$. The f functions represent heating processes, and g is a cooling function that describes dissipation in the gas. The stellar disk is expected to heat up by means of gravitational instabilities and possibly because of scattering by giant molecular clouds. The gas component should suffer heating by means of the same gravitational instabilities, but at a faster rate because its lower velocity dispersion (and smaller thickness, if we have in mind real galaxy disks) leads to a stronger reaction. Thus we expect $f_g > f_\star$, with both f functions being rapid functions of the effective stability parameter Q, so that $f_\star \sim f_g \approx 0$ for $Q > 1$ and $f_\star$ and f_g would be very large for $Q < 1$. One possible representation is given by $f \sim Q^{-\mu}$, where $\mu > 0$ is a constant (different values of μ are introduced for the gas and the stars). The cooling term g is large and essentially independent of Q. It represents turbulent dissipation, by means of inelastic cloud-cloud collisions, which is expected to act on a short time scale (for the interstellar medium in our Galaxy, this time scale is estimated to be much shorter than the rotation period, i.e., than the dynamical time scale).

The two equations are coupled by the definition of Q [see Eq. (16.14); in our simple case we have $\alpha = $ constant]. Because of the strong dependence of the f functions on Q, the system is thus self-regulated at $Q \approx 1$, and the gas cools off (i.e., Q_g decreases) to compensate for the heating in the stellar disk. The self-regulation mechanism acts in such a way that Q can remain close to its marginal value for a relatively long time. However, β decreases in time so that eventually the two components may become dynamically decoupled. In the long run, the stellar disk then becomes practically immobile, with the gas component self-regulated at $Q_g \approx 1$.

We could improve the schematic description given earlier by studying in detail and properly incorporating all the relevant sources of heating and cooling, especially the role of star-formation processes. The mechanism might be even more efficient in a disk in which an injection of cool gas takes place, for example, as a result of the acquisition of cold gas from the intergalactic environment. The focus of this section has been on the long-term evolution of galactic parameters. Another direction in which we could extend and improve the self-regulation study would be to include the evolution effects associated with the intermediate time scale of shocks induced by large-scale spiral structures. (Thus the relevant heating terms would depend on the amplitude of the spiral modes; in turn, the amplitude would have to obey an evolution equation appropriate to the mechanisms that make the modes grow.) The whole process should be studied within the global framework of the inhomogeneous disk.

16.4 Different Behaviors of Gas and Stars at Resonances

In Chapter 15 it was mentioned that the study of density waves in the vicinity of resonances requires a special treatment. Resonances show up as singularities in the relevant linearized equations (see the fluid model described in Chapter 17) or directly in the dispersion relation [see the poles of Eq. (15.9) for tightly wound waves in a disk of stars]. The special role of resonances as locations where energy and momentum are exchanged between the waves and the basic state has long been recognized.[15] Thus the mechanisms involved can be basically described in terms of Landau damping (see Chapter 12). Below we will briefly recall some of the features that characterize the behavior of a fluid disk with respect to that of a collisionless disk of stars. These properties have an important impact on the problem of large-scale spiral structure in galaxies.[16] *N*-body experiments allow us to explore the nonlinear behavior at resonances, which is very hard to follow by means of analytical techniques, yet the limited number of particles used and the lack of resolution in phase space suggest that the overall behavior simulated in many *N*-body studies of galaxy disks resembles more the behavior of a fluid disk than that of a collisionless disk of stars.

16.4.1 Lindblad Resonances

In the ordering of tightly wound spirals, at the Lindblad resonances ($\nu^2 = 1$) the long waves formally become infinitely long ($k \to 0$) in both the quadratic dispersion relation of a fluid model

[Eq. (15.11)] and the stellar dynamic dispersion relation [Eq. (15.9)]. In reality, as indicated by use of the cubic dispersion relation [Eq. (15.16)], which is better suited to describe open waves, the long waves do not manage to reach the Lindblad resonances because they are refracted back at $v^2 < 1$. Thus we can focus on the short-wave branch. Here a difference is immediately apparent in the two cases.

The fluid dispersion relation is regular, with short waves characterized by $|\hat{k}| \approx 4/Q^2$. In Chapter 17 we will recognize that the integrodifferential fluid equations for linearized perturbations show a singularity at $v^2 = 1$, but the density-wave solution can proceed regularly as a modified sound wave across the resonance, with only a modest amount of absorption.[17] In other words, in the fluid model no major effects are found on the propagation of density waves.

In contrast, study of the equations of stellar dynamics shows that no wave solutions can be found beyond the principal range across the Lindblad resonances (i.e., at $v^2 > 1$). This was realized from the very beginning[18] when the dispersion relation for tightly wound waves in a disk of stars was found to lead to infinitely short waves ($k \to \infty$) at the Lindblad resonances. This suggests that short trailing waves are fully absorbed at the Lindblad resonances.[19]

To prove that absorption occurs in the stellar disk and to calculate the appropriate wavelength of the density wave in the vicinity of the resonance, we can resort to introduction of a complex wave number,[20] so Eq. (15.9) must be interpreted as

$$s_k \frac{k_0}{k} = -2 \frac{e^{-x}}{x} \sum_{n=1}^{\infty} \frac{I_n(x)}{(\frac{v}{n})^2 - 1} = \frac{\mathcal{F}_v(x)}{1 - v^2}, \tag{16.19}$$

where $s_k = \text{sgn}[\text{Re}(k)]$. All the other quantities retain the definitions given in Chapter 15 [see also the notation in Eq. (15.42)]. Equation (16.19) is derived from integration in velocity space of the perturbed distribution function [Eq. (15.7)] under the assumption that the response is regular. This assumption fails at a resonance because a fraction of the stars would contribute a singular response. Therefore, to find the dispersion relation for neutral waves, for which ω and, consequently, v are real, we follow the discussion of Chapter 12 and calculate the dispersion relation starting from the case of growing waves [$\text{Im}(\omega) < 0$]. Then, in the vicinity of $v = s_v$, where $s_v = -1$ for the inner Lindblad resonance (ILR) [$s_v = 1$ for the outer Lindblad resonance (OLR)], we can introduce a length scale L by referring to the spatial variation of v, that is,

$$v(r) \sim s_v + s_\Omega \frac{r - r_L}{L} + i \, \text{Im}(v); \tag{16.20}$$

typical cases of interest have $s_\Omega = 1$. The initial-value approach prescribes that the dispersion relation for neutral waves be obtained by analytical continuation from $\text{Im}(v) \to 0^-$. Similarly, we may introduce a dimensionless distance from the resonant radius r_L by means of the quantity

$$z_L = s_\Omega \frac{\sqrt{2} L \kappa}{c} (v - s_v). \tag{16.21}$$

The latter quantity is a stretched coordinate because it is scaled on the size of the stellar epicyclic radius. Then we obtain the basic dispersion relation from Eq. (16.19) by separating the stellar reduction factor $\mathcal{F}_v$ into a nonresonant part,

$$\frac{\mathcal{F}_v^{nr}(x)}{1 - v^2} = -\frac{e^{-x}}{x} \sum_{n \neq s_v} \frac{n I_n(x)}{v - n}, \tag{16.22}$$

and a resonant part, which can be shown to be

$$\frac{\mathcal{F}_\nu^r(x)}{1-\nu^2} = -is_\nu s_\Omega \frac{\sqrt{2}L\kappa^3}{c^3 k^2} \int_{-\infty s_\Omega}^0 \exp[-\tau^2 + iz_L\tau - \xi]I_{s_\nu}(\xi)d\tau, \qquad (16.23)$$

where

$$\xi = \xi(\tau) = \frac{k^2 c^2}{\kappa^2} + \sqrt{2}\frac{kc}{\kappa}\tau. \qquad (16.24)$$

If we restrict these expressions to the case of ILR ($s_\nu = -1$) with $s_\Omega = 1$, we find for the nonresonant part that

$$\frac{\mathcal{F}_\nu^{nr}(x)}{1-\nu^2} = \frac{1}{x}\left[1 - e^{-x}I_0(x) + \frac{2\nu-1}{1-\nu}e^{-x}I_1(x) + 2\nu^2\sum_{n=2}^\infty \frac{e^{-x}I_n(x)}{n^2-\nu^2}\right]. \qquad (16.25)$$

Finally, by considering that the argument of the modified Bessel functions is large, we obtain for the resonant response

$$\frac{\mathcal{F}_\nu^r(x)}{1-\nu^2} \sim is_k \frac{\sqrt{2}L\kappa^4}{c^4 k^3}\frac{1}{\sqrt{2\pi}}\int_{-\infty}^0 \exp[-\tau^2 + iz_L\tau]d\tau = is_k \frac{k_0 k_L^2}{k^3}w(-z_L/2). \qquad (16.26)$$

Here we have introduced the resonant wave number $k_L^2 = \pi G\sigma\kappa^2 L/c^4$, evaluated at r_L. The w function is the well-known plasma dispersion function[21] defined as

$$w(z) = \exp(-z^2)\left(1 + \frac{2i}{\sqrt{\pi}}\int_0^z e^{s^2}\,ds\right) = \exp(-z^2)\mathrm{erfc}(-iz). \qquad (16.27)$$

The presence of the plasma dispersion function is the signature of the action of the Landau damping mechanism (see also Section 12.2).

By combining Eqs. (16.19), (16.22), and (16.23) for a given value of the pattern frequency ω/m and for given properties of the basic state, we can calculate the complex wave number $k = k(r)$ and show that in the vicinity of the Lindblad resonances (i.e., within a resonant annulus approximately of the width given by two epicycles) "both leading and trailing short waves are found to decay spatially in the direction of their respective group propagation"[22] (Fig. 16.6). The real part of the wave number is large but remains finite. Thus, with respect to short trailing wave signals propagating away from the corotation circle, the Lindblad resonances are found to act as perfect absorbers. These results can be conveniently expressed in terms of sources (actually sinks) for the wave energy and wave action.

It should be stressed that these conclusions show that a straightforward application of the original dispersion relation [Eq. (15.9)] with its singularities can be seriously misleading, whereas a regular behavior of the waves at the Lindblad resonances is expected and can be quantified in detail. In fact, on the basis of Eq. (15.9), it had been argued that rings might mark the location of the Lindblad resonances because for $k \to \infty$ the waves would result in a ringlike structure; in reality, from the preceding analysis, such a result is not expected to occur. In turn, some ringlike structures close to the Lindblad resonances may be generated by the nonlinear response of the gas.[23]

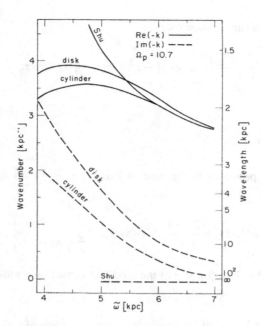

Fig. 16.6. Absorption in a stellar dynamic disk at the inner Lindblad resonance for a model of the Milky Way (from Mark, J. W.-K. "On density waves in galaxies. I: Source terms and action conservation," 1974. *Astrophys. J.*, **193**, 539; reproduced by permission of the AAS). Here the curves marked "Shu" are obtained by means of the dispersion relation (15.9).

16.4.2 Corotation Resonance

As will be discussed in Chapter 17, the linearized equations for density-wave perturbations in a fluid model have a term that becomes singular at the corotation circle, which is

$$B_{cr} = \epsilon^2 \frac{2m\Omega}{r\kappa\nu} \frac{d}{dr} \left[\ln\left(\frac{\kappa^2}{\sigma\Omega} \right) \right]. \tag{16.28}$$

Away from the corotation circle, this term is $O(\epsilon^2)$ [see Eq. (15.8)] and is thus dropped in the derivation of the local dispersion relations for both the quadratic and cubic relations given in Chapter 15. The gradient involved has been identified with the gradient of the angular-momentum distribution in the disk (sometimes called *vortensity*).[24] We recall that $\kappa^2/2\Omega$ is the vorticity of the basic rotational flow. Therefore, the quantity appearing in the gradient is, for the basic state, precisely the quantity that is conserved in a general flat, two-dimensional gas flow.[25]

In the fluid model, the corotation resonance term has been recognized to be the tool that makes bar driving possible.[26] In general, when the gradient of $\kappa^2/\sigma\Omega$ is positive, the resonant term has been found to contribute an outward transport of angular momentum across the corotation circle, and thus it is destabilizing.[27] A number of proposed amplification mechanisms for spiral structure, such as the negative-mass instability[28] and the groove instability,[29] can be traced to the corotation resonance.

The collisionless case is different in two respects.[30] First, the presence of epicyclic motions smears out the effects of the resonance, which is thus weakened.[31] Second, the resonant contribution in the collisionless case also includes a term that is proportional to the gradient of the adiabatic invariant c^2/κ (i.e., to the temperature gradient in the disk). In any case, direct calculations of global spiral modes show that the corotation resonance effects are typically weaker than those that are generally anticipated on the basis of the fluid model.[32]

16.5   Role of Near-Infrared Observations

Spiral structure in galaxy disks[33] has long been recognized to owe its existence and its appearance to a subtle interplay of actions by stars and gas. The optical morphology is generally delineated by bright young stars and dust lanes and thus is dominated by the properties of the interstellar medium; in turn, dynamical theories are aimed at determining the properties of the gravitational field, which is largely dominated by relatively old stars that give a less prominent contribution to optical images. Thus, in the context of galactic dynamics, one major goal of the observations in the past four decades has been to find an empirical way to separate the contribution of the stars to the overall spiral gravitational field from that of the gas, that is, to settle directly the question of whether spiral arms are mostly stellar or gaseous features. As pioneered by Zwicky,[34] the main means of quantitatively analyzing the properties of the underlying stellar disk has been to compare images in different wave bands and especially to move toward the long-wavelength part of the spectrum.

After the considerable progress made through studies[35] at $\approx 1\,\mu$m, it now appears that the acquired imaging capability near $2\,\mu$m is finally bringing us very close to the desired target of a direct diagnostic of the mass distribution in galaxy disks[36] (Figs. 16.7 and 16.8). A quantitative assessment for typical or specific cases of the precise amount of contamination to near-infrared (near-IR) images by relatively young objects, such as red supergiants, is still a matter of discussion. However, given the low extinction levels with respect to optical wave bands, there is little doubt that a new extragalactic perspective is growing out of the near-IR (K and K') images of spiral galaxies. These are giving us the best view of the underlying mass distribution in galaxy disks and have already provided a number of impressive and sometimes surprising results of great dynamical interest:[37]

- One crucial point of the density-wave theory, that spiral arms (at least for a grand-design structure) are stellar (i.e., associated with a significant density perturbation), is now definitely proved.
- Especially for gas-rich galaxies, spiral morphology in the near-IR range can be very different from that of optical images.
- A smooth and coherent grand-design spiral structure is rather common in the near-IR range.
- Infrared spiral structure is generally dominated by low-m features, with an $m = 1$ component showing up prominently in several cases.
- The amplitude modulation of spiral arms is found to persist in IR images, which confirms that this is a property of the underlying gravitational potential.
- Barred structure is more frequent and more prominent in the near-IR range, although not always present (e.g., in M99; even in the case of M33, in which the continuation of optical spiral arms down to the center has often been interpreted in terms of a weak bar, clear evidence of a bar is not found); long spiral arms without a dominant large-scale bar are also frequent, and in some cases (such as NGC 4622) they are very tight.

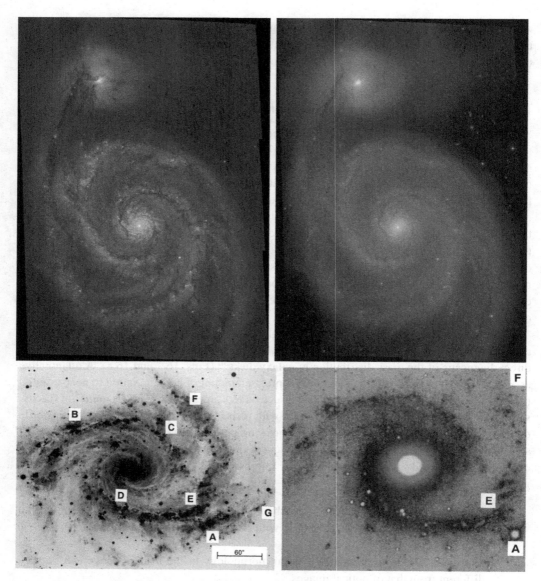

Fig. 16.7. *Top:* The grand-design galaxy M51 (NGC 5194, in a close encounter with NGC 5195), viewed by the *Hubble Space Telescope* in the blue (*left*) and in the near-infrared at ≈ 1 μm (*right*). [Credit: NASA, ESA, S. Beckwith (STScI), and The Hubble Heritage Team (STScI/AURA).] *Bottom:* NGC 2997 in the optical blue band (*left*) and in the near-IR *K'* band at 2.1 μm (*right*) (from Block, D. L., Bertin, G., et al. 1994. *Astron. Astrophys.*, **288**, 365.)

- Despite its smooth appearance, large-scale spiral structure is sometimes characterized by a strong arm-interarm contrast (values between 2 and 3 have been reported for M51 and M99), and thus it must have reached a highly nonlinear stage; however, this should not be taken as a general rule (e.g., M100 has a lower contrast, as also indicated by its kinematics; for the bisymmetric structure in M33, the contrast has been found to be ≈ 1.2).

Fig. 16.8. *Top:* The barred galaxy NGC 1300 viewed by the *Hubble Space Telescope* in the optical blue band (*left*) and in the near-infrared *I* band (*right*) [Credit: NASA, ESA, The Hubble Heritage Team (STScI/AURA); P. Knezek (WIYN)] (see also Fig. 2.2). *Bottom:* The barred galaxy NGC 1365, viewed in the visible (*left*; with the FORS1 camera on ESO's VLT) and in the near infrared (*right*; with the HAWK-I camera on ESO's VLT). [Credit: ESO/P. Grosbøl.]

- The exponential scale length of the disk may change significantly as a function of the wave band of observation, being shorter in the near-IR range.

Systematic surveys of morphological properties in the near-IR range (near 2 μm) have been carried out in the last fifteen years, confirming many of the points just made but also raising new, interesting issues.[38] Much of the emphasis of these investigations is on the frequency and properties of bars.[39] The frequency of bars in the near-IR range may not be as high as sometimes claimed.[40] Empirical studies of bars are often related to the issue of bar driving.[41] Very recent analyses complement these studies with observations from *Spitzer* (for which the shortest wavelengths that are presumed to trace the stellar mass distribution are 3.6 and 4.5 μm).[42]

At the present stage, the modal theory of spiral structure, which interprets the large-scale properties of spiral arms in terms of a quasi-stationary structure determined by the intrinsic characteristics of the individual galaxies, is definitely strengthened by these observational breakthroughs. Before the physical description of spiral structure in galaxies is addressed, in Chapter 17 the discussion of the dynamics of density waves is completed by working out the properties of global modes.

Notes

1. Lin, C. C., Shu, F. H. 1966. *Proc. Natl. Acad. Sci. USA*, **55**, 229; Lynden-Bell, D. 1967. *Lect. Appl. Math.*, **9**, 131; Graham, R. 1967. *Mon. Not. Roy. Astron. Soc.*, **137**, 25; Miller, R. H., Prendergast, K. H., Quirk, W. J. 1970. *Astrophys. J.*, **161**, 903; Quirk, W. J. 1971. *Astrophys. J.*, **167**, 7; Kato, S. 1972. *Publ. Astron. Soc. Jpn.*, **24**, 61; Jog, C. J., Solomon, P. M. 1984. *Astrophys. J.*, **276**, 114 and 127; Sellwood, J. A., Carlberg, R. G. 1984. *Astrophys. J.*, **282**, 61; Bertin, G., Romeo, A. B. 1988. *Astron. Astrophys.*, **195**, 105; Romeo, A. B. 1992. *Mon. Not. Roy. Astron. Soc.*, **256**, 307; and 1994. *Astron. Astrophys.*, **286**, 799.

2. Roberts, W. W. 1969. *Astrophys. J.*, **158**, 123. In this monograph, this topic is not covered. The reader may refer to the monograph by Bertin, G., Lin, C. C. 1996. *op. cit.*, for an outline of the shock scenario and for its important impact on the comparison between density-wave theory and observations.

3. Kalnajs, A. J. 1972. *Astrophys. Lett.*, **11**, 41, raised the problem of dissipation as a major concern for the picture of quasi-stationary spiral structure. Roberts, W. W., Shu, F. H. 1972. *Astrophys. Lett.*, **12**, 49, demonstrated the beneficial influence of shock dissipation on a self-excited spiral structure. See also Shu, F. H. 1985. In *IAU Symposium*, Vol. 106, eds. H. van Woerden et al. Reidel, Dordrecht, The Netherlands, p. 530; Lubow, S. H. 1986. *Astrophys. J. Lett.*, **307**, L39; Lubow, S. H., Balbus, S., Cowie, L. L. 1986. *Astrophys. J.*, **309**, 496.

4. Bertin, G. 1991. In *IAU Symposium*, Vol. 146, eds. F. Combes, F. Casoli. Kluwer, Dordrecht, The Netherlands, p. 93. See also Bertin, G., Romeo, A. B. 1988. *op. cit.*; Bertin, G., Lin, C. C., Lowe, S. A., Thurstans, R. P. 1989. *op. cit.* Processes that may regulate the amplitude of large-scale coherent spiral structure in a collisionless disk of stars also have been studied; see Contopoulos, G., Grosbøl, P. 1986. *Astron. Astrophys.*, **155**, 11; Patsis, P. A., Hiotelis, N., Contopoulos, G., Grosbøl, P. 1994. *Astron. Astrophys.*, **286**, 46.

5. For the overall dynamics of galaxy disks, self-regulation has often been presented in general terms; see Miller, R. H., Prendergast, K. H., Quirk, W. J. 1970. *op. cit.*; Sellwood, J. A., Carlberg, R. G. 1984. *op. cit.*; Ostriker, J. P. 1985. In *IAU Symposium*, Vol. 106, eds. H. van Woerden et al. Reidel, Dordrecht, The Netherlands, p. 638; Lin, C. C., Bertin, G. 1985. In *IAU Symposium*, Vol. 106, eds. H. van Woerden et al. Reidel, Dordrecht, The Netherlands, p. 513; Bertin, G., Lin, C. C. 1987. *Publ. Astron. Czech. Acad. Sci.*, **69**(4), 255.

6. Spitzer, L. 1968. *Diffuse Matter in Space*, Wiley, New York; 1978. *Physical Processes in the Interstellar Medium*, Wiley, New York; McKee, C. F., Ostriker, J. P. 1977. *Astrophys. J.*, **218**, 148; Kwan, J. 1979. *Astrophys. J.*, **229**, 567; Cowie, L. L. 1980. *Astrophys. J.*, **236**, 868; Jog, C. J., Ostriker,

J. P. 1988. *Astrophys. J.*, **328**, 404; self-regulation as specifically related to star-formation processes has been considered by Quirk, W. J. 1972. *Astrophys. J. Lett.*, **176**, L9; and Kennicutt, R. C. 1989. *Astrophys. J.*, **344**, 685.

7. Here we follow the article by Bertin, G., Romeo, A. B. 1988. *op. cit.*; an extension to the case of two components with finite thicknesses can be found in Romeo, A. B. 1992. *op. cit.*

8. As done by Lin, C. C., Shu, F. H. 1966. *op. cit.*

9. The intuitive prescription was indeed adopted by Bertin, G., Romeo, A. B. 1988. *op. cit.*, and by Bertin, G., Lin, C. C., Lowe, S. A., Thurstans, R. P. 1989. *op. cit.*; Bertin and Romeo (see Fig. 16.5) checked that the relevant propagation diagrams remain basically unchanged when expected.

10. Bertin, G. 1991. *op. cit.*

11. Bertin, G., Lin, C. C., Lowe, S. A., Thurstans, R. P. 1989. *op. cit.*

12. See Salo, H. 1991. *Astron. Astrophys.*, **243**, 118; Combes, F., Elmegreen, B. G. 1993. *Astron. Astrophys.*, **271**, 391; Elmegreen, B. G., Thomasson, M. 1993. *Astron. Astrophys.*, **272**, 37.

13. See Romeo, A. B. 1994. *Astron. Astrophys.*, **286**, 799.

14. Here we follow the discussion of Bertin, G., Romeo, A. B. 1988. *Astron. Astrophys.*, **195**, 105; and Bertin, G. 1991. In *IAU Symposium*, Vol. 146, eds. F. Combes, F. Casoli. Kluwer, Dordrecht, The Netherlands, p. 93.

15. Coppi, B., Rosenbluth, M. N., Sudan, R. N. 1969. *Ann. Phys.*, **55**, 207; for the galactic context, see the references to the work by Mark, and Lynden-Bell and Kalnajs given in Subsection 15.1.1.

16. Bertin, G. 1993. *Publ. Astron. Soc. Pac.*, **105**, 640.

17. This has become particularly evident when a code has been constructed for the numerical integration of the integrodifferential problem; Pannatoni, R. F. 1979. Ph.D. dissertation, Massachusetts Institute of Technology, Cambridge, MA; Pannatoni, R. F. 1983. *Geophys. Astrophys. Fluid Dyn.*, **24**, 165; Haass, J. 1982. Ph.D. dissertation, Massachusetts Institute of Technology, Cambridge, MA; Haass, J. 1983. In *IAU Symposium*, Vol. 100, ed. E. Athanassoula. Reidel, Dordrecht, The Netherlands, p. 121.

18. See Lin, C. C., Yuan, C., Shu, F. H. 1969. *Astrophys. J.*, **155**, 721; Shu, F. H. 1970. *Astrophys. J.*, **160**, 99.

19. See the physical discussion by Lynden-Bell, D., Kalnajs, A. J. 1972. *op. cit.*

20. Here we follow Mark, J. W.-K. 1971. *op. cit.*; 1974. *op. cit.*

21. Fried, B. D., Conte, S. D. 1961. *The Plasma Dispersion Function*, Academic, New York; Abramowitz, M., Stegun, I. A. 1972. *Handbook of Mathematical Functions*, Dover, New York.

22. Quoted from Mark, J. W.-K. 1974. *op. cit.*

23. See Yuan, C., Cheng, Y. 1991. *Astrophys. J.*, **376**, 104.

24. Lynden-Bell, D., Kalnajs, A. J. 1972. *op. cit.*, Feldman, S. I. 1973. Ph.D. dissertation, Massachusetts Institute of Technology, Cambridge, MA.

25. See Hunter, C. 1983. *Lect. Appl. Math.*, **20**, 179.

26. Feldman, S. I., Lin, C. C. 1973. *Studies Appl. Math.*, **52**, 1.

27. Feldman, S. I. 1973. *op. cit.*; Lau, Y. Y., Bertin, G. 1978. *Astrophys. J.*, **226**, 508.

28. Lovelace, R. V. E., Hohlfeld, R. G. 1978. *Astrophys. J.*, **221**, 51

29. See Sellwood, J. A., Kahn, F. D. 1991. *Mon. Not. Roy. Astron. Soc.*, **250**, 278, and references therein.

30. A detailed derivation with a thorough discussion is given by Mark, J. W.-K. 1976. *Astrophys. J.*, **203**, 81; **205**, 363.

31. See also Lau, Y. Y., Bertin, G. 1978. *Astrophys. J.*, **226**, 508.

32. Bertin, G., Haass, J. 1982. *Astron. Astrophys.*, **108**, 265.

33. This short discussion opens the way to the physical description of spiral structure in galaxies that will be given in Chapter 18. Much of this is taken from Bertin, G. 1996. In *New Extragalactic Perspectives in the New South Africa*, eds. D. L. Block, J. M. Greenberg. Kluwer, Dordrecht, The Netherlands, p. 227.

34. Zwicky, F. 1957. *Morphological Astronomy*, Springer, Berlin. The idea of comparing images in different wave bands was stressed by Oort, J. H. 1962. In *Interstellar Matter in Galaxies*, ed. L. Woltjer. Benjamin, New York, p. 234; and later developed by Schweizer, F. 1976. *Astrophys. J. Suppl.*, **31**, 313.

35. See Elmegreen, D. M. 1981. *Astrophys. J. Suppl.*, **47**, 229; Elmegreen, D. M., Elmegreen, B. 1984. *Astrophys. J. Suppl.*, **54**, 127; Pierce, M. J. 1986. *Astron. J.*, **92**, 285.

36. Rix, H.-W. 1993. *Publ. Astron. Soc. Pac.*, **105**, 999; Rix, H.-W., Rieke, M. J. 1993. *Astrophys. J.*, **418**, 123.

37. See Block, D. L., Wainscoat, R. J. 1991. *Nature (London)*, **353**, 48; Zaritsky, D., Rix, H.-W., Rieke, M. J. 1993. *Nature (London)*, **364**, 313; Block, D. L., Bertin, G., et al. 1994. *Astron. Astrophys.*, **288**, 365; Regan, M. W., Vogel, S. N. 1994. *Astrophys. J.*, **434**, 536; Peletier, R. F., Valentijn, E. A., et al. 1994. *Astron. Astrophys. Suppl.*, **108**, 621; Knapen, J. H., Beckman, J. E., et al. 1995. *Astrophys. J. Lett.*, **443**, L73; Rix, H.-W., Zaritsky, D. 1995. *Astrophys. J.*, **447**, 82; González, R. A., Graham, J. R. 1996. *Astrophys. J.*, **460**, 651; Thornley, M. D. 1996. *Astrophys. J. Lett.*, **469**, L45; Block, D. L., Elmegreen, B. G., Wainscoat, R. J. 1996. *Nature (London)*, **381**, 674; see also the volume *New Extragalactic Perspectives in the New South Africa*, Kluwer, Dordrecht, The Netherlands.

38. Seigar, M. S., James, P. A. 1998. *Mon. Not. Roy. Astron. Soc.*, **299**, 672 and 685, study a sample of forty-five face-on spirals, finding that the pitch-angle of spiral arms does not correlate with Hubble type and that bars do not appear to correlate with arm strength. This investigation has been followed up in the paper by Seigar, M. S., Chorney, N. E., James, P. A. 2003. *Mon. Not. Roy. Astron. Soc.*, **342**, 1, in which seventeen inclined spirals are imaged in the near infrared to test how the radial extent of grand-design spiral structure correlates with the galaxy rotation curve; the study confirms that "the spiral structure of these galaxies in the near-infrared is extremely regular."

39. Grosbøl, P., Patsis, P. A., Pompei, E. 2004. *Astron. Astrophys.*, **423**, 849, study a sample of fifty-four normal spiral galaxies, with focus on the presence of faint bars, showing that the majority of the sample possesses an inner two-armed grand-design structure, often accompanied by multiple-armed structure in the outer regions. Block, D. L., Buta, R., et al. 2004. *Astron. J.*, **128**, 183, study spiral strength and bar strength for a sample of fifteen spirals.

40. Eskridge, P. B., Frogel, J. A., et al. 2000. *Astron. J.*, **119**, 536 (followed by Eskridge, P. B., Frogel, J. A., et al. 2002. *Astrophys. J. Suppl.*, **143**, 73, who produce a *BVRJHK* imaging survey of a well-defined sample of 205 bright, nearby spiral galaxies), find that only 27 percent of a sample of 186 spirals is unbarred in the *H* band, whereas Menéndez-Delmestre, K., Sheth, K., et al. 2007. *Astrophys. J.*, **657**, 790, find a smaller fraction of barred galaxies in their sample of 151 nearby spiral galaxies taken from the 2MASS survey and studied in *J*, *H*, and *K*.

41. Buta, R. J., Knapen, J. H., et al. 2009. *Astron. J.*, **137**, 4487; Salo, H., Laurikainen, E., et al. 2010. *Astrophys. J. Lett.*, **715**, L56.

42. See Buta, R. J., Sheth, K., et al. 2010. *Astrophys. J. Suppl.*, **190**, 147. In particular, Elmegreen, D. M., Elmegreen, B. G., et al. 2011. *Astrophys. J.*, **737**, 32, study a sample of forty-six spirals, comparing flocculent, multiple-armed, and grand-design properties viewed in the optical with those viewed in the near infrared and analyzing strength of bars and spiral arms and amplitude modulations. Kendall, S., Kennicutt, R. C., Clarke, C. 2011. *Mon. Not. Roy. Astron. Soc.*, **414**, 538, focus on a sample of thirty-one spiral galaxies; among other morphological properties and correlations, the authors study the frequency of a two-armed grand design, of barred structure, of power in higher modes, and of the presence of companions.

17 Global Spiral Modes

The discussion of density waves given in Chapter 15 is only one important step in the study of the dynamics of even perturbations on an axisymmetric self-gravitating disk. As already emphasized at the end of that chapter, we should complete the analysis by taking into account the inhomogeneity of the disk and the relevant physical conditions at the boundaries. However, the focus on the study of waves is given priority because of its simplicity and its usefulness in identifying the relevant dynamical mechanisms (see Section 15.5) and, historically, because of its immediate application in testing the density-wave theory against observations of spiral structures in galaxies within a semiempirical approach.

Given the symmetry of the basic state and the radial inhomogeneity of the disk, we consider disturbances for which the perturbed density of the disk is given by $\sigma_1(r)\exp[i(\omega t - m\theta)]$ and, for given values of ω and m, look for appropriate functions $\sigma_1(r)$ that satisfy the dynamical equations and the relevant physical conditions at the boundaries. In general, this is an eigenvalue problem, and the associated spectrum is not continuous in the sense that the problem admits solutions for only selected values of the frequency. These solutions are often called *global modes* or *simply modes*. For a given model, once we have formulated the (linearized) mathematical problem, we can use numerical integration for the calculation of such modes with no need for the approximate description in terms of waves contained in a local dispersion relation [see Eq. (15.1)]. The eigenvalues of the problem define the internal notes of the disk, that is, the relevant pattern frequencies at which the modes rotate rigidly, and the eigenfunctions specify the morphologies characteristic of the self-consistent disk oscillations. If $\text{Im}(\omega) < 0$, the mode grows exponentially in time and thus is unstable or self-excited; if $\text{Im}(\omega) > 0$, the mode decays exponentially and thus is damped (therefore, it may play an important role in a driven case, e.g., as a result of tidal interactions). The presence of dissipation in the gas and of resonances in the collisionless stellar system makes the case of neutral modes [$\text{Im}(\omega) = 0$] exceptional for application to galaxies.

These concepts can be formulated clearly and easily within a linear perturbation analysis, but they also have a direct counterpart in the anticipated nonlinear evolution of the disk. Making this connection is not at all trivial because it requires a deep evaluation of the various physical aspects of the systems considered and because our knowledge of the nonlinear evolution of complex dynamical systems is rather limited. In the application of self-excited modes to observed structures in galaxies, the linear modes are expected to give a clue to the morphology of the observed structures and the value of the relevant pattern frequencies, although it is recognized that in reality the structures are nonlinear, with the linear growth expected to be saturated at finite amplitude. Discussion of the spectrum of global modes for realistic galaxy models can justify use of the hypothesis of quasi-stationary spiral structure, which has been taken as the basis for comparison between theory and observations from the very beginning. In other words, a thorough study of the properties of global modes forms the dynamical foundation of the framework for the interpretation of spiral structure in galaxies, as provided in Chapter 18. In this respect, it should be emphasized that the interpretation of spiral structure in terms of self-excited modes should not be taken literally as an indication that in the distant past galaxy disks had to be axisymmetric. Rather, much as for the case of unstable modes on Maclaurin spheroids (see Chapter 10), the modes are probably telling us under what conditions a nonaxisymmetric configuration with large-scale spiral structure is favored energetically, where galaxies may have settled naturally in the course of their evolution.

The formulation of the exact eigenvalue problem for linear global spiral modes is available for stellar dynamical[1] and fluid[2] zero-thickness models of the disk; several studies have investigated the spectrum of the global modes for a variety of basic states. If we knew the properties of the underlying basic state of galaxy disks within an adequate dynamical model described by a relatively simple set of equations, the problem of studying spiral structure could be reduced to a survey of the modes for those basic states and an evaluation of the consequences of possible mode superpositions in the context of various evolutionary scenarios. In reality, such a blind search is hopeless because we have only a few observational constraints with significant uncertainties on basic quantities such as density distributions and an incomplete picture of the relevant basic states because of the tremendous difficulties inherent in the modeling process. Already from the simple studies presented in previous chapters it should be clear that dynamical evolution is very sensitive to the detailed parameter regimes involved and the specific characteristics of the adopted models. Thus modal surveys are useful primarily in clarifying the dynamical properties of given basic states and identifying interesting parameter regimes; however, they cannot be used as tools to predict the properties of spiral structure in galaxies by means of a purely deductive approach. Instead, it has been proved that we can proceed semiempirically by trying to see whether modal morphologies match observed structures and under what circumstances this may occur. This procedure, carried out within the available observational constraints, has been found to be generally successful and to lead to predictions to be tested by new observations.

Aiming at a scheme for the derivation of all the modes for given models and at a modal survey of all possible models would bring us to investing a huge amount of effort, most of which would be related to unrealistic basic states. In turn, we may aim at a search for the relevant modes in basic states under the guidance of the available observations of spiral galaxies and our understanding of dynamical mechanisms (as developed in previous chapters by means of approximate methods that allow us to keep track of a large number of physical aspects of the problem). The physical conclusions that can be drawn from such a modal survey of realistic galaxy models will be described in Chapter 18.

To bridge the gap between what is learned in terms of waves (see Chapter 15) and what can be calculated as global modes (e.g., by numerically integrating the set of equations of Section 17.1), it is instructive to study in detail a simple case in which the physical mechanisms and processes that guarantee excitation and maintenance to the modes can be worked out analytically. Such a reference case is best exemplified by the fluid model in the regime of tightly wound spiral structure, which is the focus of this chapter. The rigorous derivation given in Sections 17.2 through 17.4 shows how to move beyond the algebraic description of waves (which is recovered, in passing, in Section 17.2.3) and allows us to check quantitatively what has been anticipated in the physical discussion of Subsections 15.5.2 through 15.5.4. In particular, the reflection process calculated at the simple turning point r_{ce} demonstrates the relevant feedback mechanism that helps to maintain the mode, and the matching of solutions at the double turning point r_{co} quantify the overreflection at corotation. Thus all the elements participating in the underlying wave cycle responsible for the self-excited discrete modes can be followed analytically. In another regime, that of open spiral structure, a similar derivation can be worked out that is fully analogous.[3] The approximate results obtained for these simple regimes have found numerical confirmation by means of direct calculations of the eigenvalue problem; these numerical studies also can cover the cases of more complex regimes for which the asymptotic analyses are bound to be insufficient.

From a physical point of view, the fluid model adopted in this chapter can be usefully explored and applied to real galaxies, provided that we keep in mind the underlying modeling issues. Thus the discussion of global spiral modes and boundary conditions must incorporate what we know from stellar dynamics and the effects associated with the three-dimensional structure of the disk and its two-component nature. In particular, a good starting point is to focus on the family of basic states identified in Section 14.4, which also was made in view of the action of self-regulation in the disk. From among these models we should be ready to discard those that are anticipated or turn out to be too violently unstable; the latter would rapidly evolve and thus would be unlikely to represent the systems presently observed.

17.1 Exact Equations for Linear Density Perturbations in a Fluid Disk Model

If we refer to the $1(ZT)$ fluid model introduced in Chapter 8 (see also the relevant family of basic states identified in Section 14.4 and the discussion of the modeling process given in Section 16.2), the exact set of linearized equations for elementary density perturbations with phase factor $\exp[i(\omega t - m\theta)]$ can be reduced to the following integrodifferential problem:[4]

$$\mathcal{L}(h_1 + \Phi_1) = -Ch_1, \tag{17.1}$$

$$\Phi_1 = \mathcal{P}(\sigma_1) = \mathcal{P}\left(\frac{\sigma}{c^2}h_1\right), \tag{17.2}$$

where

$$\mathcal{L} \equiv \frac{d^2}{dr^2} + A\frac{d}{dr} + B, \tag{17.3}$$

$$A = -\frac{d}{dr}\left\{\ln\left[\frac{(1-\nu^2)\kappa^2}{r\sigma}\right]\right\}, \tag{17.4}$$

$$B = -\frac{m^2}{r^2} - \frac{4m\Omega}{r\kappa}\frac{1}{(1-\nu^2)}\frac{d\nu}{dr} + \frac{2m\Omega}{r\kappa\nu}\frac{d}{dr}\left[\ln\left(\frac{\kappa^2}{\sigma\Omega}\right)\right], \tag{17.5}$$

$$C = -\frac{\kappa^2}{c^2}(1-\nu^2), \tag{17.6}$$

$$\nu = \frac{(\omega - m\Omega)}{\kappa}. \tag{17.7}$$

Here Φ_1 is the gravitational potential perturbation, and h_1 and σ_1 represent the enthalpy and density perturbations of the active disk. In these equations, the perturbed quantities are all meant to be functions of the radial coordinate only because the phase factor $\exp[i(\omega t - m\theta)]$ cancels out from the linear equations. The Poisson operator is integral:

$$\Phi_1(r) = \mathcal{P}(\sigma_1) = -2\pi G \int_0^\infty K(r,r')\sigma_1(r')\,dr', \tag{17.8}$$

where the kernel $K(r,r')$ is well known, that is,

$$K(r,r') = \frac{r'}{r+r'}\frac{2}{\pi}\int_0^{\pi/2}\frac{\cos(2mx)}{\sqrt{1-\zeta\cos^2 x}}\,dx, \tag{17.9}$$

with

$$\zeta = \frac{4rr'}{(r+r')^2}, \tag{17.10}$$

and exhibits a logarithmic singularity at $r = r'$.

Under the appropriate boundary conditions, the set of Eqs. (17.1) and (17.2) has been studied numerically by means of different codes[5] for determination of the exact eigenvalues and eigenfunctions. The results of these studies form the basis of the discussion presented in Chapter 18. In Sections 17.2 through 17.4, a simple regime is described for which the calculation of global spiral modes can be worked out analytically. All the dynamical mechanisms participating in the maintenance and excitation of the modes will thus be clarified.

17.2 Reduction to an Ordinary Differential Equation for Tightly Wound Perturbations

One natural way to study the preceding problem in an approximate analytical description is to consider an asymptotic expansion based on the parameter

$$\epsilon = \frac{c}{r\kappa}, \tag{17.11}$$

which for a cold disk is a small quantity (see Chapter 14). With respect to the epicyclic parameter, the coefficients entering Eq. (17.1) have the following ordering:

$$rA = O(1), \qquad r^2 B = O(1), \qquad r^2 C = O(\epsilon^{-2}). \tag{17.12}$$

This suggests that we look for solutions with large gradients in the perturbed quantities so that in Eq. (17.1),

$$r\frac{d}{dr} = O(\epsilon^{-1}). \tag{17.13}$$

Solutions of this type also would allow for an asymptotic study of the Poisson equation, as shown in the next subsection.

17.2.1 Poisson Equation for Tightly Wound Perturbations

A local (ordinary differential) relation between Φ_1 and σ_1 is expected in the limit in which the total wave number associated with σ_1 is large because in this case we can perform a stationary phase analysis of Eq. (17.8). If we refer to Eq. (17.13) and to the additional ordering [see also Eq. (17.12)]

$$m^2 = O(1), \tag{17.14}$$

appropriate for tightly wound spirals, the differential form of the Poisson equation for the reduced potential U (obtained by multiplication of the perturbed potential by $r^{1/2}$) becomes

$$4\pi G r^{1/2} \sigma_1 \delta(z) = \frac{\partial^2 U}{\partial r^2} + \frac{\partial^2 U}{\partial z^2} - \frac{(m^2 - 1/4)}{r^2} U \sim \frac{\partial^2 U}{\partial r^2} + \frac{\partial^2 U}{\partial z^2}. \tag{17.15}$$

A simple discussion of this equation under the proper boundary conditions at large values of $|z|$ and the required jump (Gauss theorem) conditions at $z = 0$ gives, to two significant orders in our asymptotic expansion,[6]

$$\frac{d\Phi_1}{dr} + \frac{\Phi_1}{2r} \sim -is_k 2\pi G \sigma_1 = -is_k \Sigma h_1, \tag{17.16}$$

where

$$s_k \equiv \text{sgn}[\text{Re}(k)], \qquad k \equiv \frac{1}{i\Phi_1} \frac{d\Phi_1}{dr}, \tag{17.17}$$

$$\Sigma = \frac{2\pi G\sigma}{c^2} = \frac{2}{r\epsilon Q}. \tag{17.18}$$

Because we are interested here in cases in which the radial variation is large and follows Eq. (17.13), we expect the natural ordering

$$r\Sigma = O(\epsilon^{-1}), \tag{17.19}$$

that is,

$$Q = O(1). \tag{17.20}$$

Note that except for the s_k index, which distinguishes trailing from leading disturbances, we can proceed from here with no explicit reference to the radial wave number k that is usually introduced in the algebraic WKBJ description.

17.2.2   A Single Ordinary Differential Equation

Based on Eqs. (17.12), (17.13), and (17.19), we can eliminate the potential Φ_1 in a systematic asymptotic expansion and find the following approximate differential equation for the perturbed enthalpy:

$$h_1'' + (-is_k \Sigma + A)h_1' + \left[C + is_k \Sigma \left(\frac{1}{2r} - \frac{\Sigma'}{\Sigma} - A \right) + q \right] h_1 = 0, \tag{17.21}$$

where $r^2 q = O(1)$. Note that, in principle, we could calculate the higher-order terms contained in q if needed. Equation (17.21) can be reduced to its normal form by means of the standard transformation

$$h_1 \equiv u \exp\left[-\frac{1}{2} \int (-is_k \Sigma + A) \, dr \right] = u \left[\frac{\kappa^2(1 - \nu^2)}{r\sigma} \right]^{1/2} \exp\left(is_k \int \frac{\Sigma}{2} \, dr \right), \tag{17.22}$$

which defines the reduced enthalpy u. The final second-order ordinary differential equation (ODE) thus obtained is

$$u'' + \frac{1}{r^2\epsilon^2}\left[\left(\frac{1}{Q^2} - 1 + v^2\right) + iO(\epsilon) + O(\epsilon^2)\right]u = 0. \tag{17.23}$$

17.2.3 The Quadratic Dispersion Relation Recovered

If we keep the leading terms in Eq. (17.21), we obtain

$$h_1'' - is_k \Sigma h_1' + Ch_1 = 0. \tag{17.24}$$

Introducing the WKBJ ansatz [see Eq. (17.17)]

$$h_1 \propto \exp\left(i\int k\,dr\right), \tag{17.25}$$

to lowest order we have

$$-k^2 + |k|\Sigma + C = 0. \tag{17.26}$$

This is equivalent to the relation

$$\kappa^2 v^2 = \kappa^2 + k^2c^2 - 2\pi G\sigma|k| \tag{17.27}$$

discussed in Chapter 15. This quadratic equation in the magnitude of the radial wave number $|k|$ may admit two real roots (short and long waves), which coincide with $|k| = \Sigma/2$ if $Q^2(1-v^2) = 1$. Thus the mathematical transformation (17.22) that leads to the reduced enthalpy u basically corresponds to subtracting the common wave number $\Sigma/2$ from the enthalpy h_1 and reminds us that in our asymptotic analysis we have to focus on either trailing ($s_k = -1$) or leading ($s_k = +1$) waves.

17.3   A Two-Turning-Point Problem

We now address the problem of finding the solution to the basic equation [see Eq. (17.23)]

$$u'' + g(r,\omega)u = 0, \tag{17.28}$$

where

$$g(r,\omega) \equiv \frac{1}{r^2\epsilon^2}\left(\frac{1}{Q^2} - 1 + v^2\right) \tag{17.29}$$

for a class of astrophysically interesting galaxy models under the appropriate boundary conditions.[7] From the relation to the algebraic WKBJ dispersion equation, it is clear that the two oscillatory solutions that are in general allowed by such a second-order ODE correspond simply to the short and long waves of the physical discussion. The advantage of proceeding with the ODE equation is that it is not subject to the limitations of the algebraic WKBJ analysis; in particular, it can properly describe the properties of the relevant turning points at which the algebraic description breaks down because of the linear wave interaction.

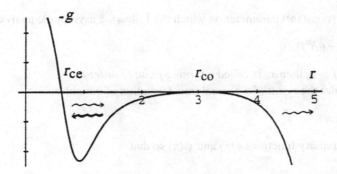

Fig. 17.1. A characterization of the two-turning-point problem relevant for spiral modes in galaxies (from Bertin, G., Lin, C. C. 1996. *Spiral Structure in Galaxies: A Density Wave Theory*, MIT Press, Cambridge, MA, p. 220; ©1996 Massachusetts Institute of Technology). With a different choice of the relevant function g and the unknown u, the same ODE and a similar description also apply to the regime of open spiral structure.

17.3.1 Boundary Conditions and the Eigenvalue Problem

We refer to the class of models for which the basic profiles $V(r)$, $\sigma(r)$, and $Q(r)$ are exemplified by those described in Section 14.4. Thus the function $g(r,\omega)$ typically behaves as shown in Fig. 17.1. Then the natural boundary conditions are those of evanescent waves at $r \to 0$ (because we are interested in regular solutions at the center) and outgoing waves (or radiation) at $r \to \infty$ (because no energy is meant to come from the outside).[8]

The specification of the two boundary conditions in our second-order linear ODE thus defines an eigenvalue problem. Note that the eigenvalue ω appears here in a more complicated manner than in standard Sturm-Liouville problems. This difficulty would have been even more evident if we had kept the higher-order terms of Eq. (17.23) that were dropped here [see Eq. (17.29)] for simplicity.

17.3.2 Two Turning Points

We now focus on the case in which the ODE presents two turning points, a simple turning point at $r = r_{ce}$ and a double turning point at $r = r_{co}$. As is easily realized from the form of the function $g(r,\omega)$, a Q profile that monotonically decreases from the center outward and reaches a constant value in the outer disk gives rise to this situation if the outer disk, where corotation is expected, is characterized by $Q = 1$; for $Q \approx 1$, the double zero at corotation splits into a pair of simple turning points[9] (which are off the real axis in the complex plane for $Q < 1$).

The study of turning points is a classical problem of applied mathematics to which the following subsection is devoted.

17.3.3 Langer's Transformation

Let us consider the second-order differential equation[10]

$$y'' + \lambda^2 q(x)y = 0, \tag{17.30}$$

where $\lambda \gg 1$ is a (constant) parameter on which the following asymptotic analysis is based. If

$$q(x) = (x - x_0)^\alpha f(x),$$ (17.31)

where $\alpha > 0$ and $f > 0$, then x_0 is called a *turning point of order* α.

We now consider the result of a general transformation of variables,

$$(x, y) \to (z, v),$$ (17.32)

defined by two auxiliary functions $\phi(x)$ and $\psi(x)$ so that

$$z \equiv \phi(x),$$ (17.33)

$$v \equiv \psi(x)y.$$ (17.34)

The function $\psi(x)$ can be chosen to be

$$\psi = \sqrt{\phi'}$$ (17.35)

so that the transformed differential equation remains in normal form, becoming

$$\frac{d^2v}{dz^2} + \frac{1}{(\phi')^2}\left[\sqrt{\phi'}\left(\frac{1}{\sqrt{\phi'}}\right)'' + \lambda^2 q\right]v = 0.$$ (17.36)

So far no approximation has been made, and we still have the choice of the function $\phi(x)$ at our disposal.

The standard WKBJ approximation is obtained by taking

$$(\phi')^2 = q(x).$$ (17.37)

In fact, this choice leads to the equation

$$\frac{d^2v}{dz^2} + [\lambda^2 + \mathcal{R}(x)]v = 0,$$ (17.38)

where

$$\mathcal{R} \equiv q^{-3/4}\left(\frac{1}{q^{1/4}}\right)''.$$ (17.39)

Note that the term $\mathcal{R}$ is $O(1)$ with respect to λ but is singular at the turning point. Thus the related WKBJ solution obtained when $\mathcal{R}$ is dropped,

$$v = \exp(\pm i\lambda z),$$ (17.40)

where

$$z = \int \sqrt{q}\, dx,$$ (17.41)

giving

$$y = \frac{1}{q^{1/4}}\exp\left(\pm i\lambda \int \sqrt{q}\, dx\right),$$ (17.42)

is also singular at the turning point, where $q = 0$. Therefore, as is well known, the standard WKBJ analysis[11] is applicable only away from turning points. Note how the preceding derivation gives in "one shot" two orders of the WKBJ expansion: the leading order (the behavior of the phase $k = \sqrt{q}$, which is the dispersion relation) and the next order (the behavior of the amplitude $\propto q^{-1/4}$, which is the conservation of wave action).

A uniform asymptotic solution applicable across a turning point can be obtained by the introduction of Langer's transformation:[12]

$$(\phi')^2 \phi^\alpha = q(x), \tag{17.43}$$

that is,

$$z = \left[(1 + \alpha/2) \int \sqrt{q}\, dx \right]^{1/(1+\alpha/2)}. \tag{17.44}$$

This choice leads to the transformed equation

$$\frac{d^2 v}{dz^2} + [\lambda^2 z^\alpha + \mathcal{S}(x)] v = 0, \tag{17.45}$$

where

$$\mathcal{S} \equiv \frac{\phi'''}{2(\phi')^3} - \frac{3(\phi'')^2}{4(\phi')^4}. \tag{17.46}$$

Note that $z \sim (x - x_0)$ for $x \to x_0$, and thus $\mathcal{S}$ is regular at the turning point. The approximate equation obtained when $\mathcal{S}$ is dropped is easily solved in terms of Bessel functions (of fractional order):[13]

$$v = z^{1/2} J_{\pm 1/(2+\alpha)} \left(\frac{2\lambda}{2+\alpha} z^{1+\alpha/2} \right) = z^{1/2} J_{\pm 1/(2+\alpha)} \left(\lambda \int \sqrt{q}\, dx \right). \tag{17.47}$$

Thus the full solution across a turning point can be expressed in terms of Bessel functions of an appropriate phase integral; by construction, far from the turning point, the solution behaves like the standard WKBJ solution.

17.4   Quantum Condition and Discrete Spectrum of Spiral Modes

We now return to the two-turning-point problem defined by Eqs. (17.28) and (17.29) and discuss the case in which the outer disk is characterized by $Q = 1$. The strategy is to focus on each of the two turning points separately and to impose the boundary condition at $r = 0$ to determine the solution around r_{ce} and the boundary condition at $r \to \infty$ to derive the solution around r_{co}; by taking advantage of the uniform solutions across the turning points provided in terms of Bessel functions, we can then obtain a representation of the solution that satisfies the relevant boundary conditions in the radial range $r_{ce} < r < r_{co}$. The two representations referred to the two turning points should be matched properly because they identify the same solution to the second-order ODE. This will give a determination of the eigenvalues of the problem.

If we refer to the inner turning point at r_{ce}, for which the order is $\alpha = 1$, the solution may be written as

$$u_{ce} = \left(\frac{\zeta}{\sqrt{g}} \right)^{1/2} Z_{1/3}(\zeta), \tag{17.48}$$

where $Z_{1/3}$ indicates a suitable combination of Bessel functions of order $1/3$, with the property that

$$u_{ce} = u_{ce}^{ev}, \qquad r < r_{ce}, \tag{17.49}$$

$$u_{ce} = u_{ce}^+ + u_{ce}^-, \quad r_{ce} < r < r_{co}. \tag{17.50}$$

The evanescent solution u_{ce}^{ev} satisfies the inner boundary condition and appears as a combination of outgoing and incoming waves outside r_{ce} (the superscripts indicate the direction of group propagation). The appropriate phase variable is

$$\zeta \equiv \int_{r_{ce}}^{r} \sqrt{g} \, dr', \tag{17.51}$$

with the branch of $\sqrt{g}$ chosen so that $\mathrm{Re}(\zeta) > 0$ for $r > r_{ce}$.

If we refer to the corotation turning point at r_{co}, for which the order is $\alpha = 2$, we may write the solution as

$$u_{co} = \left(\frac{z}{\sqrt{g}} \right)^{1/2} Z_{1/4}(z), \tag{17.52}$$

where $Z_{1/4}$ indicates a suitable combination of Bessel functions of order $1/4$, with the property that

$$u_{co} = u_{co}^+ + u_{co}^-, \qquad r_{ce} < r < r_{co}, \tag{17.53}$$

$$u_{co} = u_{co}^{st}, \quad r > r_{co}. \tag{17.54}$$

The short-trailing-wave solution u_{co}^{st} satisfies the radiation outer boundary condition and appears as a combination of outgoing and incoming waves inside r_{co}. The appropriate phase variable is

$$z \equiv - \int_{r_{co}}^{r} \sqrt{g} \, dr', \tag{17.55}$$

with the branch of $\sqrt{g}$ chosen so that $\mathrm{Re}(z) > 0$ for $r < r_{co}$.

The preceding description is implemented by the following choice of functions:

$$u_{ce}^+ = \exp\left(\frac{\pi}{6} i \right) \left(\frac{\zeta}{\sqrt{g}} \right)^{1/2} H_{1/3}^{(1)}(\zeta), \tag{17.56}$$

$$u_{ce}^- = \exp\left(-\frac{\pi}{6} i \right) \left(\frac{\zeta}{\sqrt{g}} \right)^{1/2} H_{1/3}^{(2)}(\zeta), \tag{17.57}$$

$$u_{co}^+ = \exp\left(\frac{\pi}{2} i \right) \left(\frac{z}{\sqrt{g}} \right)^{1/2} H_{1/4}^{(2)}(z), \tag{17.58}$$

$$u_{co}^- = \sqrt{2} \exp\left(\frac{3\pi}{4} i \right) \left(\frac{z}{\sqrt{g}} \right)^{1/2} H_{1/4}^{(1)}(z). \tag{17.59}$$

Here $H^{(1),(2)}$ are Hankel functions of the first and second kind, respectively.[14] Note that for large values of the argument, the Hankel functions behave as expected in the standard WKBJ analysis;

that is, for $\rho > 0$,

$$H_\beta^{(1)}(\rho) \sim \sqrt{\frac{2}{\pi\rho}} \exp[i(\rho - \pi/4)]\exp(-i\beta\pi/2), \tag{17.60}$$

$$H_\beta^{(2)}(\rho) \sim \sqrt{\frac{2}{\pi\rho}} \exp[-i(\rho - \pi/4)]\exp(-i\beta\pi/2). \tag{17.61}$$

The different sign in the ρ dependence of the phase is related to the two possible group propagation directions.

Because the problem we are dealing with is linear, for a proper matching of solutions, it is sufficient that the functions referred to the two turning points be equal within a common proportionality factor b:

$$u_{co}^+ = bu_{ce}^+, \tag{17.62}$$

$$u_{co}^- = bu_{ce}^-. \tag{17.63}$$

In the part of the complex r plane in which both $|\zeta|$ and $|z|$ are large, we can apply the asymptotic expressions for the Bessel functions and find the matching condition by elimination of b:

$$\exp(i\pi) = \sqrt{2}\exp\left(2i\int_{r_{ce}}^{r_{co}} \sqrt{g}\,dr\right); \tag{17.64}$$

hence

$$\int_{r_{ce}}^{r_{co}} \sqrt{g(\omega, r)}\,dr = \left(n + \frac{1}{2}\right)\pi + \frac{1}{4}i\ln 2, \tag{17.65}$$

with $n = 0, 1, 2, \ldots$. This is the desired global dispersion relation that gives the eigenvalues of the global problem.

A physical interpretation is better appreciated if we expand the g function for small values of the imaginary part of ω:

$$\sqrt{g(\omega, r)} \sim \sqrt{g(\omega_R, r)} + i\left(\frac{\partial\sqrt{g}}{\partial\omega}\right)_{\omega_R}\omega_I, \tag{17.66}$$

which allows us to separate the real and imaginary contributions in Eq. (17.65). If we now refer back to the notation of the algebraic WKBJ analysis (and recall that the group velocity is given by $c_g = -\partial\omega/\partial k$), these can be written as

$$\oint k(\omega_R, r)\,dr = (2n + 1)\pi, \tag{17.67}$$

which is a Bohr-Sommerfeld quantum condition, and

$$\gamma\tau \equiv \gamma\int_{r_{ce}}^{r_{co}} \frac{dr}{|c_g|} = \frac{1}{4}\ln 2, \tag{17.68}$$

which gives the value of the growth rate ($\gamma = -\omega_I$) in terms of the group propagation time τ between the two turning points. The modes are found to be self-excited as a result of overreflection at corotation.

17.5   From Linear Modes to Spiral Structures in Galaxies

This section serves the purpose of making a transition from the study of global modes to the application to spiral structures in galaxies. The main conclusions of this application will be given in Chapter 18. The key points to be considered in this transition are summarized in the following subsections.[15] The gist of what is described in the following is that in the statistical majority of galaxies, large-scale, regular spiral structure should be quasi-stationary, and this can happen if the dynamics of the disk is dominated by one mode or by a very small number of modes, with the linear theory able to identify at least the key properties of such modes. The natural occurrence of self-excited modes and the frequently observed grand-design structure in galaxies (especially in the near infrared) make the interpretation in terms of driven (damped) modes less appealing.

17.5.1   Morphology and Other Properties of Modes

The main property of modes that is considered in the application to observed galaxies is their morphology. Their growth rate is of minor importance; for practical purposes, we need to know only that the modes are moderately unstable (see also the comments given in Subsection 17.5.2). This is in line with the initial developments of the density-wave theory, when we sought an explanation for the observed patterns in terms of (neutral) waves and tried to fit them by means of the algebraic dispersion relation. Interestingly, in a large survey of thousands of galaxy models[16] of the type described in Section 14.4, indeed the morphology of the relevant modes was found to resemble that of observed large-scale structures, ranging from the case of tightly wound arms to the case of open morphologies with a two-blob structure in the inner disk, which is the signature of a bar in the linear context (some examples are shown in Fig. 17.2).

Besides the general shape of the arms, which is within the reach of the algebraic dispersion relations (see Chapter 15) whenever applicable, there are morphological aspects that can be predicted only in the context of modes. One important property is the amplitude modulation along spiral arms, which is so natural in the calculated modes because it displays the underlying wave interference at the basis of mode maintenance. This modulation is generally observed in grand-design spiral galaxies. Another crucial property is that the eigenfunctions that determine the morphology of the modes are associated with well-defined eigenvalues, that is, with a prediction of the location of the corotation circle (which determines the morphology of the overall flow patterns) and the location of the other resonances. (In the initial development of the density-wave theory, instead, the corotation circle was treated as a free parameter to be determined empirically; one of the main questions left was indeed *why* that value was selected by the underlying dynamics in a given galaxy.) In particular, normal (nonbarred) tightly wound spiral structures generally should be associated with corotation in the outer disk, whereas corotation for barred structures is predicted to occur just outside the tip of the bar, often in the middle of the optical disk.

Modal surveys also have important applications of a more dynamical interest. Once the modes are calculated numerically from the exact linearized equations, we can diagnose the modes to test the related asymptotic theories. In particular, the survey of thousands of fluid models mentioned earlier has demonstrated the general usefulness and shown the wide applicability range of the cubic dispersion relation introduced in Subsection 15.1.3. This type of test is important because most of the physical description and our understanding of dynamical mechanisms, in one way or another, are based on asymptotic theories. It is also especially important because most of the

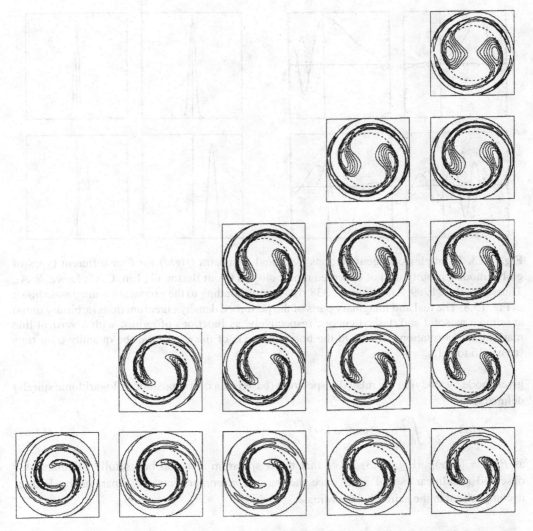

Fig. 17.2. A two-armed mode followed in a set of different models by variation of the parameters of the relevant basic state (see choice described in Section 14.4) from a wide survey of galaxy models (from Bertin, G., Lin, C. C., Lowe, S. A., Thurstans, R. P. 1989. *Astrophys. J.*, **338**, 78). These are perturbed density contours (for only the positive part). The dotted circle is the corotation circle; the outer circle corresponds to $\nu = 0.75$. The mode eigenfunctions and eigenvalues have been obtained numerically by integration of the integrodifferential problem [Eqs. (17.1) and (17.2)]. The results conform to the expectations of asymptotic analyses (outlined in the text) and are able to clarify the properties not only of tightly wound spirals but also of open and barred structures.

basic observational tests of the density-wave theory and the hypothesis of quasi-stationary spiral structure were initially carried out under the guidance of simple algebraic dispersion relations. The modes and models in the survey are generally diagnosed by means of the relevant propagation diagrams (see Subsection 15.3.2). It is often convenient to probe the wave content of a

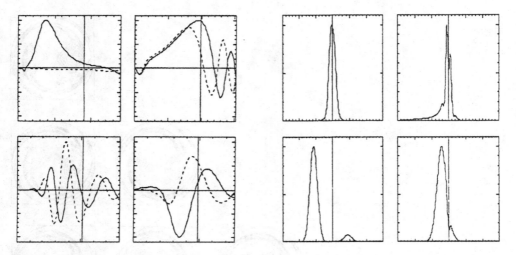

Fig. 17.3. Examples of eigenfunctions (*left*) and α spectra (*right*) for four different types of global modes, prototypical of observed morphologies (from Bertin, G., Lin, C. C., Lowe, S. A., Thurstans, R. P. 1989. *Astrophys. J.*, **338**, 104) corresponding to the propagation diagrams shown in Fig. 15.8. The real and imaginary parts of the perturbed density eigenfunctions (arbitrary units) are drawn as solid and dashed curves, respectively, as functions of radius, with a vertical line marking the corotation radius. On the horizontal axis of the α spectra, the quantity α/m runs from -15 to $+15$.

given mode by studying the related α spectrum (based on a decomposition in logarithmic spirals) defined from

$$\sigma(\alpha) = \frac{1}{2\pi} \int_0^\infty \sigma_1(r) \exp(i\alpha \ln r) \, \frac{dr}{r} \tag{17.69}$$

as $P(\alpha) = |\sigma(\alpha)|^2$ (Fig. 17.3). Note that the α spectrum also can be usefully calculated for observed (nonlinear) spiral arms once the various m components are separated by Fourier decomposition in the azimuthal coordinate.

17.5.2   Modeling and the Dynamical Window

When thinking in terms of application to real galaxies, we should not follow blindly the properties of all the modes found in given models. Many of these modes are probably to be discarded by means of physical arguments that go beyond the equations that define the linear modal analysis used. For example, the modal survey based on integration of the equations given in Section 17.1 generally leads to a variety of self-excited modes for given values of m [see Eq. (17.67)], and this would be the case for different values of m.[17] If the fluid model is used to simulate the properties of a real galaxy, we should then include other effects, in particular, those associated with the active presence of the collisionless disk of stars. We have seen that in the stellar disk, perfect absorption occurs at the inner Lindblad resonance (ILR) (see Subsection 16.4.1), whereas a fluid disk is largely transparent. Thus those fluid modes for which the ILR is exposed would be damped in a more realistic galaxy model because the relevant wave cycle would be interrupted. It is easy to check that this argument leads to discarding all the higher-n and higher-m modes, leaving very few $m = 1$ and $m = 2$ self-excited modes to be considered for a given model. This is very important for the interpretation of large-scale structure as quasi-stationary and is found to explain why large-scale spiral structure, especially in the near infrared, is associated mainly with $m = 1$ and $m = 2$.

Even if we started out with a modal survey based on the equations of stellar dynamics, we would have to reinterpret our results carefully in view of the three-dimensional and multicomponent character of the real galaxy disk that is being modeled. Besides carefully examining the role of resonances in the multicomponent disk, we should generally pay great attention to specification of the basic state. One attempt in the direction of realistic galaxy models was that given in Section 14.4, but there is no doubt that a proper modal study should be extended to many other models that might be realistic and lie outside the modal survey mentioned earlier.

A curious feature of modal surveys of one-component, zero-thickness models of the type described earlier is that because of the subtle issues involved in the modeling process (see Section 16.2), two *different* basic states may happen to be relevant for the calculation of modes for the *same* galaxy. This counterintuitive statement refers to galaxy disks that, under special circumstances, happen to support both tightly wound and open modes, which is quite natural if we recall that the Population I disk is bound to support tightly wound spirals, whereas open structures do not suffer from thickness effects so much, and for these, the gas is mostly passive. Coexistence of different spiral morphologies is indeed found among observed galaxies.

Another already mentioned important aspect of the modeling procedure is that we should restrict attention mainly to models subject to moderately unstable modes only. Violently unstable models should be discarded because they would generally be inadequate to describe currently observed galaxies.

When we go through these arguments and detailed checks in the process of matching the morphology of modes to observed structures, we come out with a number of dynamical constraints on the structural parameters that define the basic state of the galaxy that we are modeling. We have often referred to this as the *dynamical window* because it sharpens the view of galactic structure that we have from other observational windows (see Chapter 5).

17.5.3 Global Modes and Evolution

Despite the encouraging results of the modal description in interpretation of spiral structure in galaxies that are given in Chapter 18, we are at only the beginning of our understanding of galaxy evolution. This is largely due to our general lack of tools to describe the nonlinear evolution of a dynamical system, when even at the linear level its dynamics is dominated by a few modes. To a large extent, this statement also applies to many hydrodynamical systems and plasma configurations that have been studied for decades with the additional help of a number of experiments. For the galactic context, comparison with observations must rely on semiempirical arguments. For example, we do know from near-infrared observations that in some grand-design galaxies spiral structure is present at small amplitudes and that in others in which the amplitude is large the azimuthal profile across the arms is smooth and sinusoidal. All this makes it likely that the morphology of spiral arms can be usefully matched by that of linear modes. Still, in the modal picture, there are only a few plausible arguments and a few mechanisms explored quantitatively that can demonstrate how growing spiral modes may saturate at finite amplitudes.

Furthermore, the presence of nonaxisymmetric structures and other processes favor the transport and redistribution of angular momentum across the disk, and ultimately, this leads to evolution of the basic state. In Section 16.3 we briefly touched on a mechanism of self-regulation, which is expected in the presence of the dissipative gas. All these are very interesting but largely unexplored issues that await further investigation, and progress is much needed.

17.5.4   A Synopsis of Alternative Scenarios

From an even more general point of view, we may wish to explore, far better than what has been done so far, the driven case in which the galaxy disk, with its own intrinsic characteristics and modes, is significantly affected by interactions with other galaxies or just by one object passing by. We have only a few ideas and a few simulations that may help us to understand the associated evolutionary process. There are also other nonmodal interpretations, in which the large-scale spiral structure is assumed to be transient or recurrent. The study of dynamical systems prone to this nonmodal behavior is incomplete. Although it appears unlikely that these nonmodal scenarios can describe the properties of the majority of grand-design spiral galaxies, nonetheless, all these alternative views should be explored further and, especially, confronted with the growing body of observational data.

The general alternatives for the interpretation of large-scale grand-design spiral structure are conveniently summarized in a simplified table (Fig. 17.4), in which columns and rows broadly correspond to the two questions raised by Oort in his definition of the problem of spiral structure given in the 1960s.[18] The modal picture is represented in the first column and identified with the hypothesis of quasi-stationary spiral structure. It is viable if the dynamical system (i.e., a realistic model of the galaxy disk) has a discrete spectrum of perturbations dominated by one or few self-excited modes or at least some discrete damped modes that can be usefully excited by a tidal interaction. The second column represents the nonmodal interpretations, which are viable if

	Quasi-stationary spiral structure	**Rapidly evolving spiral structure**
Internal origin	Discrete spectrum; one or few dominant self-excited modes. No tides are required; tides, if present, may help to strengthen and organize spiral structure.	Continuous spectrum. Regenerative or recurrent spiral structure.
External origin	Discrete spectrum; damped modes. Tidal driving?	Continuous spectrum. One-shot " tidal encounter driving transient spiral structure.

Fig. 17.4. A synopsis of alternative scenarios (see description in Subsection 17.5.4).

the dynamical system under consideration lacks discrete modes and is instead characterized by a continuous spectrum of perturbations (or a large number of modes). The first line represents the view that, as a rule, spiral structure is of internal origin. The second line assumes that external (tidal) driving is generally required for the existence of grand-design spiral structure.

Note that in the early development of the theory, spiral structure in barred galaxies was often thought of as associated with a kind of external driving (nontidal; the bar was the driving factor, but the bar potential was generally imposed and its origin was kept aside as a different problem), whereas the current modal theory (outlined in this chapter and Chapter 18) places barred and unbarred structures on an equal footing, that is, interpreted by means of self-excited modes (of internal origin) characterized by different morphologies.

The frequent existence of grand-design structures proved by observations in the near infrared (see Chapter 18 and the comments at the end of Chapter 16) definitely supports the picture followed in this book, that, in general, grand-design spiral structure is quasi-stationary and of internal origin.

Notes

1. See the general formulation by Shu, F. H. 1970. *Astrophys. J.*, **160**, 89; the interesting matrix equation developed by Kalnajs, A. J. 1977. *Astrophys. J.*, **205**, 751; and the detailed study of self-similar models, with flat rotation curve, by Zang, T. A. 1976. Ph.D. dissertation, Massachusetts Institute of Technology, Cambridge, MA; see also Hunter, C. 1992. *Ann. N.Y. Acad. Sci.*, **675**, 22. Many other references are available in the monographs by Fridman, A. M., Polyachenko, V. L. 1984. *Physics of Gravitating Systems*, Springer-Verlag, Berlin; and Palmer, P. L. 1994. *Stability of Collisionless Stellar Systems*, Kluwer, Dordrecht, The Netherlands. Especially in relation to the origin of bar modes, a relatively recent interesting investigation is presented by Jalali, M. A., Hunter, C. 2005. *Astrophys. J.*, **630**, 804, who apply the matrix equation to study a variety of galaxy models (in terms of density profiles, Q profiles, and orbit composition), some characterized by the presence of many radial orbits; they find typically two dominant unstable bar modes, without an inner Lindblad resonance, thus at variance with the type of slow bar modes envisioned by Lynden-Bell, D. 1979. *Mon. Not. Roy. Astron. Soc.*, **187**, 101; and later investigated by Polyachenko, E. V. 2004. *Mon. Not. Roy. Astron. Soc.*, **348**, 345; 2005. *Mon. Not. Roy. Astron. Soc.*, **357**, 559.
2. Besides the work described later in this chapter, we should recall the work, among others, by Aoki, S., Noguchi, M., Iye, M. 1979. *Publ. Astron. Soc. Jpn.*, **31**, 737; see also Hunter, C. 1963. *Mon. Not. Roy. Astron. Soc.*, **126**, 299; 1965. *Mon. Not. Roy. Astron. Soc.*, **129**, 321; 1969. *Stud. Appl. Math.*, **48**, 55; and Erickson, S. A. 1974. Ph.D. dissertation, Massachusetts Institute of Technology, Cambridge, MA.
3. Bertin, G., Lin, C. C., Lowe, S. A., Thurstans, R. P. 1989. *Astrophys. J.*, **338**, 104.
4. The fluid-model equations given here essentially follow the notation and the analysis of Feldman, S. I., Lin, C. C. 1973. *Stud. Appl. Math.*, **52**, 818; then continued by Lin, C. C., Lau, Y. Y. 1975. *SIAM J. Appl. Math.*, **29**, 352; Lin, C. C., Lau, Y. Y. 1979. *Stud. Appl. Math.*, **60**, 97, and in many other papers.
5. Pannatoni, R. F. 1979. Ph.D. dissertation, Massachusetts Institute of Technology, Cambridge, MA; Pannatoni, R. F. 1983. *Geophys. Astrophys. Fluid Dyn.*, **24**, 165; Haass, J. 1982. Ph.D. dissertation, Massachusetts Institute of Technology, Cambridge, MA; Thurstans, R. P. 1987. Ph.D. dissertation, Massachusetts Institute of Technology, Cambridge, MA; Lowe, S. A. 1988. Ph.D. dissertation, Massachusetts Institute of Technology, Cambridge, MA.
6. Shu, F. H. 1970. *Astrophys. J.*, **160**, 99; Mark, J. W.-K. 1971. *Proc. Natl. Acad. Sci. USA*, **68**, 2095. Asymptotic studies of the Poisson equation have been performed in a variety of ways; the discussion by Bertin, G., Mark, J. W.-K. 1979. *SIAM J. Appl. Math.*, **36**, 407, based on a stationary phase analysis of integral representation (17.8), goes one step beyond approximation (17.16) and includes a comparison with the results of exact numerical integrations.

7. This important discussion for the case of tightly wound spirals was first made by Lau, Y. Y., Lin, C. C., Mark, J. W.-K. 1976. *Proc. Natl. Acad. Sci. USA*, **73**, 1379.

8. This is analogous to the plasma case described by Berk, H., Pearlstein, L. D., et al. 1969. *Phys. Rev. Lett.*, **22**, 876. There are also interesting analogies with the case of α decay of atomic nuclei; see Gamow, G. 1928. *Z. Phys.*, **51**, 204; Gurney, R. W., Condon, E. U. 1929. *Phys. Rev.*, **33**, 127. Another important analogy can be drawn with the so-called quasi modes of a black hole; see Chandrasekhar, S. 1983. *The Mathematical Theory of Black Holes*, Clarendon Press, Oxford, UK.

9. Within our context of self-excited spiral modes, the more general three-turning-point problem has been worked out by Nishimoto, T. 1979. *Stud. Appl. Math.*, **60**, 11.

10. The discussion follows Nayfeh, A. H. 1973. *Perturbation Methods*, Wiley, New York, chapter 7. See also Heading, J. 1962. *An Introduction to Phase-Integral Methods*, Wiley, New York.

11. This asymptotic method is also well known in the context of quantum mechanics for the discussion of the Schrödinger equation; see Messiah, A. 1958. *Quantum Mechanics*, Wiley, New York.

12. Langer, R. E. 1931. *Trans. Am. Math. Soc.*, **33**, 23.

13. See, for example, Abramowitz, M., Stegun, I. A. 1965. *Handbook of Mathematical Functions*, Dover, New York.

14. These expressions can be checked to properly implement the desired boundary conditions by considering the behavior of the Hankel functions for large values of the argument and the relevant analytical continuation formulas, as recorded, e.g., in Abramowitz, M., Stegun, I. A. 1965. *Handbook of Mathematical Functions*, Dover, New York.

15. A thorough evaluation of the physical aspects involved is given in the monograph by Bertin, G., Lin, C. C. 1996. *op. cit.*

16. Bertin, G., Lin, C. C., Lowe, S. A., Thurstans, R. P. 1989. *Astrophys. J.*, **338**, 78.

17. See Haass, J. 1983. In *IAU Symposium*, Vol. 100, ed. E. Athanassoula. Reidel, Dordrecht, The Netherlands, p. 121.

18. Oort, J. H. 1962. In *Interstellar Matter in Galaxies*, ed. L. Woltjer. Benjamin, New York, p. 234.

18 Spiral Structure in Galaxies

With most of the necessary dynamical tools already introduced, we now turn to the astrophysical problem. In this chapter we will try to highlight only some important points.[1] The reader is referred to the monograph by Bertin and Lin (1996) for the full definition of the astrophysical issues involved, a physical outline of the modal theory in the semiempirical context, and the description of many important observational tests not mentioned here.

Spiral structure in galaxies has been the focus of a vast set of observational and theoretical investigations during the entire twentieth century. A decisive turning point in the course of these studies was the realization, in the early 1960s, that density waves are at the basis of the most spectacular phenomena associated with spiral structure. In particular, this realization came as the conclusion of the pioneering work of Bertil Lindblad, who first formulated the hypothesis that the large-scale spiral structure in galaxies is quasi-stationary despite the presence of differential rotation in the disk.[2] Because of differential rotation, any material structure could not persist but would be stretched and would wrap up on a very short time scale. This is commonly known as the *winding dilemma*. In contrast, as the manifestation of waves, the density concentrations that define the spiral arms could move in the disk in a manner that is, to a large extent, decoupled from the motion of the individual particles (stars or gas clouds) that support it, much like a sound wave in an ordinary gas. Thus spiral arms, if associated with a wave phenomenon, could survive differential rotation as quasi-stationary patterns.

The hypothesis of quasi-stationary spiral structure was adopted[3] as the starting point of a density-wave theory to be worked out quantitatively (in terms of dispersion relations, much as in other fields of classical hydrodynamics) within a semiempirical approach, in which the need for a positive comparison with observations was emphasized as important well before a full dynamical theory could be put together. In particular, it was predicted that in general, being part of a wave process, the density patterns should correlate with well-defined patterns of noncircular motions, and these observational tests were immediately carried out with great success for the density-wave interpretation. Furthermore, because of the differential rotation and the resulting shocks in the cooler and dissipative interstellar medium, in this scenario, different age groups and colors were expected to be observed at different phases with respect to the peak of the spiral gravitational field, which was in line with what was known about spiral arms and later received further interesting confirmations.[4]

Since the initial formulation of the hypothesis of quasi-stationary spiral structure, it has been clear that for this to be an acceptable framework for the description of the existing morphology, the large-scale dynamics of the disk should be dominated by a very small number of global spiral modes. In practice, the resulting modal theory was developed quantitatively only starting in the mid-1970s; it has now grown into an expansive framework that is able to interpret a large number of astrophysical phenomena. Although the theory has a semiempirical foundation, it stands on a rather wide set of detailed quantitative dynamical investigations that cover the linear and some of the nonlinear aspects of the problem.

The term *mode* is sometimes (even by astronomers in this field of research) used interchangeably with *Fourier m component* or *density wave*. In the present context, it should be emphasized that the term *mode* has a much more specific dynamical meaning, as described in Chapter 17. A mode is essentially a standing wave that can be supported by the disk; thus it is associated with a spiral density perturbation and gravitational field that rotate rigidly at a given frequency around the galaxy center and do not propagate in the radial direction. It also should be emphasized that on the small scale most of the arguments in favor of a quasi-stationary structure do not apply; indeed, also by analogy with the physics of ordinary fluids, the gaseous component, which is dynamically very cold, should be prone to small-scale, transient (rapidly evolving) spiral activity. In many galaxies, the gas distribution is rather diffuse, so its dynamical impact is concentrated mainly in the outer regions, beyond two exponential scale lengths of the stellar disk. Thus the processes of self-regulation and Jeans destabilization that the gas can contribute (see Chapter 16), which are very important for relatively light disks, tend to operate mainly in the outer disk; this explains why corotation for normal (nonbarred) spiral structure is generally expected in the outer disk.

One of the most interesting aspects of the modal theory is that it can provide an explanation for why some galaxies are barred and others are not and why some galaxies have a grand design and others are flocculent. All this is traced to the relevant modal intrinsic characteristics of the basic state, as described in this chapter. This framework has been demonstrated by a study of the conditions under which a basic state may be dominated by simple mode prototypes, which are found to cover all the broad categories in the morphological classification of spiral galaxies. It should be extended to more complex models of realistic basic states in view of the detailed variety of morphologies observed, especially for barred and ringed galaxies. Without aiming at a comprehensive description of what has been done or what could be done, we focus in this chapter on a few important observational facts that are naturally explained in the modal context.

Of course, different interpretations from the quasi-stationary modal picture can be conceived, and indeed, many papers take different points of view,[5] with the spiral structure as either a recurrent or a transient phenomenon (see last subsection in Chapter 17). It seems that these alternative views, which tend to emphasize the evolutionary process, have not been worked out in coherent frameworks that are able to provide quantitative answers to the various astrophysical issues involved. In general, either in the modal context or outside it, the dynamical evolution of galaxy disks remains largely unexplored.

18.1   Quasi-Stationary Spiral Structure and Three Levels of Persistence

The hypothesis of quasi-stationary spiral structure refers to only the large-scale regular structure that is frequently observed in galaxies. Traditionally it is associated with the problem of grand-design spiral structure in nonbarred galaxies.[6] In reality, it should also be addressed to the case of barred galaxies. The reason why barred galaxies are often not considered in this context may be that it is generally taken for granted that their structure is quasi-stationary. In this regard, the modal theory has the advantage over other nonmodal approaches because it sets barred and nonbarred large-scale regular morphologies on an equal footing. Others have sometimes argued instead that barred structure is intrinsic (and thus de facto modal), whereas nonbarred structure is tidally driven and nonmodal, which is less appealing.

The hypothesis of quasi-stationary spiral structure states that for the statistical majority of galaxies, the underlying potential that is associated with the grand-design spiral structure is approximately stationary in a suitable rotating frame. The hypothesis thus makes a clear distinction between small-scale and large-scale spiral arms. There is no doubt that small-scale spiral features may coexist, as often observed, with grand-design structure and that they may be rapidly evolving. The term *quasi* is very important: Without it, the hypothesis would not make sense. The picture is that the overall morphology does not change significantly over one or two rotation periods of the disk. In other words, the morphological evolution is admitted to be present, but it is considered to be slow. The hypothesis also recognizes that in a few individual galaxies, the observed structure need not be quasi-stationary, because the arguments in its favor are of a statistical nature.

The arguments in favor of the hypothesis are mainly empirical. The Hubble classification scheme appears to correlate properties that are directly connected with the spiral morphology (e.g., the pitch angle of spiral arms) with others that define the basic state (bulge size and gas content); the latter properties are not expected to change significantly on the fast dynamical scale. From a different point of view, transient regular spiral structures of the type often advocated as alternatives to the quasi-stationary modal case, even if occurring under appropriate conditions among real galaxies, would be observed rarely simply because they are short-lived. Therefore, the statistical majority of objects should naturally fall within the quasi-stationary picture. Another semiempirical justification to the hypothesis comes from experience with the case of hydrodynamics, geophysics, and other dynamical systems, for which it is often noted that large-scale patterns are associated with quasi-stationary phenomena (e.g., the case of hurricanes; see Fig. 3.5), whereas small-scale irregular features generally evolve quickly. From a theoretical point of view, large-scale structures "see" the boundaries of the system and thus are likely to be associated with global modal behavior. This is especially natural in thin, rotating disks; in contrast, truly three-dimensional, nonrotating systems are often prone to turbulent behavior.

Quasi-stationarity is to be considered in the spirit of a working hypothesis in a semiempirical approach. A number of observational tests have been carried out on its basis. In these tests, the properties of the structure are approximated as if the structure were steady. There are impressive quantitative results from application of the hypothesis (e.g., the work on the flow patterns in M81 carried out in the 1970s[7]). More recently, a number of projects aimed at measuring the relevant pattern speed all apply the hypothesis of quasi-stationary spiral structure; the positive results that are obtained confirm that the description in terms of a very small number of modes is indeed adequate.

There are three levels at which quasi-stationarity has been questioned, and an answer has been provided for all them. The simplest level is that of the winding dilemma. We have seen that the wave concept easily resolves the apparent paradox that follows if we assume that spiral structure is made of material arms. In the late 1960s it was noted that invoking the presence of density waves by itself does not solve the persistence problem because wave packets tend to quickly propagate away.[8] Obviously, there is no such difficulty if we refer to global modes (see Chapter 17) because they do not propagate in the radial direction, with a natural feedback process able to guarantee the maintenance of the standing wave. A third problem for the persistence of spiral structure was raised and stressed especially in the 1980s. It was noted that if the mode that is argued to dominate a given spiral structure relies on the condition of marginal stability (Q very close to unity), as required for tightly wound modes (see Chapter 17), then any heating of the disk, especially that associated with spiral activity itself, would soon bring the disk outside the desired range for Q, thus making it impossible for the spiral structure associated with the mode to be long-lasting. This is a fully legitimate concern for a purely collisionless stellar disk. As an answer to the heating problem, it was pointed out that the presence of dissipative gas in the outer disk can give rise to a process of self-regulation that is able to keep the desired conditions over a long time scale (see Chapter 16). Thus gas plays an important active role in tightly wound nonbarred galaxies. For the open and barred modes, the heating problem is less severe because the precise value of Q for their support (as is already clear from inspection of marginal stability in the cubic dispersion relation) is less important.

The main focus of the modal theory is on the current morphology of spiral structure and its maintenance. The issue of excitation is given lower priority, which is natural if we notice that the detailed dynamical mechanisms at work and the specific evolution of the disk are very hard to extract from the observational data.

18.1.1 Spiral Structure in N-Body Simulations

The comments given earlier in this chapter should clarify the reasons why the investigation by means of N-body simulations of the problem of spiral structure in galaxies has remained frustratingly slow and inconclusive. The main difficulties of this approach based on numerical experiments can be traced to (1) the problem of identifying a sufficiently rich and realistic basic state for the initial conditions of the simulation and of guaranteeing an adequate representation of the relevant physical mechanisms, (2) the general trend to use the simulations to investigate long-term evolution rather than specific dynamical mechanisms, and (3) the inherent lack of reliability of the simulations if used to study the evolution of structures and basic state on very long time scales (comparable with the Hubble time). At the interpretation level, the difficulties are aggravated by frequent confusion about the concept of modes, the modal theory, and the meaning of the assumption of quasi-stationary spiral structure. A thorough review of this line of research would deserve a monograph of its own and thus is well beyond the scope of this book. In this short subsection we comment on these issues by taking advantage of two recent papers that are specifically devoted to testing some general ideas and possible alternatives for the interpretation of spiral structure in galaxies with the help of state-of-the-art simulations.

The first paper[9] addresses the question of the lifetimes of spiral patterns in disk galaxies. The main objective of the paper is to test the stability of a model taken from the survey described in Section 17.5 and used in Section 18.2 to exemplify the modal interpretation of normal, nonbarred spiral structure. As widely discussed by the authors of that survey,[10] the basic state and fluid equations used are meant to model the complex properties of a relatively light (submaximal) disk with finite thickness in which the presence of cold interstellar medium plays an essential role in enforcing the process of self-regulation, and the stellar component, being characterized by perfect absorption of waves at the inner Lindblad resonance, plays a key role in excluding modes characterized by high-m or high-n values.[11] In particular, the Q profile adopted in the survey matches the properties of the effective Q applicable to a two-component disk made of stars and gas (see Chapter 16).[12] The paper instead considers simulations of a purely stellar disk and initializes them with the formal properties of the basic state just mentioned, without ensuring the active presence of the dissipative gas and an adequate test of the ability of the simulations to handle all the relevant resonances properly, as known to be active in a collisionless stellar disk. The results presented are found to be puzzling by the author because the disk does not produce any rapidly growing bisymmetric instabilities, but rather it eventually exhibits multiple-armed spiral patterns and heats up to values of the Q parameter above 1.5 over the entire disk.

These results are not puzzling because they can be easily understood. Figure 5 in the paper confirms that the "heating of the disc, which coincides with the occurrence of visible spiral patterns, is increasingly delayed as larger numbers of particles are employed." The emergence of higher-m activity takes place at longer and longer time scales as the number of particles N increases, consistent with the interpretation that the higher-m activity is improperly present in the simulations because of the insufficient number of particles used (so that the screening role of the inner Lindblad resonances is not represented properly). Furthermore, the resulting (unrealistic) heating saturates at $Q \approx 1.7$ to 2 (as shown in fig. 4 of the paper), which is well consistent with the condition expressed by Eq. (15.34) and Fig. 15.9 of this book. In other words, the simulations are (nominally[13]) collisionless because they lack the cooling of self-regulation provided in galaxies by the dissipative gas but are improperly characterized by fluid behavior in relation to the Lindblad resonances. Finally, the entire paper (and also the second paper, which will be briefly addressed at the end of this subsection) improperly confuses the focus on moderately unstable models, characteristic of the modal theory, with the idea that the mechanism of overreflection (of long into short trailing waves) is intrinsically weak, whereas swing amplification is described as intrinsically strong and powerful. As explained in Chapter 15 and illustrated in Fig. 15.9, overreflection also can be strong, and swing is just a kind of overreflection (of leading into trailing waves) that also can be weak, depending on the parameter regime. The focus on moderately unstable models is physically justified because those are the only ones to be considered for realistic applications to observed galaxies.

For N-body simulations without a tidal perturber, some claims of the possibility of long-lived spiral structure[14] are criticized in the above-mentioned paper because long-term evolution beyond ten rotation periods has not been properly addressed, and the observed pattern in the simulation is actually changing from snapshot to snapshot. We note that independent of an assessment of whether those simulations were performed in a satisfactory manner, the criticism raised does not touch the modal interpretation nor the hypothesis of quasi-stationarity as explained in this book (in particular, in Section 18.1) and in the monograph by Bertin and Lin (1996). Similarly, the criticism of other papers,[15] because the claimed long-lived spiral structure turns out "to be the superposition of several waves having different pattern speeds" is not a criticism of the modal interpretation, given the fact that more than one mode is generally expected to operate in a dynamical system (see also discussion in Subsection 17.5.4). Another point where the paper appears to add confusion is the issue of coexistence. The coexistence of small-scale spiral activity with coherent large-scale spiral structure has been noted from the very beginning as an empirical fact and a natural expectation of the theory (see Section 18.1).[16]

The second paper[17] tests by means of N-body simulations the stability of a self-similar model described briefly in Subsection 14.3.2. The model,[18] with a perfectly flat rotation curve and a perfectly constant $Q = 1.5$, is constructed on purpose in such a way that the inner Lindblad resonance for any $m \geq 2$ is exposed, so no resonant cavity could be created to support a global self-excited mode. Thus, except for the $m = 1$ modes, the model is expected to be (linearly) stable and has been used in the tidal interpretation of spiral structure in galaxies to exemplify disks that, on the one hand, lack internally excited spiral structure and, on the other hand, are ready to accept tidal driving by the action of an occasional encounter.[19] The disk in this model is relatively heavy, in order to favor the amplification of $m = 2$ waves by overreflection of leading waves (induced by external tides) into trailing waves. This second paper focuses on the development of $m = 2$ disturbances in such a model. At early times, the stability of the model is confirmed, but the simulation shows that the model eventually heats up in the central regions and produces an internally excited global structure. It is likely that the combination of heating (i.e., the formation of an inner Q barrier) and inadequate resolution in phase space for a proper treatment of resonances effectively makes the basic state evolve into one for which a resonant cavity for self-excited modes supported by leading and trailing waves can be established. It appears that the main message of these simulations is the demonstration that choices such as that of a constant Q throughout the disk are artificial and that with a more natural Q profile, self-excited global modes are hard to avoid.[20]

18.1.2   Investigations into the Large-Scale Shock Scenario

The study of the large-scale shock scenario represented in Fig. 15.1 and briefly mentioned at the beginning of this chapter has been the focus of interest for a number of theoretical and observational investigations. Some of the issues involved have been briefly addressed in the introduction to Chapter 16. As for the topic of Subsection 18.1.1, to which the subject is related, it would be impossible here to give a proper review of the tools used in these investigations and the main results obtained. It appears that a proper understanding of the observed phenomena and the underlying physical mechanisms will depend on an improved representation of the physical properties of the interstellar medium that is undergoing the shock in galaxy disks.[21] In turn, because most of the modeling is currently based on the use of numerical simulations, it is particularly important that the simulations are set up with realistic initial conditions and carried out with the active participation of the relevant physical ingredients. The subject has recently gained a revival of interest. Obviously, this topic is related to the problem of measuring the pattern speed or, equivalently, the corotation radius in grand-design galaxies (which will be described briefly in Section 18.2).

The classical tests of the large-scale shock scenario, started in the 1960s, address the issues of enhanced star formation along the arms,[22] the formation and properties of giant molecular clouds,[23] the migration of stars away from the arms with the consequent formation of color gradients and age gradients,[24] and the general structure of the interstellar medium in relation to the observed spiral patterns. In principle, different views about the nature of large-scale spiral structure (see Subsection 17.5.4) may be tested by comparing the different predictions on the above-mentioned morphological indicators.[25] In particular, it would be interesting to compare the behavior of the interstellar medium implied by the presence of a swinging, rapidly evolving large-scale spiral[26] with that implied by the presence of one or few dominant global modes. In practice, the theoretical and observational results do not have simple interpretations.[27] Tests on specific individual galaxies, such as M51 and M81, have been particularly interesting.[28]

18.2   Dynamical Classification of Spiral Morphologies

Linear spiral modes, different from commonly known standing waves in other mechanical systems, may not only be damped (i.e., they may decay in time) but sometimes may be unstable or self-excited (i.e., they may grow exponentially in time as a global instability) at the expense of the free energy stored in the differential rotation of the disk (see Chapter 17); it is part of a nonlinear analysis to demonstrate how they can saturate (i.e., equilibrate) at finite amplitudes. Much as for other types of standing waves, the amplitude of the density perturbation associated with a global spiral mode, for the point of view of an observer moving along the arms, is in general modulated; this, like the presence of nodes for a standing wave, is due to interference of the elementary waves that can be imagined to maintain the mode. This amplitude modulation was immediately recognized[29] as an important feature for comparison with the observations.

The finding of self-excited global modes in a given galaxy model is telling us that under the physical conditions considered, spiral structure is energetically favored and expected to be realized by intrinsic mechanisms, even without external help (such as that of tidal interactions). When moderate instability is involved, the structure of the unstable modes gives an approximate description of the properties of the nonaxisymmetric state, much as the unstable $m = 2$ modes on a Maclaurin ellipsoid give an indication of the properties of the triaxial Jacobi ellipsoids (in the vicinity of the bifurcation between the two equilibrium sequences).

A given basic state is subject to a spectrum of global spiral modes (i.e., a set of pattern frequencies, with the related self-consistent spiral gravitational field), which in principle can be calculated from the relevant dynamical equations; the spectrum just reflects the intrinsic characteristics of the basic state. Thus, given a perfect knowledge of the galaxy disk, we could predict the morphology and pattern frequency (or frequencies) of its spiral structure (much as we can calculate the eigenstates of the hydrogen atom). In practice, the galaxy disk is too complex (see Section 16.2), and its basic state and general dynamical conditions are known in only an incomplete form from the data available, so a deductive approach has no realistic chance of applicability. This point becomes even more obvious if we note that the physical system owes its current state to a variety of nonlinear mechanisms and evolutionary processes that are impossible to check by direct observation. This somewhat discouraging situation is by no means specific to our problem for spiral galaxies but rather is well known to many other research fields (e.g., plasma physics, fluid dynamics, geophysics, meteorology, etc.) that also deal with complex collective macroscopic systems.

In this situation, one productive way to proceed is to take a semiempirical approach. An extensive survey of realistic models of galaxy disks[30] has shown that the morphology types of the global spiral modes that can be generated in a disk match the general morphological categories that are found along the Hubble sequence (Fig. 18.1). Depending on the parameter regime that characterizes a given galaxy disk (see Subsection 15.5.1), the dominant mode may be of the A type (with normal, nonbarred morphology and corotation generally in the outer disk) or the B type (with a generally prominent bar and corotation predicted to occur just outside the tip of the bar). Different excitation mechanisms operate for the two classes of modes: long-short or leading-trailing wave overreflection at corotation (see Subsection 15.5.3). Type A modes rely on a combined support of gas and stars, whereas type B modes are star-dominated, with the gas often playing mostly a passive role. Type A modes are rather tight. Type B modes can have a plain two-lump structure, as in many SB0 galaxies, such as NGC 2859; a pair of rather tight arms

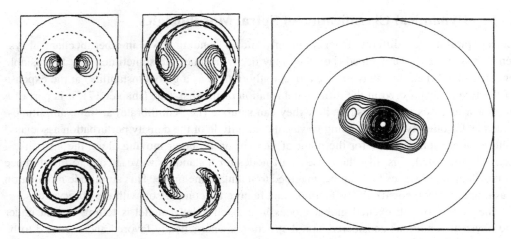

Fig. 18.1. *Left:* Four mode prototypes (represented as in Fig. 17.2) from a wide survey of galaxy models corresponding to the four propagation diagrams of Fig. 15.8 and to the eigenfunctions and α spectra of Fig. 17.3): an SB0 mode (*upper left*), an SB(s) mode (*upper right*), an S mode with moderate growth (*lower left*), and a violently unstable S mode (*lower right*). Each frame corresponds to a different galaxy model. *Right:* The superposition of a barred mode onto its axisymmetric mass density distribution. (All from Bertin, G., Lin, C. C., Lowe, S. A., Thurstans, R. P. 1989. *Astrophys. J.*, **338**, 78.)

departing from the central bar, as in NGC 1300 (see also M100); or even an almost normal but rather open spiral appearance for systems that have a significant active contribution from the gas. Because the three-dimensional geometry of the two-component disk has a different impact on tight or open waves, under special circumstances, the same galaxy disk can support large-scale modes of the two classes (see the comment in Subsection 17.5.2), thus displaying a mixed or dual type of morphology. [An interesting but different type of dual morphology may occur when spatially separate zones of the disk are marked by distinct dynamical effects (e.g., see the inner structure of NGC 4622[31] or the striking inner structure of M100[32] on a scale of one-sixth the disk's exponential scale length).]

By relation of these findings of the modal survey and the concepts of the modal theory to a discussion of the various possible physical conditions that may characterize a galaxy disk, a general framework for the dynamical classification of spiral galaxies has been proposed (Fig. 18.2) with good support from the new near-infrared (near-IR) data.[33] In this framework, the transition from SA to SB morphology is controlled by the active mass of the disk (relative to the total mass, which includes that of the bulge-halo), the transition along the Hubble sequence (from *a* to *c* galaxies) is mostly controlled by the gas content, and the transition from grand-design to flocculent galaxies (in relatively light disks) is controlled by the size of the stellar epicycles [in the sense that the stellar component, when it is too hot, becomes decoupled from the gas and remains inactive (see also Chapter 16); in this case the disk is subject to only small-scale spiral activity generated by the gas]. This framework is not intended as a rigid scheme to reduce all the observed morphological types to a simple, small set of dynamical prototypes; rather, somewhat like the original Hubble classification scheme, the modal approach offers a useful set of broad

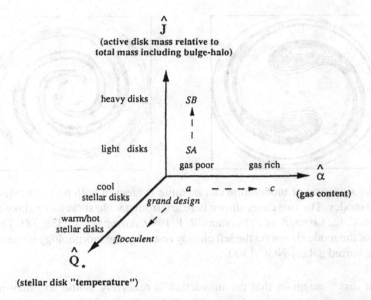

Fig. 18.2. Framework for the dynamical classification of the observed spiral structure in galaxies on the basis of their intrinsic modal characteristics (from Bertin, G., Lin, C. C. 1996. *Spiral Structure in Galaxies: A Density Wave Theory*, MIT Press, Cambridge, MA, p. 122; ©1996 Massachusetts Institute of Technology).

prototypes to guide us in understanding the rich variety of morphologies and in constructing more specific and complex models as required by the observations.

In this context it has been interesting to find that a growing body of observations of galaxies without a prominent bar point to the conclusion that many galaxy disks are likely to be submaximal (for a definition of maximum disk decomposition of rotation curves, see Subsection 4.2.1). This finding applies to the Milky Way Galaxy (as shown by the resolution of the Oort discrepancy in the solar neighborhood; see Subsection 14.1.2); is likely to be the case for galaxies such as NGC 3198, traditionally thought to exemplify the issue of disk-halo degeneracy in the decomposition of rotation curves (see discussion in Section 14.5); and also appears from the results of the Disk Mass Project, a long-term project devoted to extending the classical studies of the solar neighborhood to external galaxies.[34]

The near-IR data[35] offer an unparalleled opportunity to test many of the arguments given and, in general, the soundness of this dynamical framework because they give a direct means of probing the underlying gravitational potential and separating the contribution of the stars from that of the gas. There are four major points that give considerable support to the modal perception:

- *Regularity of the spiral structure.* Grand-design morphology is rather frequent in the near IR; it occurs even in galaxies that in the optical appear to be less regular, multiarmed, or quasi-flocculent. This shows that the conditions for the support of a regular large-scale structure are rather common and, as such, must reflect the intrinsic characteristics of galaxy disks. Even in cases such as M51, in which some interaction with another galaxy is obviously taking place, the extreme coherence observed in the underlying

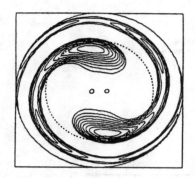 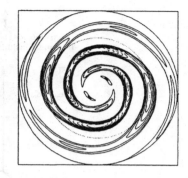

Fig. 18.3. The signature of modulation can be quite evident in both barred modes and normal, tightly wound modes. The two cases shown here are from a wide survey of galaxy models (from Bertin, G., Lin, C. C., Lowe, S. A., Thurstans, R. P. 1989. *Astrophys. J.*, **338**, 78). The modulation and structure of the mode shown on the left closely resemble the morphology observed in the near infrared in the barred galaxy NGC 1300.

stellar disk[36] suggests that the interaction is relatively gentle and may just bring out the intrinsic modal characteristics of the disk that existed before the encounter.

- *Morphology of the large-scale structure.* The morphology of the large-scale spiral structure nicely matches the morphology of the calculated spiral modes. This also has been noted for the nontrivial case of barred galaxies, such as NGC 1300.[37] This point is confirmed and actually strengthened by near-IR images; for instance, we might compare the near-IR morphology of NGC 4622 with some spiral modes.[38]
- *Amplitude modulation along the arms.* This important observed feature, which has been confirmed in the IR for several objects (see the cases of NGC 4622, M51, and NGC 1300), is a signature naturally predicted by the modal theory (Fig. 18.3).
- *Setting the Corotation Circle.* To this important point we devote a short separate subsection.

18.2.1   Locating the Corotation Circle

Any positive identification of a well-defined pattern frequency in a galaxy disk is automatically a strong case for the modal theory. The alternative interpretations of regenerative structure and tidally induced transient spiral structure, for which the large-scale spiral structure is not associated with discrete modes, do not anticipate the existence of a well-defined pattern frequency. Thus all the tests devised to measure the location of the corotation circle from the data[39] (Fig. 18.4) implicitly support the modal theory.[40] Some methods are based on the use of the observed offsets between the pattern of newly born stars and that of molecular clouds (for a short description of the ongoing discussion about the existence of these offsets, see also Subsection 18.1.2).[41]

Note that for type B modes corotation is expected to occur well inside the optical disk (at 1 to 2 exponential scale lengths), just outside the tip of the bar, which is empirically verified for objects with a prominent large-scale bar; the case of M99 (with corotation at 0.6 to 0.7 R_{25}) also appears to be within the general expectations of the modal theory. However, as emphasized earlier, a detailed prediction of corotation for an individual object would require a very laborious and difficult modeling procedure.

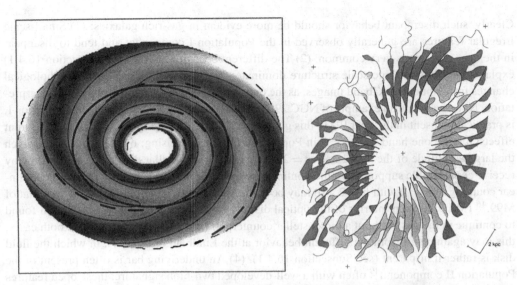

Fig. 18.4. A geometric phase method for the empirical determination of the corotation circle recognizes that the observed residual velocity map for an inclined disk, in the presence of a rigidly rotating two-armed spiral field, changes character from $m = 1$ to $m = 3$ across the corotation circle (*left*). Applied to available HI data for M81 (*right*; radio data from Visser, H. C. D. 1978. Ph.D. dissertation, University of Groningen, Groningen, The Netherlands), the observed $m = 1$ character suggests that the displayed spiral structure lies completely inside the corotation circle (from Canzian, B. "A new way to locate corotation resonances in spiral galaxies," 1993. *Astrophys. J.*, **414**, 487; reproduced by permission of the AAS).

In the picture of quasi-stationary spiral structure, the observed pattern is expected to be dominated by one or few modes but is certainly not expected to be strictly steady and associated with a unique pattern frequency (see Chapter 17). Therefore, the frequently reported finding of coexisting patterns characterized by different pattern speeds should not be regarded as an argument against the modal theory nor against the hypothesis of quasi-stationary spiral structure.[42]

18.3 Interpreting Specific Observed Features

As should be clear from the detailed description of dynamical mechanisms given in previous chapters, the cooperation of stars and gas is essential for the support of large-scale normal, nonbarred spiral structure. Stars ensure large-scale coherence of the pattern, and gas makes it possible to maintain, by means of self-regulation, a relatively cool outer disk, as required for the excitation of global modes and the responsiveness of the disk to waves. Gas also can provide a saturation mechanism for the growing linear modes at finite amplitudes. Cases in which this cooperation process is particularly evident are galaxies such as NGC 4622 and M81. For NGC 4622, the arms are very long, thin, and tight,[43] and it is extraordinary that the overall optical morphology is basically preserved in the near IR.[44] This shows empirically that the stars in NGC 4622 must be characterized by small epicycles forming a thin, light disk.

The competition between the two main components of the galaxy disk is best shown by the features in which the morphology in one component does not have a counterpart in the other.

Clearly, such discrepant behavior should be more evident in gas-rich galaxies: (1) Small-scale irregular features are generally observed in the Population I component and tend to disappear in the near IR. This is very common. (2) The different behavior at ILR (see Subsection 16.4.1) explains why low-m large-scale structure dominates the near-IR images. Thus morphological changes from optical to K-band images, as the ones shown by NGC 309, have a natural interpretation. Note the beautiful example of NGC 2997, in which a third arm, prominent in Population I, is practically absent in the near IR. In this galaxy we may witness a combination of two different effects. On the one hand, the $m = 3$ in Population I is not surprising; on the other hand, given the large amplitude of the $m = 1$ and $m = 2$ components in the stellar disk,[45] the third arm may receive considerable support from the nonlinear coupling of the low-m modes present; a nonlinear coupling of $m = 1$ with $m = 2$ also may be at the basis of other morphologies, such as that of M99.[46] (3) In the deep interior of the optical disk[47] or outside it,[48] spiral arms are often found to continue in the gas, without a direct stellar counterpart (see Subsection 18.3.1); in both cases, this may again result from the different behavior at the Lindblad resonances, in which the fluid disk is rather transparent (see Subsection 16.4.1). (4) An underlying bar is often present in the Population II component,[49] often with a well-developed two-blob signature; these open features are very natural in relatively heavy stellar disks; the frequency of this underlying bar morphology may shed light on the general issue of whether disks tend to conform to the maximum-disk ansatz (see Chapter 20). (5) It is likely that some ring structures originate in the vicinity of the Lindblad resonances,[50] and these should be associated with the gas.[51] If so, we would expect rings to be less prominent in near-IR images. A possible example of this situation may be the inner structure of NGC 309.

A rather frequent feature in near-IR images is the presence of a significant $m = 1$ component,[52] often manifested by a lopsidedness of the disk. Even before such direct views of the underlying stellar disk were available, it had already been noted that the frequency of $m = 1$ asymmetries demands an intrinsic mechanism that should explain the persistence of the observed structure.[53] This is basically the same argument that is used in general to support the modal perception for grand-design bisymmetric spiral structure. Indeed, the linear modal theory predicts that $m = 1$ modes should generally be dominated by $m = 2$ modes when available because the latter are more efficient in transporting angular momentum outward but that they should be relatively frequent, especially because they do not suffer from ILR inhibition (see Subsection 16.4.1).[54] Within the modal theory, we may argue that one-armed structures and lopsidedeness are all the nonlinear result of $m = 1$ modes; lopsidedness may be the $m = 1$ equivalent of the broad bar of the $m = 2$ context.

From the observational point of view, probably the most urgent task to be completed is a systematic investigation, by means of near-IR images of large samples of galaxies of different morphological class, of the statistical significance of points of dynamical interest of the type noted in Chapter 16 (see Section 16.5).[55] Another important issue that is currently under investigation is the identification of the epoch, at redshifts of cosmological interest, corresponding to the establishment of spiral morphology of the type characterizing galaxies in the nearby universe.[56] From a theoretical point of view, the linear analysis of the dynamics of two-component disks should be completed soon. Nonlinear processes, such as the complex mechanism of self-regulation (beyond the preliminary analysis reported in Section 16.3) or the detailed processes of mode saturation at finite amplitudes, are probably more difficult to tackle, but they should be within reach in the near future; here some work that has been carried out on the properties of orbital support to a given spiral field[57] may turn out to be of indirect help for cases in which gas cannot cooperate in the saturation process.

18.3.1 Prominent Spiral Arms in the Gaseous Outer Disk

Here we briefly focus on a recent study[58] that adds to the list of examples demonstrating that many diverse observed phenomena are naturally interpreted in the modal theory of spiral structure. As is the rule in the dynamics of galaxies, studies of this type can be seen either as a specific prediction to be tested by the observations or, alternatively, as a dynamical tool to diagnose properties of the system under investigation that may be difficult to measure directly.

Several spiral galaxies are known to exhibit coherent spiral structure in the gaseous outer disk, well outside the bright optical disk.[59] Some beautiful examples are those of NGC 2915,[60] NGC 628, NGC 1058, NGC 6946,[61] and NGC 3741.[62] In particular, the case of NGC 6946 has been the focus of recent deep investigations;[63] a spectacular set of gaseous arms can be traced all the way out, with a significant degree of regularity and symmetry, even if the outer disk is lopsided and fragmented. Such pronounced spiral structure in the outer regions has been found to be puzzling[64] because grand-design spiral structure is traditionally associated with the main body of the galaxy disk, dominated by the stars that support the density wave, in a region that is generally thought to be dominated by the visible mass. Furthermore, according to a general misconception, light disks (and the outermost gaseous disk is expected to be very light, given the dominance of the dark matter halo) would be unable to sustain regular spiral structure with a small number of arms (see also Subsection 18.1.1 and the comment on the finding of submaximal disks in Section 18.2).

In this subsection it is suggested that such coherent structure is very natural in the context of the modal theory of spiral structure. In more detail, the picture can be quantified in the following way. Consider a spiral galaxy with a reasonably well-established grand design in the bright optical disk and with a radially extended gas component. The grand design is naturally interpreted in terms of self-excited global modes of the kind described in Chapter 17. The dominant modes would all be associated with an outgoing short trailing wave signal (see Fig. 17.1). In contrast with the stellar component, the gas component is partially transparent to such signal at the outer Lindblad resonance, so short trailing waves manage to find their way into the outer regions characterized by $v > 1$. In those regions, the amplitude of the wave must simply follow the behavior dictated by the conservation of wave action. For an outgoing linear density wave of given pattern frequency in a one-component, zero-thickness fluid disk, the density amplitude must follow the relation

$$\left|\frac{\sigma_1}{\sigma_0}\right|^2 \propto G(v,Q)r^{-1}\kappa^4\sigma_0^{-4}, \tag{18.1}$$

where the proportionality constant is independent of r, and

$$G(v,Q) \equiv \frac{v^2 - 1}{Q^2\sqrt{1+(v^2-1)Q^2}}. \tag{18.2}$$

Here the notation is the same as in Subsection 15.1.2. In addition, the pitch angle i of the spiral arms should follow the behavior $\tan i = m/(rk)$ associated with the short trailing wave branch, for which

$$k = -\left(\frac{\kappa^2}{2\pi G\sigma_0}\right)\frac{2}{Q^2}\left[1+\sqrt{1+(v^2-1)Q^2}\right]. \tag{18.3}$$

Finally, the continuity and vorticity equations require that the amplitude of the perturbed radial motion u_1 follows the relation

$$\left|\frac{u_1}{r\Omega}\right| = v\left(\frac{\kappa}{\Omega}\right)\left(\frac{1}{|rk|}\right)\left|\frac{\sigma_1}{\sigma_0}\right| \propto \left(\frac{\kappa}{\Omega}\right)H(v,Q)r^{-3/2}\sigma_0^{-1}, \tag{18.4}$$

where

$$H(v,Q) \equiv \frac{vQ^2\sqrt{G(v,Q)}}{1+\sqrt{1+(v^2-1)Q^2}}. \tag{18.5}$$

The outer spiral arms are then interpreted as the natural extension in the outer disk of the outgoing short trailing waves that are responsible for exciting the global spiral structure in the star-dominated main body of the galaxy disk. Because of this conservation, the density amplitude of the outer arms is expected to increase with radius in the regions where the inertia of the disk becomes smaller and smaller, much like ocean waves can reach high amplitudes when moving close to the shore.

These predictions of the linear theory can be exemplified by means of a simple reference model characterized by a flat rotation curve, corotation at $r_{co} = 3h$ (h is the disk exponential scale), studied beyond $r = 6h$, where the unperturbed gas density profile is taken to be inversely proportional to r, and for simplicity, $Q = 1$. The results for a two-armed mode are shown in Fig. 18.5.

As for other aspects of the modal approach to the study of spiral structure in galaxies, this example shows that application of the theory can proceed quantitatively along two directions. On the one hand, a detailed test on a specific galaxy, such as NGC 6946 (other objects, such as

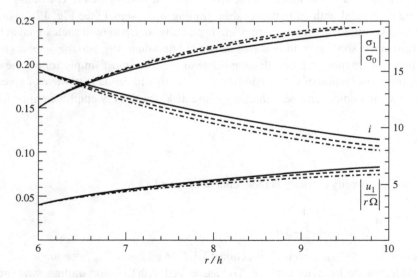

Fig. 18.5. Relative amplitude profiles of the spiral arms in terms of density σ_1/σ_0 and radial velocity $u_1/(r\Omega)$ (dimensionless, *left axis*) and pitch angle i of spiral structure (degrees, *right axis*), for three cases of the simple reference model mentioned in the text: zero-thickness disk (thick lines), finite-thickness totally self-gravitating disk (*dashed lines*), finite-thickness non-self-gravitating disk (*dash-dotted lines*) (from Bertin, G., Amorisco, N. C. 2010. *Astron. Astrophys.*, **512**, id.A17).

NGC 628 and NGC 2915, are different systems; each individual object may have its own special character), would demonstrate whether the linear theory with the assumptions made (e.g., in relation to the thickness of the disk) is satisfactory or a full nonlinear theory is required to deal with such large-amplitude structures. On the other hand, a positive test against the observations would provide one additional independent way of measuring some properties of the basic state. In particular, application of Eqs. (18.1), (18.3), and (18.4) would lead to an independent diagnostic tool to test whether the disk in the outer parts is indeed light or heavy.

18.4 **Evolution**

To a large extent, questions related to the evolution of spiral galaxies can be considered separately from the morphological issues described earlier. On the relatively short time scale, we might ask how the spiral structure observed in a given galaxy was generated in detail or, in different terms, what the observed structure should have been a few dynamical time scales ago, given its current morphological characteristics. On the long time scale, we might wish to investigate how the overall characteristics of the galaxy (not only morphology but also thickness, mass, and kinetic-energy distribution) would change as a result of internal mechanisms and interactions with the environment.

It is well known that for complex, dissipative systems it may be practically impossible to answer the first question, especially because a given current physical state may result from a variety of initial conditions. The modal theory is particularly suited to bypass our ignorance of the evolutionary process. As in other complex systems (such as those found in meteorology or plasma physics), modes tend to stand out and dominate after the rapid initial transient, independent of the exact initial conditions.

As to the long-term evolution of galaxy disks, a comprehensive predictive picture probably would require a tremendously detailed knowledge of the galaxy structure, which, even if we might imagine it to become available from the data, would most likely be impossible to handle (much as for long-term weather predictions despite the very accurate and complete sets of initial conditions now available at any given time). One more limited and still interesting question that can be tackled is determination of the time scale for which long-term evolutionary effects of a given type are expected; in particular, the effects associated with the presence of significant torques and the related angular momentum transport may be better estimated now that the amplitude of the spiral gravitational field is better diagnosed.[65]

Notes

1. Part of this chapter follows the presentation given in the symposium held at Johannesburg. 1996. *New Extragalactic Perspectives in the New South Africa*, eds. D. L. Block, J. M. Greenberg. Kluwer, Dordrecht, The Netherlands, p. 227.
2. Lindblad, B. 1963. *Stockholm Observ. Ann.*, **22**, 3.
3. Lin, C. C., Shu, F. H. 1964. *Astrophys. J.*, **140**, 646; 1966. *Proc. Natl. Acad. Sci. USA*, **55**, 229.
4. Age groups have long been noted and identified; see Lin, C. C. 1970. In *IAU Symposium*, Vol. 38, eds. W. Becker, G. Contopoulos. Reidel, Dordrecht, The Netherlands, p. 377. A convincing demonstration of the existence of significant color gradients across spiral arms has been provided only recently, for M99, by González, R. A., Graham, J. R. 1996. *Astrophys. J.*, **460**, 651; this paper also includes an interesting determination of the pattern frequency derived from the drift of the young stars away from their birthplace. From the study of age/color gradients

for a sample of nineteen spirals, Martínez-García, E. E., González-Lópezlira, R. A. 2013. *Astrophys. J.*, **765**, id.105, find frequent evidence of long-lived spiral structure. Earlier, interesting discussions based on a sample of thirteen SA and SAB spirals are given by Martínez-García, E. E., González-Lópezlira, R. A., Bruzual, G. 2009. *Astrophys. J.*, **694**, 512. For barred galaxies and the related issue of locating the corotation radius, see Martínez-García, E. E., González-Lópezlira, R. A. 2011. *Astrophys. J.*, **734**, 122, who also address the question of whether the bar and the spiral arms of a given barred galaxy are characterized by equal or different pattern speeds. Finally, Martínez-García, E. E., González-Lópezlira, R. A., Gómez, G. C. 2009. *Astrophys. J.*, **707**, 1650, address the problem of the effects of noncircular motions on the observed color gradients.

5. Starting with Lindblad, P. O. 1960. *Stockholm Observ. Ann.*, **21**, 3. Here we should mention the important papers by Goldreich, P., Lynden-Bell, D. 1965. *Mon. Not. Roy. Astron. Soc.*, **130**, 125; Toomre, A. 1981. In *The Structure and Evolution of Normal Galaxies*, eds. S. M. Fall, D. Lynden-Bell. Cambridge University Press, Cambridge, UK, p. 111.

6. See the statement of the problem by Oort, J. H. 1962. In *Interstellar Matter in Galaxies*, ed. L. Woltjer. Benjamin, New York, p. 234.

7. Following the important analysis by Visser mentioned in Chapter 15, more recent studies aimed at applying the modal theory have considered fresh new data and attempted a more advanced modeling of M81. Among these, we wish to refer to the article by Lowe, S. A., Roberts, W. W., et al. 1994. *Astrophys. J.*, **427**, 184, and some very recent work currently in progress by C. C. Feng, H.-H. Wang, and L.-H. Lin. Note that for this galaxy, the study by Kendall, S., Kennicutt, R. C., et al. 2008. *Mon. Not. Roy. Astron. Soc.*, **387**, 1007, confirms that the large-scale pattern is likely to be long-lived, with a well-defined corotation radius.

8. Toomre, A. 1969. *Astrophys. J.*, **158**, 899.

9. Sellwood, J. A. 2011. *Mon. Not. Roy. Astron. Soc.*, **410**, 1637. In the introduction it is stated that "Although many spirals in galaxies could be driven responses, the ubiquity of the spiral phenomenon suggests others are likely to be self-excited features of discs" and that "spirals driven by bars could also be long-lived patterns." In view of the discussion given in Section 18.1, we see that the paper is in general support of an intrinsic modal origin of large-scale spiral patterns in galaxies, although the author appears to favor an interpretation in terms of recurrent cycles of unstable modes; see Sellwood, J. A., Lin, D. N. C. 1989. *Mon. Not. Roy. Astron. Soc.*, **240**, 991; Sellwood, J. A., Kahn, F. D. 1991. *Mon. Not. Roy. Astron. Soc.*, **250**, 278.

10. Bertin, G., Lin, C. C., Lowe, S. A., Thurstans, R. P. 1989. *Astrophys. J.*, **338**, 78.

11. It is thus improper to call the survey by Bertin, G., Lin, C. C., Lowe, S. A., Thurstans, R. P. 1989. *op. cit.*, a study "in the hydrodynamic approximation," as Sellwood calls it on p. 1642 of his paper.

12. The important role of the cold-gas component was known from the very beginning. In particular, see first reference in Chapter 16; curiously, on p. 1638, Sellwood attributes this point to Rafikov, R. R. 2001. *Mon. Not. Roy. Astron. Soc.*, **323**, 445.

13. This point, of the caution required in the use of N-body simulations as true representations of collisionless systems, has often been emphasized in the past, in particular, by Bertin, G., Lin, C. C. 1996. *op. cit.*; for example, see the comment on "the excessive noise and relaxation often present" in N-body simulations on p. 111 of that monograph. This general concern has been reinforced by recent numerical experiments; see Sellwood, J. A. 2013. *Astrophys. J. Lett.*, **769**, id.L24.

14. Thomasson, M., Elmegreen, B. G., et al. 1990. *Astrophys. J. Lett.*, **356**, L9; Elmegreen, B. G., Thomasson, M. 1993. *Astron. Astrophys.*, **272**, 37.

15. Donner, K. J., Thomasson, M. 1994. *Astron. Astrophys.*, **290**, 475; Zhang, X. 1996. *Astrophys. J.*, **457**, 125. Zhang, X. 1998. *Astrophys. J.*, **499**, 93, notes that heavier disks are prone to bars, which is also consistent with the picture described in Section 18.2.

16. Oort, J. H. 1962. *op. cit.*; Lin, C. C. 1970. In *IAU Symposium*, Vol. 38, eds. W. Becker, G. Contopoulos. Reidel, Dordrecht, The Netherlands, p. 377. Curiously, on p. 1644, Sellwood, J. A. 2011. *op. cit.*, attributes the suggestion of coexistence to Tremaine, as a private communication.

17. Sellwood, J. A. 2012. *Astrophys. J.*, **751**, id.44.

18. Its linear stability is studied by Zang, T. A. 1976. Ph.D. dissertation, Massachusetts Institute of Technology, Cambridge, MA.

19. Toomre, A. 1981. In *The Structure and Evolution of Normal Galaxies*, eds. S. M. Fall, D. Lynden-Bell. Cambridge University Press, Cambridge, UK, p. 111. This is the picture corresponding to the lower-right box in Fig. 17.4.

20. In contrast, Sellwood argues that his simulations prove the inadequacy of linear theories and that what is observed in the simulations corresponds to yet unexplored nonlinear global instabilities.

21. See Allen, R. J. 1996. In *New Extragalactic Perspectives in the New South Africa*, eds. D. L. Block, J. M. Greenberg. Kluwer, Dordrecht, The Netherlands, p. 50, who often emphasized this point.

22. A long-debated issue is whether grand-design arms actually trigger enhanced star-formation or simply collect and organize young stars, with only little difference in star-formation rates with respect to flocculent galaxies; see Elmegreen, B. G., Elmegreen, D. M. 1986. *Astrophys. J.*, **311**, 554.

23. For recent analyses by means of dedicated simulations, see, among other papers, Shetty, R., Ostriker, E. C. 2006. *Astrophys. J.*, **647**, 997; 2008. *Astrophys. J.*, **684**, 978; Bonnell, I. A., Dobbs, C. L., et al. 2006. *Mon. Not. Roy. Astron. Soc.*, **365**, 37; Wada, K. 2008. *Astrophys. J.*, **675**, 188; Dobbs, C. L., Burkert, A., Pringle, J. E. 2011. *Mon. Not. Roy. Astron. Soc.*, **417**, 1318.

24. See González, R. A., Graham, J. R. 1996. *op. cit.*; Martínez-García, E. E., González-López lira, R. A. 2013. *op. cit.*; and the other papers cited at the beginning of this chapter.

25. Dobbs, C. L., Pringle, J. E. 2010. *Mon. Not. Roy. Astron. Soc.*, **409**, 396, consider numerical simulations of four different situations: (1) a case in which a spiral field is imposed with a given pattern speed; (2) a disk subject to a bar instability; (3) a case without a grand design, in which arms are intermittently driven by local instabilities; and (4) a disk subject to a strong external spiral field. The simulations of the gas flow are performed by means of a smoothed particle hydrodynamics code. For case (1), the self-gravity of the gas is not included; in all models, the gas is taken to be isothermal. Thus they "make no pretence of modelling star-formation, feedback, radiative processes in the interstellar medium (ISM), and so on, in any detail." Still they argue that, based on their simulations, from the observations of the locations of age-dated stellar clusters it will be possible to discriminate between the various proposed excitation mechanisms of spiral structure.

26. Such as the one illustrated in fig. 8 of the paper by Toomre, A. 1981. *op. cit.*

27. In a recent paper, in an attempt at applying the results of the simulations by Dobbs, C. L., Pringle, J. E. 2010. *op. cit.*, and Foyle, K., Rix, H.-W., et al. 2011. *Astrophys. J.*, **735**, id.101, conclude that the angular cross-correlations for different star-formation sequence tracers in a sample of twelve nearby galaxies (approximately one-half of which do not possess a well-established grand design) do not exhibit the systematic offsets that would be expected for a stationary spiral pattern of well-defined pattern speed. But in their concluding remarks the authors "caution, however, that the cross-correlation functions of the other simulated spiral structures in this study did not agree in detail with the observations. Thus, continued detailed comparisons between observations and simulations will be required in order to uncover the nature and persistence of spiral structure." Apparently, other authors are able to detect the anticipated offsets: working on a sample of fourteen galaxies, Tamburro, D., Rix, H.-W., et al. 2008. *Astron. J.*, **136**, 2872, thus determine the dominant corotation radii of the large-scale patterns to be typically ≈ 2.7 in units of the exponential scale length of the disk; Egusa, F., Kohno, K., et al. 2009. *Astrophys. J.*, **697**, 1870, detect a clear offset in five of thirteen galaxies, for which they are able to measure the relevant corotation radii.

28. For M81, Kendall, S., Kennicutt, R. C., et al. 2008. *Mon. Not. Roy. Astron. Soc.*, **387**, 1007, use different wave bands and near-IR data to make a comparison "between the phase of the stellar density wave and gas shock. The relationship between this angular offset and radius suggests that the spiral structure is reasonably long-lived and allows the position of corotation to be determined." In contrast, for M51, Dobbs, C. L., Theis, C., et al. 2010. *Mon. Not. Roy. Astron. Soc.*, **403**, 625, make simulations that convince the authors that the large-scale spiral structure in that galaxy is not a quasi-steady density wave; see also the observational (kinematic) study by Shetty, R., Vogel, S. N., et al. 2007. *Astrophys. J.*, **665**, 1138.

29. Bertin, G., Lau, Y. Y., Lin, C. C., Mark, J. W.-K., Sugiyama, L. 1977. *Proc. Natl. Acad. Sci. USA*, **74**, 4729.

30. Bertin, G., Lin, C. C., Lowe, S. A., Thurstans, R. P. 1989. *op. cit.*

31. Block, D. L., Bertin, G., et al. 1994. *Astron. Astrophys.*, **288**, 365.

32. Knapen, J. H., Beckman, J. E., et al. 1995. *Astrophys. J. Lett.*, **443**, L73.

33. Block, D. L., Bertin, G., et al. 1994. *op. cit.*

34. For a general description of the project, see Verheijen, M. A. W., Bershady, M. A., et al. 2004. *Astron. Nachr.*, **325**, 151; for the galaxy UGC 463, see Westfall, K. B., Bershady, M. A., et al. 2011. *Astrophys. J.*, **742**, 18; for a general statement on a sample of thirty intermediate-to-late-type galaxies, see Bershady, M. A., Martinsson, T. P. K., et al. 2011. *Astrophys. J. Lett.*, **739**, L47. In the last paper, six galaxies that are classified as barred galaxies do not appear to behave differently from the other galaxies of the sample, with respect to the general statement that galaxies possess submaximal disks. This appears to contradict the expectation that barred galaxies should be associated with heavier disks and thus would require a separate discussion; for each individual galaxy of the sample considered in that study we should inspect the role of thickness and the detailed morphology present.

35. In particular, see Block, D. L., Bertin, G., et al. 1994. *op. cit.*; see also Puerari, I., Block, D. L., et al. 2000. *Astron. Astrophys.*, **359**, 932, and several following related articles.

36. Rix, H.-W., Rieke, M. J. 1993. *Astrophys. J.*, **418**, 123.

37. See fig. 1 in the article by Bertin, G. 1993. *Publ. Astron. Soc. Pac.*, **105**, 640.

38. Like that shown by fig. 8 in the article by Bertin, G., Lin, C. C., et al. 1989. *op. cit.*; for a near-infrared survey of barred galaxies, see Elmegreen, D. M., Elmegreen, B. G., et al. 1996. *Astron. J.*, **111**, 1880.

39. See the set of papers introduced by Allen, R. J., Canzian, B., Lubow, S. H. 1993. *Publ. Astron. Soc. Pac.*, **105**, 638. See also Tremaine, S., Weinberg, M. D. 1984. *Astrophys. J. Lett.*, **282**, L5; Elmegreen, B. G., Elmegreen, D. M., Montenegro, L. 1992. *Astrophys. J. Suppl.*, **79**, 37; Canzian, B. 1993. *Astrophys. J.*, **414**, 487; Oey, M. S., Parker, J. S., et al. 2003. *Astron. J.*, **126**, 2317.

40. In this respect, some notable cases are NGC 936; see Kent, S. 1987. *Astron. J.*, **93**, 1062; M81, see Lowe, S. A., Roberts, W. W., et al. 1994. *Astrophys. J.*, **427**, 184, and references therein, and Westpfahl, D. J. 1998. *Astrophys. J. Suppl.*, **115**, 203; M99; see González, R. A., Graham, J. R. 1996. *Astrophys. J.*, **460**, 651. The method devised by Canzian was applied to a study of NGC 4321 (M100) by Sempere, M. J., García-Burillo, S., et al. 1995. *Astron. Astrophys.*, **296**, 45 (together with a second method based on numerical simulations of the molecular cloud hydrodynamics), and by Canzian, B., Allen, R. J. 1997. *Astrophys. J.*, **479**, 723. The method devised by Tremaine and Weinberg assumes that the adopted observed tracer follows the continuity equation. It has been applied frequently; for NGC 936, Merrifield, M. R., Kuijken, K. 1995. *Mon. Not. Roy. Astron. Soc.*, **274**, 933; for NGC 4596, Gerssen, J., Kuijken, K., Merrifield, M. R. 1999. *Mon. Not. Roy. Astron. Soc.*, **306**, 926; for NGC 7079, Debattista, V. P., Williams, T. B. 2004. *Astrophys. J.*, **605**, 714; for M51, M83, and NGC 6946 (in comparison with a determination of corotation based on CO data), Zimmer, P., Rand, R. J., McGraw, J. T. 2004. *Astrophys. J.*, **607**, 285.

41. Egusa, F., Sofue, Y., Nakanishi, H. 2004. *Publ. Astron. Soc. Jpn.*, **56**, L45; Egusa, F., Kohno, K., et al. 2009. *Astrophys. J.*, **697**, 1870.

42. Meidt, S. E., Rand, R. J., et al. 2008. *Astrophys. J.*, **676**, 899, extend the method of Tremaine and Weinberg to the situation in which the pattern frequency is determined locally and, in general, is not constant with radius. Applications of this method to M51, M101, IC 342, NGC 3938, and NGC 3344 indicate that multiple pattern speeds may often be present; Meidt, S. E., Rand, R. J., et al. 2008. *Astrophys. J.*, **688**, 224; Meidt, S. E., Rand, R. J., Merrifield, M. R. 2009. *Astrophys. J.*, **702**, 277. Given the complex morphology of IC 342, NGC 3938, and NGC 3344, this result is quite natural.

43. See Strom, S. E., Strom, K. M. 1978. In *IAU Symposium*, Vol. 77, eds. E. M. Berkhuijsen, R. Wielebinski. Reidel, Dordrecht, The Netherlands, p. 69; Byrd, G. G., Thomasson, M., et al. 1989. *Celest. Mech.*, **45**, 31; Buta, R., Crocker, D. A., Byrd, G. G. 1992. *Astron. J.*, **103**, 1526.

44. Block, D. L., Bertin, G., et al. 1994. *op. cit.*

45. As noted in the near-IR image; see fig. 2 of Block, D. L., Bertin, G., et al. 1994. *op. cit.*

46. González, R. A., Graham, J. R. 1996. *op. cit.*; see also Iye, M., Okamura, S., et al. 1982. *Astrophys. J.*, **256**, 103.

47. For M51, see Zaritsky, D., Rix, H.-W., Rieke, M. J. 1993. *Nature (London)*, **364**, 313; see also the case of NGC 2997.

48. For example, for M101, NGC 6946, and NGC 628, and Sancisi, R. 1990. In *Windows on Galaxies*, eds. G. Fabbiano et al. Kluwer, Dordrecht, The Netherlands, p. 199; Kamphuis, J. 1993. Ph.D. dissertation, Groningen University, Groningen, The Netherlands.

49. See NGC 309 and NGC 1637 in the article by Block, D. L., Bertin, G., et al. 1994. *op. cit.* For M100, see Pierce, M. J. 1986. *Astron. J.*, **92**, 285; Knapen, J. H., Beckman, J. E., et al. 1995. *Astrophys. J. Lett.*, **443**, L73; see also the interesting related observations of molecular spiral structure by Rand, R. J. 1995. *Astron. J.*, **109**, 2444.

50. See discussion in the papers introduced by Allen, R. J., Canzian, B., Lubow, S. H. 1993. *Publ. Astron. Soc. Pac.*, **105**, 638. For a recent investigation, see Kim, W.-T., Seo, W.-Y., Kim, Y. 2012. *Astrophys. J.*, **758**, id.14.

51. See Bertin, G. 1993. *Publ. Astron. Soc. Pac.*, **105**, 640.

52. See NGC 2997 and NGC 1637 in the article by Block, D. L., Bertin, G., et al. 1994. *op. cit.*; Rix, H.-W., Zaritsky, D. 1995. *Astrophys. J.*, **447**, 82; see also Zaritsky, D., Salo, H., et al. 2013. *Astrophys. J.*, **772**, id.135, and references therein.

53. Baldwin, J. E., Lynden-Bell, D., Sancisi, R. 1980. *Mon. Not. Roy. Astron. Soc.*, **179**, 23.

54. See also Zang, T. A. 1976. Ph.D. dissertation, Massachusetts Institute of Technology, Cambridge, MA.

55. See Elmegreen, D. M., Elmegreen, B. G., et al. 2011. *Astrophys. J.*, **737**, id.32.

56. For example, see Sheth, K., Elmegreen, D. M., et al. 2008. *Astrophys. J.*, **675**, 1141; Elmegreen, D. M., Elmegreen, B. G., et al. 2009. *Astrophys. J.*, **701**, 306.

57. See Patsis, P. A., Hiotelis, N., et al. 1994. *Astron. Astrophys.*, **286**, 46, and references therein. As noted in Subsection 18.1.1, some numerical simulations have been claimed to provide evidence for nonlinear saturation of regular self-excited spiral structures in collisionless disks; see also Khoperskov, A. V., Just, A., et al. 2007. *Astron. Astrophys.*, **473**, 31.

58. This subsection basically summarizes the results of the paper by Bertin, G., Amorisco, N. C. 2010. *Astron. Astrophys.*, **512**, id.A17, in which the calculations are extended to the case of a disk with finite thickness.

59. Shostak, G. S., van der Kruit, P. C. 1984. *Astron. Astrophys.*, **132**, 20; Dickey, J. M., Hanson, M. M., Helou, G. 1990. *Astrophys. J.*, **352**, 522; Kamphuis, J. J. 1993. Ph.D. dissertation, Groningen University, Groningen, The Netherlands.

60. Meurer, G. R., Carignan, C., et al. 1996. *Astron. J.*, **111**, 1551.

61. Ferguson, A. M. N., Wyse, R. F. G., et al. 1998. *Astrophys. J. Lett.*, **506**, L19.

62. Begum, A., Chengalur, J. N., Karachentsev, I. D. 2005. *Astron. Astrophys. Lett.*, **433**, L1.

63. Boomsma, R. 2007. Ph.D. dissertation, Groningen University, Groningen, The Netherlands; Boomsma, R., Oosterloo, T., A., et al. 2008. *Astron. Astrophys.*, **490**, 555.

64. Sancisi, R., Fraternali, F., et al. 2008. *Astron. Astrophys. Rev.*, **15**, 189.

65. Bertin, G. 1983. *Astron. Astrophys.*, **127**, 145; Gnedin, O. Y., Goodman, J., Frei, Z. 1995. *Astron. J.*, **110**, 1105. Zhang, X. 1996. *Astrophys. J.*, **457**, 125; these analyses still have to be extended to the regime of open spiral structure.

19 Bending Waves

The gas layer of our Galaxy is known to be warped in a rather asymmetric way.[1] Outside the solar circle, the HI disk reaches ≈ 3 kpc above the plane on one side (in the North), whereas it extends down to only 1 kpc below the plane on the other side (in the South); in the outermost regions, at a galactocentric radius of ≈ 30 kpc, the warp is found to reach the height of 6 kpc. Many galaxies are affected by similar global distortions of the disk[2] (Fig. 19.1; see also Fig. 3.4 in Part I). In a survey of twenty-six edge-on galaxies, twenty were found to be warped in HI, often in an asymmetric way, with the warp usually starting at the edge of the optical disk.[3] Warps also have been noted (although often only as marginal evidence) for the stellar disk.[4] The fact that stars and gas both participate in the warp is an indication that the phenomenon is probably dominated by gravitational forces. The presence of significant warps only in the outer regions, where the stellar disk ends, may simply be due to the small inertia of the disk in the outer parts compared with the heavier inner regions.

Besides the frequent integral-sign appearance (i.e., dominance of an $m = 1$ relatively open distortion of the disk), the morphology of warps is not as well documented. Some asymmetries may be interpreted as the result of a significant $m = 2$ component. Corrugations (i.e., small-amplitude, higher-m components) are also present (as observed in the HI layer of our Galaxy and confirmed by recent investigations[5]). The inclination of the warp, as measured by the local value of the ratio of the out-of-plane displacement H to the radial distance r, is sometimes small but often becomes very large,[6] making small-amplitude analyses questionable. The warp tends to be rather straight where it begins, at the edge of the optical disk, and then it may develop into some open-spiral form in the outer regions.[7] To obtain a three-dimensional description of the phenomenon, the large-scale distortions of galaxy disks are often analyzed by means of simple tilted-ring models;[8] note that these models have little dynamical foundation and are clearly inadequate to represent asymmetric warps such as those observed in our Galaxy.

If we refer to the simple description in terms of tilted rings, it is clear that the main concern about the phenomenon is the presence of differential precession, which would make, much as for the winding dilemma in the context of spiral structure, any material distortion of the disk of the type observed eventually become tightly corrugated (but see further comments in the following paragraph). The ubiquity of warps thus poses a problem of not only generation but also persistence. This naturally leads to an interpretation in terms of bending waves, for which we should ascertain the excitation and propagation properties together with a discussion of the establishment of global modes and the nature of their spectrum. Certainly it is tempting to look for explanations based on the internal dynamics of the disk.[9]

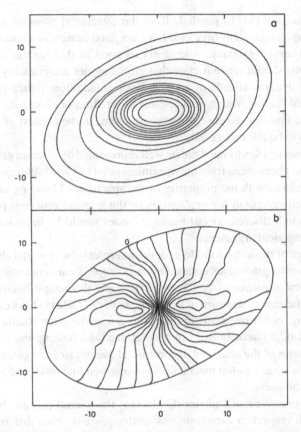

Fig. 19.1. Tilted-ring model (*a*) and warp model for the HI isovelocity contours (*b*) for the inclined spiral galaxy NGC 5055 (from Bosma, A. 1978. Ph.D. dissertation, University of Groningen, Groningen, The Netherlands, p. 83). The model contours closely resemble those obtained from the HI radio data.

Despite the many analogies that can be drawn, the dynamical problem of warps is different from that of spiral structure in galaxies. As indicated earlier, warps are found to affect the outer parts of the disk (typically they begin between R_{25} and the so-called Holmberg radius, where the B surface brightness reaches 26.5 mag arcsec^{-2}). This simple fact has important consequences: (1) In the outer disk, the dynamical clock runs slowly, which generally makes the differential precession not as dramatic (see also the later comment on the role of dark halos); in addition, because the clock is slow, the group propagation time for bending-wave packets is long (longer than the corresponding propagation time of density waves in the disk). (2) In the outer disk, the internal forces of the galaxy are weak. (3) So far we have no evidence for correlations between the morphology of warps and other aspects of galactic structure related to galaxy classification (in contrast with the well-known correlations between properties of spiral structure, bulge size, and gas content along the Hubble sequence). Because of point (3), from an empirical point of view, there is thus no compelling reason to argue that warps are quasi-stationary; we recall that the hypothesis of quasi-stationary spiral structure was the main foundation for the modal theory in that case (see Chapter 18). In parallel, from the dynamical point of view, because of the relatively long group propagation time, we have no good reason to expect modal behavior to be established in observed galaxies, at least not as good as that for the spiral-structure problem. These comments should make it clear that global modes are probably not the best way to represent the actual dynamical state of warped disks. In addition, points (1) and (2) make the possibility of external driving, which could be due to satellites or close galaxy encounters, much more appealing here than for the case of spiral structure. In the context of warps, any external interaction is bound to be felt more easily.

In conclusion, observed warps need not be well represented by discrete global modes and may be relatively transient phenomena from the dynamical point of view (although, in practice, slowly evolving). Thus concerns with the properties of the spectrum of bending modes[10] probably go beyond what is strictly required for explanation of the relevant observed phenomena. Finding that some models admit discrete global bending modes would be interesting but is not really suggested by semiempirical arguments.

The general nature of theoretical models for the interpretation of warps changed considerably in the late 1970s after the gradual acceptance by the astrophysical community of the role of dark halos in the dynamics of galaxies (see Chapter 20). Early theoretical interpretations (generally focused on the specific case of the warp of the gas layer of our Galaxy) have explored the possibility of external driving, either by means of interactions with the intergalactic medium[11] or, more simply, by means of tidal interactions.[12] In general, the tidal forcing was found to be too weak to justify the amplitudes of the observed distortions. However, isolated galaxy disks were recognized to lack an intrinsic excitation mechanism because bending modes of fully self-gravitating disks were found to be stable.[13]

The presence of massive halos obviously changes the overall picture, both with respect to persistence and with respect to excitation. As to the concern about differential precession, it was soon noted that in the presence of a sufficiently massive spherical halo, the problem of differential precession would not exist.[14] The same idea was basically followed later, by detailed investigations in the more general context of axisymmetric halos.[15] In addition, the halo may have a beneficial influence on excitation of the observed warps. Triaxial halos, by means of a kind of parametric instability, can be a source for excitation,[16] but surprisingly, even axisymmetric halos can amplify or excite bending waves because of the relative velocity between disk and halo material.[17]

As often is the case in the study of dynamical mechanisms, added interest in quantifying the processes involved derives from the possibility of setting dynamical constraints on the mass distribution of galaxies from given observed morphologies (see Chapter 20). Additional confidence in the study of bending waves derives from investigations of planetary rings,[18] for which in situ measurements can give direct quantitative tests of some theoretical arguments.

The discussion of warps is intimately related with that of the dynamics of the so-called polar rings around galaxy disks.[19] Much of the analysis carries the same ingredients. For the study of polar rings, numerical investigations are often performed in terms of hydrodynamical simulations, with the aim of assessing whether and on what time scale lumps of infalling gas are bound to settle as polar rings in one of the principal planes of the mass distribution of the host galaxy.[20] The outcome depends rather sensitively on the dissipation properties of the gas material that is considered and the amount of figure rotation of the underlying mass distribution (for the general triaxial case).

19.1 Bending Waves in a Simple Slab Model

If we consider a slab model of a disk (homogeneous in the horizontal plane and inhomogeneous in the vertical direction, symmetric with respect to the $z = 0$ plane; e.g., see Subsection 14.1.1), the dispersion relation for odd (i.e., bending) waves in the long-wavelength regime (wavelength longer than the thickness of the slab; see the similar limit in Section 19.4) is found to be[21]

$$\omega^2 = 2\pi G\sigma |k| + (c_\perp^2 - c_\parallel^2)k^2, \tag{19.1}$$

where $c_\perp$ is the vertical velocity dispersion. Here we recognize that for bending waves, self-gravity provides a restoring force and is stabilizing, in contrast with the Jeans instability term for density waves [see Eq. (15.11) in Chapter 15]. The gravity-independent term associated with pressure anisotropy has the well-known character of a fire-hose instability[22] but was found to be not particularly interesting given the numbers characterizing the pressure anisotropy in the solar neighborhood.[23]

Much as in the case of density waves, we can derive a similar relation as a WKBJ approximation for bending waves in an inhomogeneous differentially rotating disk (see Section 19.3) by considering elementary displacements of the thin disk of the form $h(r, \theta, t) = \text{Re}\{H(r) \exp[i(\omega t - m\theta)]\}$ with

$$H(r) = \alpha \exp\left[i \int^r k(r') \, dr'\right]. \tag{19.2}$$

For a cold, fully self-gravitating disk, the reference dispersion relation thus would be

$$(\omega - m\Omega)^2 = 2\pi G\sigma |k|, \tag{19.3}$$

where k is the local radial wave number, and ω of Eq. (19.1) is replaced with the relevant Doppler-shifted frequency.

Following the standard treatment of dispersive waves (see Section 15.3), we may then proceed to evaluate a number of interesting properties of bending waves. In particular, the equations for the radial group velocity c_g and angular-momentum density $\mathcal{G}$ are

$$c_g = -\frac{\partial \omega}{\partial k} = -\frac{\pi G\sigma}{m(\Omega_p - \Omega)} s_k, \tag{19.4}$$

$$\mathcal{G} = \frac{1}{2}m^2(\Omega_p - \Omega)\sigma |H|^2, \tag{19.5}$$

where we have introduced the pattern frequency Ω_p and the sign s_k for identification of trailing and leading bending waves. Thus trailing-bending-wave trains move away from the corotation circle. Note the sign change of $\mathcal{G}$ across corotation.

For the case in which possible amplification mechanisms (e.g., see Section 19.2) are weak, we can set up an equation governing the conservation of wave action[24] by requiring $rc_g\mathcal{G}$ to be constant, that is,

$$|r^{1/2}\sigma(r)H(r)|^2 = \text{constant}. \tag{19.6}$$

For a given disk density profile, this relation predicts how the amplitude of the warp should increase in the outer regions of the galaxy disk. In turn, for an observed amplitude profile of a galaxy warp, the relation provides a dynamical constraint on the disk density profile. This argument has been applied to several galaxies, including our own, to show that the observed warps support the picture in which a significant dark halo is indeed present[25] (see Chapter 20).

19.2   Disk-Halo Interaction and Related Two-Stream Instability

The presence of a halo changes the picture in important aspects. In this section we focus on the relative motion between disk and halo stars. Such relative streaming provides a natural reservoir of free energy for the possible amplification (or excitation) of waves with appropriate pattern frequency, that is, of waves rotating at a velocity between that of the disk and the (smaller) velocity of the halo. This phenomenon is well known in hydrodynamics, giving rise to the Kelvin-Helmholtz instability,[26] but also may take place in collisionless environments, such as high-temperature plasmas, in the form of a two-stream instability.[27] In the collisionless context, the process occurs by means of a kind of inverse Landau damping, that is, through wave-particle linear interaction; in galactic dynamics, this latter process is often associated with a sort of dynamical friction and is sometimes visualized in terms of wakes in the distribution of stars. We thus anticipate that a halo can make a galaxy disk flap, much like a flag in the wind.

If, for simplicity, the halo is taken to be nonrotating, the local growth rate of bending waves associated with the disk-halo interaction is estimated to be given by

$$\gamma \sim \frac{1}{m(\Omega - \Omega_p)} \left(\frac{4\sqrt{2}\pi^{3/2}G^2\sigma\rho_h}{mc_h\Omega_p} \right) N(\zeta_0), \tag{19.7}$$

where

$$N(\zeta_0) = 1 - \frac{\exp(-\zeta_0^2)}{\zeta_0} \int_0^{\zeta_0} \exp(t^2)\, dt, \tag{19.8}$$

$$\zeta_0 = \left| \frac{\omega}{\sqrt{2}c_h k} \right|. \tag{19.9}$$

In these expressions, c_h and ρ_h denote the velocity dispersion and density, respectively, of the halo stars (which here are assumed to be isotropic). Note the obvious relation of the positive function $N(\zeta_0)$ (Fig. 19.2) with the plasma dispersion function [e.g., see Eq. (16.27)]. The result for γ, with the sign condition on $\Omega_p(\Omega_p - \Omega)$, confirms the intuitive expectation that bending waves can be amplified (or excited) by disk-halo interaction only if the pattern speed of the wave falls between the angular velocity of the disk and that of the halo (which is taken here to be vanishingly small). A similar mechanism is also known to operate for even (density) waves.[28] The important role of disk-halo interaction has been confirmed by a more recent analysis, precisely in the terms outlined earlier.[29]

The mechanism discussed in this section may help to amplify signals produced otherwise; for example, it has been argued that it might provide the additional amplification needed to make the tidal interaction by the Large Magellanic Cloud on our Galaxy effective in explaining the observed warp.[30] Basically, the mechanism can operate on wave packets by means of convective amplification by a factor

$$g(r_0, r) = \exp\left[\int_{r_0}^{r} \gamma(r') \frac{dr'}{c_g(r')} \right], \tag{19.10}$$

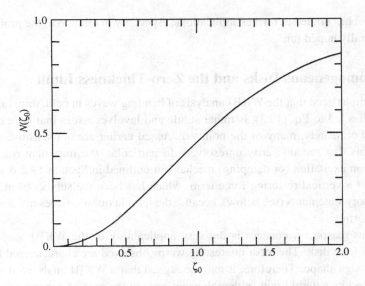

Fig. 19.2. The function N for disk-halo interaction (Bertin, G., Mark, J. W.-K. 1980. *Astron. Astrophys.*, **88**, 289).

which represents the gain in amplitude during (group) propagation of a wave packet from r_0 to r. In the presence of a suitable feedback and a wave cycle, such convective amplification might even be organized in the growth rate of a global mode (see Chapter 17); however, the process may give interesting amplifications even when global modes are not involved or do not have time to be established. If the condition on $\Omega_p(\Omega_p - \Omega)$ is not met, the process introduces a damping. Obviously, in the presence of dynamical friction, neutral perturbations are very difficult to produce.

Because the analysis involved in the derivation of the dynamical friction effects of disk-halo interaction is rather complicated, and because the instability is in reality of a fluid character, a very simple fluid model has been investigated in which disk and halo are both treated as incompressible fluids of finite constant thickness streaming through one another.[31] The idea here is to simplify the description of disk-halo interaction and thus does not pretend to address the issue of the dynamics of the gas components of the galaxy.[32] The two fluid inviscid layers couple only through gravity, in particular, through the distorted boundaries. The relevant dispersion relation for bending waves then can be written as

$$\left(D_d - \frac{\rho_h}{\rho_d}\right)\left(\frac{\rho_h}{\rho_d}D_h - \frac{z_d}{z_h}\right) - \frac{\rho_h}{\rho_d}D_c = 0. \tag{19.11}$$

Here z and ρ denote thickness and density, respectively, and subscripts identify disk (d) or halo (h) quantities. The one-component dispersion relation for bending waves (in the absence of the halo layer) would be $D_d = 0$; thus D_h is the same function obtained when the disk quantities are replaced with halo quantities. The function D_c represents the coupling between the two components that is due to gravitational forces. The D functions involve relatively simple combinations of hyperbolic functions. From Eq. (19.11), the sign condition for instability on $\Omega_p(\Omega_p - \Omega)$ can be recovered, and the qualitative picture of the role of disk-halo interaction is confirmed. So far this fluid model has not led to further applications, especially in relation to the problem of global modes, as initially hoped for.

19.3   Inhomogeneous Disks and the Zero-Thickness Limit

It should be emphasized that the WKBJ analysis of bending waves in cold, thin, inhomogeneous disks, which has led to Eq. (19.3), is rather subtle and involves issues that have not been fully clarified yet. Fortunately, many of the points discussed earlier are not sensitive to the answer to the questions that remain partly unresolved. In particular, the important role of disk-halo interaction as an excitation (or damping) mechanism outlined in Section 19.2 does not depend on inclusion of a vertical restoring force term, which has been the subject of many confusing and contradictory statements (see below), because the term in question does not alter the relevant transport properties.

One obvious reason for concern is the very applicability of the WKBJ ansatz to realistic perturbations of the disk. The most interesting warps observed are characterized by $m = 1$ and by an integral sign shape. Therefore, it may be argued that a WKBJ analysis in this context is stretched beyond its natural limits, although experience in the case of density waves shows that some asymptotic analyses work better than we would expect. But even for bending waves that satisfy the WKBJ conditions, the situation is not completely clear. Most likely, as is already apparent from the nontrivial study of the plane-parallel slab (see also Section 19.4), the main subtlety is traced to the zero-thickness limit itself. Note, for example, that the linear study of bending waves on an incompressible fluid layer (as briefly referred to in Section 19.2) considers ripples of vanishingly small amplitude; this demonstrates that in the limit of zero thickness, any bending wave of given amplitude is to be considered a nonlinear perturbation. The various difficulties that are noted in this limiting process have a natural counterpart in singularities that occur in the relevant equations, especially those describing the potential theory when we try to start directly from an analysis of bending a zero-thickness disk:

$$\rho(r,z,\theta,t) = \sigma(r)\delta\left(z - h(r,\theta,t)\right). \tag{19.12}$$

Although the analysis has been carried out in careful detail, some confusion still remains.[33] In a review of the theories of warps, Toomre[34] emphasized the role of restoring forces felt by rings of test particles and then argued strongly that because of this, dispersion relation (19.3) should include an additional kinematic restoring term.[35]

However, the blind use in the equations governing the dynamics of a bent infinitesimally thin disk of the kinematic restoring force felt by a ring of test particles in a three-dimensional configuration is clearly wrong. For example, if we consider test particles very close to the equatorial plane of a self-gravitating disk (see Section 14.1.2), the force involved would be dominated by the local term proportional to $G\rho(z = 0)$; alternatively, just outside a very thin disk, a test particle would feel a constant vertical field that is approximately independent of the height of the particle above the equatorial plane. In these two examples, the restoring force felt by the test particles has very little to do with the restoring force on a displaced (bent) disk. Therefore, some commonly reported derivations, such as the one in section 6.6 of the monograph by Binney and Tremaine,[36] are simply incorrect.

However, the presence of a separate component, such as a spheroidal halo, obviously requires a modification of the dispersion relation (19.3). Thus a term ν_z^2 corresponding "to the restoring force of the spheroidal halo in the direction of the disk plane"[37] must be added:

$$(\omega - m\Omega)^2 = v_z^2 + 2\pi G\sigma |k|. \tag{19.13}$$

The justification for v_z^2 is trivial. (In passing, we note here that through this term, an oblate and a prolate halo would act differently on bending waves.) In addition, this term guarantees the proper natural limit when the dynamics is fully dominated by a spherical halo, so $v_z^2 \sim \Omega^2$. In fact, in this latter case, the expected[38] neutral tilt mode (characterized by $m = 1$, $k \to 0$) can be recovered from the dispersion relation, much as Lindblad's kinematic density waves can be recovered from Eq. (15.11). It is reassuring to find that a later, independent derivation of the properties of bending waves confirms these conclusions.[39]

Another difficult issue is that of defining the appropriate boundary conditions for bending modes in realistic galaxy models. Some important insights have been gathered by considering the dynamics of truncated models.[40] Nonetheless, as mentioned in the introduction of this chapter, it might well be that a resolution of the issues related to the spectrum of global bending modes will not really be crucial to an explanation of the observed warps in galaxies.

In conclusion, some elements of the discussion are well understood, but the problem of determining a safe representation of the dynamics of bending waves in inhomogeneous disks remains. We may argue, as mentioned at the beginning of this section, that dealing directly with the dynamics of a zero-thickness disk will be the source of unwelcome singularities in the problem. Here we should add that when a disk with finite thickness and finite (anisotropic) velocity dispersion is considered, this does change the relevant transport properties of waves [e.g., see Eq. (19.1)]. Unfortunately, a rigorous analysis of the consequences of all these effects is not yet available.

19.4 Bending Waves on a Current Sheet and the Solar Sectors

A curious analogy has been found with the problem of bending waves in a completely different context: in the dynamics of the heliosphere.[41] Here a short account is given of some interesting aspects of the plasma phenomenon.

In the 1970s, in situ measurements allowed us to gather direct information on the plasma properties of the solar wind, particularly on the overall magnetic configuration associated with it. These measurements made it clear that the classical spherically symmetric model of the solar wind as an expansion of the coronal plasma[42] should be considered as only a first idealization of the actual state of the heliosphere. One interesting feature noted by space probes was that the solar wind, as observed in the vicinity of the plane of the ecliptic, is characterized by magnetic sectors. These are regions of well-defined polarity (where the field points typically outward or inward with respect to the Sun) separated by narrow regions where the polarity reverses rather abruptly. Normally, the observations report two sectors, but the cases of four and six sectors also occur. Above a certain heliographic latitude, the sector structure is found to disappear.[43] We can explain these findings by considering that the space probe measuring the magnetic field may be either above or below a thin current sheet extending approximately on the plane of the ecliptic.[44] In fact, models of the heliosphere involving the presence of a large-scale thin equatorial current sheet had been proposed and studied for a long time.[45] Attempts at reconstructing a three-dimensional configuration consistent with the observed solar sectors (such as that illustrated by the data secured by the *Pioneer 10* interplanetary mission) led to empirical models[46] (Fig. 19.3) that

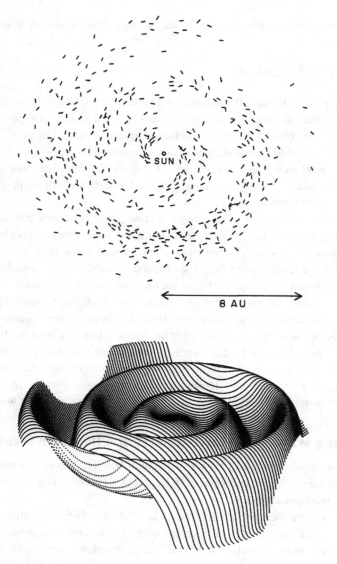

Fig. 19.3. The bending of the current sheet in the solar heliosphere (from Jokipii, J. R., Thomas, B. "Effects of drift on the transport of cosmic rays. IV: Modulation by a wavy interplanetary current sheet," 1981. *Astrophys. J.*, **243**, 1115; reproduced by permission of the AAS). The data (*above*) are projected magnetic-field unit vectors in the plane of the ecliptic (in the frame of reference corotating with the Sun) from *Pioneer 10*, shown only when the field is found to point toward the Sun; two solar sectors are then measured in this case. The measurements are interpreted as a result of a bending of the heliospheric current sheet, shown here on a different scale (25 AU across), as might be seen by an observer at 30° above the plane and 75 AU from the Sun.

clearly suggest the presence of a bending wave on the current sheet. In qualitative terms, the phenomenon is thus related to the ballerina effect argued by Alfvén.[47]

Much as in the case of galaxy warps, once the phenomenon is associated with the presence of bending waves, we should try to work out the dynamics of the wave process in order to get quantitative predictions to be compared with the observations. Conceptually, the simplest model for the current sheet that can be analyzed in detail, in close analogy with what is described in Section 9.1, is that of a simple plane-parallel slab.

In Cartesian coordinates, consider a configuration characterized by

$$\mathbf{B} = B_0 \mathcal{O}(z) \mathbf{e}_y, \tag{19.14}$$

where $\mathcal{O}(z)$ is an odd, monotonically increasing function of z such that for $z/\delta_z \to \pm\infty$, $\mathcal{O}(z) \to \pm 1$, and, for $|z/\delta_z| \ll 1$, $\mathcal{O}(z) \sim z/\delta_z$. Thus δ_z represents the thickness of the current sheet described by

$$J_x = J_0 \frac{d\mathcal{O}}{dz}. \tag{19.15}$$

The pressure-balance vertical equilibrium condition requires a pressure profile of the form

$$p_\perp = \frac{B_0^2}{8\pi} [1 - \mathcal{O}^2(z)] + p_\infty, \tag{19.16}$$

which is assumed to be associated with a mass-density profile of the form

$$\rho = \rho_0 \mathcal{E}(z) + \rho_\infty, \tag{19.17}$$

where $\mathcal{E}(z)$ is an even, decreasing (for $z > 0$) function, with $\mathcal{E}(0) = 1$. Thus the Alfvén velocity v_A changes across the current sheet following the relation

$$v_A^2 = \frac{B^2}{4\pi\rho} = \frac{B_0^2 \mathcal{O}^2(z)}{4\pi [\rho_0 \mathcal{E}(z) + \rho_\infty]}. \tag{19.18}$$

We then consider magnetohydrodynamic (MHD) perturbations on the current sheet. Given the assumed symmetry, at the linear stage, even and odd perturbations can be studied separately. Even perturbations in the presence of a small amount of resistivity may give rise to magnetic reconnection.[48]

Odd bending waves of the current sheet are characterized by a restoring force provided by a $\mathbf{J} \times \mathbf{B}$ term, which is fully analogous[49] with the self-gravity restoring force, leading to stable dispersion relation (19.3). A study of the MHD equations governing odd perturbations of the form

$$\mathbf{B}_1 = \tilde{\mathbf{B}}(z) \exp[i(-\omega t + k_\parallel y + k_x x)] \tag{19.19}$$

leads to a type of shear Alfvén waves described by the second-order differential equation

$$\frac{d}{dz} \left(g \frac{d\xi}{dz} \right) = g k^2 \xi, \tag{19.20}$$

where

$$g = \rho\omega^2 - \frac{k_\parallel^2 B^2}{4\pi}. \tag{19.21}$$

Here $k^2 = k_x^2 + k_\parallel^2$, and ξ represents the vertical component of the relevant plasma displacement. The preceding analysis is carried out by applying the frozen-in condition and the incompressibility constraint div $\boldsymbol{\xi} = 0$ on a sheet characterized by isotropic pressure. The long-wavelength limit is defined by the condition $|k\delta_z| \ll 1$.

We then complete this linear analysis by imposing as a boundary condition the evanescence of the solution for $\xi(z)$ away from the current sheet and by properly matching this outer solution to the inner solution across the current sheet. This leads to the desired dispersion relation. The zero-thickness limit is nontrivial. In order to derive a dispersion relation similar to Eq. (19.3), we should impose the condition $\rho_\infty/\rho_0 < |k\delta_z|$; if also $|k_x| \ll |k_\parallel|$, we finally obtain

$$\omega^2 \approx \frac{|k_\parallel| B_0^2}{4\pi \langle \rho \rangle \delta_z}. \tag{19.22}$$

An interesting and subtle aspect of the analysis, which also has emerged in the analogous study of the collisionless self-gravitating slab,[50] is the dissipation introduced by the vertical resonance. In the MHD plasma slab, resonance occurs (at two locations symmetrically placed with respect to the $z = 0$ plane) where the function g vanishes (Alfvén resonance). This is found to occur at the edge of the current sheet. Curiously, a damping also exists in the limit in which physical dissipation, which may be induced by resistivity or by ion viscosity, is vanishingly small.[51] For a thin sheet, when $\rho_\infty/\rho_0 < |k\delta_z|$, the Alfvén resonance introduces a damping rate $O(|k\delta_z|^2)$.

Notes

1. Burke, B. F. 1957. *Astron. J.*, **62**, 90; Kerr, F. J. 1957. *Astron. J.*, **62**, 93; Westerhout, G. 1957. *Bull. Astron. Inst. Neth.*, **13**, 201; Kerr, F. J. 1969. *Annu. Rev. Astron. Astrophys.*, **7**, 39; Henderson, A. P. 1979. *Astron. Astrophys.*, **75**, 311; Henderson, A. P., Jackson, P. D., Kerr, F. J. 1982. *Astrophys. J.*, **263**, 116; Kulkarni, S. R., Blitz, L., Heiles, C. 1982. *Astrophys. J. Lett.*, **259**, L63. A more modern study of the vertical structure of the HI disk of our Galaxy is presented by Levine, E. S., Blitz, L., Heiles, C. 2006. *Astrophys. J.*, **643**, 881.
2. Sancisi, R. 1976. *Astron. Astrophys.*, **53**, 159; Newton, K., Emerson, D. T. 1977. *Mon. Not. Roy. Astron. Soc.*, **181**, 573; Bosma, A. 1978. Ph.D. dissertation, University of Groningen, Groningen, The Netherlands; Briggs, F. 1990. *Astrophys. J.*, **352**, 15.
3. García-Ruiz, I., Sancisi, R., Kuijken, K. 2002. *Astron. Astrophys.*, **394**, 769.
4. Tsikoudi, V. 1977. Ph.D. dissertation, University of Texas, Austin, TX; Strom, S. E., Strom, K. M. 1979. *Sci. Am.*, April issue, p. 72; van der Kruit, P. C. 1979. *Astron. Astrophys. Suppl.*, **38**, 15; Tsikoudi, V. 1980. *Astrophys. J. Suppl.*, **43**, 365; Sandage, A., Humphreys, R. M. 1980. *Astrophys. J. Lett.*, **236**, L1; van der Kruit, P. C., Searle, L. 1981. *Astron. Astrophys.*, **95**, 105; Innanen, K. A., Kamper, K. W., et al. 1982. *Astrophys. J.*, **254**, 515; Sasaki, T. 1987. *Publ. Astron. Soc. Jpn.*, **39**, 849; Sánchez-Saavedra, M. L., Battaner, E., Florido, E. 1990. *Mon. Not. Roy. Astron. Soc.*, **246**, 458.
5. Levine, E. S., Blitz, L., Heiles, C. 2006. *op. cit.*, note that the corrugations have a local character, whereas the large-scale warp is dominated by $m = 0, 1, 2$ perturbations.
6. In NGC 3198, for example, the warp is gentle; see Begeman, K. 1987. Ph.D. dissertation, University of Groningen, Groningen, The Netherlands. In NGC 4013, it reaches 30°; see Bottema, R., Shostak, G. S., van der Kruit, P. C. 1987. *Nature (London)*, **328**, 401; for a recent study of this unusual galaxy, see Comerón, S., Elmegreen, B. G., et al. 2011. *Astrophys. J. Lett.*, **738**, L17.
7. Kerr, F. J. 1985. In *IAU Symposium*, Vol. 106, eds. H. van Woerden et al. Reidel, Dordrecht, The Netherlands, p. 501; Briggs, F. 1990. *op. cit.*
8. Rogstad, D. H., Lockhart, I. A., Wright, M. C. H. 1974. *Astrophys. J.*, **193**, 309; Rogstad, D. H., Wright, M. C. H., Lockhart, I. A. 1976. *Astrophys. J.*, **204**, 703; Reakes, M. L., Newton, K. 1978. *Mon. Not. Roy. Astron. Soc.*, **185**, 277.

9. Starting with Lynden-Bell, D. 1965. *Mon. Not. Roy. Astron. Soc.*, **129**, 299.

10. As was addressed in detail by Hunter, C., Toomre, A. 1969. *Astrophys. J.*, **155**, 747, and often emphasized in some reviews; the interested reader is referred to Saar, E. M. 1978. In *IAU Symposium*, Vol. 84, ed. W. B. Burton. Reidel, Dordrecht, The Netherlands, p. 513; Toomre, A. 1983. In *IAU Symposium*, Vol. 100, ed. E. Athanassoula. Reidel, Dordrecht, The Netherlands, p. 177; Binney, J. 1992. *Annu. Rev. Astron. Astrophys.*, **30**, 51. Bertin, G., Mark, J. W.-K. 1980. *Astron. Astrophys.*, **88**, 289, also speculated as to the possibility of discrete bending modes at the basis of observed warps.

11. Kahn, F. D., Woltjer, L. 1959. *Astrophys. J.*, **130**, 705.

12. For the case of our Galaxy, the driver might be the Magellanic Clouds; see Elwert, G., Hablick, D. 1965. *Z. Astrophys.*, **61**, 273; Avner, E. S., King, I. R. 1967. *Astron. J.*, **72**, 650; Fujimoto, M., Sofue, Y. 1976. *Astron. Astrophys.*, **47**, 263; Fujimoto, M., Sofue, Y. 1977. *Astron. Astrophys.*, **61**, 199; Spight, L., Grayzeck, E. 1977. *Astrophys. J.*, **213**, 374; see also Murai, T., Fujimoto, M. 1980. *Publ. Astron. Soc. Jpn.*, **32**, 581; and Lin, D. N. C., Lynden-Bell, D. 1982. *Mon. Not. Roy. Astron. Soc.*, **198**, 707.

13. Hunter, C. 1969. *Stud. Appl. Math.*, **48**, 55; Hunter, C., Toomre, A. 1969. *Astrophys. J.*, **155**, 747.

14. Tubbs, A. D., Sanders, R. H. 1979. *Astrophys. J.*, **230**, 736.

15. Petrou, M. 1980. *Mon. Not. Roy. Astron. Soc.*, **191**, 767; Sparke, L. S., Casertano, S. 1988. *Mon. Not. Roy. Astron. Soc.*, **234**, 873.

16. Binney, J. 1978. *Mon. Not. Roy. Astron. Soc.*, **183**, 779; Binney, J. J. 1981. *Mon. Not. Roy. Astron. Soc.*, **196**, 455.

17. Bertin, G., Mark, J. W.-K. 1980. *op. cit.*; Bertin, G., Casertano, S. 1982. *Astron. Astrophys.*, **106**, 274.

18. See, for example, Shu, F. H., Cuzzi, J. N., Lissauer, J. J. 1983. *Icarus*, **53**, 185.

19. A conference has been devoted to this topic; see Casertano, S., Sackett, P., Briggs, F., eds. 1991. *Warped Disks and Inclined Rings around Galaxies*, Cambridge University Press, Cambridge, UK. See also Moiseev, A. V., Smirnova, K. I., et al. 2011. *Mon. Not. Roy. Astron. Soc.*, **418**, 244, and references therein.

20. Tohline, J. E., Simonson, G. F., Caldwell, N. 1982. *Astrophys. J.*, **252**, 92; Habe, A., Ikeuchi, S. 1988. *Astrophys. J.*, **326**, 84; Pearce, F. R., Thomas, P. A. 1991. *Mon. Not. Roy. Astron. Soc.*, **248**, 688; Christodoulou, D. M., Katz, N., et al. 1992. *Astrophys. J.*, **395**, 113; Katz, N., Rix, H.-W. 1992. *Astrophys. J. Lett.*, **389**, L55; Steiman-Cameron, T. Y., Kormendy, J., Durisen, R. H. 1992. *Astron. J.*, **104**, 1339.

21. The analysis of the collisionless plane-parallel slab has been carried out in detail by Mark, J. W.-K. 1971. *Astrophys. J.*, **169**, 455, who recovered [in his eq. (77)] the limit of Eq. (19.1) for the case $c_\perp^2 \ll c_\parallel^2$; the problem has been revisited by Weinberg, M. 1991. *Astrophys. J.*, **373**, 391. For a derivation and a discussion of most of the results given here and in the following two sections, the reader is referred to Bertin, G., Mark, J. W.-K. 1980. *op. cit.*

22. See Schmidt, G. 1979. *Physics of High Temperature Plasmas*, 2nd ed., Academic, New York.

23. Toomre, A. 1966. In *Notes from the Geophysical Fluid Dynamics Summer Program*, Woods Hole Oceanographic Institute, Woods Hole, MA, p. 111; Kulsrud, R. M., Mark, J. W.-K., Caruso, A. 1971. *Astrophys. Space Sci.*, **14**, 52.

24. Bertin, G., Mark, J. W.-K. 1980. *op. cit.*

25. Lake, G., Mark, J. W.-K. 1980. *Nature (London)*, **287**, 705; Rohlfs, K., Kreitschmann, J. 1981. *Astrophys. Space Sci.*, **79**, 289; Blitz, L., Mark, J. W.-K., Sinha, R. P. 1981. *Nature (London)*, **290**, 120; Bertin, G., Casertano, S. 1982. *op. cit.*

26. Chandrasekhar, S. 1961. *Hydrodynamic and Hydromagnetic Stability*, Oxford University Press, Oxford, UK (reprinted in 1981 by Dover, New York).

27. Buneman, O. 1958. *Phys. Rev. Lett.*, **1**, 8; Buneman, O. 1959. *Phys. Rev.*, **115**, 503.

28. Mark, J. W.-K. 1976. *Astrophys. J.*, **206**, 418; see also Marochnik, L. S., Suchkov, A. A. 1969. *Astron. Zh.*, **46**, 319 and 524.

29. Nelson, R. W., Tremaine, S. 1995. *Mon. Not. Roy. Astron. Soc.*, **275**, 897; see also Binney, J., Jiang, I.-G., Dutta, S. 1998. *Mon. Not. Roy. Astron. Soc.*, **297**, 1237.

30. Bertin, G., Mark, J. W.-K. 1980. *op. cit.* A more recent investigation that includes the beneficial role of the dark halo to make the tidal interaction by the LMC more effective is given by Weinberg, M. D.,

Blitz, L. 2006. *Astrophys. J. Lett.*, **641**, L33, based on a general model envisioned by Weinberg, M. D. 1998. *Mon. Not. Roy. Astron. Soc.*, **299**, 499; but see also the negative results by García-Ruiz, I., Kuijken, K., Dubinski, J. 2002. *Mon. Not. Roy. Astron. Soc.*, **337**, 459. Interestingly, to describe the lively character of the tidally driven warp, on p. L34 of their paper Weinberg and Blitz use the expression "The image looks rather like a flag flapping in the breeze," which is very similar to the one used in the first paragraph of this subsection and precisely what inspired the analysis by Bertin, G., Mark, J. W.-K. 1980. *op. cit.*, on p. 294: "disks can be driven to warp much like a flag waving in the wind."

31. Bertin, G., Casertano, S. 1982. *op. cit.*; the fluid model is very similar to the one used by Goldreich, P., Lynden-Bell, D. 1965. *Mon. Not. Roy. Astron. Soc.*, **130**, 97.

32. See, e.g., Nelson, A. H. 1976. *Mon. Not. Roy. Astron. Soc.*, **174**, 661; Waxman, A. M. 1978. *Astrophys. J.*, **222**, 61; Waxman, A. M. 1979. *Astrophys. J. Suppl.*, **41**, 635 and 647.

33. Hunter, C., Toomre, A. 1969. *op. cit.*, admit that the separation of the vertical self-force into two parts (i.e., their F_1 and F_2) is a subtle formal step [see their sections II(b) and II(d); in their section IV(c) they note that their models were chosen because "the frequency $\nu^{(N)}$ predicted by equation (46) and that due to non-local restoring forces presumably add like squares" (pp. 760–761)] but basically leave the point as a question to be investigated further; in this respect, some issues are clarified by the analysis provided by Hunter, C. 1995. *Ann. N.Y. Acad. Sci.*, **773**, 111. Bertin, G., Mark, J. W.-K. 1980. *op. cit.*, prove that the leading-order WKBJ analysis in the limit in which the spheroidal component can be omitted (see their appendix D on p. 296), supports formula (19.3) for the dispersion relation.

34. Toomre, A. 1983. In *IAU Symposium*, Vol. 100, ed. E. Athanassoula. Reidel, Dordrecht, The Netherlands, p. 177, with the attention focused on the outer parts of a galaxy, writes, "To understand that sensitivity in modern terms, start with a ring of test particles in orbit near the periphery of a galaxy, each bobbing up and down with frequency $\kappa_z(r)$ about the plane $z = 0$, etc."

35. Toomre, A. 1983. *op. cit.*, in relation to the discussion of possible global bending modes, criticizes the article by Bertin, G., Mark, J. W.-K. 1980. *op. cit.*, because "their reasoning rested on a blunder. In their dispersion relation (II-3) they simply forgot the frequency κ_z due to distant matter" (his p. 181); it appears that no issue is raised by Toomre in that paper against the disk-halo amplification mechanism itself. Binney, J. 1992. *Annu. Rev. Astron. Astrophys.*, **30**, 51, tries to restate the point, but he introduces an expression of the alleged missed term that only adds confusion to the discussion; his eq. (12a) on p. 65 is called the "true" dispersion relation (and credited to Hunter, C. 1969. *op. cit.*), and he identifies the vertical frequency with the radial epicyclic frequency (without distinction of disk and halo contributions)!

36. Binney, J., Tremaine, S. 1987. *Galactic Dynamics*, Princeton University Press, Princeton, NJ; in section 6.6, the authors describe the kinematic evolution of warps by means of the single-particle vertical frequency ν [see their eq. (3.66) and eq. (6.77c)], which is applicable to star motions inside the disk, and include the same kinematic term (ν^2) in the dispersion relation for bending waves [see their eq. (6.87)].

37. Bertin, G., Mark, J. W.-K. 1980. *op. cit.*, appendix D, p. 296; after eq. (D7) it is also noted that "the new term ν_z^2 does not depend on ω or k and therefore it does not modify the transport properties of bending waves and the interaction with the spheroidal component."

38. Tubbs, A. D., Sanders, R. H. 1979. *op. cit.*

39. Nelson, R. W., Tremaine, S. 1995. *op. cit.*, repeat the calculations of Bertin, G., Mark, J. W.-K. 1980. *op. cit.*, and confirm that analysis. In particular, Nelson and Tremaine check that the contribution of the disk to the dispersion relation in the WKBJ limit is given by $2\pi G\sigma |k|$ [see their eq. (42)], so their dispersion relation eq. (43) (on p. 904) coincides with eq. (19.13), as derived by Bertin and Mark.

40. As, for example, done by Hunter, C., Toomre, A. 1969. *op. cit.*

41. This section is based on the article by Bertin, G., Coppi, B. 1985. *Astrophys. J.*, **298**, 387.

42. Parker, E. N. 1958. *Astrophys. J.*, **128**, 664.

43. Smith, E. J. 1979. *Rev. Geophys. Space Sci.*, **17**, 610.

44. Smith, E. J., Tsurutani, B. T., Rosenberg, R. L. 1978. *J. Geophys. Res.*, **83**, 717. For a recent discussion based on the three-dimensional results collected by the polar mission *Ulysses*, see Erdős, G., Balogh, A. 2010. *J. Geophys. Res.*, **115**, A01105, and references therein.

45. Pneuman, G. W., Kopp, R. A. 1971. *Solar Phys.*, **18**, 258; Schulz, M. 1973. *Astrophys. Space Sci.*, **24**, 371.

46. See, e.g., the clear representation given by Jokipii, J. R., Thomas, B. 1981. *Astrophys. J.*, **243**, 1115.

47. Alfvén, H. 1977. *Rev. Geophys. Space Sci.*, **15**, 271.

48. Dungey, J. W. 1953. *Philos. Mag.*, **44**, 725; Furth, H. P., Killeen, J., Rosenbluth, M. N. 1963. *Phys. Fluids*, **6**, 459. See also Schmidt, G. 1979. *op. cit.*

49. Bertin, G., Coppi, B. 1985. In *IAU Symposium*, Vol. 107, eds. M. R. Kundu, G. D. Holman. Reidel, Dordrecht, The Netherlands, p. 491, outlined a derivation for an infinitesimally thin current sheet.

50. Mark, J. W.-K. 1971. *op. cit.*

51. A clarification of this apparently paradoxical result can be found in the article by Bertin, G., Einaudi, G., Pegoraro, F. 1986. *Comments Plasma Phys. Cont. Fusion*, **10**, 173.

20   Dark Matter in Spiral Galaxies

From what has been described so far, it should be clear that the structure and evolution of spiral galaxies depend largely on the amount and distribution of dark matter actually present in these systems. (This statement would apply in even stronger terms to evolutionary scenarios in which galaxy-galaxy interactions play a major role.) In modeling of the basic state (e.g., see Sections 14.4 and 14.5) and discussion of several dynamical mechanisms (e.g., see Section 18.2 or Chapter 19), we have recognized that, in general, a dark halo may be present. However, we have left open the possibility that in some galaxies the halo may dominate, whereas in others it may be relatively small or even insignificant from the dynamical point of view. In this chapter we briefly review the arguments that have brought us to believe that dark matter is indeed present in spiral galaxies. This will also show that many issues remain far from being fully understood.

Starting with the studies by Kapteyn and Oort of the stellar motions in the solar neighborhood (see Section 14.1.2) and the studies by Zwicky and Smith of the galaxy motions inside clusters,[1] it has long been suspected that the Universe contains much more matter than meets the eye, that is, through the telescopes at our disposal. On the scale of galaxies and clusters of galaxies, objects were found to move faster than would be expected on the basis of the gravitational forces associated with visible matter. The most natural explanation is that some matter that is present and contributes to the gravitational field simply escapes detection.[2] Still the issue of dark matter was not a major concern to astrophysicists until the early 1970s, when detailed studies of the motions of atomic hydrogen around spiral galaxies led to the suggestion that dark halos are probably ubiquitous. Earlier, in the 1960s, we find discussions related to the possible existence of dark halos, but only in the late 1970s and early 1980s did the concept of dark halos around galaxies become established. Much of this belief is derived from the pressure of arguments from the cosmological context (see the brief description in Chapter 4). However, the amounts of dark matter required for explaining kinematics in galaxies, where direct convincing evidence for dark matter was first discovered and is strongest, are practically insignificant from the cosmological point of view.

From the methodological point of view, it is important to recall that we observe through various windows in the electromagnetic spectrum (see Chapter 2), from which we infer the mass of the astronomical sources of the observed radiation. Much of the optical light comes from the stars. Theories of stellar structure now allow us to estimate the mass of a star when its type is properly identified. However, the light from a region of a galaxy disk comes from the contribution of stars of many different types, that is, a whole stellar population, and a model is needed to give an estimate of the relevant mass-to-light ratio M/L, which, in turn, is needed to convert observed photometric profiles into mass profiles. The model clearly depends on the number of low-luminosity stars assumed to be present in a typical population, and this number is poorly known. Radio emission comes mainly from cold gas present in the disk. (For elliptical galaxies and for clusters of galaxies, X-ray emission may come from hot coronal plasma or from the intracluster medium; see Chapter 24.) The theory of emission processes usually gives us a way to estimate the total amount of gas responsible for the emission. In the case of the cold gas in spiral galaxies, this is just a fraction of the total mass of gas because large amounts of molecules, often not directly involved in the primary emission, may be present. For most types of galaxies, gas provides only a small fraction of the total mass. In addition, our view is complicated by the presence of dust because it obscures some of the light coming from the stars, even if it contributes little to the total mass; therefore, the amount of mass present in the form of stars is best diagnosed by observations in the near infrared (IR).

In view of these problems, the total mass of an extended object such as a galaxy (or a cluster of galaxies) is best measured dynamically. This is done as follows: We first assume some geometric properties of the system being considered. For spiral galaxies, axisymmetry is often assumed for the basic state and, to some extent, is justified by the data. Then we adopt a model that is usually characterized by simple kinematics (such as circular motions in a disk galaxy) and by a condition of overall equilibrium (for the hot X-ray-emitting gas in ellipticals or in clusters, we may refer to *hydrostatic equilibrium* or, for a group of stars or galaxies, *virial equilibrium*). Then we interpret and fit the available kinematic data and thus obtain a dynamical measurement of the total mass that determines the gravitational field. Recently, some alternative methods have become available that also might be considered dynamical measurements based on the detection of gravitational lensing; in principle, these are applicable even to systems that are not in dynamical equilibrium (see Chapter 26). The dynamical model thus provides an independent estimator of the mass. For example, the simple use of the third law of Keplerian motion $GM = \Omega^2 R^3$ applied to orbits of individual stars in the central region of our Galaxy (see Fig. 1.3) has led to the most accurate measurement of the mass of a compact dark object associated with the radio source Sgr A*, which we believe to be a supermassive black hole.

When the dynamically estimated mass exceeds the mass that we can reasonably associate with the visible components, we may be forced to resort to the presence of unseen material, although it may simply be that the model used is too naive and that some of the assumptions of the model are unjustified. Sometimes we are entitled to take a fairly conservative attitude and conclude that, strictly speaking, the evidence for dark matter is just not there; given the complexity of the factors that are involved and the limitations of the observations, if the mass discrepancy between dynamical measurements and directly identified mass is not too strong, some caution is advisable (see the description of the study of the solar neighborhood reported in Subsection 14.1.2). Still, this kind of inference argument in astronomy has been marked by extraordinary successes in the past if we think of the discovery of Neptune in 1846 based on the anomalous kinematics of Uranus. By similar arguments, we have been brought to believe that black holes have been discovered[3] or that planets have been found around a pulsar.[4]

In the context of galaxies, the claim that dark matter is present is usually based on two factors that generally coexist for the strongest cases. One is the magnitude of the dynamically inferred mass-to-light ratio. Normal populations of stars, such as those found in the solar neighborhood, can easily be reconciled with mass-to-light ratios (in the blue band) up to a few in solar units. If the inferred value turns out to exceed normal values by some significant factor, we are forced to admit that dark matter is present. The second factor is a gradient in the mass-to-light ratio that may be too strong with respect to the observed degree of homogeneity of the underlying stellar population. For galaxies, this is typically noted as an increase in the mass-to-light ratio with radius, indicating that substantial amounts of matter may be present in their outer parts.

20.1 Rotation Curves of External Galaxies and Their Decomposition

The material in the disks of spiral galaxies rotates around the galaxy center on approximately circular orbits. The motion of this material can be measured by means of spectroscopic studies on the basis of the Doppler effect on emission or absorption lines, thus providing the rotation curve $V = V(r)$ and indirectly a measure of the gravitational force that keeps the material on such orbits. There is good evidence, especially for spirals without a prominent bar, that the underlying gravitational field is axisymmetric to a high degree of approximation. These spectroscopic studies can be done at optical as well as radio wavelengths. In the case of the stellar disk, we should rely on the absorption lines from stellar atmospheres. But these are hard to measure. Therefore, optical rotation curves are usually obtained from the emission lines of hot gas surrounding the regions of recent star formation. As a consequence, optically determined rotation curves generally sample a relatively small radial range. It is no surprise that the most decisive observations pointing to the existence of dark halos come from radio studies of cold atomic hydrogen (at 21 cm) because cold atomic hydrogen is often found to extend very far out, sometimes even out to a radius larger than twice the radius of the easily visible disk.

Broadly speaking, the luminosity profile of the disk of spiral galaxies declines exponentially with radius (see Chapter 4) as $\exp(-r/h_\star)$, so at 4 to $5h_\star$ the optical disk practically ends (more than 90 percent of the light is contained within $4h_\star$). If the stellar light traced the mass distribution, as might be naively anticipated, we could easily calculate, apart from an overall scale factor, the gravitational pull that is due to the mass distributed in the disk as a function of radius [see Eq. (14.32) in Chapter 14]; beyond the bright optical disk, this would rapidly become well approximated by the inverse-square law. In other words, for a mass-to-light ratio M/L that is constant with radius, we can show that the expected rotation curve $V(r)$ should be characterized by an initial rise out to a radius of $\approx 2h_\star$, followed by a rapid Keplerian decline $V \sim r^{-1/2}$. If the light distribution is not exactly exponential, or if the assumption of a constant M/L ratio is only approximately justified, we may have some leverage in explaining the observed rotation curve, but eventually, by probing the gravitational field of a galaxy sufficiently far out, we would expect to observe the Keplerian decline. In turn, the observed lack of such a decline is the main direct evidence for the existence of dark halos around spiral galaxies.

One of the earliest indications for the existence of relatively flat rotation curves in external galaxies arises from the study of M31 in Andromeda[5] (Fig. 20.1). The impression is that at that time not much thought was given to the possibility that the encountered discrepancy between expected and dynamically measured mass would become a serious problem in astrophysics. For example, the Schmidt model of our Galaxy (see Subsection 14.1.2) is characterized by a rapid Keplerian decline of the rotation curve in the region outside the solar circle.

If we argue that an essentially spherical dark halo is responsible for the observed flat rotation curve so that (at large radii) $V^2 \approx GM(r)/r$, its associated density profile should be approximately of the form $\rho \sim r^{-2}$. This is precisely the form expected for an isothermal distribution of matter. Even though a dark halo has no reason to have relaxed into an isothermal sphere,

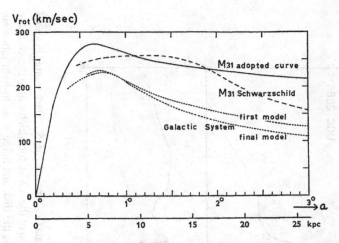

Fig. 20.1. A comparison of the rotation curve for the Andromeda galaxy with that considered for the Milky Way, after the first HI observations of M31 were made at Dwingeloo (from van de Hulst, H. C., Raimond, E., van Woerden, H. 1957. *Bull. Astron. Inst. Neth.*, **14**, 1). The curve marked as adopted refers to that suggested by the radio data.

this is a strong indication that the dark-matter distribution required in this context would not be ad hoc, as would actually be appealing from the physical point of view. Note that the distribution of stars in elliptical galaxies inside the sphere that contains half the total luminosity of the system is very close to $\rho_\star \sim r^{-2}$ (for a constant M/L ratio; see Chapter 22). The properties of the rotation curves of disks embedded in self-consistent isothermal halos were described in Section 14.5.

In the 1970s, the issue of flat rotation curves rapidly became the focus of various observational and theoretical studies.[6] Initially, emphasis was placed on the flatness of the rotation curve, with no special qualification placed on its radial range. In other words, the existence of dark halos was often inferred just on the basis of optically derived rotation curves.[7] However, it was soon pointed out that kinematic studies restricted to the optical disk actually would be unable to provide convincing evidence for the existence of dark halos; in fact, by including the role of the bulge when present, we could show that in general the photometric profile of the disk could be converted, with a constant M/L ratio, into a mass distribution reasonably consistent with the optically derived rotation curves[8] (Fig. 20.2). The decisive evidence for dark halos comes from the radially extended radio rotation curves, as proved by the clear-cut case of the spiral galaxy NGC 3198[9] (Fig. 20.3). The study of high-quality radio rotation curves shows that dark matter is indeed required even for the rare cases in which the rotation curve shows a declining trend in the outer parts.[10] It should be emphasized that to convert the measurements of rotation curves into meaningful statements on the overall gravitational field, the velocity field of the investigated galaxy should be highly symmetric so that we are not misled by distortions in the field that are known to be induced by bars or warps.

If we admit that dark halos exist, we should try to find constraints on their intrinsic shapes. First, we should ask whether their symmetry conforms to that of the disk, that is, whether they can be considered as essentially axisymmetric. Then we should ask whether they are fairly round or instead are very flat. The case of our Galaxy would suggest axisymmetry and argues against

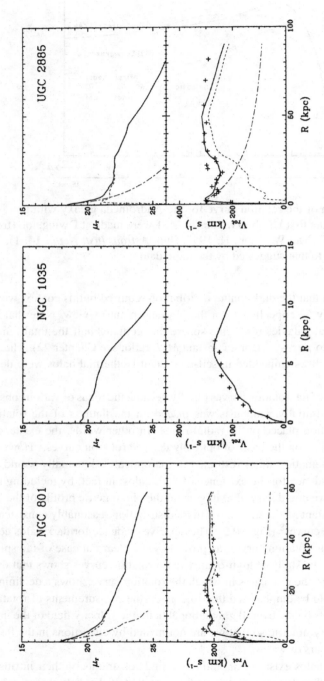

Fig. 20.2. Three cases (one small and two large Sc spirals) from a study of thirty-seven late-type galaxies (from Kent, S. "Dark matter in spiral galaxies with optical rotation curves," 1986. *Astron. J.*, **91**, 1301; reproduced by permission of the AAS), for which a fit to the available optical rotation curve is made with a bulge-disk decomposition without a dark halo. In each galaxy, the disk and bulge mass-to-light ratios are treated as constant free parameters (NGC 1035 is taken as having no bulge).

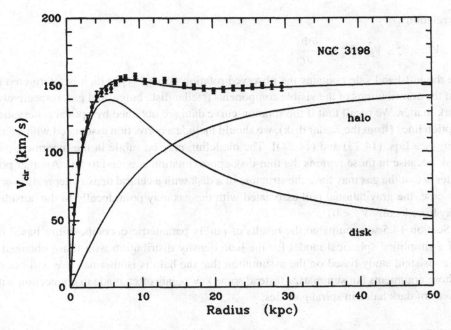

Fig. 20.3. The disk-halo decomposition with maximum disk for the HI rotation curve of the Sc galaxy NGC 3198 (from van Albada, T. S., Bahcall, J. N., Begeman, K., Sancisi, R. "Distribution of dark matter in the spiral galaxy NGC 3198," 1985. *Astrophys. J.*, **295**, 305; reproduced by permission of the AAS). The scale length of the exponential luminosity profile (see also Fig. 4.1) has been taken to be 2.7 kpc.

the distribution of dark matter in a thin disk. The study of radially extended rotation curves in spirals shows indeed that in many cases the underlying gravitational potential should be essentially axisymmetric. It also has been argued that the small scatter in the luminosity-velocity relation would be incompatible with halos dominated by triaxiality.[11] Studies of external galaxies based on the observations of polar rings around galaxy disks suggest that dark halos should be oblate and fairly flat.[12] The issue of the shape of dark halos has an interesting counterpart in the problem of the possible candidates for dark matter. Very flat halos would be difficult to reconcile with nonbaryonic dark matter, which would lack the dissipation that can make a system collapse to a disk. Here we should recall that one line of investigations[13] has pursued the possibility that a large fraction of dark matter is in the form of very cold molecular hydrogen. In Subsection 18.3.1 we showed how a study of the prominent spiral arms often observed in the cold gas beyond the bright optical disk may provide a way to measure the disk density in the outermost regions.

In conclusion, for spiral galaxies, rotation curves provide the natural diagnostics of the gravitational field in the plane of the disk. The most convincing results are best obtained by studying the simplest objects, particularly those that do not exhibit significant deviations from axial symmetry, such as major warps of the disk or bars or high-amplitude spiral arms. In Subsection 4.2.1 we described in some detail the steps that are taken to decompose an observed rotation curve into the contributions of the various visible components and the dark halo. In particular, we referred to the relation

$$V^2 = V_\star^2 + V_b^2 + V_g^2 + r\frac{\partial \Phi_h}{\partial r}, \tag{20.1}$$

where the left-hand side contains the observed rotation curve $V(r)$, which is interpreted as the sum of the contributions of the visible components (stellar disk, bulge, and gas, respectively) and the dark matter. We recall that if the rotation-curve data are obtained by optical observations of absorption lines (from the stellar disk), we should include a correction associated with asymmetric drift [see Eqs. (14.23) and (14.24)]. The modeling may be subtle in the innermost parts of the disk because in those regions the thin-disk approximation is bound to fail. Another point to consider is that the gas may have the structure of a disk with a central density depression or hole; in this case, the gravitational pull associated with the gas may point locally to the outside (i.e., we may have locally $V_g^2 < 0$).

In Section 14.5 we compared the results of purely parametric decompositions based on the use of a simplified spherical model for the halo density distribution with those obtained from a self-consistent study based on the assumption that the halo is isothermal. We will not repeat here those arguments but summarize instead some key points often made in connection with the discovery of dark halos in spiral galaxies.

20.1.1 The Maximum-Disk Decomposition

If we take the conservative attitude of invoking the minimum amount of dark matter required by the data, we are brought to adopt a disk-bulge-halo decomposition (of an observed rotation curve) that maximizes the contribution of the visible material. This is usually called the *maximum-disk solution* because it selects the maximum value for the M/L of the disk compatible with the observed rotation curve; lower values of this mass-to-light ratio also would be acceptable and would correspond to larger amounts of dark matter present. Maximum-disk models thus tend to emphasize that dark matter is not required by optically derived rotation curves. In principle, we might leave the halo contribution in Eq. (20.1) completely free. In practice, as an additional physical constraint, the density distribution for the dark matter is generally taken to be monotonically decreasing with radius, as occurring for the parametric choice, expressed by Eq. (14.54). The ratio of dark-to-luminous mass within a sphere enclosing the optical disk of the galaxy is of the order of unity. Beyond the optical disk, the main contribution to the rotation curve comes from the invisible halo.

Studies of the vertical dynamics of spiral galaxies, much as the study of the solar neighborhood outlined in Chapter 14, can shed further light on the problem of dark matter. In principle, these efforts should be able to tell whether the maximum-disk solution is the one preferred in nature. The first results from the Disk Mass Project, a long-term project devoted to extending the classical studies of the solar neighborhood to external galaxies, appear to point to the presence of submaximal disks.[14]

20.1.2 Conspiracy

From this line of reasoning, one puzzling aspect of the flatness of rotation curves emerged and was emphasized, especially in the 1980s, under the name of *conspiracy*.[15] This is the consideration that in the previously described picture the inner part of the rotation curve is supported almost exclusively by the disk, whereas the outer part is halo-dominated. Still, over the whole radial range, the rotation curve is often found to be flat and essentially featureless. (Of course, this is meant to be only a zeroth-order description because even flat rotation curves are not entirely featureless.)[16] What makes the properties of the dark halo and the stellar disk so finely tuned and coupled to allow them to enforce such smooth, flat behavior of the rotation curve, even if the dynamical behavior of the disk and that of the halo are expected to be, in principle, decoupled from each other? A similar kind of conspiracy also has been noted, although in not such strong terms, in the discussion of the mass distribution of elliptical galaxies (see Chapter 24).

The puzzle is reinforced by the existence of a well-defined luminosity-velocity relation with very small dispersion (see Subsection 4.3.1 and Fig. 4.6), given the fact that dark and visible components are presumably characterized by largely independent formation and evolution histories and thus are not necessarily bound to contribute in a homologous way to the dynamics of galaxies. Note that the luminosity of a galaxy is determined by only visible matter, whereas the support of the rotation curve is shared by both the dark and visible components. This is to be recognized as a type of conspiracy even if galaxies do not conform to the maximum-disk hypothesis.

Recently it has been noted that there is one additional, even more surprising aspect of the conspiracy problem. For the maximum-disk solution to be always acceptable, at least in principle, as demonstrated even in the case of low-surface-brightness galaxies, the scale that defines the central slope of the rotation curve should correlate with the scale of the distribution of the luminous disk. In turn, if dark matter is indeed present and partly contributes to the field in the inner regions of the galaxy, its density distribution must have a length scale compatible with that of the scale of the luminous component. This correlation is confirmed empirically (see Subsection 4.3.2 and Fig. 4.7).

Finally, it has been noted that features present in the observed rotation curves are generally found in correspondence with well-defined features in the distribution of visible matter.[17]

20.1.3 Degeneracy

The decomposition of rotation curves is affected by a degeneracy problem, in the sense that for disks that are lighter than those identified by the maximum-disk solution, the rotation curves can be equally well fitted by adding a heavier halo. A lower limit to the lightness of the disk would be posed by studies of stellar populations. This ambiguity is obvious if the halo contribution in Eq. (20.1), $r\partial\Phi_h/\partial r$, is taken to be completely free to explain the part of the rotation curve not accounted for by the presence of visible matter. Surprisingly, the degeneracy remains and is actually emphasized in decompositions based on parametric models of the dark matter density distributions, often assumed to be spherical, as described in Section 14.5. An investigation of the rotation curve of NGC 3198 and of a fiducial rotation curve suggested by a sample of thirty-five pure disk galaxies (about half of which are low-surface-brightness galaxies) confirms that parametric models tend to overemphasize the apparent degeneracy, which is removed if self-consistent, physically based models are used in the rotation-curve decomposition.[18]

20.2   **Dynamical Arguments**

In the 1970s, much of the emphasis from the theoretical side on the need for a dark halo came from general stability considerations. Based on study of the properties of classical ellipsoids (see Chapter 10), a few linear modal analyses, and some N–body simulations of galaxy disks, it was argued[19] that self-gravitating disks, with an excess of kinetic energy in the form of systematic rotation K_{ord} relative to the gravitational binding energy $|W|$, with $K_{ord}/|W| > 0.14$, would be prone to a violent bar instability; furthermore, one way to stabilize the disk against such bar modes could be by means of the presence of a sufficiently massive halo (which would contribute to an increase in $|W|$ without any significant change in K_{ord}. Curiously, the estimate of the dark-to-luminous mass inside the optical radius based on stability requirements is similar to that required by the study of rotation curves. Thus the empirical need for dark matter was linked to a deeper theoretical motivation, and this strongly pushed in the direction of a general consensus on the existence of dark halos around spiral galaxies.

Later a number of counterexamples cast doubts on the validity of the general stability criterion used to invoke the presence of dark halos, which soon became less popular as an argument. Remarkably, a thorough investigation of the stability of galaxy disks aimed at interpreting spiral structure in galaxies (see Chapters 17 and 18) now confirms the view that a large-scale nonbarred spiral structure can be supported and justified in general only if the active disk is relatively light, as would occur if the disk were embedded in a massive halo. Put in different terms, the conclusion of these studies on the morphology of large-scale spiral modes in galaxies suggests that if the maximum-disk solution were the rule (i.e., if only a minimal amount of dark matter were present), then most galaxy disks, if sufficiently thin, would indeed be prone to bar modes.

There is nothing dramatic in the occurrence of a barred structure by itself. The key point is that the application of the Ostriker-Peebles criterion tacitly assumed that many disks in spiral galaxies are nonbarred, starting with the case of our own Galaxy, and this led to the requirement of a sizable dark halo. Now it appears that the presence of a bar is not so uncommon. Through the recently acquired imaging capability in the near infrared (near 2.2 μm) it has become easier to detect a bar in the underlying evolved stellar disk. Indeed, many objects (such as NGC 309 or NGC 1637) do show an underlying bar even when the optical picture does not. Other objects (such as NGC 4622) show no evidence of a bar. Wide surveys have tested the frequency of occurrence of the bar morphology, showing that a significant fraction also remains nonbarred in the near infrared (some references are given in Chaptesr 3 and 16). If most spiral galaxies turned out to be barred in their underlying stellar disk, then the maximum-disk decomposition would have been likely to be the one preferred in nature (see the discussion in Chapter 18).

Warps often affect the outer regions of galaxy disks and thus may put some dynamical constraints on dark halos on the large scale. In Chapter 19, aspects of the dynamics of bending waves, which are thought to be associated with the phenomenon of warped disks, were described. Some studies have noted that if we interpret warps as steady (neutrally stable), nonpropagating kinds of distortions of the galactic plane, then we may argue that the halo is dominant so as to make the gravitational field spherical[20] or at least within a given range of flattening depending on the shape of the observed warp.[21] We might object that the strict assumption of persistence implicit in these studies is not justified by the long time scale at the periphery of the galaxy. In addition, neutrally stable perturbations are hard to justify [see comment following Eq. (19.10)]. Fortunately, if we look at the problem from the point of view of excitation and wave propagation, we can identify other interesting constraints. In particular, the amplitude profile of the warp $H(r)$ as a result of conservation of wave action (see Section 19.1) should follow the relation $H(r) \sim r^{-1/2}\sigma^{-1}$; the observed behavior suggests that a sizable dark halo must be present.[22] In the absence of a dark halo, bending appears to lack an adequate source of amplification essentially because the Jeans instability term associated with density waves is replaced by a restoring-force term for bending waves. In contrast, a nonrotating or slowly rotating spheroidal dark halo would provide a natural source of excitation by means of a kind of two-stream instability for the disk to flap like a flag in the wind (see Section 19.2).

20.3 **Dark Matter in Our Galaxy**

In Chapter 14 we discussed the basic arguments related to the issue of dark matter for our Galaxy, especially in connection with modeling of the thickness of the disk in the solar neighborhood. Here we just recall that the program started by Oort in the 1930s to measure the density of the disk by studying its vertical structure, which initially appeared to suggest the possibility of a relatively heavy disk, was further developed in the 1980s and gradually changed into a controversy. Finally, at the turn of the past century, accurate measurements from *Hipparcos* resolved the controversy in favor of the picture that in the solar neighborhood there is no evidence for a mass discrepancy, the disk of our Galaxy is light, and dark matter is present in the form of a round halo. We also recall that if the controversy had been resolved in the opposite way, in favor of the presence of a dark disk, this would have posed a serious challenge to the currently accepted picture that dark matter is made of collisionless material.

In the recent years, long-term observational investigations have been undertaken with the goal of setting further constraints on the dark halo of our Galaxy. One project focuses on the stars in the direction of the center of our Galaxy; others, on the stars in the Magellanic Clouds and M31. The general idea is to monitor millions of stars for a long period of time with the expectation of catching some microlensing events that would be produced by the dark objects of the halo passing along the line of sight in front of the monitored stars. From the theory of general relativity, we can calculate the properties of the light curve associated with the temporary brightening of the microlensed star under the assumption that the objects that make the dark halo have a mass in a given range. In particular, the expected magnification factor (for the flux density from a given pointlike source) induced by a lensing object along the line of sight is given by[23]

$$A = \frac{1 + 2\theta_E^2/\theta^2}{\sqrt{1 + 4\theta_E^2/\theta^2}}, \tag{20.2}$$

where θ measures the angular distance of the source from the lens along the bent light-ray path, and $\theta_E \propto \sqrt{m}$ is the angular width of the so-called Einstein ring for a lens in the given geometric position and mass m. A few events were soon identified with the appropriate signature of microlensing in the direction of the Large Magellanic Cloud.[24] These were interpreted as due to intervening dark objects with mass $\approx 0.1\, M_\odot$. The very small number of events observed appeared to be consistent with the number expected if a sizable fraction of the dark halo of our galaxy is dominated by objects of this type. In turn, the statistics of microlensing events in the sky region in the direction of the Galaxy center[25] seemed to indicate a significant amount of matter in the disk. The projects based on microlensing, although interesting and promising, involve major difficulties in relation to the small number of events that are expected, the statistical analysis involved in the selection of positive events from other causes of variability, the uncertainty on the distance of the lensing object in individual events, and the interpretation in terms of models for the dark halo. Recent assessments of the entire program[26] indicate that events are caused by the lensing of known populations of stars and that there is no need for introducing the existence of dark-matter compact objects to explain the observed event rates.

One major question that should be addressed when a dark halo is invoked concerns its overall shape. Spherical or spheroidal symmetry is favored by the presence of observed objects, such as the globular cluster system, that show an approximately spherical distribution. (Curiously, globular clusters are distributed in the Galaxy with an $r^{-3.5}$ density distribution;[27] see Chapter 22 for a discussion of the density distribution of spherical stellar systems, especially Subsection 22.4.4). Another argument for it is the implied density profile of the dark halo that would resemble that of an isothermal sphere (see Chapter 22). Of course, an analysis of the type made for the kinematics of the solar neighborhood would be unable to distinguish between a quasi-spherical dark halo and a highly flattened spheroid (or rather thick disk). One shape factor that may be checked is the overall axisymmetry of the gravitational field (and thus of the dark halo). Arguments have been put forward in favor of a significant triaxiality,[28] but a local examination of the Oort constants[29] and the large-scale kinematics of carbon stars strongly suggest that the large-scale gravitational potential is axisymmetric to a high degree of approximation.

20.4   Cosmological Arguments

As noted at the beginning of this chapter, the amounts of dark matter required for explaining kinematics in galaxies are practically insignificant from a cosmological point of view. In this section we briefly comment on two topics suggested by cosmological simulations that have attracted considerable interest.

20.4.1   Central Cusps

One focus of interest is the structure of dark halos as emerging from cosmological simulations.[30] Apparently the structures formed in the simulations of an expanding universe made of cold dark matter are characterized by halos of different scales but with a universal density profile associated with a central cusp. Extensive investigations have been done to test the presence of such cuspy halos on the very small scale of individual galaxies. The fact that in many cases evidence is found for cored distributions, against the cosmological predictions, is generally ascribed to the fact that the simulations suggesting such universal density profiles did not include the presence of baryonic matter. Curiously, the models developed to calculate the effects of the infall of baryonic matter into the potential wells dominated by dark-matter halos, the so-called models of adiabatic growth,[31] would make the resulting central density distributions even more cuspy. It has been noted that the problem may be partly alleviated by dynamical friction, which is a slow but nonadiabatic mechanism that is able to soften the final density profiles.[32]

In Section 14.5 we described the properties of self-consistent equilibrium configurations made of a disk embedded in an isothermal halo. In view of the preceding remarks, it would be interesting to study self-consistent models in which the dark-matter distribution follows the predictions of cosmological models, which, in addition to the central cusp, definitely favor a density profile with asymptotic behavior at large radii $\sim r^{-3}$. To carry out this study, we would require a physically justified form of the distribution function for the dark halo from which to start. Then it would be interesting to test to what extent the cosmologically predicted distributions could be reconciled with the existence of cases in which the rotation curve remains very flat in a wide radial range.

20.4.2 Substructures

Another focus of interest is a noted discrepancy between the observed relatively small number of small satellites around galaxies and the large number of substructures expected from cosmological simulations.[33] This corresponds to the low-mass end of dark halos approximately described by a power-law mass function.[34] Ways to resolve the apparent discrepancy have been discussed.[35] All this clearly demands a better understanding of the formation and evolution of globular clusters, dwarf spheroidals, and other low-luminosity galaxies.[36] Curiously, although they span a similar range in optical luminosity, globular clusters and dwarf spheroidals have distinct properties.[37] In particular, globular clusters are believed to be devoid of dark matter, whereas dwarf spheroidals are believed to be characterized by the highest values of the mass-to-light ratio.[38]

Many investigations have addressed directly the problem as posed by data and models of our Galaxy. But the problem is general and has attracted the attention of several different lines of research, in particular, those related to gravitational lensing.[39]

20.5 Modified Newtonian Dynamics?

The discussion so far has assumed that the laws of physics, as derived from experiments in the laboratory, are applicable to the scale of galaxies and beyond. Progress in astrophysics gives us continual assurance of the basic correctness of this assumption. Yet, when a serious discrepancy or an unresolved puzzle arises, it is not unreasonable to wonder whether we are witnessing a failure of the physical laws as formulated on our small planetary scale. In this respect, there are several notable examples from the past, among which we should recall some reactions to the anomalous precession of the orbit of Mercury before the proper explanation could be provided within the framework of general relativity. In the general context of galactic dynamics, we even find a violation of the law of gravitational attraction proposed by Jeans.[40] As to the specific problem of dark matter, one early suggestion of solution in terms of a modification of the gravity law was given in the 1960s.[41]

In this section one line of thinking is briefly described that manages to explain phenomena such as flat rotation curves without invoking the presence of unseen material at the cost of modifying Newtonian dynamics. Starting from semiempirical grounds, it has been able to face quite successfully probably all the observations that are generally quoted as evidence for the presence of dark matter.[42] Modified Newtonian Dynamics (MOND), initially formulated in a rudimentary way,[43] has now been cast in a more formal framework of an alternative relativistic gravitation theory,[44] commonly known as *TeVeS*.

If we decide to break the laws of Newtonian dynamics, there are various routes that we can follow. Of course, the guiding principle is to propose modifications that are able to explain the outstanding puzzles first. The most natural way, which was indeed initially followed in the 1980s, is to imagine that the breakdown occurs at some scale length r_0 of the order of a few kiloparsecs so that flat rotation curves are observed on the scale to which such new dynamics applies. Among others, one specific proposal was a correction to the standard Newtonian field in terms of a Yukawa-type potential.[45] All these attempts were not feasible for two main reasons.

1. The desired correlation between galaxy diameter (at the outermost available kinematic data point) and mass discrepancy is not observed. In particular, some very large galaxies such as UGC 2885 do not demand dark matter (in the framework of Newtonian dynamics) on the scale of 70 kpc, whereas other very small objects, such as UGC 2259, do so below the scale of 10 kpc.[46]
2. These modifications lead to a relation between asymptotic velocity of rotation and mass of the type $M \sim V^2 r_0$, which is contradicted by the existence of the relation between luminosity and velocity $L \sim V^4$.

The most promising approach,[47] which resolves the two difficulties just mentioned, is to set the breakdown of the Newtonian dynamics at a threshold acceleration a_0 of the order of 10^{-8} cm s^{-2}. (One curious numerical coincidence has been noted in a relation with the Hubble constant H_0 and the speed of light c, that is, $a_0 \approx H_0 c$.) Two options can be considered. We can assume that the modification affects the second law of dynamics; that is, we argue that the relation between force and acceleration is changed in low-acceleration regimes. This has some dramatic and undesired consequences for the basic principles of classical mechanics. The second preferred approach is to think in terms of a modification to the force of gravity, retaining the validity of the second law of dynamics. A simple way to formulate the law of modified dynamics is to say that the true gravitational field $\mathbf{g}$ replaces the standard Newtonian field $\mathbf{g_N}$ following the rule $\mu(g/a_0)\mathbf{g} = \mathbf{g_N}$, with the specification $\mu(x) \sim 1$ for $x \gg 1$ (limit of the standard Newtonian field) and $\mu(x) \sim x$ for $x \ll 1$ (modified gravity field). The relation for the asymptotic velocity $V^4 \sim GMa_0$ that follows is consistent with the luminosity-velocity relation.

This modified dynamics prescription can be formulated in the framework of a modified Poisson equation. In a nonrelativistic Lagrangian theory,[48] the standard Poisson equation is replaced by the field equation

$$\nabla \left[\mu \left(\frac{\|\nabla\phi\|}{a_0} \right) \nabla\phi \right] = 4\pi G\rho, \tag{20.3}$$

so the MOND acceleration field is provided by $\mathbf{g} = -\nabla\phi$. The relevant boundary condition for a system of finite mass is $\nabla\phi \to 0$ as $\|\mathbf{x}\| \to \infty$. The connection with the Newtonian field described earlier involves the presence of a solenoidal field $\mathbf{S}$, that is, $\mu(g/a_0)\mathbf{g} = \mathbf{g_N} + \mathbf{S}$; the field $\mathbf{S}$ depends on the specific density distribution ρ that is considered. On this basis, numerical codes have been constructed capable of performing N-body simulations in MOND dynamics.[49] Furthermore, it has been possible to study the properties of separable triaxial density-potential pairs in MOND.[50]

Based on a simple, often-used form for the transition function $\mu(x) = x(1+x^2)^{-1/2}$, it has been argued that current data on the planetary orbits in the solar system set an upper limit[51] to the threshold acceleration: $a_0 \lesssim 3.8 \; 10^{-7} \; \text{cm s}^{-2}$. Related to this preliminary study, a rather intriguing result comes from a long-term analysis of data from the probes *Pioneer 10* and *Pioneer 11*, tracked over a period of more than 10 years, which point to the presence of an anomalous acceleration in the direction of the Sun of the order of a_0.[52] Another anomaly that has sparked curiosity in relation to our understanding of gravitation is the one relative to Earth flybys by several spacecraft probes, one of which was *Rosetta*.[53]

Extensive comparisons among the predictions of these modified gravity laws and various observational constraints appear to be quite encouraging. In particular, a single value of the acceleration parameter $a_0 = 1.2 \; 10^{-8} \; \text{cm s}^{-2}$ appears to be able to provide a good fit to all the galaxies of a sample of eleven selected on the basis of the quality and symmetry of the available kinematic data[54] (Fig. 20.4); the sizes of these objects span a range between 7 and 40 kpc (for the outermost available kinematic data point). The fit becomes very good if the distance of the galaxy is left somewhat free, which is not unreasonable given the large uncertainties involved in distance determination. In contrast, the fit to the same objects by means of models based on classical Newtonian dynamics requires the specification of various amounts of dark matter in amounts and density profiles that may be judged to be ad hoc to the extent that unseen material is invoked. Thus an attractive aspect of the modified theory is its ability to make predictions that can be compared with observations.

Two topics may reveal weaknesses of the MOND interpretation. On the theoretical side, it appears that MOND would predict more efficient relaxation and dynamical friction[55] than standard Newtonian gravity. Therefore, MOND may run into difficulties in interpreting some cases where lack of relaxation is apparent (see Chapter 7). On the observational side, optical and X-ray

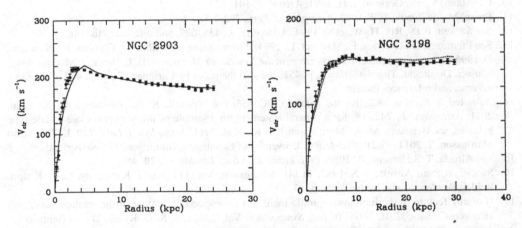

Fig. 20.4. Two cases from a study of a sample of spiral galaxies (from Begeman, K. G., Broeils, A. H., Sanders, R. H. "Extended rotation curves of spiral galaxies: Dark halos and modified dynamics," 1991. *Mon. Not. Roy. Astron. Soc.*, **249**, 523; by permission of Oxford University Press, on behalf of the Royal Astronomical Society) in which the observed rotation curve is fitted without dark matter by considering a Modified Newtonian Dynamics. Two fits have been performed, a one-parameter fit (with M/L as free parameter), and a two-parameter fit, for which the distance to the galaxy also can be adjusted slightly; the slightly discrepant fit for NGC 3198 is the one-parameter fit. The value of the threshold acceleration is taken to be fixed at $a_0 = 1.21 \times 10^{-8} \; \text{cm s}^{-2}$.

data, combined with lensing analyses, have shown that in some clusters of galaxies, the total mass distribution is peaked at points away from the peaks of the visible mass distribution, thus making it impossible for a theory such as MOND to interpret the observations simply by changing the laws of gravity.[56]

Notes

1. Zwicky, F. 1933. *Helv. Phys. Acta*, **6**, 110; Smith, S. 1936. *Astrophys. J.*, **83**, 23. As already noted at the beginning of Section 4.4, the discussions of dark matter presented in Chapters 20 and 24 partly follow the article "Materia oscura," by Bertin, G., van Albada, T. S. 1998. In *Enciclopedia del Novecento*, Vol. 11, Istituto della Enciclopedia Italiana, Rome, p. 149.
2. See the reference to Henri Poincaré given at the beginning of Section 4.4.
3. For a giant black hole in the galaxy NGC 4258, see Miyoshi, M., Moran, J., et al. 1995. *Nature (London)*, **373**, 127.
4. Wolszczan, A., Frail, D. A. 1992. *Nature (London)*, **355**, 145.
5. van de Hulst, H. C., Raimond, E., van Woerden, H. 1957. *Bull. Astron. Inst. Neth.*, **14**, 1.
6. See Roberts, M. S. 1976. *Comments Astrophys.*, **6**, 105.
7. Especially through the work of Rubin and collaborators; see Rubin, V. C. 1987. In *Dark Matter in the Universe*, eds. J. Kormendy, G. R. Knapp. Reidel, Dordrecht, The Netherlands, p. 51, and references therein.
8. Kalnajs, A. 1983. In *Internal Kinematics and Dynamics of Galaxies*, ed. E. Athanassoula. Reidel, Dordrecht, The Netherlands, p. 87; Kent, S. M. 1986. *Astron. J.*, **91**, 1301.
9. van Albada, T. S., Bahcall, J. N., Begeman, K., Sancisi, R. 1985. *Astrophys. J.*, **295**, 305; earlier this point was made by the survey by Bosma, A. 1978. Ph.D. dissertation, University of Groningen, Groningen, The Netherlands.
10. Casertano, S., van Gorkom, J. H. 1991. *Astron. J.*, **101**, 1231.
11. Franx, M., de Zeeuw, P. T. 1992. *Astrophys. J. Lett.*, **392**, L47.
12. See Sackett, P. D., Rix, H.-W., et al. 1994. *Astrophys. J.*, **436**, 629, and references therein.
13. See Pfenniger, D., Combes, F., Martinet, L. 1994. *Astron. Astrophys.*, **285**, 79; Combes, F., Pfenniger, D. 1996. In *New Extragalactic Perspectives in the New South Africa*, eds. D. L. Block, J. M. Greenberg. Kluwer, Dordrecht, The Netherlands, p. 451; see also the paper by Carignan, C., on p. 447 of the same volume, and references therein.
14. As noted in Section 18.2; for the galaxy UGC 463, see Westfall, K. B., Bershady, M. A., et al. 2011. *Astrophys. J.*, **742**, 18; for a general statement on a sample of thirty intermediate-to-late-type galaxies, see Bershady, M. A., Martinsson, T. P. K., et al. 2011. *Astrophys. J. Lett.*, **739**, L47; see also Martinsson, T. 2011. Ph.D. dissertation, University of Groningen, Groningen, The Netherlands.
15. van Albada, T. S., Sancisi, R. 1986. *Phil. Trans. Roy. Soc. London A*, **320**, 447.
16. Sancisi, R., van Albada, T. S. 1987. In *IAU Symposium*, Vol. 117, eds. J. Kormendy, G. R. Knapp. Reidel, Dordrecht, The Netherlands, p. 67.
17. "For any feature in the luminosity profile there is a corresponding feature in the rotation curve and vice versa"; Sancisi, R. 2004. In *IAU Symposium*, Vol. 220, eds. S. D. Ryder, D. J. Pisano, et al. Astronomical Society of the Pacific, San Francisco, p. 233.
18. This point was made in the analysis by Amorisco, N. C., Bertin, G. 2010. *Astron. Astrophys.*, **519**, A47, briefly summarized in Section 14.5.
19. Ostriker, J. P., Peebles, P. J. E. 1973. *Astrophys. J.*, **186**, 467.
20. Tubbs, A. D., Sanders, R. H. 1979. *Astrophys. J.*, **230**, 736.
21. Sparke L. S., Casertano, S. 1988. *Mon. Not. Roy. Astron. Soc.*, **234**, 873.
22. Bertin, G., Mark, J. W.-K. 1980. *Astron. Astrophys.*, **88**, 289.
23. See Schneider, P., Ehlers, J., Falco, E. E. 1992. *Gravitational Lenses*, Springer-Verlag, New York.
24. Alcock, C., Akerlof, C. W., et al. 1993. *Nature (London)*, **365**, 620; Aubourg, E., Bareyre, P., et al. 1993. *Nature (London)*, **365**, 623.

25. See the program described by Udalsky, A., Szymański, M., et al. 1993. *Acta Astron.*, **43**, 289.

26. For the Optical Gravitational Lensing Experiment (OGLE), see Wyrzykowski, L., Skowron, J., et al. 2011. *Mon. Not. Roy. Astron. Soc.*, **416**, 2949, who combine the results of monitoring stars in the direction of the Magellanic Clouds.

27. See Djorgovski, S., Meylan, G. 1994. *Astron. J.*, **108**, 1292.

28. Blitz, L., Spergel, D. N. 1991. *Astrophys. J.*, **370**, 205.

29. Kuijken, K., Tremaine, S. D. 1991. In *Dynamics of Disk Galaxies*, ed. B. Sundelius. Göteborg University, Sweden, p. 71.

30. Navarro, J. F., Frenk, C. S., White, S. D. M. 1996. *Astrophys. J.*, **462**, 563; Moore, B., Governato, F., et al. 1998. *Astrophys. J. Lett.*, **499**, L5; Ghigna, S., Moore, B., et al. 2000. *Astrophys. J.*, **544**, 616; Navarro, J. F., Hayashi, E., et al. 2004. *Mon. Not. Roy. Astron. Soc.*, **349**, 1039; and many following papers.

31. Blumenthal, G. R., Faber, S. M., et al. 1986. *Astrophys. J.*, **301**, 27; Mo, H. J., Mao, S., White, S. D. M. 1998. *Mon. Not. Roy. Astron. Soc.*, **295**, 319; Kochanek, C. S., White, M. 2001. *Astrophys. J.*, **559**, 531; Gnedin, O. Y., Kravtsov, A. V., et al. 2004. *Astrophys. J.*, **616**, 16.

32. El-Zant, A., Shlosman, I., Hoffman, Y. 2001. *Astrophys. J.*, **560**, 636; Bertin, G., Liseikina, T., Pegoraro, F. 2003. *Astron. Astrophys.*, **405**, 73; El-Zant, A., Hoffman, Y., et al. 2004. *Astrophys. J. Lett.*, **607**, L75; Ma, C.-P., Boylan-Kolchin, M. 2004. *Phys. Rev. Lett.*, **93**, 021301; Nipoti, C., Treu, T., et al. 2004. *Mon. Not. Roy. Astron. Soc.*, **355**, 1119; Arena, S. E., Bertin, G., et al. 2006. *Astron. Astrophys.*, **453**, 9; see also Chapter 7.

33. Klypin, A., Kravtsov, A. V., et al. 1999. *Astrophys. J.*, **522**, 82; Moore, B., Ghigna, S., et al. 1999. *Astrophys. J. Lett.*, **524**, L19. The issue has been revisited by means of more advanced simulations; see Springel, V., Wang, J., et al. 2008. *Mon. Not. Roy. Astron. Soc.*, **391**, 1685; Diemand, J., Kuhlen, M., et al. 2008. *Nature (London)*, **454**, 735.

34. See Tormen, G., Diaferio, A., Syer, D. 1998. *Mon. Not. Roy. Astron. Soc.*, **299**, 728; Ghigna, S., Moore, B., et al. 2000. *Astrophys. J.*, **544**, 616; De Lucia, G., Kauffmann, G., et al. 2004. *Mon. Not. Roy. Astron. Soc.*, **348**, 333; Gao, L., White, S. D. M., et al. 2004. *Mon. Not. Roy. Astron. Soc.*, **355**, 819.

35. For example, see Okamoto, T., Gao, L., Theuns, T. 2008. *Mon. Not. Roy. Astron. Soc.*, **390**, 920, and references therein.

36. See also Helmi, A. 2008. *Astron. Astrophys. Rev.*, **15**, 145, and references therein.

37. For example, see van den Bergh, S. 2008. *Mon. Not. Roy. Astron. Soc.*, **390**, L51.

38. For example, see Lokas, A. L., Mamon, G. A., Prada, F. 2005. *Mon. Not. Roy. Astron. Soc.*, **363**, 918.

39. See Mao, S., Schneider, P. 1998. *Mon. Not. Roy. Astron. Soc.*, **295**, 587; Metcalf, R. B., Madau, P. 2001. *Astrophys. J.*, **563**, 9; Bradač, M., Schneider, P., et al. 2002. *Astron. Astrophys.*, **388**, 373; Chiba, M. 2002. *Astrophys. J.*, **565**, 17.

40. Jeans, J. 1923. *Mon. Not. Roy. Astron. Soc.*, **84**, 60.

41. Finzi, A. 1963. *Mon. Not. Roy. Astron. Soc.*, **127**, 21.

42. For an early review of this unorthodox point of view, see Sanders, R. H. 1990. *Astron. Astrophys. Rev.*, **2**, 1. For a more recent discussion, see Sanders, R. H. 2010. *The Dark Matter Problem: A Historical Perspective*, Cambridge University Press, Cambridge, UK.

43. Milgrom, M. 1983. *Astrophys. J.*, **270**, 365; Bekenstein, J. D., Milgrom, M. 1984. *Astrophys. J.*, **286**, 7. A discussion of Modified Newtonian Dynamics in the cosmological context was given by Sanders, R. H. 1998. *Mon. Not. Roy. Astron. Soc.*, **296**, 1009, and later reviewed by the same author in 2006. *EAS Publ. Ser.*, **20**, 231.

44. Bekenstein, J. D. 2004. *Phys. Rev. D*, **70**, id.083509; with an erratum 2005. *Phys. Rev. D*, **71**, id.069901.

45. Sanders, R. H. 1984. *Astron. Astrophys.*, **136**, L21.

46. Carignan, C., Sancisi, R., van Albada, T. S. 1988. *Astron. J.*, **95**, 37.

47. Milgrom, M. 1983. *op. cit.*; and following papers.

48. Bekenstein, J. D., Milgrom, M. 1984. *op. cit.*

49. Brada, R., Milgrom, M. 1999. *Astrophys. J.*, **519**, 590; Ciotti, L., Londrillo, P., Nipoti, C. 2006. *Astrophys. J.*, **640**, 741; Nipoti, C., Londrillo, P., Ciotti, L. 2007. *Astrophys. J.*, **660**, 256; Tiret, O.,

Combes, F. 2007. *Astron. Astrophys.*, **464**, 517. For MOND cosmological simulations, see Llinares, C., Knebe, A., Zhao, H. 2008. *Mon. Not. Roy. Astron. Soc.*, **391**, 1778.

50. Ciotti, L., Zhao, H., de Zeeuw, T. 2012. *Mon. Not. Roy. Astron. Soc.*, **422**, 2058.

51. Bracci, L., Fiorentini, G. 1989. *Il Nuovo Cimento*, **12C**, 121.

52. Anderson, J. D., Laing, P. A., et al. 2002. *Phys. Rev. D*, **65**, id.082004. See Hellemans, A. 2005. *Scientific American*, October 10; but see also Turyshev, S. G.,Toth, V. T., et al. 2012. *Phys. Rev. Lett.*, **108**, id.241101; Anderson, J. D., Morris, J. R. 2012. *Phys Rev. D*, **85**, id.084017.

53. Anderson, J. D., Campbell, J. K., et al. 2008. *Phys. Rev. Lett.*, **100**, 091102.

54. Begeman, K. G., Broeils, A. H., Sanders, R. H. 1991. *Mon. Not. Roy. Astron. Soc.*, **249**, 523. A similar study for a sample of thirty spiral galaxies in the Ursa Major cluster is reported by Sanders, R. H., Verheijen, M. A. W. 1998. *Astrophys. J.*, **503**, 97, bringing up to approximately eighty the number of galaxy rotation curves fitted by Modified Newtonian Dynamics.

55. Ciotti, L., Binney, J. 2004. *Mon. Not. Roy. Astron. Soc.*, **351**, 285; Nipoti, C., Ciotti, L., et al. 2008. *Mon. Not. Roy. Astron. Soc.*, **386**, 2194.

56. The most important cases are those of the Bullet Cluster; Clowe, D., Bradač, M., et al. 2006. *Astrophys. J. Lett.*, **648**, L109; and the Ring Cluster; Jee, M. J., Ford, H. C., et al. 2007. *Astrophys. J.*, **661**, 728; but see the complex case of the cluster A520, studied by Jee, M. J., Mahdavi, A., et al. 2012. *Astrophys. J.*, **747**, id.96.

Part IV
Elliptical Galaxies

21 Orbits

Chapter 13 started out by noting the importance of the study of orbits as a key step in the study of the dynamics of galaxies. Here the reason why it is important is briefly summarized. Even better than other stellar systems, elliptical galaxies may be modeled as purely collisionless collections of stars. It is well known that for the distribution function in phase space that is used to describe these systems, the evolution operator dictated by the mean field is the same as that for single-particle orbits. The resulting Jeans theorem (see Chapter 9) thus reduces the problem of finding the most general stationary solution to the collisionless Boltzmann equation to identification of the relevant integrals of the motion. Furthermore, the frequencies characterizing single-particle orbits play an important role in the stability of the stellar system with respect to internal or driven perturbations. Finally, we should recall that closed periodic orbits are often used to model relatively smooth and cold flow patterns that are sometimes observed around galaxies, thus offering a very interesting diagnostic tool for the underlying potential (see also Subsection 13.6.1 and the discussion of dark matter in Chapters 20 and 24). Unfortunately, when we deal with the dynamics of elliptical galaxies and with the related orbits, we have to face two very difficult issues.

The first important point is that ellipticals are dynamically hot stellar systems. Therefore, even if we focus our attention on axisymmetric basic states, approximations such as those that have led to the discussion of quasi-circular orbits in the context of spiral galaxies are bound to be of little use. This makes it difficult to set up semianalytical investigations of dynamical mechanisms, in contrast to the simpler case of galaxy disks. In this regard, we should recall that the possibility of separating the discussion of vertical and horizontal equilibria for a disk (in the simple intuitive way described in Chapter 14) and the possibility of setting up a local description of waves (see Chapters 15 and 16) depend crucially on the coldness of the galaxy disk under investigation.

The second major difficulty is that if we consider obviously nonspherical systems, such as E3 or later types, we have no firmly established symmetry that we may associate with the basic state of elliptical galaxies. It is not clear to what extent ellipticals can be thought of as generically axisymmetric, although strong triaxialities should not be involved.[1] Triaxiality probably occurs rather frequently. In some cases, for example when a twisting of the isophotes is observed, the galaxy image itself proves that axisymmetry is not applicable.[2] Some kinematic features also detect departures from axisymmetry, which are naturally involved, for example, when minor-axis rotation is observed. Despite this, it may well be that for some purposes a large fraction of ellipticals can be approximated as quasi-axisymmetric systems, much as the gravitational field can be taken to be approximately axisymmetric (even though spiral arms and bars are often prominent) for the purpose of describing many aspects of the dynamics of galaxy disks. For ellipticals of the type E0 to E2, we may adopt spherical symmetry as a natural starting point, in the spirit of Chapter 9; however, for any specific observed round object, we should be ready to accept the possibility of a pole-on view of an intrinsically flatter system (the prolate possibility is also to be considered but is physically less appealing).

Because symmetry determines the available integrals of the motion that characterize the particle orbits, the search for reasonable basic states (i.e., stationary equilibrium configurations) to be associated with the underlying structure of elliptical galaxies, with the above-noted uncertainties on the symmetry involved, immediately leads to an exploration of the integrability properties of potentials of possible interest in galactic dynamics. This is done in view of application to the Jeans theorem. The concepts of integral of the motion and the Jeans theorem require some clarification[3] (see Section 21.2). In a general three-dimensional potential, the maximum number of meaningful integrals of the motion is expected to be three (others would be related to resonances of the phase variables and thus would disappear at the slightest perturbation[4]). Spherical symmetry is a somewhat degenerate case for which all the meaningful integrals of the motion are available. Axisymmetry is in general characterized by two classical integrals (E and J_z). A three-dimensional potential with no specified symmetry would have only one classical integral (E).

Empirically, we are led to the study of potentials with isolating integrals (beyond the classical ones) because evidence has grown in the direction of stellar systems being often characterized by anisotropic velocity dispersions. Broadly speaking, one conserved quantity in single-particle orbits leads to the identification of one axis in the velocity ellipsoid for a collection of stars, so the realization of a fully anisotropic pressure tensor should require the presence of three meaningful integrals of the motion. Thus we may provide natural guidelines to the construction of equilibrium distribution functions, that is, of suitable superpositions of stellar orbits, compatible with a given underlying potential. In many analyses, the potential is imposed, and consequences are thus drawn on the orbits and the resulting choices for the distribution function. However, for elliptical galaxies, much more than for the case of spiral galaxies, in which separate populations are empirically identified (disk, bulge, and halo; stars and gas within the disk), it is of great interest to proceed to the fully self-consistent case in which a single distribution function supports the field by means of the Poisson equation. This step will be considered in some detail in Chapter 22, but it is mentioned here to emphasize the underlying motivation for many important orbital studies.

In this situation, there have been basically two complementary lines of work. Broadly speaking, these reflect two different approaches in facing the self-consistent problem in stellar dynamics (see Section 22.1). One approach is rather ambitious because it is particularly aimed at resolving the very difficult case in which the departure from spherical symmetry is significant. Usually in this line of research, priority is given to the density distribution, so the potential is fixed in advance; we then ask whether there is a suitable superposition of orbits, that is, an appropriate distribution function in phase space, that is able to support it. In these terms, there is no guarantee that physically acceptable solutions to this question exist. For example, in the case of galaxy disks dominated by strong bars, it is not obvious that exact equilibria involving a steadily rotating bar are available and realized in nature; it may well be that only slowly evolving bars are realized, and these would lack the integrals of the motion that some steady triaxial configurations demand to exist a priori. Even if some steady bars do exist, this might be realized under only very special conditions (in particular, on the mass distribution and pattern speed characterizing the bar) or under special requirements on the underlying axisymmetric basic state (see the modal point of view given in Chapter 17).

The second approach gives priority to the distribution function in phase space. It postpones the study of significantly nonspherical ellipticals and examines what are the natural physical requirements that a distribution of stellar orbits should obey in the simpler case in which the field is approximately spherical. Once the form of the distribution function is thus constrained, the Poisson equation (as a nonlinear equation in Φ) is studied to determine the resulting potential and, a posteriori, the associated mass distribution. This approach has two important points of merit. One is simply the fact that many arguments can be easily followed semianalytically. The other, more interesting point is that this route naturally leads to explaining, at least in the quasi-spherical case, one basic empirical property of elliptical galaxies, that is, the $R^{1/4}$ character of the luminosity profile (see Chapter 4). This route, despite its limitations, has a clear physical basis. Indeed, there are general arguments related to the statistical mechanics of incomplete violent relaxation (and thus to formation processes; see Chapter 25) that further justify the models eventually identified. The description is also naturally extended to the case of models that include diffuse dark halos, and indeed, it has been found to lead to interesting results in determination of the amount and distribution of dark matter in elliptical galaxies (see Chapter 24). In other words, examining quasi-spherical models is not a trivial exercise; they can indeed teach us much about the dynamics of elliptical galaxies.

In any case, a demonstration of the existence of collisionless triaxial models is very interesting per se. For this purpose, we start out with a very symmetric potential (i.e., a potential with good integrals of the motion; see Sections 21.2 and 21.3). This has been done and has demonstrated that indeed anisotropic stellar systems can have shape and rotation that, in principle, are fully decoupled from each other. Unfortunately, these attempts have immediately shown that the easiest route to triaxial models makes use of density distributions that are different from those suggested by the observed luminosity profiles. In particular, the triaxial models produced tend to have a central core that is too wide. However, it might simply be that nonrealistic density distributions arise from an insistence on a complete decoupling between shape and rotation; some figure rotation might be involved in significantly nonspherical ellipticals, and this might be the result of the relevant formation process. In the following, although it is recognized that the study of all possible potentials is rewarding and valuable per se, higher priority is given to quasi-spherical systems because they seem to be preferred in the semiempirical context. We will thus mostly focus on the spherical case and try to clarify the way such symmetry is broken.

21.1 Spherical Potentials

For a spherical potential, the orbits are characterized in the same way as was done in Chapter 13 for orbits in the plane of an axisymmetric galaxy disk. The only difference is that the plane identified by the angular-momentum vector $\mathbf{J}$ can have any orientation in space. Thus interpreted, the formulas provided in Section 13.1 for action and angle variables and in Section 13.2 for quasi-circular orbits are directly applicable to the spherical case. As noted earlier, the case of quasi-circular orbits appears to be of lesser interest here.[5]

If we refer to Eqs. (13.10) and (13.18) as definitions of the circular frequency and the epicyclic frequency (applicable to the limit of quasi-circular orbits) and recall the expression of the Poisson equation in the spherical case, that is,

$$\frac{1}{r^2}\frac{d}{dr}\left(r^2\frac{d\Phi}{dr}\right)=4\pi G\rho,$$

(21.1)

we find the interesting relation

$$\kappa^2=\Omega^2+4\pi G\rho.$$

(21.2)

A few special potentials are worth a short digression.

21.1.1 Keplerian Potential

The $\Phi=-GM_\bullet/r$ potential generated by a point mass is characterized by $\kappa=\Omega$. Indeed, in this degenerate case, there is only one orbital frequency [see Eqs. (13.15) and (13.16)], which depends on the energy E only

$$\Omega_r=\Omega_\theta=(1/GM_\bullet)(-2E)^{3/2};$$

(21.3)

the radial action [see Eq. (13.13)] is given by

$$J_r=\frac{GM_\bullet}{\sqrt{-2E}}-J.$$

(21.4)

Circular orbits are characterized by $J_r\to0$, so for them there is a relation between energy and angular momentum given by

$$J_{circ}=\frac{GM_\bullet}{\sqrt{-2E}}.$$

(21.5)

21.1.2 Harmonic Potential

The harmonic potential $\Phi=\omega_0^2r^2/2$ is associated with a constant density distribution, with $4\pi G\rho=3\omega_0^2$. Thus $\kappa=2\Omega=2\omega_0$. Here

$$\Omega_r=2\Omega_\theta=2\omega_0,$$

(21.6)

$$J_r=\frac{E}{2\omega_0}-\frac{J}{2}.$$

(21.7)

Thus, for circular orbits,

$$J_{circ}=\frac{E}{\omega_0}.$$

(21.8)

21.1.3 Logarithmic Potential

The logarithmic potential $\Phi=V_0^2\ln(r/b)$ is supported by a singular density distribution, with $4\pi G\rho=V_0^2/r^2$, and generates a flat rotation curve with $\kappa=\sqrt{2}\Omega=\sqrt{2}V_0/r$. For circular orbits, we have

$$J_{circ}=\frac{bV_0}{\sqrt{e}}\exp(E/V_0^2).$$

(21.9)

A reasonable approximation for the action[6] is found to be

$$J_r \simeq \sqrt{\frac{e}{2\pi}} \left[J_{\text{circ}}(E) - J \right].$$

(21.10)

An even more accurate expression, with relative error below 0.004, is

$$J_r \simeq \sqrt{\frac{e}{2\pi}} J_{\text{circ}}(E) \left[1 - (J/J_{\text{circ}})^{\sqrt{\pi/e}} \right].$$

(21.11)

Relation (21.11) also reproduces the exact radial period of orbits for $J = 0$ and for $J = J_{\text{circ}}(E)$.

21.1.4 Isochrone Potential

The isochrone potential[7] is defined as

$$\Phi(r) = -\frac{GM}{b + \sqrt{b^2 + r^2}}.$$

(21.12)

For simplicity, we refer to dimensionless units by setting $GM = 1$ and $b = 1$ in Eq. (21.12). Then the supporting density distribution (with total mass equal to 4π and half-mass radius $r_M \approx 3.06$) is

$$\rho(r) = \left(1 + 2\sqrt{1 + r^2} \right) \left(\sqrt{1 + r^2} \right)^{-3} \left(1 + \sqrt{1 + r^2} \right)^{-2}.$$

(21.13)

There is also a simple expression for the epicyclic frequency:

$$\kappa(r) = (1 + r^2)^{-3/4}.$$

(21.14)

The isochrone potential is characterized by the fact that the ratio Ω_r / Ω_θ depends on only angular momentum (and thus is independent of energy). The explicit formulas are

$$\Omega_r = (-2E)^{3/2} = 2\Omega_\theta \left(1 + \frac{J}{\sqrt{4 + J^2}} \right)^{-1},$$

(21.15)

$$J_r = \frac{1}{\sqrt{-2E}} - J/2 - \sqrt{1 + J^2/4}.$$

(21.16)

The relation between energy and angular momentum in circular orbits is given by

$$E = -\frac{1}{2 \left(J_{\text{circ}}/2 + \sqrt{1 + J_{\text{circ}}^2/4} \right)^2};$$

(21.17)

hence

$$J_{\text{circ}} = \frac{1 + 2E}{\sqrt{-2E}}.$$

(21.18)

An explicit expression is available for the function $r_0(J)$ [the radius of circular orbits, defined implicitly by Eq. (13.11)]:

$$r_0(J) = \left[2J^2(1 + J^2/4) + J(J^2 + 2)\sqrt{1 + J^2/4} \right]^{1/2}.$$

(21.19)

21.1.5   The Perfect Sphere

A density distribution that is quite similar to that of the isochrone case and that also has some
interesting properties (it is the spherical limit of a very symmetric ellipsoidal density distribution;
see Subsection 21.3.4) is given by

$$\rho(r) = \frac{M}{\pi^2} \frac{b}{(b^2 + r^2)^2}. \tag{21.20}$$

The associated potential is

$$\Phi(r) = -\frac{2GM}{\pi} \frac{\arctan(r/b)}{r}. \tag{21.21}$$

The half-mass radius thus occurs at $r_M \approx 2.264b$.

21.2   Classification of Potentials in Relation to the Jeans Theorem

"Only isolating integrals should be used in Jeans theorem."[8] An integral of the motion $I(\mathbf{x}, \mathbf{v})$
is isolating if the phase-space sets defined by $I(\mathbf{x}, \mathbf{v}) = k_1$ and $I(\mathbf{x}, \mathbf{v}) = k_2$, for almost any $k_1 \neq
k_2$, are disjoint. For the spherical problem, except for the special cases of the Keplerian and
harmonic potentials, there are four independent isolating integrals: the energy E and the angular-
momentum vector $\mathbf{J}$; the fifth integral is nonisolating.

The concept of an isolating integral is best introduced by means of the simple example of a
two-dimensional (2D) anisotropic harmonic oscillator.[9]

21.2.1   The Concept of Isolating Integral Exemplified by the Two-Dimensional
Anisotropic Harmonic Oscillator

If ω_x and ω_y are the relevant frequencies, the energy integral of the 2D anisotropic harmonic
oscillator can be thought of as the sum of two independent integrals of the motion $E = I_x + I_y$,
where

$$I_x = \frac{1}{2}\dot{x}^2 + \frac{1}{2}\omega_x^2 x^2, \tag{21.22}$$

$$I_y = \frac{1}{2}\dot{y}^2 + \frac{1}{2}\omega_y^2 y^2. \tag{21.23}$$

These are isolating integrals of the motion in the sense that the conditions $I_x(x, \dot{x}, y, \dot{y}) = I_1$ and
$I_y(x, \dot{x}, y, \dot{y}) = I_2$ (where I_1 and I_2 are constants) isolate hypersurfaces in phase space. A projection
in the configuration (x, y) space shows that the orbit is indeed confined to a rectangle identified
by the straight lines $x = \pm\sqrt{2I_1}/\omega_x$ and $y = \pm\sqrt{2I_2}/\omega_y$.

For the 2D problem, we know that there must exist a third time-independent integral of the
motion $I = I(x, \dot{x}, y, \dot{y})$, which may, in principle, be required for the most general solution to the
collisionless Boltzmann equation [$f = f(I_x, I_y, I)$; Jeans theorem]. In this example, an analytical
expression for this third integral can be given because we know the explicit solution for the orbit:

$$y = \frac{\sqrt{2I_2}}{\omega_y} \sin\left[(-1)^k \frac{\omega_y}{\omega_x} \operatorname{Sin}^{-1}\left(\frac{\omega_x x}{\sqrt{2I_1}}\right) + \frac{\omega_y}{\omega_x} k\pi - \frac{I_3}{\omega_x}\right]. \tag{21.24}$$

Here we have introduced a phase constant I_3 and made use of the relation

$$\arcsin z = (-1)^k \operatorname{Sin}^{-1}(z) + k\pi, \tag{21.25}$$

where k is a relative integer number that defines the principal value Sin^{-1} within the $[-\pi/2, \pi/2]$ interval of the arcsin function.

When the ratio ω_y/ω_x is rational, the third integral I is isolating because the condition

$$I(x, \dot{x}, y, \dot{y}) = I_3, \tag{21.26}$$

which is obtained directly from Eq. (21.24) by solving for I_3, isolates points in phase space. This is shown in configuration space by the fact that the related orbit is closed. [When the frequency ratio is an integer, the orbit of Eq. (21.24) can be used to define the Chebyshev polynomials.]

In contrast, if ω_y/ω_x is not a rational number, the orbit fills ergodically the whole rectangle, thus demonstrating that I is, in this latter case, a nonisolating integral. By means of this simple example, it should now be clear why only isolating integrals are to be used in the Jeans theorem.

21.2.2 Local Integrals, Exceptional Integrals, and Quasi-Integrals

Once the differences among the integrals of the motion in relation to the Jeans theorem are recognized, it is important to devise tools to identify and classify the potentials accordingly. One possibility is to consider generalizations of the well-known fact that the existence of an ignorable coordinate can be seen as counterpart to the existence of an underlying symmetry and thus is associated with the existence of an isolating integral of the motion.[10] This idea has led to a classification of potentials on the basis of the existence of local integrals.[11] In practice, the potentials that admit the so-called local integrals have these conserved on all orbits, which may be an excessive restriction for some applications. In any case, this class of integrals should play an important role in stellar dynamics given the fact that the potentials involved leave one or more free functions and thus might be considered to be generic potentials for classes of stellar systems or stages in the slow evolution of individual stellar systems. In contrast, the fifth isolating integral for the Kepler potential or isotropic harmonic oscillator appears to result from a hidden underlying symmetry,[12] which is a singular property associated with the exact radial dependence of the potential. Exceptional integrals of this type, even if found to exist in special nonspherical potentials, would be less appealing from the astrophysical point of view.

There is another approach that can be taken in this context. Broadly speaking, in this approach it is recognized from the beginning that physical stellar systems often have an age of the order of 100 (in terms of the relevant dynamical time). For most applications, there is thus no real need to require strict conservation of some quantities, or in other words, approximately conserved quantities would behave much like exact integrals of the motion. Therefore, it is interesting to study the existence of quasi-integrals of the motion for realistic classes of potentials, possibly within a simple analytical description. There have been several asymptotic investigations in this direction, which we will not follow directly here.[13] The construction of equilibrium models relevant to elliptical galaxies that we will provide in Chapter 22 (Section 22.4), although referring to potentials with local integrals (to be described in the next section), is very much in the spirit of the study of quasi-integrals.

21.3   Nonspherical Potentials with Isolating Integrals

The best-known class of nonspherical potentials with three isolating integrals of the motion is the one that goes under the name of *Eddington* or *Stäckel potentials*.[14] These are also in the list of potentials with local integrals, which can be traced to invariance properties associated with the ellipsoidal coordinates (λ, μ, ν).[15]

21.3.1   Triaxial Case

The ellipsoidal coordinates (λ, μ, ν) are defined as the roots for τ of the equation

$$\frac{x^2}{\tau+\alpha} + \frac{y^2}{\tau+\beta} + \frac{z^2}{\tau+\gamma} = 1, \tag{21.27}$$

where we consider $\alpha < \beta < \gamma < 0$ to be constants (thus we may visualize a long x axis, an intermediate y axis, and a short z axis; see Subsection 21.3.4). The coordinates are related to standard Cartesian coordinates by the expressions

$$x^2 = \frac{(\lambda+\alpha)(\mu+\alpha)(\nu+\alpha)}{(\alpha-\beta)(\alpha-\gamma)}, \tag{21.28}$$

$$y^2 = \frac{(\lambda+\beta)(\mu+\beta)(\nu+\beta)}{(\beta-\alpha)(\beta-\gamma)}, \tag{21.29}$$

$$z^2 = \frac{(\lambda+\gamma)(\mu+\gamma)(\nu+\gamma)}{(\gamma-\alpha)(\gamma-\beta)}. \tag{21.30}$$

The coordinates are taken to vary within the following ranges:

$$-\gamma \leq \nu \leq -\beta \leq \mu \leq -\alpha \leq \lambda. \tag{21.31}$$

The surfaces at constant λ are ellipsoids, so λ can be regarded as a radial coordinate. The surfaces at constant μ are one-sheet hyperboloids around the long x axis, and the surfaces at constant ν are two-sheet hyperboloids, symmetric with respect to the (x,y) plane (Fig. 21.1). The metric element can be expressed in terms of these coordinates as

$$ds^2 = \mathcal{P}^2(\lambda, \mu, \nu)d\lambda^2 + \mathcal{Q}^2(\lambda, \mu, \nu)d\mu^2 + \mathcal{R}^2(\lambda, \mu, \nu)d\nu^2. \tag{21.32}$$

Here $4\mathcal{P}^2 = (\lambda-\mu)(\lambda-\nu)/[(\lambda+\alpha)(\lambda+\beta)(\lambda+\gamma)]$; the expressions for $\mathcal{Q}^2$ and $\mathcal{R}^2$ are obtained by permutation with respect to the variables (λ, μ, ν).

Under these definitions, the class of potentials defined in terms of three free functions ζ, η, and ξ,

$$\Phi = \frac{\zeta(\lambda)}{\mathcal{P}^2} + \frac{\eta(\mu)}{\mathcal{Q}^2} + \frac{\xi(\nu)}{\mathcal{R}^2} = \frac{\hat{\zeta}(\lambda)}{(\lambda-\mu)(\lambda-\nu)} + \frac{\hat{\eta}(\mu)}{(\mu-\nu)(\mu-\lambda)} + \frac{\hat{\xi}(\nu)}{(\nu-\lambda)(\nu-\mu)}, \tag{21.33}$$

can be shown to admit two isolating integrals of the motion in explicit form in addition to the energy integral E. The existence of such a complete set of isolating integrals allows us to give a complete classification of orbits. In each of the three ellipsoidal coordinates, the motion can be either an oscillation between turning points or a rotation. Box orbits are those that involve

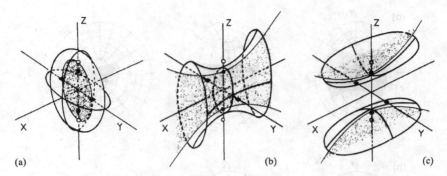

Fig. 21.1. Ellipsoidal coordinates (from de Zeeuw, P. T. "Elliptical galaxies with separable potentials," 1985. *Mon. Not. Roy. Astron. Soc.*, **216**, 273; by permission of Oxford University Press, on behalf of the Royal Astronomical Society). Surfaces of constant λ are ellipsoids (a), surfaces of constant μ are one-sheet hyperboloids (b), surfaces of constant ν are two-sheet hyperboloids (c). Black symbols and the pair of open circles represent the three pairs of foci.

oscillation for all three coordinates; they can come arbitrarily close to the center. Long-axis tubes have no turning point in the ν direction, and the motion is a rotation around the long axis. Two types of long-axis tubes are realized, inner or outer, depending on whether the crossing of the (x, y) plane occurs inside or outside the foci of the ellipsoidal coordinate system. Short-axis tubes have no turning point in the μ direction (Fig. 21.2).

21.3.2 Axisymmetric Case

For the axisymmetric case, it is convenient to refer to spheroidal coordinates (λ, μ, ϕ) defined in terms of the cylindrical polar coordinates (R, z, ϕ) by means of the transformations

$$\lambda = \left[\frac{(R^2 + z^2 + \beta^2)}{2} + \sqrt{\frac{(R^2 + z^2 + \beta^2)^2}{4} - \beta^2 z^2} \right]^{1/2} , \tag{21.34}$$

$$\mu = \pm \left[\frac{(R^2 + z^2 + \beta^2)}{2} - \sqrt{\frac{(R^2 + z^2 + \beta^2)^2}{4} - \beta^2 z^2} \right]^{1/2} , \tag{21.35}$$

where $\lambda \geq \beta$, $-\beta \leq \mu \leq \beta$. The foci are separated by a distance 2β. The coordinate ϕ is the same in the two systems of coordinates and is defined as the azimuthal angle around the symmetry axis. The surfaces at constant λ are prolate spheroids, and those at constant μ are hyperboloids (Fig. 21.3). The coordinates (λ, μ) can be seen as the roots for τ of

$$\frac{z^2}{\tau^2} + \frac{R^2}{\tau^2 - \beta^2} = 1. \tag{21.36}$$

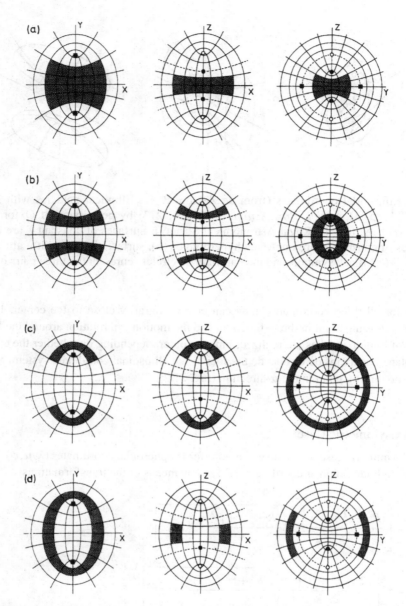

Fig. 21.2. Volumes allowed to the main families of orbits (from de Zeeuw, P. T. 1984. Ph.D. dissertation, University of Leiden, The Netherlands, p. 32), with the pairs of foci marked as in Fig. 21.1. The critical hyperbola and the critical ellipse are given as dashed curves: (*a*) box, (*b*) inner long-axis tube, (*c*) outer long-axis tube, (*d*) short-axis tube.

Under these definitions, the class of potentials defined in terms of two free functions ζ and η, that is,

$$\Phi = \frac{\zeta(\lambda) + \eta(\mu)}{\lambda^2 - \mu^2},\tag{21.37}$$

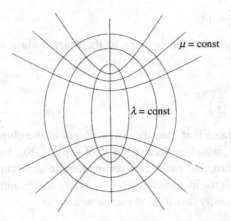

Fig. 21.3. Spheroidal coordinates. The surfaces at constant λ and μ are obtained by rotation around the vertical symmetry axis.

has been shown to admit a third isolating integral of the motion in addition to E and J_z, which can be written as

$$I_3 = \frac{1}{2}[(\mu^2-\beta^2)v_\lambda^2+(\lambda^2-\beta^2)v_\mu^2+(\lambda^2+\mu^2-2\beta^2)v_\phi^2] + \frac{(\mu^2-\beta^2)\zeta(\lambda)+(\lambda^2-\beta^2)\eta(\mu)}{\lambda^2-\mu^2},$$

$$(21.38)$$

where v_λ, v_μ, and v_ϕ are the components of the velocity vector. Here the orbital families are reduced with respect to the triaxial case. In the prolate case, only long-axis tubes are left, whereas in the oblate case, only short-axis tubes remain. The barrier provided by J_z conservation eliminates the box orbits (the ones that would have gone arbitrarily close to the center).

21.3.3 Asymptotic Behavior at Large Radii

Sufficiently far from the center, the ellipsoidal coordinates reduce asymptotically to spherical coordinates (r,θ,ϕ). The expressions for the Stäckel potentials and the related integrals become more intuitive. In particular, the potentials of Eq. (21.33) can be written as

$$\Phi(r,\theta,\phi) \sim \Phi_0(r) + \frac{\eta(\theta)}{r^2} + \frac{\xi(\phi)}{r^2\sin^2\theta}, \qquad (21.39)$$

where Φ_0, η, and ξ are arbitrary functions. The two relevant integrals of the motion (in addition to E) become

$$I_2 \sim \frac{J_z^2}{2} + \frac{\xi(\phi)}{\sin^2\theta}, \qquad (21.40)$$

$$I_3 \sim \frac{J^2}{2} + \eta(\theta). \qquad (21.41)$$

These expressions show directly the desired axisymmetric and spherical limits.

21.3.4 The Perfect Ellipsoid

One interesting triaxial potential that belongs to the Stäckel class is that associated with the density distribution:[16]

$$\rho = \frac{\rho_0}{\left(1 + \frac{x^2}{a^2} + \frac{y^2}{b^2} + \frac{z^2}{c^2}\right)^2},$$ (21.42)

where a, b, and c are constants that identify the geometry of the ellipsoid. One often considers $a > b > c$. The distribution, which generalizes that of Eq. (21.20), has one length scale, which gives the size of a broad core, and two shape parameters, in addition to a mass scale (the total mass is $M = \pi^2 abc\rho_0$). With the identification $\alpha = -a^2$, $\beta = -b^2$, and $\gamma = -c^2$ and the notation of Subsection 21.3.1, the density distribution can be written as

$$\rho = \frac{\rho_0 a^4 b^4 c^4}{(\lambda\mu\nu)^2}.$$ (21.43)

From the theory of classical ellipsoids[17] (see Chapter 10), we find that

$$\Phi = -\frac{(\lambda+\alpha)(\lambda+\gamma)}{(\lambda-\mu)(\lambda-\nu)}\mathcal{G}(\lambda) - \frac{(\mu+\alpha)(\mu+\gamma)}{(\mu-\nu)(\mu-\lambda)}\mathcal{G}(\mu) - \frac{(\nu+\alpha)(\nu+\gamma)}{(\nu-\lambda)(\nu-\mu)}\mathcal{G}(\nu),$$ (21.44)

where the basic function $\mathcal{G}(s)$ is defined by the integral

$$\mathcal{G}(s) = \frac{GM}{\pi}\int_0^\infty \frac{\sqrt{u-\beta}}{\sqrt{(u-\alpha)(u-\gamma)}}\frac{du}{u+s}.$$ (21.45)

This demonstrates that the potential can indeed be separated in the form of Eq. (21.33). It has been shown that the perfect ellipsoid is the only (nonsingular) distribution associated with an Eddington potential in which the density is stratified on similar ellipsoids.[18]

21.4 More General Potentials

Numerical studies of the structure of star orbits in more general potentials have shown that the main classes of regular orbits often reproduce the same orbital families found in the study of Stäckel potentials. In particular, one thorough investigation has considered the orbits in the potential associated with a density distribution of a Hubble profile modified by suitable quadrupoles:[19]

$$\rho(x,y,z) = \frac{1}{(1+r^2)^{3/2}} - g(r)\frac{(2z^2-x^2-y^2)}{2r^2} + h(r)\frac{3(x^2-y^2)}{r^2},$$ (21.46)

where the functions of the spherical radial coordinate $g(r)$ and $h(r)$ are chosen to give the equidensity surfaces fixed axial ratios 1:1.25:2. The majority of the orbits computed (more than 1,000 orbits) turn out to be regular box or tube orbits, that is, to possess three effective integrals of the motion (two nonclassical integrals in addition to E). A suitable superposition of these orbits demonstrates the existence of self-consistent collisionless triaxial models with shape determined by pressure anisotropy.

In a later investigation,[20] the study of the same density distribution was extended to include slow figure rotation (so slow that the main resonances fall outside the model) and some rotational internal streaming. In general, these results would encourage the use of Stäckel potentials as representative of regular orbits in wide classes of potentials. However, potentials associated with sufficiently large central density concentrations have shown the existence of boxlets, which are nonsymmetric orbits with respect to the principal planes that tend to replace the standard box orbits in regions of high central concentration.[21]

The study of orbits in general potentials, much as in the case of spiral galaxies (see the end of Chapter 13), has undergone great progress. This line of work is not pursued any further here.[22]

Notes

1. Franx, M., Illingworth, G. D., de Zeeuw, P. T. 1991. *Astrophys. J.*, **383**, 112. Part IV is partly based on the review by Bertin, G., Stiavelli, M. 1993. *Rep. Prog. Phys.*, **56**, 493.

2. See Ryden, B. S. 1991. *Mon. Not. Roy. Astron. Soc.*, **253**, 743.

3. Lynden-Bell, D. 1962. *Mon. Not. Roy. Astron. Soc.*, **124**, 1 and 95, made a significant move beyond the original formulation of the Jeans theorem (Jeans, J. H. 1915. *Mon. Not. Roy. Astron. Soc.*, **76**, 71) by sharpening the concept of isolating integrals and classifying the relevant potentials. Previously, several authors had investigated important aspects of these issues; e.g., see Eddington, A. S. 1915. *Mon. Not. Roy. Astron. Soc.*, **76**, 37; Chandrasekhar, S. 1939. *Astrophys. J.*, **90**, 1; 1940. *Astrophys. J.*, **92**, 441; Kurth, R. 1955. *Astron. Nach.*, **282**, 241; Kuzmin, G. G. 1956. *Tartu Teated*, **2**; Bernstein, I. B., Greene, J. M., Kruskal, M. D. 1957. *Phys. Rev.*, **108**, 546; Contopoulos, G. 1958. *Stockholm Obs. Ann.*, **20** (5), 1; Contopoulos, G. 1960. *Z. Astrophys.*, **49**, 273.

4. This would bring us to the KAM theorem; see Arnol'd, V. I. 1976. *Les Méthodes Mathématiques de la Mécanique Classique*, Mir, Moscow.

5. For the purpose of constructing spherical models with tangentially biased pressure tensor, a method similar to that described in Section 14.2, which is based on the epicyclic expansion, is given by Bertin, G., Leeuwin, F., et al. 1997. *Astron. Astrophys.*, **321**, 703.

6. See Cipollina, M., Bertin, G. 1994. *Astron. Astrophys.*, **288**, 43.

7. Hénon, M. 1959. *Ann. Astrophys.*, **22**, 126 and 491. This potential has been considered in many subsequent articles. One description can be found in a paper by Gerhard, O. E., Saha, P. 1991. *Mon. Not. Roy. Astron. Soc.*, **251**, 449.

8. The statement is taken from the title of the article by Lynden-Bell, D. 1962. *Mon. Not. Roy. Astron. Soc.*, **124**, 1.

9. This discussion is taken from the article by Woltjer, L. 1967. In Vol. 9 of AMS *Lectures in Applied Mathematics*, ed. J. Ehlers. American Mathematical Society, Providence, RI, p. 1.

10. An important theorem is at the foundation of this point; Noether, E. 1918. See *Collected Papers 1983*, ed. N. Jacobson. Springer-Verlag, Berlin.

11. Lynden-Bell, D. 1962. *Mon. Not. Roy. Astron. Soc.*, **124**, 95.

12. $O(4)$ invariance for the Kepler problem associated with the Lenz vector and $SU(3)$ invariance for the harmonic oscillator.

13. For example, see Contopoulos, G. 1958. *op. cit.*; 1960, *op. cit.*; 1967. In Vol. 9 of AMS *Lectures in Applied Mathematics*, ed. J. Ehlers. American Mathematical Society, Providence, RI, p. 98; more recent attempts relevant to the issues discussed in the next chapter are those by Petrou, M. 1983. *Mon. Not. Roy. Astron. Soc.*, **202**, 1195 and 1209; Lupton, R. H., Gunn, J. F. 1987. *Astron. J.*, **93**, 1106; see also Binney, J. J. 1982. *Mon. Not. Roy. Astron. Soc.*, **201**, 15.

14. Stäckel, P. 1890. *Math. Ann.*, **35**, 91; Eddington, A. S. 1915. *Mon. Not. Roy. Astron. Soc.*, **76**, 37. These potentials have been the subject of a thorough analysis by de Zeeuw, P. T. 1984. Ph.D. dissertation, Leiden University, Leiden, The Netherlands; 1985. *Mon. Not. Roy. Astron. Soc.*, **215**, 731; **216**, 273; **216**, 599; and by Bishop, J. 1986. *Astrophys. J.*, **305**, 14; and Statler, T. S. 1986. Ph.D. dissertation, Princeton University, Princeton, NJ; 1987. *Astrophys. J.*, **321**, 113.

15. See Lynden-Bell, D. 1962. *op. cit.*

16. Kuzmin, G. G. 1956. *Astron. Zh.*, **33**, 27; 1973. Translated in 1987. *IAU Symposium*, Vol. 127, ed. T. de Zeeuw. Reidel, Dordrecht, The Netherlands, p. 553.

17. In particular, see theorem 12 in chapter 3 of Chandrasekhar, S. 1969. *Ellipsoidal Figures of Equilibrium*, Yale University Press, New Haven, CT.

18. de Zeeuw, P. T., Lynden-Bell, D. 1985. *Mon. Not. Roy. Astron. Soc.*, **215**, 713.

19. Schwarzschild, M. 1979. *Astrophys. J.*, **232**, 236.

20. Schwarzschild, M. 1982. *Astrophys. J.*, **263**, 599. Apparently, the known Stäckel potentials cannot be generalized to include rotation. In two dimensions, one case of rotating Stäckel potential has been discovered (see Contopoulos, G., Vandervoort, P. O. 1992. *Astrophys. J.*, **389**, 118), which is the only one known to admit a second integral of the motion quadratic in the momenta, in addition to the Jacobi integral, besides the case of the rotating anisotropic harmonic oscillator (Freeman, K. C. 1966. *Mon. Not. Roy. Astron. Soc.*, **134**, 1 and 15). Unfortunately, its generalization to the three-dimensional case is not feasible.

21. Lees, J. F., Schwarzschild, M. 1992. *Astrophys. J.*, **384**, 491; see also the study of the semiclassical Rydberg states of hydrogen atoms in the presence of an external magnetic field by Delos, J. B., Knudson, S. K., Noid, D. W. 1983. *Phys. Rev. A*, **28**, 7.

22. Among other studies, one interesting technique is that developed by McGill, C., Binney, J. J. 1990. *Mon. Not. Roy. Astron. Soc.*, **244**, 634. As general references, see Boccaletti, D., Pucacco, G. 1996. *Theory of Orbits*, Vol. 1: *Integrable Systems and Nonperturbative Methods*, Springer-Verlag, Berlin; 1999. *Theory of Orbits*, Vol. 2: *Perturbative and Geometrical Methods*, Springer-Verlag, Berlin; Contopoulos, G. 2002. *Order and Chaos in Dynamical Astronomy*, Springer-Verlag, Berlin.

22 Stellar Dynamical Models

The process of modeling elliptical galaxies (or constructing models of interest for elliptical galaxies) can be interpreted in very different ways depending on the goals we have in mind. Some of this variety of approaches was already implicit in Chapter 14 in the context of the modeling of galaxy disks. There much of the emphasis was on the construction of realistic basic states as a prerequisite for an appropriate stability analysis; thus the discussion focused on a number of physical arguments (see especially Section 14.4) aimed at identifying general and flexible classes of models with realistic properties. The main goal was to produce a sound physical basis for a dynamical study of some outstanding morphological aspects of galaxy disks, especially spiral structure. Nevertheless, in that chapter we found it instructive to describe a few different models; for example, in Section 14.1 we took a close look at the vertical equilibrium of the disk, which might have been explored even further if we had in mind a deeper analysis of the problem of dark matter in the solar neighborhood. In addition, we also introduced a few exact models (Section 14.3) that offer useful analytical tools and clarify some of the issues related to the support of equilibrium configurations.

There are a number of important dynamical questions related to elliptical galaxies:

For bright elliptical galaxies, why is there basically one universal luminosity profile (the $R^{1/4}$ law) on the global scale?

What is such universality telling us about the formation and the long-term evolution of these stellar systems?

Can we explain it, that is, can we propose a simple physical interpretation for it?

How can we reconcile the variety of observed kinematic profiles with the existence of a relatively fixed photometric structure?

To what extent can pressure anisotropy be blamed for the variety of observed kinematic profiles?

For low-luminosity elliptical galaxies, why do the luminosity profiles deviate from the $R^{1/4}$ law, and correspondingly, why does rotational support apparently become more important?

For the very low end of the luminosity distribution of ellipsoidal stellar systems (dwarf spheroidals and globular clusters), why do we find so diverse structures among small stellar systems of comparable luminosity?

How can the diversity of structural properties noted between bright ellipticals and low-luminosity ellipticals be reconciled with the apparent continuity in the observed scaling laws, which is sometimes claimed to hold all the way down to the case of very small stellar systems?

Given the fact that spiral galaxies appear to be embedded in massive dark halos, how would a dark halo influence the dynamics of an elliptical galaxy?

Would dark matter just follow the luminous component so that only negligible M/L gradients occur, or should we expect, by analogy with spiral galaxies, that rather diffuse dark halos coexist with more compact stellar distributions?

How can we measure the amount and distribution of dark matter around elliptical galaxies?

Is it true that some ellipticals (such as NGC 3379) and classes of small stellar systems (such as globular clusters) can be thought of as practically devoid of dark matter, whereas other galaxies and classes of small stellar systems (such as dwarf spheroidals) of comparable luminosity are dark-matter-rich?

Given the range of observed shapes, can we infer the relevant distribution of intrinsic shapes among galaxies?

How are the various shapes dynamically supported?

Are the shapes dominated by pressure anisotropy or produced with significant contributions from internal streaming and figure rotation?

In other words, how frequent and how significant is the presence of pressure anisotropy?

Can we obtain clues on how the shapes of ellipticals may be generated in terms of plausible formation processes?

On the smaller scale, can we obtain a satisfactory description of the core and nuclear regions of ellipticals?

What can the resulting models tell us about the possible ubiquity of massive black holes at the center of elliptical galaxies?

Many of these issues can be addressed by constructing and studying stellar dynamical equilibrium configurations. Clearly, there are also important issues, such as the existence and role of scaling laws, and, in general, questions related to evolution of the elliptical galaxies on the Hubble time scale, that most likely escape a pure dynamical discussion because they depend on more complex physical mechanisms, the detailed star-formation processes, and stellar evolution. In this respect, some points of interest will be considered in Part V.

In this chapter the properties of stellar dynamical models are described that may help us to answer some of the questions just raised. We will make use of the Jeans theorem, considered a useful working tool. We will try to see whether, based on guidelines dictated by physical arguments, the theorem can lead us to models that match and clarify the most significant observed properties of elliptical galaxies. If the process works, we will consider it as an indication of the plausibility of the physical arguments made. If the process fails, we should try to revise our physical interpretation, and we might even question whether the Jeans theorem, in the simple form that is usually adopted, should be applied at all to the physical systems under investigation. In particular, some difficulties that are encountered in the construction of realistic triaxial dynamical models may be telling us that fully triaxial stellar systems cannot be approximated by exactly integrable models. In practice, for reasons illustrated later, we will emphasize a modeling process in which we start from physically justified distribution functions.

Within the same basic framework of stellar dynamical tools (i.e., orbits, integrals of the motion, and the Jeans theorem), very different alternative approaches can be taken (see Section 22.1). Apart from some technical aspects that are involved, two major lines of thought characterize these efforts. On one side, some studies, in line with those of many other branches of physics, try to be predictive and thus focus on the consequences of physical assumptions in relation to the observations. In a semiempirical approach, this process receives continual inspiration from the data. These efforts try to produce a theory, which may be disproved by existing data or by new observations. In a second line of thinking, which is essentially descriptive, other studies aim at extracting from the data as much as possible before making physical assumptions, with the hope that the physical aspects can be discussed later when all the structural parameters are secured directly from the data. Almost by definition, the descriptive stage is unable to provide explanations. Probably at the root of this other point of view, in contrast with the semiempirical approach, is the hope of deriving the evolution of stellar systems starting from directly measured initial conditions. It has often been emphasized that such an inherently deductive approach is bound to suffer from severe limitations. In practice, because the data points sample only limited spatial regions with finite accuracy, projected along the line of sight, the descriptive inversion process also must rely on a set of assumptions (in the best cases, on the geometry and the mass-to-light ratio profile), and it is known to be inherently unstable and thus leads easily to nonphysical solutions.

22.1   Four Approaches

It may be convenient to distinguish four different ways of working on the construction of stellar dynamical models. It should be noted that in some cases the differences are more in the perception than in the actual methods, and a given model may be reached from different directions.

22.1.1   Search for Interesting Exact Models

One simple way to proceed is to collect all possible self-consistent models. Usually here the emphasis is on the exactness or the analytical simplicity of the models.[1] By producing a larger and larger set of models potentially applicable to galaxies, we hope to find some that are able to reasonably match the observed profiles (see Chapter 4), others that are best suited for the numerical simulation of a specific dynamical mechanism (e.g., see Chapter 23), or models that are convenient to test or demonstrate the occurrence of special features, such as the structure of the line-of-sight velocity profiles (see Subsection 22.1.3).

This blind search operates between two extremes. In one limit, spherical symmetry, with all the desired classical integrals of the motion readily available, appears to be a degenerate case. There are an infinite number of ways of specifying a distribution function and constructing the related self-consistent model. Even here we may note that full self-consistency puts strong limitations on the set of solutions allowed by the Jeans theorem alone. In particular, much as for the classification of quantum states, the Jeans theorem would, in principle, allow any $f = f(E, J^2, J_z)$. In practice, if we require that the spherical potential Φ in E be generated by the distribution function itself, then we have to drop the use of J_z; that is, we should work with $f = f(E, J^2)$. (In principle, an odd dependence on J_z would not break the symmetry of the mass distribution, but the resulting internal streaming would correspond to the existence of a preferred axis, with no natural counterpart in the assumed symmetry of the problem.) Still, within this framework we may construct models with a wide variety of density and pressure anisotropy profiles, and in many different ways we may match the main observed characteristics. Some notable models will be mentioned later in the context of distribution-function priority models.

At the other extreme, the general triaxial case seems to be an almost impossible task, at least from the analytical point of view. If we anticipate the possibility of three good integrals of the motion (to allow for a fully anisotropic pressure tensor), almost any combination $f = f(E, I_2, I_3)$ is going to lead to a density distribution that is incompatible with the symmetry required for the existence of I_2 and I_3. Stäckel potentials have three isolating integrals of the motion (see Chapter 21) and offer a very wide choice of options (specified in terms of three free functions). They thus may be considered the natural environment for the construction of self-consistent tri-axial models.[2] Unfortunately, a discussion of Stäckel-type density-potential pairs has shown the undesired feature that the associated density profile does not have sufficiently concentrated central regions (with the empirically suggested r^{-2} density behavior). The application of a theorem[3] that guarantees the construction of a Stäckel potential with a desired set of axial ratios starting from a specified density profile along the z axis has shown that realistic density profiles can be obtained only at the cost of introducing a relatively massive disk.[4] The issue of construction of triaxial models with central cusps has been the focus of much work.[5] We should also recall that Stäckel potentials have been proved to be free of isophotal twisting,[6] so they are unable to reproduce the strongest empirical argument in favor of triaxiality.

More general triaxial models can be constructed by numerical superposition of stellar orbits in a density-priority approach in line with the studies mentioned in Section 21.4.

22.1.2 Density-Priority Models

In the construction of models giving priority to the density distribution, we specify in advance the potential-density pair (which may be fully or partially self-consistent) and then try to invert the integral relation

$$\rho = \int f d^3 v. \tag{22.1}$$

One motivation at the basis of this point of view is very simple. The quantity that is best constrained by the observations under the assumption of a constant mass-to-light ratio is the projected density profile. If we add an assumption about the three-dimensional (3D) geometry of the system, we thus have an empirical determination of the volume-density profile that appears on the left-hand side of Eq. (22.1) (further discussion of this point in connection with the empirical $R^{1/4}$ law is postponed to Subsection 22.4.4). It is thus natural to ask for various geometric config-urations whether or not a (positive-definite) function of the integrals of the motion can be found that is compatible with the imposed density profile suggested by the observations. Furthermore, we may ask whether or under what conditions such a solution is unique.

In the case of spherical symmetry, various options can be taken. Because the density profile depends on only the radial coordinate, the possibility of a unique inversion prescription is natu-rally expected when the distribution function (or the part of the function that is left free to vary) depends on one integral only. For example, if we take $f = f(E)$, the solution is provided by an Abel inversion[7] [see Eq. (14.15)]:

$$f(E) = \frac{1}{2\pi^2 \sqrt{2}} \left[\int_E^0 \frac{d^2\rho}{d\Phi^2} \frac{d\Phi}{\sqrt{\Phi - E}} - \frac{1}{\sqrt{-E}} \left(\frac{d\rho}{d\Phi} \right)_{\Phi=0} \right]. \tag{22.2}$$

In this inversion formula, the density profile is assumed to be monotonic, so the radial coor-dinate can be eliminated to give $\rho = \rho(\Phi)$. The function $f(E)$ resulting from the inversion should be checked to ensure that it is positive-definite. The model thus produced is characterized by isotropic pressure. Another possibility is to consider a dependence of f on a combination

$Q = Q(E,J^2)$ specified in advance, thus producing anisotropic models according to some desired criteria. In this respect, the case where Q is a linear combination of E and J^2 has often been used, especially to produce models with radially biased anisotropy profiles,[8] with some difficulties appearing when used for the construction of tangentially biased anisotropic models.[9] Clearly, many other options can be considered and lead to several interesting models.[10] Some separations with a rather intuitive basis can be adopted, focusing on a circularity function[11] [which measures the distance from the relation $J_{\rm circ}(E)$, given for a few potentials in Chapter 21] or even on the epicyclic energy for the case in which our goals are tangentially biased models.[12]

For the case of axial symmetry, a very elegant generalization of the Abel inversion has been devised[13] that can yield $f(E,J_z^2)$ for a given analytical axisymmetric potential; the method is based on a suitable contour integral in the complex plane. The most general triaxial case appears to require the use of direct numerical superposition of stellar orbits, as briefly indicated at the end of Chapter 21.[14]

22.1.3   Deprojecting the Data

It was mentioned earlier that one motivation of the density-priority approach is to impose the quantity that is best constrained by the observations. This viewpoint may be generalized in procedures in which all the available observational constraints are imposed. In other words, models can be constructed basically by deprojection of the available photometric and spectroscopic profiles. In this respect, one aspect that has recently attracted much interest is the information contained in the line-of-sight velocity profiles;[15] this is reminiscent of analogous interest in the 21-cm line profiles in radio observations of spiral galaxies.[16]

To define the deprojection process, we thus may work with a mathematically convenient basis of distribution functions and try to determine empirically the combination that leads to a best fit for all the available data. A simple class of these techniques involves the use of distributions that are polynomials[17] in E and J.

We may even bypass the explicit use of distribution functions and refer to numerical combinations of stellar orbits, with the additional guidance of cost functions or entropy functions.[18] An even more abstract, nonparametric method has been proposed that has the important feature of ensuring that the derived distributions in phase space are smooth.[19] In practice, all these methods based on the superposition of stellar orbits can be traced back to the general concepts described in Section 21.4; they are generally referred to as *Schwarzschild methods* and have found wide applications in the study of stellar systems. The main goal of these descriptive methods is to identify a positive-definite distribution function that is able to give a reasonable fit to the available data. On the one hand, the methods often suffer from the imperfect and incomplete data available for a given stellar system. On the other hand, it has been pointed out that, curiously, these orbit-based methods would actually fail even in the opposite limit, that of the presence of perfect data.[20]

An alternative method resorts to the use of the moment fluid equations (see Chapter 8). The modeling can be carried through relatively easily in the axisymmetric case.[21] The moment equations can be simplified considerably if $f = f(E,J_z)$ (without assuming any specific form for the underlying f) and become (with cylindrical coordinates R,z)

$$\frac{1}{\rho}\frac{\partial}{\partial z}\left(\rho \langle v_R^2 \rangle\right) + \frac{\partial \Phi}{\partial z} = 0, \tag{22.3}$$

$$\frac{1}{\rho}\frac{\partial}{\partial R}\left(\rho\langle v_R^2\rangle\right)+\frac{\langle v_R^2\rangle-\langle v_\phi^2\rangle}{R}+\frac{\partial\Phi}{\partial R}=0. \tag{22.4}$$

We can close the equations by making some assumptions about the pressure anisotropy profile (often taken to be constant) and on the part of $\langle v_\phi^2\rangle$ that is associated with systematic streaming motions. The photometric profiles taken at various position angles are fitted to provide information on the 3D structure of the galaxy, and the density-potential pair thus inserted into the fluid equations gives the resulting kinematic profiles that are then compared with the spectroscopic data. In practical cases, the method may show whether reasonable fits can be provided with constant mass-to-light ratios and whether significant pressure anisotropies are involved. It is basically by means of these tools that it has been established[22] that most ellipticals are probably not too far from oblate systems with $f=f(E,J_z)$. It is emphasized that there is no guarantee that the solutions thus found admit a positive-definite supporting distribution function. In addition, even when the solution selected by the data exists, it may require a rather artificial distribution of stellar orbits. In fact, we have already noted that descriptive methods of this kind tend to postpone the discussion of their physical justification.

As noted earlier in this chapter, we should be aware that the data points sample only limited spatial regions with good accuracy and that the inversion process is generally unstable, so we may often be led to unrealistic or physically implausible distributions of stellar orbits.

22.1.4 Distribution-Function Priority Models

From the physical point of view, the methods mentioned so far remain unsatisfactory. For a collisionless stellar system, all the physics is contained in the distribution function in phase space, and most methods mentioned earlier leave the resulting distribution function out of control and without critical examination. To some extent it is as if, in making models of stellar structure, we first assigned the density profile inside the star, and then we asked whether there is an equation of state able to justify the imposed density distribution, limiting the physical discussion to a study of uniqueness or to checking that the pressure profile is positive-definite.

The fourth approach (i.e., the line of work better developed in this chapter) tries to overcome this criticism by attacking the problem at its roots. The idea is thus to proceed through a physically based choice of distribution function. This approach is predictive in the sense that it tries to look for implications, in terms of both density profiles and kinematic profiles, that should follow as a result of some physical scenarios. Current or future data may disprove the arguments put forward in this modeling process.

The main difficulty of this line of thinking is that the statistical mechanics of self-gravitating systems is poorly known, and there are very few empirical clues that we may take advantage of. Thus, so far the most successful solutions are found for cases in which the system is supposed to have relaxed significantly (see Section 22.3) or for partially relaxed models of ellipticals in which we exploit important clues offered by numerical simulations of collisionless collapse (see Section 22.4). Note that in general the physical assumptions that offer guidance in the choice of the relevant distribution function are unable to determine its form uniquely. This is also true for the choice of models described by Eq. (22.24), which represents only one available option within the physical picture that leads to isotropic quasi-Maxwellian truncated models.

The truncated Maxwellian distribution functions described in Section 22.3 have found wide applications in describing the structure and evolution of globular clusters, for many of which indeed relaxation processes occur on a relatively fast time scale. Even in their simplest form, that is, in the spherical nonrotating case, the models define an interesting (one-parameter) family of equilibrium configurations. To give a stronger physical justification for this and other models invoked to describe weakly collisional systems, the distribution functions have been compared directly with self-similar solutions of the Fokker-Planck equation.[23] Although the application of these models to elliptical galaxies is less justified, it is sometimes considered.

For a description of collisionless stellar systems, such as elliptical galaxies, it would be important to produce, at least when the system is not too far from spherical symmetry, a model or classes of models under physical justification with the capability of realistic luminosity and kinematic profiles. Because we operate between the two extremes (spherical degeneracy and triaxial impossibility) mentioned in Subsection 22.1.1, the case of quasi-spherical symmetry is simple enough to lead to some workable solutions and yet can teach us something beyond the spherical degenerate case. The models described in Section 22.4 go in the desired direction, but so far they have not been extended to cover cases for which significant departures from spherical symmetry are involved.

Interesting attempts at constructing nonspherical models have been carried out, some focusing on the algorithm for numerical solution of the Poisson equation[24] and others on the more interesting aspect of including approximate integrals of the motion.[25] These attempts, giving priority to the distribution function, follow the most natural procedure of a physical approach, as is generally done in plasma physics.[26] In Subsections 22.3.2 through 22.3.4, some recent developments will be described in the construction of nonspherical models for quasi-relaxed stellar systems that are applicable to globular clusters and hopefully of interest for some low-luminosity elliptical galaxies.

22.2 Local Description of Relaxed Systems

We now consider the spherical analogue of the isothermal slab solution described in Section 14.1 (in the context of the vertical equilibrium of a disk). That calculation was especially instructive as a very simple case of a stellar dynamical study with distribution-function priority. We recall that the isothermal slab implied a constant gravity field at large distances from the plane [see Eq. (14.9)], so as a model of a galaxy disk its relevance was mainly for a self-consistent description of the vicinity of the equatorial plane.

Similarly, besides its general interest as indicative of the divergences intrinsic to the hypothesis of a fully relaxed stellar system, the spherical solution described in the following subsections provides a useful representation of the local behavior of many stellar systems in their central regions. It can clarify some of the properties of the more physical King models (which have been applied to the interpretation of global luminosity profiles of globular clusters; see Section 22.3). In addition, the isothermal solution also turns out to describe much of the structure inside the half-mass radius of the partially relaxed, finite-mass models that will be found to incorporate the $R^{1/4}$ profile (described in Section 22.4). Finally, as we outlined in Section 14.5, the isothermal model, flattened by the presence of an embedded disk, may provide a physical picture of the distribution of dark matter in spiral galaxies, at least out to the largest radius for which rotation curves can be measured.

22.2.1 The Isothermal Sphere

If we take the distribution function

$$f = A \exp(-aE),$$ (22.5)

with no restrictions on the star velocities, we obtain an isothermal equation of state given by

$$p = \rho/a, \tag{22.6}$$

where

$$\rho = \hat{A}\exp(\psi) = \hat{A}\hat{\rho}(\psi). \tag{22.7}$$

Here we have introduced the dimensionless potential $\psi = -a\Phi(r)$. Note that the two constants A and a are dimensional parameters that may be used to match some physical scales of the system under investigation (e.g., the core radius and the central velocity dispersion in the core of a galaxy). In contrast with the models described in Section 22.3, there are no free dimensionless parameters available here. (These relations are analogous to those of Subsection 14.1.1 and should be compared with the corresponding equations of Subsection 22.3.1.) The fully self-consistent problem requires the solution of the Poisson equation

$$\frac{1}{\zeta^2}\frac{d}{d\zeta}\left(\zeta^2\frac{d\psi}{d\zeta}\right) = -\exp(\psi). \tag{22.8}$$

The scale length λ is defined by means of the relation [see Eq. (22.31)]

$$\lambda^2 = \frac{1}{4\pi G\hat{A}a}. \tag{22.9}$$

If we look for a solution with finite central density, the natural boundary conditions are $\psi(0) = \Psi$ and $\psi'(0) = 0$. Because no free dimensionless parameters are available, here we can choose the value of Ψ in such a way that $\lambda = r_0$ [see Eq. (22.33)]. As is well known in studies of polytropic stars, the isothermal solution beyond the scale r_0 becomes asymptotically close to $\rho \sim r^{-2}$, thus leading to an integrated mass that increases linearly with radius and making it inapplicable for describing the global profiles of stellar systems.

In the central regions [basically at radii $r = O(r_0)$], the density profile that is obtained numerically, when projected along the line of sight, is found to match the general behavior of galaxy cores. The projected density (here R denotes the projected radius) is very close to the so-called modified Hubble profile

$$\Sigma = \frac{\Sigma_0}{1 + R^2/r_0^2}, \tag{22.10}$$

which can be traced to a volume-density profile[27] of the form

$$\rho = \frac{\rho_0}{(1 + r^2/r_0^2)^{3/2}}. \tag{22.11}$$

Obviously, the latter expressions find frequent applications because of their analytical simplicity.

One special solution with a singularity at the center is obtained when the condition of finite central density is dropped. This singular isothermal sphere corresponds to $\psi = -2\ln(\zeta/\sqrt{2})$, which solves Eq. (22.8) exactly. Here the model loses the characterization in terms of the central density and becomes self-similar. The constants A and a are related to the constants b and V_0 of the logarithmic spherical mass model of Subsection 21.1.3 by means of the relations $a = 2/V_0^2$ and $b = \sqrt{2}\lambda$, with λ given by Eq. (22.9). The singular isothermal sphere presents close analogies to the isothermal self-similar disk described in Subsection 14.3.2 in relation to the logarithmic potential that is found and to the type of divergences occurring at small and large radii.[28] For the associated orbit structure, we may thus refer to Subsection 21.1.3.

Historically, study of the isothermal sphere has its roots in the theory of stellar structure in relation to the Lane-Emden equation for polytropic star models.[29] We recall that the isotropic distribution function $f = A(-E)^s$ for $E < 0$ (and $f = 0$ otherwise), with $\Phi = 0$ defining the surface of the associated stellar system, generates a density distribution $\rho(r) = B[-\Phi(r)]^{s+3/2}$. Here A and B are constants. The Poisson equation thus becomes the same as the equation for a polytropic sphere[30] with index $n = s + 3/2$. For the latter equation, analytical solutions are known for $n = 0, 1, 5$. The isothermal limit of the polytropic spheres $(n \to \infty)$ is nontrivial.[31] The case with $n = 5$ (with infinite radius) is often considered for its simple analytical structure.[32] In particular, the associated Plummer density–potential pair is given by

$$\rho(r) = \frac{3Mb^2}{4\pi} \frac{1}{(b^2 + r^2)^{5/2}}, \tag{22.12}$$

$$\Phi(r) = -\frac{GM}{(b^2 + r^2)^{1/2}}. \tag{22.13}$$

The integrated mass is given by

$$M(r) = M \frac{r^3}{(b^2 + r^2)^{3/2}}, \tag{22.14}$$

so the half-mass radius occurs at $r_M \approx 1.3\, b$. The mean-square velocity is exactly $\langle v^2 \rangle = -\Phi(r)/2$, and the projected density is given by

$$\Sigma(R) = \frac{M}{\pi} \frac{b^2}{(b^2 + R^2)^2}. \tag{22.15}$$

22.2.2 Galaxy Cores and Anisotropic Models Obtained by Means of Adiabatic Growth

We noted at the beginning of this section that different emphasis can be placed on a global or a local application of stellar dynamical models. On a small scale $(r \ll r_M)$, many models, such as the King models or the ones that will be introduced in Section 22.4, are characterized by a quasi-isothermal core in the sense that they behave approximately like the (regular) isothermal sphere. However, nuclei of galaxies are believed to frequently host massive pointlike density concentrations (possibly black holes). It is thus of great interest to see whether adequate self-consistent models can be constructed for which the field is due to the sum of the contributions of the stellar system and a central point mass. Clearly, a comparison of the properties of these models with high-resolution photometric and spectroscopic observations should eventually disclose which galaxies host a central massive black hole and which do not. Major projects have been undertaken that focus on these issues.[33] Either by means of stellar dynamical studies alone (in particular, for disk galaxies such as M31, NGC 3115, NGC 4594, our Galaxy, and ellipticals[34] such as M32 and NGC 3377) or with the help of kinematic data from gas orbiting close to the central nucleus (see the spectacular cases of the giant elliptical M87 and the spiral galaxy NGC 4258 = M106; Fig. 22.1), convincing evidence has been accumulating for central black holes with masses in the range 10^6 to $10^{10} M_\odot$.

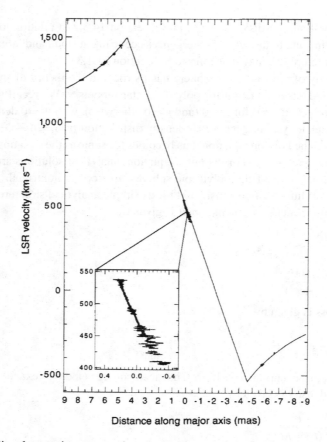

Fig. 22.1. A radio observation suggesting the presence of a massive black hole at the center of the galaxy M106 (from Miyoshi, M., Moran, J., et al. 1995. *Nature (London)*, **373**, 127; © 1995, Rights Managed by Nature Publishing Group). The plot gives the line-of-sight velocity as a function of radial distance (in milliarcseconds) from the center along the major axis of the observed molecular disk, with data superimposed on a model curve described in the cited paper. At the distance of the object (estimated to be 6.4 Mpc), 8 mas define a length of 0.25 pc.

Kinematic data are probably crucial to decide whether a central point mass exists. Still it has been noted that in the innermost regions (where the $R^{1/4}$-like profile is usually smeared out by seeing in observations from the ground[35]), in some cases the luminosity profile, when studied at sufficiently high spatial resolution, does flatten out into a core structure, whereas in others the profile increases basically as a power law,[36] all the way insofar as it is possible to observe. With the aid of the *Hubble Space Telescope*, for a sample of dozens of ellipticals (and bulges), this behavior has been traced down to a scale of 1 pc from the center. Curiously, some of the parameters that characterize the galactic nuclear regions are found to correlate strongly with more global parameters.

In this short subsection we do not go further into this fascinating subject; instead, we restrict our attention to some modeling issues that are found to be interesting because of their simple physical motivation. In contrast to the case without a central point mass, for which a core structure is naturally expected, in the presence of a $1/r$ singularity in the potential, the stellar system should display a central cusp in its density distribution. In a study of this type, one important parameter is clearly the ratio $M_\bullet/M_c$ of the mass of the black hole to the mass of the core of the large-scale stellar density distribution.

Originally, much of the attention was drawn to the case of a black hole inside a collisional environment, such as that at the center of a globular cluster. Steady-state solutions of the Fokker-Planck equation were sought and constructed, and these are applicable to time scales longer than the relevant two-body relaxation time.[37] In the regime of small masses of the central black hole, under the assumption that binary stars are unimportant, the cusp that is found is approximately given by $\rho \sim r^{-7/4}$. In the following years, motivated by some of the observational studies mentioned earlier, the focus of attention changed to the case of the more collisionless galaxy cores. A relatively recent revival of interest in collisional steady-state solutions[38] is motivated by the fact that we may imagine even the center of a galaxy to produce a collisional cusp by means of merging of small star clumps.

The collisionless case suggests the likely presence of significant anisotropies very close to the central point mass. There radial orbits would feel selectively the influence of the central singularity. We can make a first simple study, by focusing on the region outside such a black-hole-dominated environment, that can be obtained in terms of loaded polytropes.[39] The basic idea is to determine how the Lane-Emden equation is modified in such an outer region, thus reducing the discussion to that of a modified (isotropic) hydrostatic equilibrium with an assumed equation of state. One interesting finding of this much-simplified description has been to show that a cusp structure forms with properties that depend little on the polytropic index.

A more elegant construction of self-consistent stellar dynamical models can be made under the assumption that the central black hole grows slowly (on a time scale much longer than the relevant dynamical time scale) inside a preexisting collisionless stellar system (thus over a time shorter than the relaxation time). In a spherical adiabatic growth,[40] the orbits associated with the distribution function are modified but conserve their angular momentum and their radial action. The process thus shows in a very intuitive way how anisotropy can be generated out of an initially isotropic system by growth of the central mass.

Without going into detail, the structure of the equations that describe spherical adiabatic evolution is given here. Suppose that we separate the potential into two parts so that the energy integral is given by

$$E = \frac{v^2}{2} + \Phi^0(r) + \Phi^0_{\text{ext}}(r). \tag{22.16}$$

Here Φ^0_{ext} is an *external* potential, and Φ^0 is the self-consistent potential generated by the density associated with the distribution function $f_0(E, J)$ that describes the stellar system. We recall that the radial action J_r is defined as

$$J_r = \oint v_r(r, E, J)\, dr; \tag{22.17}$$

the function v_r and the integral between the two turning points depend on the sum of the external and self-consistent potentials.

Now we can assume that the external potential is varied slowly by changing one parameter λ from its initial value λ_0 so that it becomes $\Phi_{\text{ext}}(r)$. This change will induce a change in the self-consistent potential from $\Phi^0(r)$ to $\Phi(r)$. In this slow process, the radial action and the angular momentum are conserved, but the energy is not. Assume that we start from an orbit characterized by $(E^{(0)}, J)$ in the potential $\Phi^0(r) + \Phi^0_{\text{ext}}(r)$ and arrive at an orbit characterized by (E, J) in the potential $\Phi(r) + \Phi_{\text{ext}}(r)$ with the condition

$$J_r(E, J; \Phi + \Phi_{\text{ext}}) = J_r(E^{(0)}, J; \Phi^0 + \Phi^0_{\text{ext}}). \tag{22.18}$$

This relation should be interpreted as a mapping between the initial value of the energy $E^{(0)}$ and the final value of the energy E required for the radial action to remain constant. In other words, this implicitly gives a relation $E^{(0)} = E^{(0)}(E, J)$. If we start from a distribution of stellar orbits f_0, the final distribution function will thus be given by

$$f(E, J) = f_0(E^{(0)}, J), \tag{22.19}$$

where $E^{(0)}$ is determined from Eq. (22.18). Self-consistency requires that

$$\Delta\Phi = 4\pi G \int f \, d^3v. \tag{22.20}$$

These nonlinear equations can be solved numerically (e.g., by iteration procedures) to derive the effects of adiabatic evolution. In particular, we consider the situation in which $\Phi^0_{\text{ext}} = 0$ and $f_0 = f_0(E)$ describes a nonsingular isothermal sphere [see Eq. (22.5)] and ask what will be the final distribution function $f(E, J)$ for $\Phi_{\text{ext}} = -GM_\bullet/r$. The influence radius of the black hole is therefore given by $r_\bullet = aGM_\bullet$, and the mass of the core can be defined from $r_0 = aGM_c$. The density cusp is found to be characterized by $\rho \sim r^{-3/2}$ and associated with a kinematic cusp $\langle v^2 \rangle \sim GM_\bullet/r$. For a small black hole, the density cusp emerges from the core region, whereas for large black holes, the transition to the outer $\sim r^{-2}$ isothermal behavior occurs by means of an intermediate $\sim r^{-5/2}$ profile, with no distinct core left (Fig. 22.2). The cusp is quasi-isotropic, with a small tangential bias at intermediate radii. These conclusions are changed if we start from initial functions significantly different from that of the nonsingular isothermal sphere. The extension to the case in which some rotation and flattening are present is not trivial.

Part of the structure of the solutions obtained by adiabatic growth can be clarified analytically. For this purpose, we refer to regions of phase space where the actions in key mapping relation (22.18) can be approximated analytically (see Chapter 21 for the Keplerian, logarithmic, and harmonic potentials). Another limit within which we can proceed semianalytically is the linear regime, in which the departures $f_1 = f - f_0$ and $\Phi_1 = \Phi - \Phi^0$ from the initial configuration are small because the change in the external potential $\delta\Phi_{\text{ext}}$ is small. The linearized equations for adiabatic changes are[41]

$$\Delta\Phi_1 = 4\pi G \int f_1 \, d^3v, \tag{22.21}$$

$$f_1 = \left(\frac{\partial f_0}{\partial E}\right)_J \left(\Phi_1 + \delta\Phi_{\text{ext}} - \langle\Phi_1 + \delta\Phi_{\text{ext}}\rangle\right), \tag{22.22}$$

where E is given by Eq. (22.16), and the angle brackets denote a bounce-orbit average over the unperturbed orbits:

$$\langle h \rangle = \frac{1}{\tau_r} \oint h(r) \frac{dr}{v_r}. \tag{22.23}$$

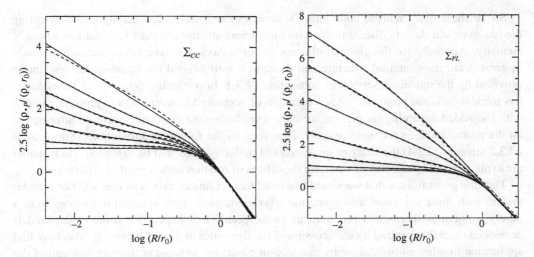

Fig. 22.2. Projected density cusps for models with a central black hole grown adiabatically onto an initial isothermal sphere are shown as solid curves [the central black hole mass values are 0, 0.03, 0.1, 0.2, 0.3, 0.5, 1 (*left*) and 0, 0.03, 0.1, 0.3, 1, 3, 10 (*right*) in units of the core mass M_c]. These dynamical model curves are shown here fitted by analytically simple formulas (*dashed curves*) introduced by Crane, P., Stiavelli, M., et al. 1993. *Astron. J.*, **106**, 1371, to fit the central photometric profiles of elliptical galaxies obtained from *Hubble Space Telescope* observations (from Cipollina, M., Bertin, G. 1994. *Astron. Astrophys.*, **288**, 43).

Here τ_r is the bounce time between the radial turning points [see Eq. (13.15)]. These equations are a somewhat subtle zero-frequency limit of the linear stability equations (see Chapter 23); they are also well known within the plasma community.[42]

It is not clear how these tools can give a realistic representation of galactic nuclei (see also Subsection 25.2.4). They have been shown here mainly because they define an interesting and physically guided device to construct self-consistent equilibrium distribution functions with a central point mass and because they represent an instructive case. A scenario of this type and similar techniques may turn out to be useful to describe other situations of interest in stellar dynamics.

22.3 Global Description of Quasi-Relaxed Stellar Systems

For systems such as globular clusters, it is natural to consider the role of a truncation introduced by the tidal field of the host galaxy. Tides obviously impose a nonspherical external field, but given the observed shapes of globular clusters, we may start from spherically symmetric models. Then it is also natural to look for a description in terms of a distribution function that depends on only the energy $E = v^2/2 + \Phi(r)$; for many of these small stellar systems, the relevant relaxation times are relatively short, so it is plausible that during their lifetime they have moved significantly in the direction of a quasi-Maxwellian distribution function. This is the basic physical motivation for solutions of the type described in Subsection 22.3.1. It should be emphasized that the main success of these truncated models is that they can indeed provide a physically based description of the global structure of an important class of stellar systems.

So far these quasi-relaxed nonrotating spherical models have been mainly tested by fitting the observed star density distributions, with insufficient attention to their kinematic structure.[43] Actually, especially for the globular clusters of our Galaxy, we have the opportunity to study in great detail their internal structure and dynamics, well beyond the zeroth-order description provided by the models described in Subsection 22.3.1. In particular, the study of their shapes has made significant progress.[44] The collection of accurate kinematic data is growing rapidly, with the added capability, at present for only few clusters, of collecting accurate information on the proper motions for many stars.[45] Therefore, in the following subsections (Subsections 22.3.2 through 22.3.4) the study of quasi-relaxed stellar systems will be extended[46] to consider the explicit triaxial geometry of tides and the effects of rotation (either rigid or differential).

Thus the general issue that we would like to address in this section is as follows: For a stellar system with finite mass and active internal relaxation, what types of small departures from a purely isothermal distribution function can be expected? It is hoped that some of the models developed recently, inspired by a discussion of the dynamics of globular clusters, also may find application to other ellipsoidal stellar systems, in particular, to those at the very low end of the luminosity distribution.

22.3.1 Spherical Isotropic (King) Models

Consider a spherical distribution of stars $f = f(E)$ limited to a sphere of radius r_t. The collection of stellar orbits described by f is thus subject to the condition that f vanishes for $E > \Phi(r_t)$. Suppose that the distribution function has the form

$$f_K = A\{\exp(-aE) - \exp[-a\Phi(r_t)]\}, \tag{22.24}$$

for $E < \Phi(r_t)$ (and $f_K = 0$ otherwise). Here A, a, and r_t are free constants, defining two scales and one dimensionless parameter. The function f_K is continuous at the truncation energy. Note that the truncation is suggested by physical arguments related to the tidal interaction with the host galaxy, but mathematically, the truncation radius r_t may be smaller than the relevant tidal radius r_T; in this case, the model is said to underfill the tidal volume.

We introduce the dimensionless escape energy ψ, defined as

$$\psi = a[\Phi(r_t) - \Phi(r)]. \tag{22.25}$$

Thus the condition $E < \Phi(r_t)$ can be written as $av^2/2 < \psi$. The density profile (for $r < r_t$, i.e., $\psi > 0$) associated with the distribution function is given by

$$\rho(r) = 4\pi A \int_0^{\sqrt{2\psi/a}} v^2\, dv\{\exp(-av^2/2)\exp[-a\Phi(r)] - \exp[-a\Phi(r_t)]\}. \tag{22.26}$$

By a simple change of integration variables, the density can be written as

$$\rho = \hat{A}\frac{3}{2}\int_0^\psi \sqrt{x}\, dx[\exp(-x)\exp(\psi) - 1] = \hat{A}\hat{\rho}(\psi), \tag{22.27}$$

where

$$\hat{\rho}(\psi) = \exp(\psi)\int_0^\psi x^{3/2}\exp(-x)\, dx = \exp(\psi)\,\gamma\left(\frac{5}{2}, \psi\right), \tag{22.28}$$

where γ is the incomplete gamma function.[47] In this isotropic model, the density depends on the radial coordinate only implicitly through the potential $\Phi(r)$. For large values of ψ, we have $\hat{\rho} \sim (3\sqrt{\pi}/4)\exp(\psi)$; that is, it behaves like the density of nontruncated isothermal models (see Subsection 22.2.1).

In passing, we note that the one-dimensional velocity dispersion can be written as

$$\langle v_r^2 \rangle = \langle v_\phi^2 \rangle = \langle v_\theta^2 \rangle = \frac{2}{5a} \frac{\gamma\,(7/2,\psi)}{\gamma\,(5/2,\psi)}. \tag{22.29}$$

By a suitable rescaling of the radial coordinate $r \to \zeta = r/\lambda$, the self-consistency relation implied by the Poisson equation thus can be written as

$$\frac{1}{\zeta^2} \frac{d}{d\zeta} \left(\zeta^2 \frac{d\psi}{d\zeta} \right) = -\hat{\rho}(\psi), \tag{22.30}$$

which is to be integrated subject to the boundary conditions $\psi(0) = \Psi > 0$ and $\psi'(0) = 0$. Thus the scale length λ can be defined in terms of the central density ρ_0 and the dimensionless depth of the central potential well Ψ by means of the relation

$$\lambda^2 = \frac{\hat{\rho}(\Psi)}{4\pi G \rho_0 a}. \tag{22.31}$$

Close to the center, the potential well can be approximated by a parabola with

$$\psi \sim \Psi - \frac{1}{6}\hat{\rho}(\Psi)\zeta^2. \tag{22.32}$$

Integrating Eq. (22.30) determines the radius r_t as the location where ψ vanishes. Thus a one-to-one correspondence is found between Ψ and r_t/λ. A commonly used scale length for this problem is

$$r_0 = \sqrt{\frac{9}{4\pi G \rho_0 a}} = \frac{3}{\sqrt{\hat{\rho}(\Psi)}}\lambda. \tag{22.33}$$

The one-parameter family of models, usually called *King models*[48] (Figs. 22.3 through 22.6), is thus identified either by the value of Ψ (frequently indicated as W_0) or, more often, by the value of the concentration index

$$C = \log(r_t/r_0). \tag{22.34}$$

As a model for elliptical galaxies,[49] the required concentration parameter is in the range $2 < C < 2.35$, although the $R^{1/4}$ profile is not reproduced in detail. Highly concentrated King models (formally $C \to \infty$) go in the direction of the isothermal sphere, which is described in Subsection 22.2.1.

It should be emphasized that despite a's being a constant, the velocity dispersion associated with the distribution function of Eq. (22.24) is not a constant. The velocity dispersion monotonically decreases with radius and vanishes at the truncation radius [see Eq. (22.29) and Fig. 22.6]. If we refer to definition (7.7) and to Eqs. (22.28) and (22.29), we see that the relaxation time increases with radius, so eventually the outer envelope of King models becomes collisionless (in particular, the relaxation time diverges at the truncation radius). This is further evidence that the

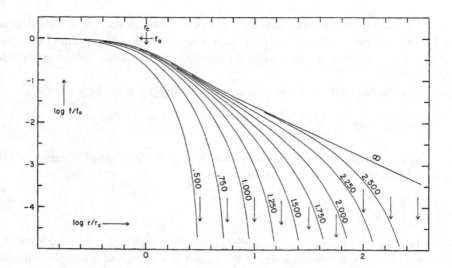

Fig. 22.3. Projected density profiles for the King models (from King, I. R. "The structure of star clusters. III: Some simple dynamical models," 1966. *Astron. J.*, **71**, 64; reproduced by permission of the AAS). The curves show the logarithm of the projected density (normalized to the central value) for selected values of the concentration parameter (marked along the curves). For each case, an arrow provides the location of the relevant truncation radius r_t.

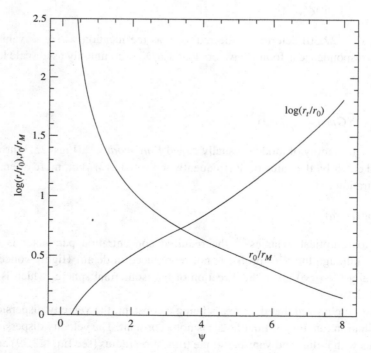

Fig. 22.4. Concentration parameter C and scale r_0 relative to the half-mass radius r_M as a function of the central dimensionless potential Ψ for King models (adapted from Vesperini, E. 1994. Ph.D. dissertation, Scuola Normale Superiore, Pisa, Italy).

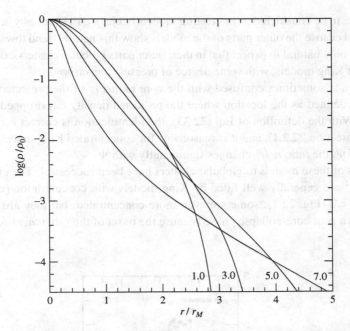

Fig. 22.5. Density profiles for selected values of the central dimensionless potential ($\Psi = 1, 3, 5, 7$) for King models (adapted from Vesperini, E. 1994. Ph.D. dissertation, Scuola Normale Superiore, Pisa, Italy).

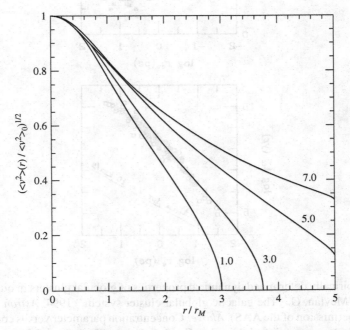

Fig. 22.6. Velocity-dispersion profiles for selected values of the central dimensionless potential ($\Psi = 1, 3, 5, 7$) for King models (adapted from Vesperini, E. 1994. Ph.D. dissertation, Scuola Normale Superiore, Pisa, Italy).

King models are to be considered as interesting models but lack a rigorous physical justification. In other words, because the outer parts of the models show this natural trend toward collisionless conditions, it is only natural to expect that in their outer parts globular clusters exhibit significant deviations from King models, with some degree of pressure anisotropy.

The radius r_0 is sometimes confused with the core radius r_c of the projected density distribution, usually defined as the location where the projected density has dropped to one-half its central value. With the definition of Eq. (22.33), the identification is correct for the isothermal sphere (see Subsection 22.2.1), and it is reasonable for concentrated King models,[50] whereas at lower values of Ψ, the ratio r_0/r_c changes significantly with Ψ.

Applications of these models to globular clusters have been successful. The globular clusters of our Galaxy[51] are generally well fitted by King models with concentration parameter in the range $0.5 < C < 2$ (Fig. 22.7); some are even more concentrated, but they are generally considered to be in a post-core-collapse phase because the onset of the *gravothermal catastrophe*[52]

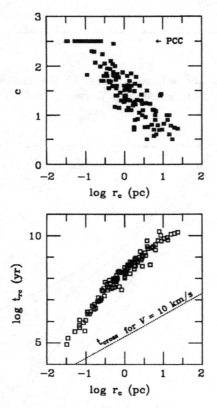

Fig. 22.7. Empirically determined dynamical properties of globular clusters in our Galaxy (from Djorgovski, S., Meylan, G. "The galactic globular cluster system," 1994. *Astron. J.*, **108**, 1292; reproduced by permission of the AAS). *Above:* Concentration parameter versus core radius; post-core-collapse clusters are assigned the value $C = 2.5$. *Below:* The value of the central relaxation time t_{rc} estimated on the basis of the empirically determined central properties of the clusters (see Chapter 7). The apparent strong correlation with the core radius simply reflects the operational definition of t_{rc}.

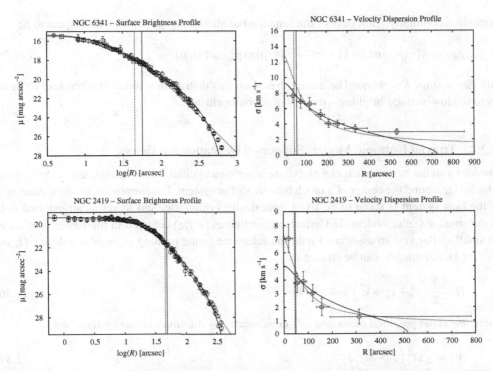

Fig. 22.8. Dynamical fits to selected globular clusters by spherical King models (*solid lines*) and by nontruncated $f^{(\nu)}$ models (*dotted lines*; see Subsection 25.2.2 for definition of the latter models); from Zocchi, A., Bertin, G., Varri, A. L. 2012. *Astron. Astrophys.*, **539**, A65, to which the reader is referred for a full description. *Above:* The more relaxed cluster NGC 6341. *Below:* The less relaxed cluster NGC 2419.

is known to take place at $\Psi \approx 7.4$ (see also Chapter 25). Unfortunately, so far the comparison between models and data has been mainly limited to a fit to the available photometric profiles.[53] Only recently has attention been drawn to the need for a combined test on photometric and kinematic data (Fig. 22.8).[54]

We can take full advantage of the equilibrium sequence by arguing that the identifying parameters (i.e., total mass, central velocity dispersion, and concentration) can change as a result of evolutionary processes (such as evaporation and disk shocking) while the underlying model retains its King appearance.[55] By this representation it is possible to follow the evolution of a whole population of globular clusters in a galaxy with a rather handy algorithm,[56] thus clarifying many of the interesting correlations in the relevant parameter space observed for the clusters of our Galaxy.

One issue that has been revisited recently as a result of current interest in the structure of the outermost regions of globular clusters and the origin of the so-called extratidal light is the truncation of the distribution function. Discontinuities may be erased by evolution in the presence of weak collisionality or collective modes. Other spherical models consider modifications of an isothermal distribution function with sharper or milder truncations. In particular, we may consider a plain truncation, $f_{PT} = A \exp(-aE)$, with a discontinuity at $E = 0$, or a truncation that guarantees continuity[57] not only of the function but also of the derivative with respect to E:

$$f_{WT} = A\{\exp(-aE) - [1 - aE + a\Phi(r_t)]\exp[-a\Phi(r_t)]\},$$
(22.35)

with $f_{WT} = 0$ for $E > \Phi(r_t)$. The models based on this distribution function often lead to better fits to the low-surface-brightness regions of globular clusters.[58]

22.3.2 Triaxial Isotropic Models Supported by Stationary Tides

Consider a stellar system such as a globular cluster on a circular orbit,[59] characterized by angular velocity Ω, around the center of a much heavier stellar system. For simplicity, the field generated by the host system is taken to be fixed, spherically symmetric, and to act as an external field on the smaller stellar system. In Cartesian coordinates (x, y, z) centered in the center of mass of the small stellar system associated with the reference frame rotating at angular velocity Ω, the relevant Jacobi integral can be written as

$$H = \frac{1}{2}\left(v_x^2 + v_y^2 + v_z^2\right) + \Phi_C + \Phi_T,$$
(22.36)

where Φ_C is the potential generated self-consistently by the small stellar system, and

$$\Phi_T = \frac{1}{2}\Omega^2\left(z^2 - \nu x^2\right)$$
(22.37)

is the external tidal potential. Here we have assumed that the center of the orbit of the stellar system is along the x axis and that the orbit is on the (x, y) plane, so $\nu = 4 - \kappa^2/\Omega^2$ is a generally positive coefficient, and the tidal potential implies a vertical compression and a stretching along the x axis.

Therefore, to construct a self-consistent model in which the external tides are taken into account in their full 3D geometry, we may start from the function at the basis of the King models, $f_K(E)$, and replace E by the Jacobi integral:[60]

$$f_K^t = A[\exp(-aH) - \exp(-aH_0)],$$
(22.38)

for $H < H_0$ and $f_K^t = 0$ otherwise. Then the boundary of the system is defined as the surface where the dimensionless escape energy

$$\psi(x, y, z) = a\{H_0 - [\Phi_C(x, y, z) + \Phi_T(x, z)]\}$$
(22.39)

vanishes. The function ψ must be determined self-consistently from the relevant Poisson equation. The truncation energy H_0 must be chosen in such a way that the boundary defined by $\psi = 0$ is a closed surface; that is, it is inside the available Roche volume (Fig. 22.9).

By analogy with the spherical King models, we introduce the scale length r_0, defined as in Eq. (22.33), and the concentration parameter $\Psi = \psi(0, 0, 0)$. The strength of the tidal field can be measured by means of the dimensionless parameter

$$\epsilon = \frac{\Omega^2}{4\pi G \rho_0},$$
(22.40)

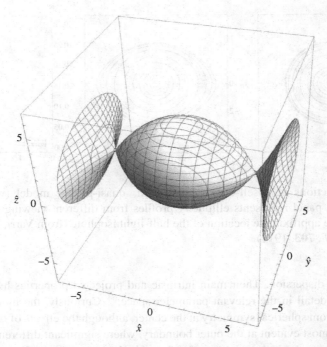

Fig. 22.9. The Hill surface defining the Roche volume for a self-consistent triaxial model with $\Psi = 2$, $\epsilon = 7.043 \times 10^{-4}$ hosted by a Keplerian potential ($\nu = 3$) (from Bertin, G., Varri, A. L. 2008. *Astrophys. J.*, **689**, 1005).

where ρ_0 is the central density. Then it can be shown that for $\psi \geq 0$, the Poisson equation can be written in dimensionless form as

$$\hat{\nabla}^2 \psi = -9\left[\frac{\hat{\rho}(\psi)}{\hat{\rho}(\Psi)} + \epsilon(1-\nu)\right], \tag{22.41}$$

whereas for $\psi < 0$ it becomes

$$\hat{\nabla}^2 \psi = -9\epsilon(1-\nu). \tag{22.42}$$

Here the dimensionless density $\hat{\rho}(\psi)$ is defined as in Eq. (22.28). As for the spherical King models, the natural boundary conditions are the two regularity conditions (on ψ and its gradient) at the origin and the condition of vanishing Φ_C at large radii. Note that because the surface of the system $\psi = 0$ is not known a priori, this is a free-boundary problem. The solution has been found and studied in the (Ψ, ϵ) parameter space by two different methods,[61] that is, by applying the (mainly analytical) method of matched asymptotic expansions[62] or the (mainly numerical) method of iteration.[63] For a given value of the tidal strength, there is a maximum value of the concentration parameter that can be attained by the models; for lower values, the models underfill the associated Roche volume. Alternatively, for a given value of the concentration parameter, there is a maximum value of the tidal strength for which equilibrium configurations can be found.

By construction, the resulting models are triaxial (the shape is approximately that of a prolate spheroid pointing in the direction of the galaxy center; Fig. 22.10) and characterized by

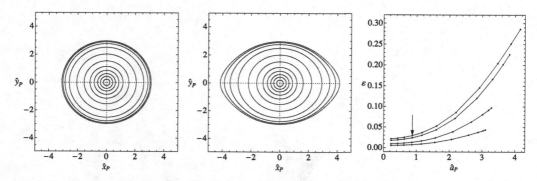

Fig. 22.10. Projections of a self-consistent triaxial, quasi-prolate model (with $\Psi = 2$ and $v = 3$). The right panel represents ellipticity profiles from different viewing angles, with the arrow marking the approximate location of the half-light isophote (from Varri, A. L., Bertin, G. 2009. *Astrophys. J.*, **703**, 1911).

isotropic velocity dispersion. Their main intrinsic and projected properties have been studied systematically in detail in the relevant parameter space.[64] Curiously, the models also exhibit finite deviations from spherical symmetry at the center, although the effects of tides on the shape of the cluster are most evident at the outer boundary, where significant differences with respect to the corresponding spherical models are also found in the kinematic profiles.

22.3.3 Rigidly Rotating Axisymmetric Models

If the stellar system has finite total angular momentum, statistical arguments suggest that relaxation should lead to a modified Maxwell-Boltzmann distribution function, with the single-particle specific energy E replaced by the function $E - \omega J_z$, where J_z represents the specific angular-momentum component in the direction of the total angular momentum of the system, and the constant ω represents the rigid angular velocity at which the system is required to rotate around the z axis. Therefore, we are brought to extend the King models to models supported by the following distribution function:[65]

$$f_K^r = A[\exp(-aH) - \exp(-aH_0)], \tag{22.43}$$

for $H > H_0$, and $f_K^r = 0$ otherwise. Here H is the energy integral in the frame of reference rotating at angular velocity ω, that is,

$$H = \frac{1}{2}\left(v_x^2 + v_y^2 + v_z^2\right) + \Phi_C - \frac{1}{2}\omega^2(x^2 + y^2). \tag{22.44}$$

The construction and discussion of the self-consistent models can be carried out in full analogy with the triaxial models of Subsection 22.3.2. The parameter space is identified by the concentration parameter Ψ and by the rotation-strength parameter

$$\chi = \frac{\omega^2}{4\pi G \rho_0}. \tag{22.45}$$

For a given value of the rotation-strength parameter there exists a maximum value of the allowed concentration parameter; alternatively, for a given value of the concentration parameter there exists a maximum value of the allowed rotation-strength parameter. This corresponds to the existence of a breakup radius; the surface that defines the outer boundary of the model cannot extend beyond such radius. The resulting models are axisymmetric and oblate and characterized by isotropic velocity dispersion. Their main intrinsic and projected properties have been studied systematically in detail in the relevant parameter space.[66] The models also exhibit finite flattening at the center, although the effects of rotation on the shape of the cluster are most evident at the outer boundary, where significant differences with respect to the corresponding spherical models are found also in the kinematic profiles. The general appearance of the models is disky, with a flattening that increases with radius.

22.3.4 Differentially Rotating Axisymmetric Models with Pressure Anisotropy

The case of rigid rotation is unsatisfactory. On the one hand, it is difficult to believe that for applications to real systems, solid-body rotation should be enforced all the way to the outer boundary, given the fact that the outer parts are expected to be collisionless [see comments shortly after Eq. (22.34)]. On the other hand, from an empirical point of view, stellar systems are known to often exhibit solid-body rotation in their central regions, changing into differential rotation in their outer regions. These considerations may have been part of the physical motivation for the construction of some well-known models, often considered as a starting point for the study of rotating ellipsoidal stellar systems.[67]

Here a new class of models is briefly presented, inspired by the study of globular clusters.[68] The idea is to refer to an integral of the motion I for an axisymmetric system with the property that $I \sim E - \omega J_z$ at low values of J_z, that is, close to the symmetry axis, and $I \sim E$ at large values of J_z, that is, far from the symmetry axis. Given the success of the truncated models described in Subsections 22.3.2 and 22.3.3, we suggest a distribution function of the form

$$f_{WT}^d = A[\exp(-aI) - (1 - aI + aE_0)\exp(-aE_0)], \tag{22.46}$$

for $E < E_0$, and $f_{WT}^d = 0$ otherwise; A and a are positive constants. This choice guarantees a smooth truncation in energy. We then take

$$I = E - \frac{\omega J_z}{1 + bJ_z^{2c}}, \tag{22.47}$$

where ω, b, and $c > 1/2$ are positive constants. As is generally the case in the distribution-function-priority approach, the choice of distribution function is guided by physical arguments, but it is not unique.

In addition to the two scales defined by A and a and the two dimensionless parameters Ψ and $\chi = \omega^2/(4\pi G\rho_0)$ setting concentration and rotation strength, the distribution function has two new dimensionless parameters (c and a parameter proportional to b) that can be varied to characterize the shape of the differential rotation. The self-consistent models have been computed by solving the relevant Poisson equation by the iteration method.[69]

The models are characterized by a generally boxy appearance and a variety of (projected) ellipticity profiles. In the case of moderate rotation, the ellipticity profiles are nonmonotonic,

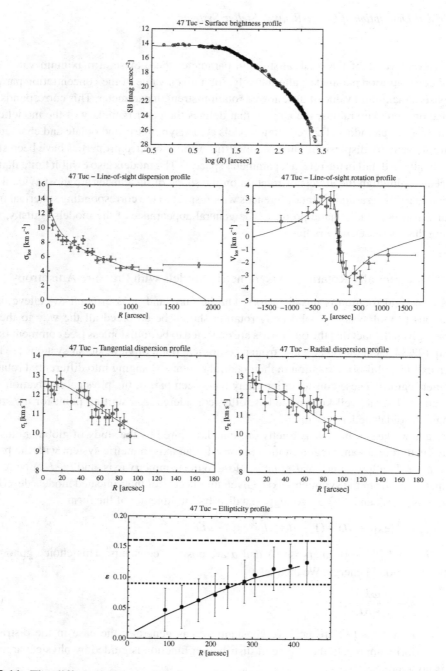

Fig. 22.11. The differentially rotating cluster 47 Tuc. *Top:* Surface brightness profile and line-of-sight kinematic profiles (measured along the projected major axis; vertical bars represent the measured errors and horizontal bars indicate the size of the bins). Solid lines represent the model profiles and open circles the observational data points; these fits determine the three physical scales of the model. *Middle:* Fit to the proper-motion dispersion profiles along the projected tangential (σ_t) and radial (σ_R) directions; this fit determines the dynamical distance to the cluster. *Bottom:* Ellipticity profile. The solid line represents the profile predicted by the model identified by the above-mentioned fits; the black dots mark the observed ellipticities (White, R. E., Shawl, S. J. 1987. *Astrophys. J.*, **317**, 246). Dotted and dashed horizontal lines indicate the average values reported in the literature (from Bianchini, P., Varri, A. L., Bertin, G., Zocchi, A. 2013. *Astrophys. J.*, **772**, id.67).

whereas models in rapid rotation have monotonically decreasing profiles. This behavior is thus complementary to that of rigidly rotating models. Models in rapid rotation exhibit a central toroidal structure.[70]

These differentially rotating models also show a variety of velocity-dispersion profiles. In addition, the pressure tensor is typically isotropic in the central regions and then becomes radially biased in the intermediate regions. Surprisingly, in the outer parts the models are characterized by tangentially biased anisotropy. It has been especially rewarding to find that globular clusters with significant rotation and large amounts of kinematic measurements (in particular, M15, 47 Tuc, and ω Cen) tend to fall rather well and naturally within the reach of the models presented in this subsection (Fig. 22.11).[71]

22.4   Global Description of Partially Relaxed Stellar Systems

In a distribution-function priority approach, one interesting family of equilibrium models has been identified by trying to match key qualitative features at small and large radii that characterize the scenario of galaxy formation by means of collisionless collapse. The relevant distribution function[72] has been called f_∞ as a reminder that its very simple analytical form originates from two important asymptotic aspects: the requirement that the associated potential at large radii be approximately a Stäckel potential and the fact that the function can be seen as the limit for $n \rightarrow \infty$ of a sequence f_n also compatible with the general picture of collisionless collapse. In addition, in the spherical limit the nontruncated models are a one-parameter (Ψ) equilibrium sequence for which the global structure remains basically unchanged for $\Psi > 7$, with the associated projected density profile remarkably close (over a range of 10 magnitudes) to the observed $R^{1/4}$ luminosity profile.[73] The f_∞ distribution function turns out to also have an interesting interpretation in terms of statistical arguments for partially relaxed stellar systems,[74] and this will be briefly outlined in Chapter 25.

As has often been emphasized, the spherical case is a degenerate case, so insight into identification of the models is best gained by starting from the case of small departures from spherical symmetry.

22.4.1   Physical Basis

In the formation picture of collisionless collapse, an elliptical galaxy has most of the stars already formed before virialization, so the collapse and violent relaxation leading to the final quasi-equilibrium state take place through the action of the mean field collectively generated by the stars (see also Chapter 7 and Fig. 25.3). Note that this picture also would apply to the case in which the initial configuration results from the merging of a number of star clumps or protogalaxies. N-body simulations[75] have shown that starting from a variety of initial conditions, the result of a relatively violent collisionless collapse is a quasi-spherical stellar system for which the projected density profile is realistic, in the sense that it resembles the luminosity profiles observed in elliptical galaxies. These simulations, besides demonstrating that this picture is astrophysically acceptable, offer important clues that observations are unable to provide. The stellar system so formed has a signature of efficient violent relaxation[76] in its central regions that turn out to be characterized by isotropic pressure, whereas it remains almost unrelaxed in the outer parts, with a radially dominated pressure tensor (Fig. 22.12).

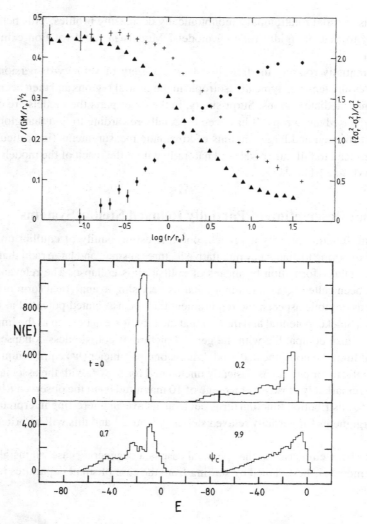

Fig. 22.12. Dynamical effects of collisionless collapse observed in numerical simulations. *Above:* Velocity dispersion and pressure anisotropy for the final configuration of a collapse model, with initial virial ratio $2K/|W| = 0.1$; the quantity shown as filled circles, with the scale given on the right vertical axis, is the quantity called α in the text. *Below:* Evolution of the energy distribution $N(E)$ (see also Fig. 22.15) in the same collapse model (from van Albada, T. S. "Dissipationless galaxy formation and the R to the 1/4-power law," 1982. *Mon. Not. Roy. Astron. Soc.*, **201**, 939; by permission of Oxford University Press, on behalf of the Royal Astronomical Society).

The pressure anisotropy can be described in terms of a function

$$\alpha = 2 - \frac{\langle v_\phi^2 \rangle + \langle v_\theta^2 \rangle}{\langle v_r^2 \rangle} \tag{22.48}$$

[where we refer to spherical coordinates (r, θ, ϕ) and we consider the case for which no internal streaming is present]. Here angle brackets denote the average in velocity space. In an N-body simulation, these averages are performed over the discrete number of particles used, properly binned. For a quasi-spherical system, the quantity α is basically a function of the radial coordinate. We can thus summarize the result of the numerical simulations, in relation to the pressure-anisotropy profile, by stating that the collisionless collapse leads generically to systems characterized by $\alpha \sim 0$ at small radii and by $\alpha \sim 2$ at large radii. We introduce the anisotropy radius r_α as the radius where $\alpha = 1$. The simulations show that $r_\alpha \approx r_M$; that is, the transition to mainly radial pressure occurs around the half-mass radius.

Can we find distribution functions capable of reproducing this apparently simple qualitative behavior? In the strictly spherical case, this can be done in an infinite number of ways. But the simulations show that this qualitative behavior occurs even when the system is appreciably far from spherical symmetry.

22.4.2 Selection Criterion and Identification of the Distribution Function

Consider the axisymmetric nonrotating case where the departure from spherical symmetry is small. If we try to construct a distribution function with a radially biased pressure anisotropy at large radii by using only the known classical integrals of the motion E and $J_z^2 = r^2 \sin^2 \theta \, v_\phi^2$, we see that we are forced to the condition

$$\alpha = 1 - \frac{\langle v_\phi^2 \rangle}{\langle v_r^2 \rangle} \leq 1 \tag{22.49}$$

because for any $f = f(E, J_z^2)$ we have $\langle v_\theta^2 \rangle = \langle v_r^2 \rangle$. Therefore, to reproduce the behavior $\alpha \sim 2$ at large radii within the Jeans theorem, we must invoke the presence of a third isolating integral of the motion.

From Chapter 21 we know that only special classes of potentials admit a third integral. The only known case of astrophysical interest is that described in Subsections 21.3.2 and 21.3.3. For our purposes, we need to assume only that the third integral exists at large radii, and for that we try the condition [see Eqs. (21.39) and (21.41)]

$$\Phi(r, \theta) \sim \Phi_0(r) + \frac{\eta(\theta)}{r^2}, \tag{22.50}$$

with

$$I_3 \sim \frac{J^2}{2} + \eta(\theta). \tag{22.51}$$

Because we are considering the fully self-consistent case, in which Φ is supported by f by means of the Poisson equation, in order to guarantee the appropriate radial behavior of the non-axisymmetric part of the potential $\Phi(r, \theta) - \Phi_0(r)$, we have to require that the density profile generated by $f(E, J_z^2, I_3)$ falls as r^{-4} at large radii. In passing, we note that indeed this is realized by the density profile of the perfect ellipsoid (see Subsection 21.3.4), but here we do not impose any specific density profile except for the asymptotic condition at large radii.

The most natural distribution function dependent on the three integrals, $f = A \exp(-aE - bJ_z^2/2 - cI_3)$ (i.e., a generalization of a Maxwellian distribution), would fail to produce the desired asymptotic density profile. A simple distribution function consistent with the r^{-4} behavior at large radii is

$$f_\infty = A(-E)^{3/2} \exp(-aE - bJ_z^2/2 - cI_3). \tag{22.52}$$

Here we take $f = 0$ for $E > 0$ (i.e., for unbound stars), and we require that A, a, and c be positive constants (the choice $c > 0$ indeed leads to radially biased velocity-dispersion profiles; negative temperature models, with $a < 0$, also have been considered,[77] but they are unphysical and violently unstable; see Chapter 23). The self-consistent problem thus can be carried out analytically for small departures from spherical symmetry, that is, for $\eta(\theta) = O(b/c) \ll 1$. The resulting models are spherical in the center and become progressively slightly oblate or prolate in the outer regions.

Using the Laplace approximate integration method for the v_ϕ and v_θ variables, we can easily show that at large radii the density associated with f_∞ has the following behavior:

$$\rho \sim \frac{\hat{A} \exp(-c\eta)\Phi^2}{r^2 \sqrt{1 + (b/c)\sin^2\theta}},$$
(22.53)

and we can also prove that $\langle v_r^2 \rangle \sim -\Phi/3$ and

$$\alpha \sim 2 - \frac{6}{(-a\Phi)(1 + cr^2/a)}.$$
(22.54)

Because the potential becomes Keplerian in the outer parts, we see that indeed the function f_∞ satisfies the outer boundary condition imposed by the picture of collisionless collapse, with the desired density behavior.

The spherical limit of the preceding distribution function is given by

$$f_\infty = A(-E)^{3/2}\exp(-aE - cJ^2/2).$$
(22.55)

Note that the nontrivial factor $(-E)^{3/2}$ has a simple orbital interpretation, being characteristic of the Keplerian frequency [see Eq. (21.3)]; note also that a similar dependence characterizes Ω_r for isochrone models [see Eq. (21.15)]. This feature has stimulated interesting discussions on the statistical mechanics of incomplete violent relaxation (see also Chapter 25).

The arguments that have led to identification of the f_∞ distribution exclude many possibilities but do not lead to a unique distribution function. In practice, the arguments can be summarized by the following selection criterion:[78] *The distribution function for elliptical galaxies should depend on three integrals of the motion in such a way that at large radii the pressure anisotropy parameter α tends to 2 and the mass density decreases as r^{-4}. In addition, in the central regions the distribution function should be very close to an isotropic Maxwellian (i.e., $\alpha \approx 0$).* Note that the density behavior at large radii allows for models with finite total mass. This selection criterion has been shown to be satisfied by other forms of distribution function, for example, by a whole sequence

$$f_n = A \frac{(-E)^{3/2}\exp(-aE)}{[1 + (bJ_z^2 + cI_3)/n]^n}.$$
(22.56)

The analysis of the resulting models has proved that the approach displays good structural stability in the sense that the interesting properties of the f_∞ models are found to reflect more the physical picture adopted in their construction than the specific analytical implementation used.

In Chapter 25 we will introduce another family of models consistent with the selection criterion just stated, based on a distribution function called $f^{(\nu)}$,[79] which is characterized by interesting properties in relation to statistical mechanics[80] and fits in detail the final products of simulations of collisionless collapse.[81] The family is characterized by an $R^{1/4}$ projected-density profile for large values of Ψ; in its plain (nontruncated) form it has also been used to fit globular clusters,[82] with some success for less relaxed clusters such as NGC 2419.

22.4.3 Properties of the Self-Consistent Nontruncated Models

Consider the spherical limit of Eq. (22.55), and refer to the natural units for radius, energy, and velocity given by $\sqrt{a/c}$, $1/a$, and $1/\sqrt{a}$, respectively. Let $\psi = -a\Phi$ and $\Psi = \psi(0)$, and introduce the dimensionless index

$$\gamma = \frac{ac}{4\pi GA}. \tag{22.57}$$

Then the Poisson equation in dimensionless form becomes [see Eq. (22.30)]

$$\frac{1}{\zeta^2}\frac{d}{d\zeta}\left(\zeta^2\frac{d\psi}{d\zeta}\right) = -\frac{1}{\gamma}\hat{\rho}(\psi,\zeta), \tag{22.58}$$

which should be solved under the natural boundary conditions $d\psi/d\zeta = 0$ at $\zeta = 0$ and $\psi \sim \hat{M}/\zeta$ for $\zeta \to \infty$. The dimensionless density $\hat{\rho}(\psi,\zeta)$ has an explicit dependence on ζ because the model is anisotropic.[83] For a given value of γ, we may look for the value (or values) of Ψ that guarantees that the relevant boundary conditions are satisfied. Conversely, we may overdetermine the problem by assigning a third boundary condition [i.e., $\Psi = \psi(0)$] and then looking at Eq. (22.58) as an eigenvalue problem for γ. The accuracy of numerical solutions can be assessed in terms of the deviations from the Keplerian behavior of the potential at large radii or, independently, by a test of the virial theorem over the self-consistent model.

For each value of Ψ, one solution for γ is found that is consistent with the imposed boundary conditions. For low values of Ψ, the relation between γ and Ψ is approximately linear up to $\Psi \approx 4$, when γ reaches the maximum $\gamma_{max} \approx 52.5$. Between 4 and 7, the function $\gamma = \gamma(\Psi)$ makes a transition and then connects to the horizontal line $\gamma \approx 18$ for larger values of Ψ (Fig. 22.13).

We recall that once a solution to Eq. (22.58) is found [i.e., a pair of values (Ψ, γ) and the associated eigenfunction $\psi(\zeta)$], the models we obtain by inserting the relevant potential in Eq. (22.55)

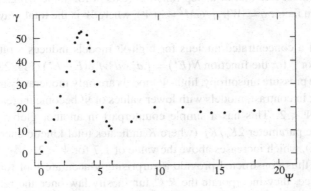

Fig. 22.13. The relation between γ and Ψ for the f_∞ models (from Stiavelli, M., Bertin, G. 1985. *Mon. Not. Roy. Astron. Soc.*, **217**, 735).

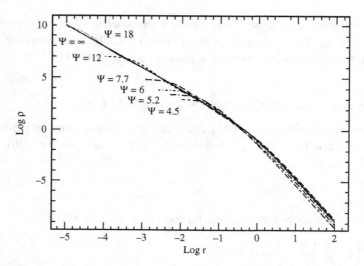

Fig. 22.14. Behavior of the density profiles for the f_∞ models for different values of the central dimensionless potential Ψ (from Bertin, G., Stiavelli, M. 1989. *Astrophys. J.*, **338**, 723).

have all their phase-space properties and all the possible observable profiles fully determined; the only freedom left is that of two scales, for example, the choice of the total mass M and the half-mass radius r_M of the model, in physical units. At large values of Ψ, the global properties and various profiles of the self-consistent models stay practically unchanged; the only variation with Ψ is associated with the development of a nucleus with higher and higher central density but smaller and smaller mass. For $\Psi \to \infty$, the models converge toward a singular f_∞ model for which the nucleus is characterized by $\rho \sim r^{-2}$ all the way in (Fig. 22.14).

In relation to the density profile, low-Ψ models have a wide core; in fact, their mass distribution is well approximated by that of the perfect sphere or the isochrone potential (see Section 21.1). For relatively high values of Ψ, instead, the density profile outside the nucleus (say, $r > 0.1 r_M$) is characterized by $\rho \sim r^{-2}$ inside the half-mass radius and by $\rho \sim r^{-4}$ outside; the slope transition occurring at $r \approx r_M$ is rather sharp. The change in the mass distribution from a wide-core structure at low Ψ to a concentrated distribution at high Ψ has a simple counterpart in the transition, almost like that of a step function (except for some wiggles), from ≈ 0.35 to ≈ 0.50, for the form factor $q = |W| r_M / G M^2 = q(\Psi)$, where W is the total gravitational energy of the model.

The presence of a concentrated nucleus for high-Ψ models induces a rather wide range of exponential behavior[84] for the function $N(E^\star) = \int d^3x \, d^3v f \, \delta(E - E^\star)$ (Fig. 22.15).

In relation to the pressure anisotropy, high-Ψ models are only moderately anisotropic because for them $r_\alpha \approx 3 r_M$; in contrast, models with lower values of Ψ become increasingly anisotropic, with $r_\alpha < r_M$ for $\Psi < 2$. This has a simple counterpart in another global indicator of pressure anisotropy: the parameter $2 K_r / K_T$ (where K indicates total kinetic energy in the radial or tangential direction), which increases above the value of 1.7 for $\Psi < 2$.

The f_∞ models thus constructed provide a surprisingly accurate tool for fitting the observations.[85] In practice, they incorporate the $R^{1/4}$ luminosity law once the relevant photometric profile is obtained from the model with the assumption of a constant M/L ratio. For the cases for which the photometric profile is best known, such as that of NGC 3379,[86] an excellent fit is

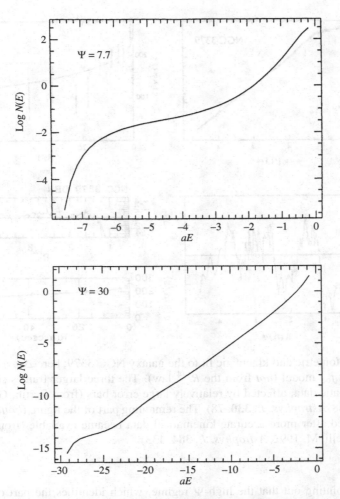

Fig. 22.15. Behavior of the energy distribution $N(E)$, defined in the text, for two concentrated f_∞ models of the sequence (from Bertin, G., Stiavelli, M. 1989. *Astrophys. J.*, **338**, 723).

found over a range of 11 magnitudes; for this galaxy, the corresponding kinematic profile (i.e., the velocity dispersion projected along the line of sight) is also well fitted out to the outermost available kinematic data point around R_e (Fig. 22.16). Fits of this kind provide a dynamical measurement of the mass-to-light ratio based on a global modeling. As indicated earlier, the fit is not sensitive to the value of Ψ (provided that $\Psi > 7$), which may be tuned only in an attempt at also fitting the possibly present small nucleus (but here other problems should be faced; see Subsection 22.2.2). This generic adequacy of the f_∞ models appears to indicate that the global properties of elliptical galaxies are indeed consistent with the picture of collisionless collapse, which is probably the explanation of the universality of the $R^{1/4}$ luminosity profile. The dependence on Ψ of the global profiles in the transition range $\Psi = 4$ to 7 might be used to parameterize the departures from the $R^{1/4}$ law (i.e., nonhomology) that have sometimes been noted in relatively small ellipticals.[87]

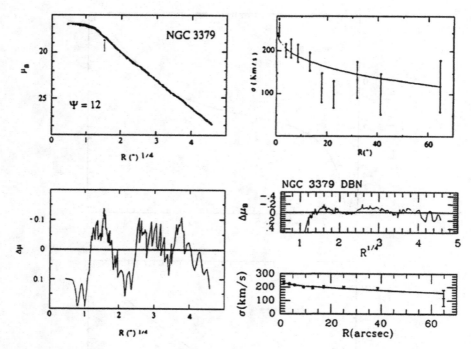

Fig. 22.16. Photometric and kinematic fit to the galaxy NGC 3379; here $\Delta\mu$ indicates residual from the best-fit f_∞ model (*not* from the $R^{1/4}$ law). The three larger frames show a fit based on earlier kinematic data, affected by relatively large error bars (from Bertin, G., Saglia, R. P., Stiavelli, M. 1988. *Astrophys. J.*, **330**, 78). The remaining part of the figure (*bottom right*) refers to a fit performed after more accurate kinematical data became available (from Saglia, R. P., Bertin, G., Stiavelli, M. 1992. *Astrophys. J.*, **384**, 433).

It is worth pointing out that the high-Ψ regime, which identifies the part of the sequence in which the models are characterized by projected density profiles well fitted by the $R^{1/4}$ law, starts at $\Psi \approx 7$: This happens to coincide with the condition for the onset of the gravothermal catastrophe briefly mentioned in Subsection 22.3.1. The interest in this occurrence is further strengthened when we inspect the detailed behavior of the function $q(\Psi)$ or the quantity GMa/r_M as a function of Ψ along the f_∞ equilibrium sequence. In Chapter 25 the argument will be better formalized by means of the $f^{(\nu)}$ family of models.

This discussion can be extended, at least in part, to nonspherical systems. Some asymptotic analyses can be carried out[88] under the ordering $\eta(\theta) = O(b/c) \ll 1$. At large radii, self-consistency leads to an inhomogeneous equation for η in the variable θ, which can be solved in terms of polynomials (in $\cos\theta$). The analysis shows the possibility of models characterized by either boxy or disky isophotes. Because the asymptotic analysis of the self-consistent nonspherical f_∞ models is nontrivial, much insight has been gained by initializing an N-body code with the asymptotic expressions of f_∞ with values of b well beyond their expected range of applicability.[89] The simulated systems are found to relax quickly to equilibrium configurations with properties that are not far from those of the approximate equilibrium states (Fig. 22.17).

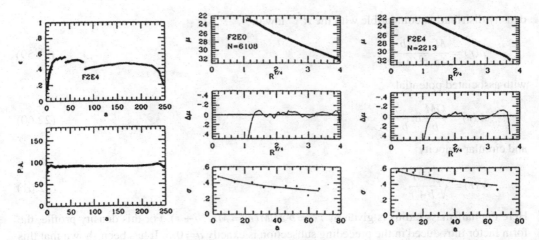

Fig. 22.17. A test for the use of spherical f_∞ models (from Saglia, R. P., Bertin, G., Stiavelli, M. 1992. *Astrophys. J.*, **384**, 433). The F2 model is a highly flattened model (actually, with a dark component included; see discussion in Subsection 22.4.6) constructed by initializing an N-body simulation with a nonspherical f_∞ distribution function with parameters well beyond the allowed limit of quasi-spherical symmetry. The N-body system quickly relaxes to an equilibrium configuration, which, from some viewing angles, is as flat as an E4 galaxy (see the ellipticity and position angle of the projected density contours in the left frames, shown on the radial range out to $4R_e$). In a simulated observation, this model is observed from two viewing angles, one at which it resembles an E0 galaxy (central frames) and one at which it appears as an E4 (right frames). The figure shows the result of a photometric and kinematic fit performed with the analytically simpler spherical (two-component) f_∞ models. The results are then used to evaluate the reliability of spherical model fits for flat ellipticals.

22.4.4 Density Behavior Associated with the $R^{1/4}$ Law

For a very long time it has been believed that the simplest description of a density profile $\rho(r)$ compatible with the observed luminosity distribution and thus with the $R^{1/4}$ law would be $\rho \sim r^{-3}$. Curiously, this belief persisted[90] well after concrete evidence had been accumulated against the r^{-3} behavior. To some extent this may have been inspired by the popularity of the so-called modified Hubble profile (see Subsection 22.2.1).

If we assume that the $R^{1/4}$ empirical law for the luminosity profile holds exactly from the center out to infinity, under the assumption of spherical symmetry, it is possible to carry out an inversion into a volume-density profile $\rho(r)$ that, once projected, gives precisely such a law.[91] The numerical solution is tabulated. But there is no physical reason to require that the law hold exactly and at every radius because the data show systematic deviations[92] from the $R^{1/4}$ profile and sample up to no more than 11 magnitudes. Thus we should compare models directly with the data and use the $R^{1/4}$ law only as a zeroth-order reference case.

It is in this light that we should consider the luminosity profiles associated with the f_∞ models, which we discussed earlier as being characterized by a density profile with two slopes, r^{-2} inside and r^{-4} outside, with a rather sharp transition around r_M. At the time when the self-consistent anisotropic f_∞ models were constructed, it was realized independently[93] that indeed a simple density distribution compatible with the $R^{1/4}$ law is

$$\rho(r) = \frac{M}{4\pi} \frac{b}{r^2(b+r)^2}, \tag{22.59}$$

with associated potential

$$\Phi(r) = \frac{GM}{b} \ln \frac{r}{b+r} \tag{22.60}$$

and circular velocity

$$V(r) = \sqrt{\frac{GM}{b+r}}. \tag{22.61}$$

Here the half-mass radius is given by $r_M = b$ as $M(r) = Mr/(b+r)$. For this density profile, the form factor introduced in the preceding subsection is exactly $q = 0.5$. It has been shown that this density distribution is qualitatively similar to that of the singular f_∞ models ($\Psi = \infty$) but quantitatively different, with relative deviations of ≈ 10 percent in the radial range $10^{-5}r_M < r < 10^2 r_M$. Thus, although by itself the preceding density-potential pair indicates little about the physical origin of the luminosity profile of elliptical galaxies, it nonetheless serves as a very handy analytical descriptive tool.

Another simple analytical model for the density distribution is[94]

$$\rho(r) = \frac{M}{2\pi} \frac{b}{r(b+r)^3}, \tag{22.62}$$

with associated potential

$$\Phi(r) = -\frac{GM}{b+r} \tag{22.63}$$

and circular velocity

$$V(r) = \frac{\sqrt{GMr}}{b+r}. \tag{22.64}$$

Here the half-mass radius is given by $r_M = (1 + \sqrt{2})b$ as $M(r) = Mr^2/(b+r)^2$. Variations on the same theme, with more freedom on the relevant exponents, also have been considered, especially with the goal of giving a better parametric description of the properties of the inner structure of galaxy cores and cusps.[95] These models are generally used with a preference for the r^{-4} decline at large radii.

22.4.5   Self-Consistent Truncated Models

The models described so far in this section are characterized by the interesting property of having finite mass even if they are of infinite radial extent. In practice, it would be important to check the properties of truncated models, in line with some of the arguments given in Section 22.3. In each one-parameter family of models, this would introduce a second dimensionless parameter, related to the truncation radius; in addition, as described in Section 22.3, various degrees of smoothness can be considered in the way the truncation is performed. A discussion of truncated models is likely to be of little relevance to applications to elliptical galaxies but may turn out to be physically interesting in the context of globular clusters. In fact, some globular clusters appear to be in a fully collisionless regime, and thus the paradigm of incomplete violent relaxation might be applicable to them.

22.4.6 Two-Component Models (with Dark Matter)

Despite their simplicity and limitations, the f_∞ models appear to capture much of the structure of elliptical galaxies. Except for the small variations associated with the precise value of Ψ for the physically interesting part of the sequence ($\Psi > 7$), and except for the freedom in the choice of the two dimensional scales, all the phase-space properties of the models are fixed. This proves that the models are of physical interest because they are found to be realistic a posteriori, with no parameter tuning. Still, such a rigid structure of the models may be embarrassing for two basic reasons. One point of concern is that, as is well known, the universal photometric structure of elliptical galaxies is curiously accompanied by a variety of kinematic profiles (see Chapter 4). Thus the very success of the f_∞ models in fitting galaxies such as NGC 3379 automatically implies a failure to fit other ellipticals with flatter velocity-dispersion profiles (e.g., NGC 4472). The second reason for concern is that, at this stage, the success of the f_∞ models supports a picture in which there is no need for dark matter; we recall that the realistic photometric profile is obtained by converting mass density into luminosity under the assumption of a constant M/L ratio. This would be fine from a methodological point of view. However, we do have evidence for the presence of massive dark halos around spiral galaxies (see Chapter 20). From the physical point of view, it would be hard to believe that ellipticals have no dark matter (see Chapter 24).

From the very beginning it has been clear that the observed variety of kinematic profiles could be ascribed basically to two physical factors: In particular, a relatively flat velocity-dispersion profile might result from the presence of a dark halo or from the dominance of tangential orbits.[96]

Following the approach emphasized in this chapter (and in general in this book), we may leave aside as unphysical the idea of populating stellar orbits in an ad hoc manner in order to produce desired velocity-dispersion profiles. In doing so, we are also encouraged by general stability arguments, which suggest that significant departures from quasi-Maxwellian distributions of stellar orbits are probably a source of collective modes that go in the direction of removing such peculiarities in phase space, as often shown in the context of plasma physics. An additional important semiempirical argument also confirms this viewpoint. If the variety of kinematic profiles corresponded to the existence of arbitrary distributions of stellar orbits, we would expect to observe some kinematic profiles that are flatter and others that are steeper than those predicted by the f_∞ models; instead, it appears that the steepest observed profiles are those consistent with the one-component f_∞ models, for which the drop in velocity dispersion projected along the line of sight from the central regions to R_e is by less than a factor of ≈ 2. Furthermore, it is difficult to imagine a physical formation process as leading to basically nonrotating spheroidal systems and to a strong bias of the pressure tensor in the tangential directions. In fact, collisionless collapse leads to a bias, but in the radial direction. Therefore, the natural option left is to explore the possibility that the observed variety of kinematic profiles results from a variety of situations associated with the presence of dark halos. In other words, we continue to follow the physical scenario of collisionless collapse and ask what would be the impact of the presence of a diffuse halo in the models.

Note that this attempt goes against one intuitive expectation: that the impressive accuracy of photometric fits based on one-component f_∞ models might be spoiled by the presence of a second component. In particular, the influence of a second component also would appear in the relevant virial constraint for the luminous matter that now becomes

$$2K_L + W_L + W_{LD} = 0, \tag{22.65}$$

where the W_{LD} term represents the interaction integral. In the spherically symmetric case, the self-interaction term can be written as $W_L = -q_L GM_L^2/r_L$, and

$$W_{LD} = -4\pi G \int_0^\infty r\rho_L(r)M_D(r)\, dr, \tag{22.66}$$

so the no-dark-matter case corresponds to $W_{LD} = 0$, $q_L \approx 0.5$.

A dark halo, if present, is also likely to follow the picture of collisionless collapse, even if we assume it to be made of baryonic matter. The simplest way to proceed is thus to devise a two-component analysis in which one component contributes to the light and observable velocity-dispersion profiles by means of a constant mass-to-light ratio, and the other contributes as dark matter to only the underlying gravitational field. Encouraged by the success of the f_∞ function in characterizing the framework of collisionless collapse, we may describe each component by the same form of distribution function, but with independent sets of parameters:

$$f_L = A_L(-E)^{3/2}\exp(-a_L E - c_L J^2/2), \tag{22.67}$$

$$f_D = A_D(-E)^{3/2}\exp(-a_D E - c_D J^2/2). \tag{22.68}$$

As usual, these expressions hold for $E < 0$; the functions are taken to vanish for $E > 0$. The potential Φ entering in the definition of the energy in these functions is the total gravitational potential, generated by the sum of the two contributions, $\rho_L + \rho_D$. The parameter space here involves four dimensionless quantities. In addition to a concentration parameter (Ψ_L), we may set three relative scales: the mass ratio M_L/M_D, the length-scale ratio r_L/r_D of the corresponding half-mass radii, and the temperature ratio a_L/a_D. In an extensive survey,[97] $\approx 3,000$ models have been computed, covering a wide grid in parameter space [especially in the plane $(r_L/r_D, M_L/M_D)$], mostly addressing the physically plausible case of diffuse halos (i.e., models with $r_L/r_D < 1$).

The main result of the survey of two-component models is that despite the variety of kinematic profiles generated, there is a natural conspiracy for the concentrated models to support realistic luminosity profiles, consistent with the $R^{1/4}$ law. A diffuse dark halo thus may alter significantly the velocity-dispersion profile with only minor effects on the density distribution of the luminous component. This comes as a surprise and adds confidence to the overall physical picture at the basis of this model construction. The concept of minimum halo can be developed by analogy with the maximum-disk decomposition for spiral galaxies (see Chapter 20). The detailed quantitative properties of the fully self-consistent two-component global models have been used to study the presence of dark halos around elliptical galaxies[98] (Figs. 22.17 and 22.18). The resulting luminous–dark-matter decomposition leads to different values of M_L/M_D (in some galaxies, on the basis of the available kinematic data, which even in the best cases reach out only to $\approx R_e$, there is no need to invoke the presence of a dark halo) but to a rather well-defined value of M_L/L.

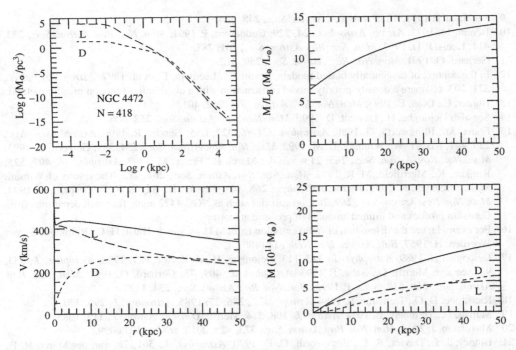

Fig. 22.18. Intrinsic properties for the best-fit model of NGC 4472 based on two-component f_∞ models: density profiles for the dark (D) and the luminous (L) components, cumulative total mass-to-light ratio, circular velocity profile decomposition, and cumulative mass decomposition (from Saglia, R. P., Bertin, G., Stiavelli, M. 1992. *Astrophys. J.*, **384**, 433).

The relevant two-component models appear to be more isotropic than the one-component models. For some reason, the presence of the dark halo tends to increase the value of r_α/r_M for the luminous component. A posteriori, this is in line with recent observational determinations of the line-of-sight velocity profiles, which generally show only modest departures from a Gaussian.

Interesting density-priority studies of two-component systems also have been carried out[99] and may be compared with the results of the preceding survey of distribution-function priority models.

Notes

1. See Dejonghe, H. 1986. *Phys. Rep.*, **133**, 217.
2. See discussion at the end of Chapter 21; Bishop, J. L. 1986. *op. cit.*; Statler, T. S. 1987. *op. cit.*
3. Kuzmin, G. G. 1956. *Astron. Zh.*, **33**, 27.
4. de Zeeuw, P. T., Peletier, R., Franx, M. 1986. *Mon. Not. Roy. Astron. Soc.*, **221**, 1001.
5. See Merritt, D. 1997. *Astrophys. J.*, **486**, 102; Sridhar, S., Touma, J. 1997. *Mon. Not. Roy. Astron. Soc.*, **292**, 657.
6. Franx, M. 1988. *Mon. Not. Roy. Astron. Soc.*, **231**, 285.
7. Eddington, A. S. 1916. *Mon. Not. Roy. Astron. Soc.*, **76**, 572; see also Tricomi, F. G. 1985. *Integral Equations*, Dover, New York.
8. Osipkov, L. P. 1979. *Pis'ma Astron. Zh.*, **5**, 77; Merritt, D. 1985. *Astron. J.*, **90**, 1027; Merritt, D. 1985. *Mon. Not. Roy. Astron. Soc.*, **214**, 25p.

9. See Saha, P. 1991. *Mon. Not. Roy. Astron. Soc.*, **248**, 464.

10. Hénon, M. 1973. *Astron. Astrophys.*, **24**, 229; Cuddeford, P. 1991. *Mon. Not. Roy. Astron. Soc.*, **253**, 414; Louis, P. D. 1993. *Mon. Not. Roy. Astron. Soc.*, **261**, 283.

11. Gerhard, O. 1991. *Mon. Not. Roy. Astron. Soc.*, **250**, 812.

12. In the context of tangentially biased models, Bertin, G., Leeuwin, F., et al. 1997. *Astron. Astrophys.*, **321**, 703, compare a density-priority model construction with a distribution-function priority method.

13. Hunter, C., Qian, E. 1993. *Mon. Not. Roy. Astron. Soc.*, **262**, 401.

14. See also Dejonghe, H., Laurent, D. 1990. *Mon. Not. Roy. Astron. Soc.*, **252**, 606.

15. Franx, M., Illingworth, G. 1988. *Astrophys. J. Lett.*, **327**, L55; Bender, R. 1990. *Astron. Astrophys.*, **229**, 441; Rix, H.-W., White, S. D. M. 1992. *Mon. Not. Roy. Astron. Soc.*, **254**, 389; Gerhard, O. 1993. *Mon. Not. Roy. Astron. Soc.*, **265**, 213; van der Marel, R., Franx, M. 1993. *Astrophys. J.*, **407**, 525; Kuijken, K., Merrifield, M. R. 1993. *Mon. Not. Roy. Astron. Soc.*, **264**, 712. The results of Winsall, M. L., Freeman, K. 1993. *Astron. Astrophys.*, **268**, 443; Bender, R., Saglia, R. P., Gerhard, O. 1994. *Mon. Not. Roy. Astron. Soc.*, **269**, 785, on galaxies such as NGC 4472 argue for small departures from Gaussian profiles and limited amounts of pressure anisotropy.

16. For example, see the discussion of radio emission from M31 by van de Hulst, H. C., Raimond, E., van Woerden, H. 1957. *Bull. Astron. Inst. Neth.*, **14** (480), 1.

17. Dejonghe, H. 1989. *Astrophys. J.*, **343**, 113; Dejonghe, H., de Zeeuw, P. T. 1988. *Astrophys. J.*, **333**, 90; see also Merritt, D., Saha, P. 1993. *Astrophys. J.*, **409**, 75; Gerhard, O. 1991. *Mon. Not. Roy. Astron. Soc.*, **250**, 512; Saha, P. 1992. *Mon. Not. Roy. Astron. Soc.*, **254**, 132.

18. Richstone, D. O., Tremaine, S. 1984. *Astrophys. J.*, **286**, 27; 1985. *Astrophys. J.*, **296**, 370.

19. Merritt, D., Tremblay, B. 1994. *Astron. J.*, **108**, 514; Merritt, D. 1996. *Astron. J.*, **112**, 1085.

20. Magorrian, J. 2006. *Mon. Not. Roy. Astron. Soc.*, **373**, 425; 2013. arXiv:1303.6099.

21. Binney, J. J., Davies, R. L., Illingworth, G. D. 1990. *Astrophys. J.*, **361**, 78; van der Marel, R. P., Binney, J. J., Davies, R. L. 1990. *Mon. Not. Roy. Astron. Soc.*, **245**, 582; van der Marel, R. P. 1991. *Mon. Not. Roy. Astron. Soc.*, **253**, 710, and many following papers.

22. See Franx, M., Illingworth, G. D., de Zeeuw, P. T. 1991. *Astrophys. J.*, **383**, 112.

23. Spitzer, L., Härm, R. 1958. *Astrophys. J.*, **127**, 544; Hénon, M. 1961. *Ann. Astrophys.*, **24**, 369; King, I. R., 1965. *Astrophys. J.*, **70**, 376.

24. Prendergast, K. H., Tomer, E. 1970. *Astron. J.*, **75**, 674; Wilson, C. P. 1975. *Astron. J.*, **80**, 175.

25. Petrou, M. 1983. *Mon. Not. Roy. Astron. Soc.*, **202**, 1195 and 1209; Lupton, R. H., Gunn, J. F. 1987. *Astron. J.*, **93**, 1106.

26. Harris, E. G. 1962. *Il Nuovo Cimento*, **23**, 115; see Section 9.1 in Part II.

27. See Rood, H. J., Page, T. L., et al. 1972. *Astrophys. J.*, **175**, 627.

28. The axially symmetric general case, also in the presence of rotation and of an embedded disk, has been studied by Richstone, D. O. 1980. *Astrophys. J.*, **238**, 103; 1982. *Astrophys. J.*, **252**, 496; Monet, D. G., Richstone, D. O., Schechter, P. L. 1981. *Astrophys. J.*, **245**, 454; Miller, R. H. 1982. *Astrophys. J.*, **254**, 75; Toomre, A. 1982. *Astrophys. J.*, **259**, 535.

29. Eddington, A. S. 1926. *Internal Constitution of Stars*, Cambridge University Press, Cambridge, UK; Chandrasekhar, S. 1939. *An Introduction to the Theory of Stellar Structure*, University of Chicago, Chicago (reprinted in 1967 by Dover, New York); Saslaw, W. C. 1985. *Gravitational Physics of Stellar and Galactic Systems*, Cambridge University Press, Cambridge, UK; see also Betti, E. 1880. *Il Nuovo Cimento*, **7**, 26.

30. See also Vandervoort, P. O. 1980. *Astrophys. J.*, **240**, 478.

31. Hunter, C. 2001. *Mon. Not. Roy. Astron. Soc.*, **328**, 839.

32. Plummer, H. C. 1915. *Mon. Not. Roy. Astron. Soc.*, **76**, 107. Curiously, the $n = 3/2$ case is characterized by a distribution function that is constant up to $E = 0$, which makes it a sort of water-bag model.

33. See, e.g., Kormendy, J., Richstone, D. 1995. *Annu. Rev. Astron. Astrophys.*, **33**, 581; Crane, P., Stiavelli, M., et al. 1993. *Astron. J.*, **106**, 1371; Faber, S. M., Tremaine, S., et al. 1997. *Astron. J.*, **114**, 1771, and the many papers cited therein. In particular, for M32, see van der Marel, R. P., de Zeeuw, P. T., et al. 1997. *Nature (London)*, **385**, 610; for M87, see Harms, R. J., Ford, H. C., et al. 1994. *Astrophys. J. Lett.*, **435**, L35; for M106, see Miyoshi, M., Moran, J., et al. 1995. *Nature (London)*, **373**, 127.

34. Many other objects not mentioned here are also likely to possess a massive central black hole; e.g., for NGC 1399, see Stiavelli, M., Møller, P., Zeilinger, W. W. 1993. *Astron. Astrophys.*, **277**, 421.

35. The failure of the $R^{1/4}$ profile in the innermost regions was already noted by Lauer, T. 1985. *Astrophys. J.*, **292**, 104, and by others.

36. The classical case is that of M87; see Young, P. J., Westphal, J. A., et al. 1978. *Astrophys. J.*, **221**, 721.

37. Peebles, P. J. E. 1972. *Astrophys. J.*, **178**, 371; Bahcall, J. N., Wolf, R. A. 1976. *Astrophys. J.*, **209**, 214.

38. Evans, N. W., Collett, J. L. 1997. *Astrophys. J. Lett.*, **480**, L103, show that a steady-state cusp with $\rho \sim r^{-4/3}$ is a self-consistent solution of the collisional Boltzmann equation.

39. Huntley, J. M., Saslaw, W. C. 1975. *Astrophys. J.*, **199**, 328.

40. Peebles, P. J. E. 1972. *Gen. Rel. Grav.*, **3**, 63; Young, P. H. 1980. *Astrophys. J.*, **242**, 1232; Goodman, J., Binney, J. J. 1984. *Mon. Not. Roy. Astron. Soc.*, **207**, 511; Lee, M. H., Goodman, J. 1989. *Astrophys. J.*, **343**, 594; Cipollina, M., Bertin, G. 1994. *Astron. Astrophys.*, **288**, 43; Quinlan, G. D., Hernquist, L., Sigurdsson, S. 1995. *Astrophys. J.*, **440**, 554. The effects of a slowly growing (and also a fast-growing) black hole inside an initially triaxial model have been studied with N-body simulations by Merritt, D., Quinlan, G. D. 1998. *Astrophys. J.*, **498**, 625.

41. Cipollina, M. 1992. Tesi di Laurea, Pisa University, Pisa, Italy; Cipollina, M., Bertin, G. 1994. *Astron. Astrophys.*, **288**, 43.

42. See Antonsen, T. M., Lee, Y. C. 1982. *Phys. Fluids*, **25**, 132.

43. See Zocchi, A., Bertin, G., Varri, A. L. 2012. *Astron. Astrophys.*, **539**, A65, and references therein.

44. After the pioneering work by White, R. E., Shawl, S. J. 1987. *Astrophys. J.*, **317**, 246; see Chen, C. W., Chen, W. P. 2010. *Astrophys. J.*, **721**, 1790, and references therein.

45. For the cluster ω Cen, see van Leeuwen, F., Le Poole, R. S., et al. 2000. *Astron. Astrophys.*, **360**, 472; and Anderson, J., van der Marel, R. P. 2010. *Astrophys. J.*, **710**, 1032; for 47 Tuc, see Anderson, J., King, I. R. 2003. *Astron. J.*, **126**, 772; and McLaughlin, D. E., Anderson, J., et al. 2006. *Astrophys. J. Suppl.*, **166**, 249.

46. The work presented here largely follows Varri, A. L. 2012. Ph.D. dissertation, Università degli Studi di Milano, Milano, Italy.

47. See Abramowitz, M., Stegun, I. A. 1970. *op. cit.*, definition 6.5.2.

48. King, I. R. 1965. *Astron. J.*, **70**, 376; 1966. *Astron. J.*, **71**, 64; Michie, R. W. 1963. *Mon. Not. Roy. Astron. Soc.*, **125**, 127, and Michie, R. W., Bodenheimer, P. H. 1963. *Mon. Not. Roy. Astron. Soc.*, **126**, 269, considered the more general anisotropic distribution function, where $E \rightarrow E + cJ^2$.

49. Kormendy, J. 1978. *Astrophys. J.*, **218**, 333; King, I. R. 1978. *Astrophys. J.*, **222**, 1.

50. Peterson, C. J., King, I. R. 1975. *Astron. J.*, **80**, 427.

51. Djorgovski, S. G., Meylan, G. 1994. *Astron. J.*, **108**, 1292; Harris, W. E. 1996. *Astron. J.*, **112**, 1487, revised and updated 2010 [arXiv:1012.3224]; Meylan, G., Heggie, D. C. 1997. *Astron. Astrophys. Rev.*, **8**, 1.

52. Antonov, V. A. 1962. *Vestn. Leningr. Univ.*, **7**, 135, translated in 1985. *IAU Symposium*, Vol. 113, eds. J. Goodman, P. Hut. Reidel, Dordrecht, The Netherlands, p. 525; Lynden-Bell, D., Wood, R. 1968. *Mon. Not. Roy. Astron. Soc.*, **138**, 495. The gravothermal catastrophe is an interesting phenomenon related to the fact that self-gravitating systems are characterized by negative specific heat. In the context of the dynamics of globular clusters, an extensive description of the mechanisms involved is given by Spitzer, L., Jr. 1987. *Dynamical Evolution of Globular Clusters*, Princeton University Press, Princeton, NJ. See also Katz, J. 1980. *Mon. Not. Roy. Astron. Soc.*, **190**, 497; Wiyanto, P., Kato, S., Inagaki, S. 1985. *Publ. Astron. Soc. Jpn.*, **37**, 715.

53. Trager, S. C., King, I. R., Djorgovski, S. 1995. *Astron. J.*, **109**, 218; McLaughlin, D. E., van der Marel, R. P. 2005. *Astrophys. J. Suppl.*, **161**, 304.

54. See Zocchi, A., Bertin, G., Varri, A. L. 2012. *op. cit.*, and references therein. The paper demonstrates that fits to the available kinematic profiles are difficult to obtain independent of the relaxation condition of the clusters; in addition, it shows that "it would be highly desirable to acquire larger numbers of accurate kinematic data-points, well distributed over the cluster field."

55. King, I. R. 1966. *Astron. J.*, **71**, 64; Prata, S. W. 1971. *Astron. J.*, **76**, 1017 and 1029; Chernoff, D. F., Kochanek, C. S., Shapiro, S. L. 1986. *Astrophys. J.*, **309**, 183; Chernoff, D. F., Shapiro, S. L. 1987. *Astrophys. J.*, **322**, 113. Numerical simulations (see Chernoff, D. F., Weinberg, M. 1990. *Astrophys. J.*, **351**, 121) have indeed demonstrated that the dynamical evolution of a cluster is well described in terms of evolution along a King sequence.

56. Vesperini, E. 1997. *Mon. Not. Roy. Astron. Soc.*, **287**, 915. See also Vesperini, E. 1998. *Mon. Not. Roy. Astron. Soc.*, **299**, 1019.

57. These can be seen as the spherical limit of the models introduced by Wilson, C. P. 1975. *Astron. J.*, **80**, 175.

58. See McLaughlin, D. E., van der Marel, R. P. 2005. *op. cit.* For a study of the problem of the extratidal light in the globular cluster system of NGC 5128, see Harris, W. E., Harris, G. L. H., et al. 2002. *Astron. J.*, **124**, 1435. The connection between the outermost density profile and the truncation of the supporting distribution function was noted in general terms by Hunter, C. 1977. *Astron. J.*, **82**, 271.

59. In a recent paper it has been shown that for clusters on noncircular orbits, "instead of the limiting radius being imposed at perigalacticon, it more nearly traces the instantaneous tidal radius of the cluster at any point in the orbit"; see Webb, J. J., Harris, W. E., et al. 2013. *Astrophys. J.*, **764**, id.124.

60. This is indeed the step envisioned by Weinberg, M. D. 1993. In *The Globular Cluster-Galaxy Connection*, Vol. 48, eds. G. H. Smith, J. P. Brodie. Astronomical Society of the Pacific Conference Series, p. 689, and taken by Heggie, D. C., Ramamani, N. 1995. *Mon. Not. Roy. Astron. Soc.*, **272**, 317, and Bertin, G., Varri, A. L. 2008. *Astrophys. J.*, **689**, 1005. A full discussion of the intrinsic and projected properties of the models is given by Varri, A. L., Bertin, G. 2009. *Astrophys. J.*, **703**, 1911.

61. Bertin, G., Varri, A. L. 2008. *op. cit.*

62. For the general concepts, see Van Dyke, M. 1975. *Perturbation Methods in Fluid Mechanics*, Parabolic Press, Stanford, CA. For an application to the analogous problem of rigidly rotating fluid polytropes, see Smith, B. L. 1975. *Astrophys. Sp. Sci.*, **35**, 223.

63. Similar to that used in an axisymmetric problem by Prendergast, K. H., Tomer, E. 1970. *Astron. J.*, **75**, 674.

64. Varri, A. L., Bertin, G. 2009. *op. cit.*

65. See appendix B in Bertin, G., Varri, A. L. 2008. *Astrophys. J.*, **689**, 1005; and section 2 in Varri, A. L., Bertin, G. 2012. *Astron. Astrophys.*, **540**, A94. The models considered by Kormendy, J., Anand, S. P. S. 1971. *Astrophys. Sp. Sci.*, **12**, 47, follow a similar argument but are characterized by a plain, discontinuous truncation.

66. Varri, A. L., Bertin, G. 2012. *op. cit.*

67. Prendergast, K. H., Tomer, E. 1970. *op. cit.*, considered as possible models of elliptical galaxies systems based on a distribution function proportional to $\exp(E - \omega J_z)$ with a sharp truncation in E; the models are then characterized by differential rotation and by a nonmonotonic ellipticity profile. Wilson, C. P. 1975. *op. cit.*, proposed models of elliptical galaxies that include differential rotation and pressure anisotropy by means of adjustable parameters, β and ζ; the supporting distribution function is constructed by multiplying the function f_{WT}, with its smooth truncation in energy, of Eq. (22.35) by a factor $\exp[\beta J_z - (1/2)\zeta^2 J_z^2]$. Other studies of rotating models include Lynden-Bell, D. 1962. *Mon. Not. Roy. Astron. Soc.*, **123**, 447; Lupton, R. H., Gunn, J. E. 1987. *Astron. J.*, **93**, 1106; Lagoute, C., Longaretti, P.-Y. 1996. *Astron. Astrophys.*, **308**, 441. In the context of the study of galaxy bulges, we should mention Jarvis, B. J., Freeman, K. C. 1985. *Astrophys. J.*, **295**, 314, who adopted the distribution function later considered by Lagoute and Longaretti, that is, $f = f_K(E)\exp(\beta J_z)$; Rowley, G. 1988. *Astrophys. J.*, **331**, 124, who used a distribution function proportional to $\exp\{-a[E - \omega J_z + (1/2)\zeta^2 J_z^2]\}$ with a truncation in $[E - \omega J_z + (1/2)\zeta^2 J_z^2]$ and provided an interesting discussion in terms of the so-called Lindblad diagram.

68. Varri, A. L., Bertin, G. 2012. *op. cit.*

69. Following the strategy outlined by Prendergast, K. H., Tomer, E. 1970. *op. cit.*, and adapted by Bertin, G., Varri, A. L. 2008. *op. cit.*, and Varri, A. L., Bertin, G. 2012. *op. cit.*

70. A toroidal structure has been noted as a possibility in other distribution-function priority models, in particular, in those studied by Lynden-Bell, D. 1962. *Mon. Not. Roy. Astron. Soc.*, **123**, 447, and

Prendergast, K. H., Tomer, E. 1970. *op. cit.* Among more recent investigations, see the models studied by Ciotti, L., Bertin, G., Londrillo, P. 2004. In *Plasmas in the Laboratory and in the Universe: New Insights and New Challenges*, eds. G. Bertin, D. Farina, R. Pozzoli. AIP Conference Proceedings, Vol. 703, Melville, NY, p. 322, and by Monari, G., Ciotti, L., et al. 2013. In preparation; these models may include the presence of a central massive black hole and solid-body rotation.

71. Bianchini, P. 2012. Tesi di Laurea, Università degli Studi di Milano, Milano, Italy; Bianchini, P., Varri, A. L., et al. 2013. *Astrophys. J.*, **772**, id.67.

72. Bertin, G., Stiavelli, M. 1984. *Astron. Astrophys.*, **137**, 26; Stiavelli, M., Bertin, G. 1985. *Mon. Not. Roy. Astron. Soc.*, **217**, 735; Bertin, G., Stiavelli, M. 1989. *Astrophys. J.*, **338**, 723.

73. This was realized from the beginning, but a detailed comparison with the observations was made only in a second stage; see Bertin, G., Saglia, R. P., Stiavelli, M. 1988. *Astrophys. J.*, **330**, 78.

74. Stiavelli, M., Bertin, G. 1987. *Mon. Not. Roy. Astron. Soc.*, **229**, 61.

75. van Albada, T. S. 1982. *Mon. Not. Roy. Astron. Soc.*, **201**, 939. These results have been confirmed by means of more modern simulations; in particular, see Trenti, M., Bertin, G., van Albada, T. S. 2005. *Astron. Astrophys.*, **433**, 57.

76. Lynden-Bell, D. 1967. *Mon. Not. Roy. Astron. Soc.*, **136**, 101; Shu, F. H. 1978. *Astrophys. J.*, **225**, 83.

77. Merritt, D., Tremaine, S., Johnstone, D. 1989. *Mon. Not. Roy. Astron. Soc.*, **236**, 829.

78. Bertin, G., Stiavelli, M. 1989. *op. cit.*

79. Stiavelli, M., Bertin, G. 1987. *Mon. Not. Roy. Astron. Soc.*, **229**, 61.

80. Bertin, G., Trenti, M. 2003. *Astrophys. J.*, **584**, 729.

81. Trenti, M., Bertin, G., van Albada, T. S. 2005. *Astron. Astrophys.*, **433**, 57.

82. Zocchi, A., Bertin, G., Varri, A. L. 2012. *op. cit.*

83. See appendix A of Bertin, G., Pegoraro, F., et al. 1994. *Astrophys. J.*, **434**, 94.

84. Binney, J. J. 1982. *Mon. Not. Roy. Astron. Soc.*, **200**, 951.

85. See Bertin, G., Saglia, R. P., Stiavelli, M. 1988. *op. cit.*

86. de Vaucouleurs, G., Capaccioli, M. 1979. *Astrophys. J. Suppl.*, **40**, 699.

87. See Sersic, J. L. 1968. *Astron. J.*, **73**, 892; Ciotti, L. 1991. *Astron. Astrophys.*, **249**, 91; Caon, N., Capaccioli, M., D'Onofrio, M. 1993. *Mon. Not. Roy. Astron. Soc.*, **265**, 1013; Graham, A., Lauer, T. R., et al. 1996. *Astrophys. J.*, **465**, 534; Prugniel, P., Simien, F. 1997. *Astron. Astrophys.*, **321**, 111; Bertin, G., Ciotti, L., Del Principe, M. 2002. *Astron. Astrophys.*, **386**, 149, and references therein.

88. Stiavelli, M., Bertin, G. 1985. *op. cit.*

89. See Bertin, G., Stiavelli, M. 1989. *op. cit.*; Saglia, R. P., Bertin, G., Stiavelli, M. 1992. *Astrophys. J.*, **384**, 433.

90. See statements on pp. 940 and 945 in van Albada, T. S. 1982. *op. cit.* and fig. 5 in the article by Binney, J. J. 1982. *Mon. Not. Roy. Astron. Soc.*, **200**, 951; Smith, B. F., Miller, R. H. 1986. *Astrophys. J.*, **309**, 522; White, S. D. M. 1987. In *IAU Symposium*, Vol. 127, ed. T. de Zeeuw. Reidel, Dordrecht, The Netherlands, p. 339 and especially pp. 345–346; White, S. D. M., Narayan, R. 1987. *Mon. Not. Roy. Astron. Soc.*, **229**, 103.

91. Young, P. J. 1976. *Astron. J.*, **81**, 807.

92. For example, see van Albada, T. S. 1982. *op. cit.*

93. Jaffe, W. 1983. *Mon. Not. Roy. Astron. Soc.*, **202**, 995.

94. Hernquist, L. 1990. *Astrophys. J.*, **356**, 359. Jaffe's and Hernquist's density-potential pairs are special cases of the pairs studied by Dehnen, W. 1993. *Mon. Not. Roy. Astron. Soc.*, **265**, 250.

95. See Carollo, C. M. 1993. Ph.D. dissertation, Ludwig-Maximilians University, Munich; Dehnen, W. 1993. *Mon. Not. Roy. Astron. Soc.*, **265**, 250; Tremaine, S., Richstone, D. O., et al. 1994. *Astron. J.*, **107**, 634; and following papers.

96. Illingworth, G. D. 1983. In *IAU Symposium*, Vol. 100, ed. E. Athanassoula. Reidel, Dordrecht, The Netherlands, p. 257; Tonry, J. L. 1983. *Astrophys. J.*, **266**, 58.

97. Saglia, R. P. 1990. Ph.D. dissertation, Scuola Normale Superiore, Pisa, Italy; Bertin, G., Saglia, R. P., Stiavelli, M. 1992. *Astrophys. J.*, **384**, 423; see also Bertin, G., Saglia, R. P., Stiavelli, M. 1989. In *Third ESO-CERN Symposium*, eds. M. Caffo et al. Kluwer, Dordrecht, The Netherlands, p. 303.

98. Saglia, R. P., Bertin, G., Stiavelli, M. 1992. *Astrophys. J.*, **384**, 433; Saglia, R. P., Bertin, G., et al. 1993. *Astrophys. J.*, **403**, 567; Bertin, G., Bertola, F., et al. 1994. *Astron. Astrophys.*, **292**, 381.

99. Ciotti, L., Pellegrini, S. 1992. *Mon. Not. Roy. Astron. Soc.*, **255**, 561; Ciotti, L. 1996. *Astrophys. J.*, **471**, 68. An interesting application of two-component models to investigate dynamical interpretations of the tilt of the fundamental plane of elliptical galaxies (see Subsection 4.3.3) is provided by Ciotti, L., Lanzoni, B., Renzini, A. 1996. *Mon. Not. Roy. Astron. Soc.*, **282**, 1.

23 Stability

The problem of stability in the context of elliptical galaxies is completely different from the studies of stability described in Part III. For disks, there are several specific morphological properties, in particular, spiral structures, bars, warps, and corrugations, that we may match in terms of appropriate modes over more symmetric equilibrium configurations. In the case of elliptical galaxies, there are no outstanding morphological features to be addressed in a similar manner. Some features do exist, for example, isophotal twisting, shells, and peculiar kinematics, that might in principle be considered, but their three-dimensional (3D) spatial structure is not known, and it is not clear whether their origin can be traced to simple low-amplitude regular patterns. Linear stability analyses also would be of little use for addressing one obvious question related to the departures from spherical symmetry: What intrinsic shapes are realized? In fact, natural modes to be considered (and in some cases found to be unstable) are those with $\ell = 2$, but they are a degenerate class that includes oblate, prolate, and triaxial perturbations. Thus one general motivation at the basis of many modal analyses (see Chapter 9) is simply not present in the case of elliptical galaxies.

There remain two major physical reasons to carry out stability investigations. One is the need for a good knowledge of the intrinsic modes of a dynamical system as a prerequisite to the proper description of the driven problem (e.g., for the study of tidal interactions). Here, not unlike the case of disk galaxies, much work remains to be done.[1] The other important reason is to check the physical plausibility of the equilibrium models. Models that turn out to be violently unstable should not be used to fit the observations because the associated dynamical system would be subject to rapid evolution in the direction of new configurations with different observable characteristics.

From the technical point of view, the case here is also very different from that of Part III. On the one hand, it seems that local analyses in terms of a dispersion relation for waves (see Chapter 15) have very limited applicability. This is due not only to the 3D character of the underlying basic state but also to the fact that most stellar orbits are not localized in radius as they are in disks and the fact that the system is very inhomogeneous on the scales sampled by the orbits. On the other hand, there is a subtle difference in the stability analysis, which is characteristic of spherical geometry. We will see that a spherical harmonic analysis leads to nontrivial separations of variables.

We would like to have simple criteria for the possible sources of instability and a feeling for the resulting modes and dynamical mechanisms involved. These would then give guidelines for the construction of physically plausible equilibrium models. For example, nonmonotonic density profiles or strong pressure anisotropies are among the obvious candidate sources of free energy for unstable modes. In practice, although much progress has been made, many issues remain unresolved.[2] The only simple criterion of practical use is that for the onset of the radial-orbit instability,[3] although other theorems exist but have lesser impact on the astrophysical context.[4] The mechanisms involved still need clarification. Even the simple case related to $\ell = 0$ modes is far from being under control. Simulations are of great help, especially because they allow us to follow the stage of nonlinear evolution, but they cannot clarify by themselves the many theoretical questions that can be raised.

The discussion that follows is limited to a presentation of the basic equations for linear modes and a brief discussion of what is known about the radial-orbit instability. In Part III we focused less on the derivation aspect because it was simpler and much was to be discussed on the properties of the relevant dispersion relations. Here less is known on the general properties of modes and instabilities, whereas the derivation of the equations is rather complex and, in many respects, instructive.

23.1   Basic Equations for Linear Modes

The derivation of the basic equations for linear modes[5] can be summarized as follows.

23.1.1   Linearized Equations

Let us consider linear perturbations with time dependence $\exp(i\omega t)$ over a spherically symmetric basic state associated with a distribution function $f_0(E, J^2)$ and an unperturbed potential $\Phi(r)$ that satisfy the collisionless Boltzmann-Poisson equations

$$\frac{\partial f_1}{\partial t} + \{f_1, H\} = \{\Phi_1, f_0\},$$

(23.1)

$$\nabla^2 \Phi_1 = 4\pi G \int f_1 \, d^3v.$$

(23.2)

Here standard Poisson brackets are used, and H is the unperturbed Hamiltonian.

We refer to spherical coordinates in physical space (r, θ, ϕ) and cylindrical polar coordinates in velocity space $(v_r, v_\perp, \alpha)$, where $\tan\alpha = v_\phi/v_\theta$. Then the angle-dependent part of the unperturbed orbit operator can be separated as

$$\frac{\partial}{\partial t} + \{, H\} = \frac{\partial}{\partial t} + \frac{v_\perp}{r}\widehat{L} + \widehat{D}$$

(23.3)

where the angle-dependent part $\widehat{L}$ is defined by

$$\widehat{L} = \cos\alpha \frac{\partial}{\partial\theta} + \frac{\sin\alpha}{\sin\theta}\frac{\partial}{\partial\phi} - \sin\alpha \cot\theta \frac{\partial}{\partial\alpha},$$

(23.4)

and the operator $\widehat{D}$ is defined by

$$\widehat{D} = v_r \frac{\partial}{\partial r} - \frac{v_r v_\perp}{r}\frac{\partial}{\partial v_\perp} + \left(\frac{v_\perp^2}{r} - \frac{d\Phi}{dr}\right)\frac{\partial}{\partial v_r};$$

(23.5)

when applied to functions of E, J^2, and r, the operator $\widehat{D}$ reduces to

$$\widehat{D} = s_v \left\{ 2 \left[E - \Phi_{\text{eff}}(J^2, r) \right] \right\}^{1/2} \frac{\partial}{\partial r} = v_r(E, J, r) \frac{\partial}{\partial r}, \tag{23.6}$$

where $s_v = \text{sgn}(v_r)$ identifies the direction of the radial motion.

At this stage it is convenient to subtract the so-called adiabatic response [which in reality is only one part of the proper adiabatic response; see Eq. (22.22)] and write

$$h_1 = f_1 - \Phi_1 \frac{\partial f_0}{\partial E} \tag{23.7}$$

so that

$$\frac{\partial h_1}{\partial t} + \{h_1, H\} = -\frac{\partial f_0}{\partial E} \frac{\partial \Phi_1}{\partial t} + \frac{\partial f_0}{\partial J} \widehat{L} \Phi_1. \tag{23.8}$$

23.1.2 Generalized Spherical Harmonics

The expansion of $h_1 = h_1(E, J, r, \theta, \phi, \alpha)$ into generalized spherical harmonics is given by

$$h_1 = \sum_{\ell, m, s} h_{ms}^\ell(E, J, r) T_{ms}^\ell(\phi, \theta, \alpha), \tag{23.9}$$

$$\Phi_1 = \sum_{\ell, m} \Phi^\ell(r) T_{m0}^\ell(\phi, \theta), \tag{23.10}$$

where

$$T_{ms}^\ell(\phi, \theta, \alpha) = \exp[-i(m\phi + s\alpha)] P_{ms}^\ell(\cos\theta), \tag{23.11}$$

$\ell = 0, 1 \ldots$ is the index labeling the multipole expansion of the perturbation, and m and s are integers, $-\ell \leq (m, s) \leq \ell$. The generalized Legendre polynomials P_{ms}^ℓ are defined as[6]

$$P_{ms}^\ell(z) = \ell^{-m-s} \left[\frac{(\ell - m)! \, (\ell - s)!}{(\ell + m)! \, (\ell + s)!} \right]^{1/2} \left(\frac{1 + z}{1 - z} \right)^{(m+s)/2}$$

$$\times \sum_{j = \max(m, s)}^{\ell} \frac{(\ell + j)! \, \ell^{2j}}{(\ell - j)! \, (j - m)! \, (j - s)!} \left(\frac{1 - z}{2} \right)^j, \tag{23.12}$$

and they are such that, for $s = 0$,

$$\exp(im\phi) \, P_{m0}^\ell(\cos\theta) = Y_m^\ell(\phi, \theta), \tag{23.13}$$

with $Y_m^\ell(\phi, \theta)$ the standard spherical harmonics. Under the action of $\widehat{L}$, the generalized spherical harmonics T_{ms}^ℓ transform as

$$\widehat{L} T_{ms}^\ell = \frac{1}{2i} \left(\alpha_{s+1} T_{m, s+1}^\ell + \alpha_s T_{m, s-1}^\ell \right), \tag{23.14}$$

where

$$\alpha_s = \sqrt{(\ell + s)(\ell - s + 1)}. \tag{23.15}$$

Equations (23.3), (23.4), and (23.14) show that as a consequence of the spherical symmetry of the equilibrium configuration, the collisionless Boltzmann-Poisson equations are diagonal with respect to ℓ and m and independent of m.

Choosing a value of ℓ and fixing arbitrarily $m = 0$, we obtain for a perturbation with frequency ω

$$\left[-\omega + iv_r(E, J, r)\frac{\partial}{\partial r}\right]h_s^\ell + \frac{J}{2r^2}\left(\alpha_s h_{s-1}^\ell + \alpha_{s+1}h_{s+1}^\ell\right)$$

$$= \omega\frac{\partial f_0}{\partial E}\Phi^\ell \delta_{s0} + \frac{1}{2}(\delta_{s+1,0} + \delta_{s-1,0})\sqrt{\ell(\ell+1)}\frac{\partial f_0}{\partial J}\Phi^\ell, \quad (23.16)$$

$$\left[\frac{1}{r^2}\frac{d}{dr}\left(r^2\frac{d}{dr}\right) - \frac{\ell(\ell+1)}{r^2}\right]\Phi^\ell = 8\pi G\int v_\perp\,dv_\perp dv_r\left(h_0^\ell + \Phi^\ell\frac{\partial f_0}{\partial E}\right), \quad (23.17)$$

where the index $m = 0$ has been dropped for the sake of simplicity. We note that although modes with different values of m have the same linear frequency and/or growth rate, they are not in general geometrically equivalent for $\ell \geq 2$. For example, for $\ell = 2$, we have either biaxial or triaxial perturbations, and the former can be either oblate or prolate. This degeneracy is removed by nonlinear terms in the case of finite-amplitude modes.

23.1.3   Diagonalization

System (23.16) consists of $2\ell + 1$ coupled equations. A simplification arises from the fact that Eqs. (23.3) and (23.16) are invariant for $(\alpha, \phi) \to (-\alpha, -\phi)$. Thus, because we have chosen $m = 0$, we can split the vector h_s^ℓ and system (23.16) into an even part ($\ell + 1$ components) and an odd part (ℓ components) under the transformation $s \to -s$. Only the $s = 0$ term in Eq. (23.9) contributes to the perturbed density, as used explicitly on the right-hand side of Eq. (23.17), and thus the odd part of the vector h_s^ℓ can be disregarded because it corresponds to perturbed motions that do not lead to a perturbed density and potential. This separation is manifested when we diagonalize system (23.16) in terms of the eigenfunctions of the unperturbed Hamiltonian H by defining the new amplitudes

$$\widehat{h}_s^\ell = \sum_{s'}O_{ss'}h_{s'}^\ell, \quad (23.18)$$

where O is an appropriate orthogonal matrix. If we introduce the vector A_s^ℓ as

$$A_s^\ell = \sum_{s'}O_{ss'}\delta_{s'0} = O_{s0}, \quad (23.19)$$

we obtain from system (23.16)

$$\left[-\omega + iv_r(E, J, r)\frac{\partial}{\partial r} + \frac{sJ}{r^2}\right]\widehat{h}_s^\ell = \left(\omega\frac{\partial f_0}{\partial E} + s\frac{\partial f_0}{\partial J}\right)\Phi^\ell A_s^\ell. \quad (23.20)$$

Here $A_s^\ell = (1/\sqrt{2})(1, 0, -1)$ and $(1/4)(\sqrt{6}, 0, -2, 0, \sqrt{6})$ for $\ell = 1$ and $\ell = 2$, respectively, and for the same values of ℓ, O is given by

$$O^{(1)} = \frac{1}{2}\begin{pmatrix} 1 & \sqrt{2} & 1 \\ -\sqrt{2} & 0 & \sqrt{2} \\ 1 & -\sqrt{2} & 1 \end{pmatrix}, \quad (23.21)$$

$$O^{(2)} = \frac{1}{4} \begin{pmatrix} 1 & 2 & \sqrt{6} & 2 & 1 \\ -2 & -2 & 0 & 2 & 2 \\ \sqrt{6} & 0 & -2 & 0 & \sqrt{6} \\ -2 & 2 & 0 & -2 & 2 \\ 1 & -2 & \sqrt{6} & -2 & 1 \end{pmatrix}. \tag{23.22}$$

The ℓ zeros of A_s^ℓ correspond to the odd part of h_s^ℓ. The remaining $(\ell+1)$ independent equations in Eq. (23.20) are coupled by the Poisson equation, which involves $h_0^\ell = \sum_s {}^tO_{0s}\widehat{h}_s^\ell = \sum_s A_s^\ell \widehat{h}_s^\ell$. For compactness, in the following, A_s^ℓ is defined as a vector with $(\ell+1)$ components only (i.e., $s = \ell, \ell-2, \ldots, -\ell$).

23.1.4 Bounce Average and Resonance Conditions

The unperturbed orbit operator on the left-hand side of Eq. (23.20) can be expressed in terms of the angle variable of the radial motion defined by

$$\Omega_r \, d\tau = \frac{2\pi}{\tau_r(E,J)} \frac{dr}{v_r(E,J,r)}. \tag{23.23}$$

It is then convenient to expand the perturbed amplitudes $\widehat{h}_s^\ell$ and the potential Φ^ℓ in a Fourier series in the angle variable $\Omega_r \tau$. This procedure is equivalent to the bounce average technique that is used extensively in plasma physics[7] (see also the end of Chapter 12). We write

$$\widehat{h}_s^\ell(E,J,r) = \sum_{p=-\infty}^{+\infty} \bar{h}_{sp}^\ell(E,J) \exp\left(ip\Omega_r \tau + is \int_0^\tau \tilde{\Omega} \, d\tau' \right), \tag{23.24}$$

$$\Phi^\ell(r) = \sum_{p=-\infty}^{+\infty} \bar{\Phi}_{sp}^\ell(E,J) \exp\left(ip\Omega_r \tau + is \int_0^\tau \tilde{\Omega} \, d\tau' \right), \tag{23.25}$$

where Ω_r follows the same notation as in Chapters 13 and 21, and the residual frequency $\tilde{\Omega}$ is defined as

$$\tilde{\Omega} = \tilde{\Omega}(E,J,r) = \frac{J}{r^2} - \Omega_\theta(E,J). \tag{23.26}$$

The term $(is \int_0^\tau \tilde{\Omega} \, d\tau')$ in the exponential function has been introduced in order to remove the τ-dependent part of sJ/r^2 in Eq. (23.20),

$$\bar{h}_{sp}^\ell(E,J) = \left\langle \widehat{h}_s^\ell \exp\left(-ip\Omega_r \tau - is \int_0^\tau \tilde{\Omega} \, d\tau' \right) \right\rangle, \tag{23.27}$$

and similarly for $\bar{\Phi}^\ell(E,J)$, and the angle brackets denote bounce average as defined by Eq. (22.23). With this expansion, system (23.20) becomes algebraic, and the amplitudes $\bar{h}_{sp}^\ell$

obey the equation

$$\bar{h}_{sp}^{\ell} = \frac{\omega \frac{\partial f_0}{\partial E} + s \frac{\partial f_0}{\partial J}}{-\omega - p\Omega_r(E,J) + s\Omega_\theta(E,J)} A_s^{\ell} \bar{\Phi}_{sp}^{\ell}. \tag{23.28}$$

The vanishing of the denominator in Eq. (23.28) singles out the stars that resonate with a mode of frequency ω. This resonance condition is characterized by a radial number p, which runs from $-\infty$ to $+\infty$, and by an azimuthal number s, which takes the $(\ell + 1)$ values $\ell, \ell - 2, \ldots, -\ell$. The remaining values of s are excluded by the selection rule arising from the symmetry of the equilibrium under inversion of angle ϕ. When the velocity space integral is performed, the denominator in Eq. (23.28) is to be defined according to the standard Landau prescription[8] (see Chapter 12).

The linearized-mode dispersion equation is obtained by combining Eqs. (23.17), (23.18), (23.24), and (23.28) and has the form of an integrodifferential equation.

23.1.5 Conditions at Large Radii

For large radii, the high-frequency limit $|J/r^2\omega| \ll 1$, $(v_r/\omega)\partial/\partial r = O(|J/r^2\omega|)$, leads to

$$h_0^{\ell} \simeq -\Phi^{\ell} \frac{\partial f_0}{\partial E} + \frac{v_r}{\omega} \frac{\partial}{\partial r} \left(\frac{\partial f_0}{\partial E} \frac{v_r}{\omega} \frac{\partial \Phi^{\ell}}{\partial r} \right) - \frac{J^2 \ell(\ell+1)}{2\omega^2 r^4} \left(\frac{\partial f_0}{\partial E} + \frac{r^2}{J} \frac{\partial f_0}{\partial J} \right) \Phi^{\ell}, \tag{23.29}$$

which when inserted into Eq. (23.17) leads to the ordinary differential equation

$$\frac{d}{dr} \left[r^2 \left(1 + \frac{4\pi G\rho}{\omega^2} \right) \frac{d\Phi^{\ell}}{dr} \right] - \ell(\ell+1) \left(1 + \frac{4\pi G\rho}{\omega^2} \right) \Phi^{\ell} = 0. \tag{23.30}$$

This equation does not apply to $\ell = 0, 1$ perturbations because it would lead to violations of the mass and center-of-mass conservation that are instead automatically satisfied by the collisionless Boltzmann equation. For $\ell = 0, 1$, the perturbations must decay faster and the dispersion equation must maintain its integrodifferential character for all values of r where $\Phi_1 \neq 0$. For $\ell \geq 2$, the regular solution is thus $\Phi^{\ell} \propto r^{-(\ell+1)}$, which we may adopt as the natural boundary condition at large radii. These comments and the following discussion are best suited for the study of a basic state that smoothly extends to infinity, that is, of models that are not radially truncated.[9]

23.1.6 Subsidiary Eigenvalue Problem and Choice of a Set of Basis Functions

To solve the system given by Eqs. (23.20) and (23.17) numerically, we may refer to the following subsidiary eigenvalue problem:

$$\wp^{\ell}[\varphi] = \lambda^{\ell} g(r)\varphi, \tag{23.31}$$

where $\wp^{\ell}[\varphi]$ is the Poisson operator defined on the left-hand side of Eq. (23.17) on the function $\varphi = \varphi(r)$, and $g(r)$ is a smooth function to be suitably chosen with properties similar to those of the unperturbed density $\rho(r)$.

Equation (23.31) has a discrete spectrum of negative real eigenvalues λ_n^{ℓ}. Then each eigenfunction $\varphi_n^{\ell}(r)$, (1) has good properties with respect to the Poisson operator, (2) incorporates the appropriate boundary conditions for the problem ($\ell \geq 2$), and (3) operates on the scale defined by the equilibrium model. Therefore, $\{\varphi_n^\ell\}$ and $\{g(r)\varphi_n^\ell\}$ are convenient bases for expanding the eigenfunctions $\Phi^\ell(r)$ and $\rho^\ell(r)$ [the right-hand side of Eq. (23.17) is $4\pi G\rho^\ell$]. The elements of the basis φ_n^ℓ can be normalized to unity:

$$\int_0^\infty r^2 g(r)\varphi_n^\ell(r)\varphi_{n'}^\ell(r)\,dr = \delta_{nn'}. \tag{23.32}$$

23.1.7 Global Dispersion Relation

Let $\rho_n^\ell(r)$ be the density response obtained by integration over velocity space of the solution of Eq. (23.20) with Φ^ℓ replaced with φ_n^ℓ. By expanding $\rho_n^\ell(r)$ onto the basis $\{g(r)\varphi_n^\ell\}$, we define the collisionless Boltzmann matrix $\Pi_{nm}^\ell(\omega)$ such that

$$4\pi G\rho_n^\ell(r) = \sum_m \Pi_{nm}^\ell(\omega)g(r)\varphi_m^\ell(r). \tag{23.33}$$

Then the desired eigenvalue equation for the mode frequency ω can be cast in the algebraic form

$$\| \lambda_n^\ell \delta_{nm} - \Pi_{nm}^\ell(\omega) \| = 0. \tag{23.34}$$

Therefore, the problem is now reduced to calculation of the collisionless Boltzmann matrix $\Pi_{nm}^\ell(\omega)$. Following Eq. (23.7), we separate two contributions

$$\Pi_{nm}^\ell(\omega) = \Pi_{nm}^{\ell(ad)} + \Pi_{nm}^{\ell(h)}(\omega). \tag{23.35}$$

Their general form is

$$\Pi_{nm}^{\ell(ad)} = 8\pi^2 G \int_0^\infty J\,dJ \int_{E_{\min}(J)}^0 dE \frac{\partial f_0}{\partial E}$$

$$\times \oint \frac{\varphi_m^\ell(r)\varphi_n^\ell(r)}{\{2[E - \Phi_{\mathrm{eff}}(J^2,r)]\}^{1/2}}\,dr, \tag{23.36}$$

$$\Pi_{nm}^{\ell(h)}(\omega) = \sum_{s,p} \mathcal{R}_{sp}^\ell(\omega) \left[\Phi_m^{\ell sp} \Phi_n^{\ell sp} \right], \tag{23.37}$$

where p is an integer; the integral operator $\mathcal{R}$ acting on the quantity inside the square brackets is defined as

$$\mathcal{R}_{sp}^\ell(\omega)\,[\cdots] = 8\pi^2 G \left(A_s^\ell\right)^2 \int_0^\infty J\,dJ \int_{E_{\min}(J)}^0 dE$$

$$\times \tau_r(E,J) \left(\frac{\omega \frac{\partial f_0}{\partial E} + s\frac{\partial f_0}{\partial J}}{-\omega - p\Omega_r + s\Omega_\theta} \right) [\cdots] \tag{23.38}$$

and

$$\Phi_m^{\ell sp}(E,J) = \frac{2}{\tau_r} \int_{r_{\min}}^{r_{\max}} \frac{dr}{\{2[E - \Phi_{eff}(J^2,r)]\}^{1/2}}$$

$$\times \varphi_m^\ell(r)\cos\left[\Theta_{ps}(E,J,r)\right], \tag{23.39}$$

where

$$\Theta_{ps}(E,J,r) = \int_{r_{\min}}^{r} \frac{p\Omega_r(E,J) + s\tilde{\Omega}(E,J,r')}{\{2[E - \Phi_{eff}(J^2,r')]\}^{1/2}} dr'. \tag{23.40}$$

23.2 Radial-Orbit Instability

The rather detailed outline of the linear stability problem given in Section 23.1 should provide a sufficiently clear indication of the many difficulties intrinsic to a stability analysis of spherical stellar systems. The final eigenvalue equation [see Eq. (23.34)] requires nontrivial numerical investigations,[10] practically without any guidance from physical or semianalytical considerations. It is thus not surprising that our knowledge in this research area is still very sketchy and far from satisfactory.[11] Other help and insights naturally derive from *N*-body simulations, best suited for a description of the nonlinear evolution and able to address initially nonspherical cases. Nonetheless, the semianalytical approach of investigating the linear modal equations has led to a remarkable simple discovery (see below) that probably summarizes most of what is currently known of astrophysical interest in this context. This is a criterion for the onset of the so-called radial-orbit instability.

23.2.1 A Conjectured Criterion for Instability

If a spherical stellar system contains too much of its kinetic energy in the radial direction, then the system is bound to develop an $\ell = 2$ instability[12] called *radial-orbit instability*. To quantify the amount of pressure anisotropy sufficient to drive the instability, a criterion has been proposed:[13]

$$\frac{2K_r}{K_T} \geq 1.7 \pm 0.25. \tag{23.41}$$

Here we have in mind a stellar system with negligible internal streaming, for which the integrated kinetic-energy content $K = K_r + K_T$ is in the form of random motions (in the radial or tangential direction; see also Subsection 22.4.3); for a system supported by scalar pressure (isotropic case), we obviously have $K_T = 2K_r$.

Some efforts have been made to show the existence of counterexamples[14] or to pinpoint the precise threshold value[15] of $2K_r/K_T$. Most likely the preceding should be considered only as a handy instability indicator, with the specific numbers to be determined for each class of model under investigation depending on the density and pressure-anisotropy profiles. For example, as mentioned in Subsection 22.4.3, for the f_∞ models, another global pressure-anisotropy indicator is given by the ratio r_α/r_M; this parameter generally correlates with $2K_r/K_T$ (Fig. 23.1). For the f_∞ models, the radial-orbit instability threshold can be expressed as $r_\alpha/r_M \approx 1$. There is no reason for the existence of a simple universal criterion of the type given earlier.

Indeed, some models constructed as a result of collisionless collapse have been found[16] to be quasi-spherical and without evidence for radial-orbit instability, although they are characterized by values of $2K_r/K_T$ as high as 2.75 and values of r_α/r_M as low as 0.2 (Fig. 23.2). These models

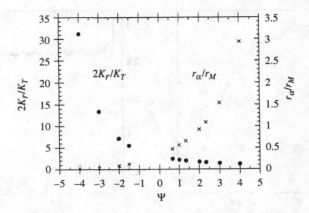

Fig. 23.1. Global radial anisotropy of the f_∞ models measured in two ways (from Bertin, G., Pegoraro, F., Rubini, F., Vesperini, E. 1994. *Astrophys. J.*, **434**, 94). We recall that the radius r_α marks the transition from the inner isotropic region of the model to the outer parts dominated by radial pressure. Also shown is the part of the sequence $\Psi < 0$ considered by some authors. This part is characterized by high radial anisotropy. For example, for the $\Psi = -3/2$ model, the value of $2K_r/K_T$ is 5.5, significantly above the threshold of radial-orbit instability, and indeed, most of the model is dominated by radial pressure because $r_\alpha = 0.12\, r_M$.

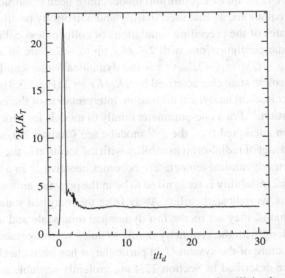

Fig. 23.2. Evolution of the ratio $2K_r/K_T$ for a simulation of collisionless collapse under very symmetric initial conditions (from Trenti, M., Bertin, G. 2006. *Astrophys. J.*, **637**, 717). Note the high value of the anisotropy parameter ($2K_r/K_T > 20$) reached during the collapse; the final equilibrium state is quasi-spherical (with axial ratios ≈ 0.97) and is characterized by $2K_r/K_T \approx 2.75$.

come from numerical experiments of collisionless collapse characterized by highly symmetric initial conditions, in which phase mixing is inefficient (in contrast to those at the basis of the analysis described in Subsection 22.4.1), under cold initial conditions. It appears that the physical factor at the basis of the observed partial suppression of the radial-orbit instability in these models

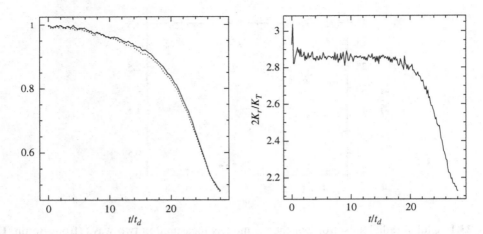

Fig. 23.3. Axial ratios (*left*) and anisotropy parameter (*right*) for the evolution of a metastable model ($2K_r/K_T = 2.92$ at $t = 0$; from Trenti, M., Bertin, G. 2006. *Astrophys. J.*, **637**, 717).

is the existence of an almost perfectly isotropic central region formed during collapse, with a sharp transition to an outer envelope characterized by radially biased pressure anisotropy. To confirm this picture, N-body quasi-equilibrium models have been constructed a posteriori, with the help of the Jeans equations, so that their density and anisotropy profiles qualitatively match those found as end states of the preceding simulations of collisionless collapse. It thus has been demonstrated that some configurations with $2K_r/K_T$ up to ≈ 2.9 are in a metastable state for a period of $\approx 20t_d$ [$t_d = GM^{5/2}/(-2E_{tot})^{3/2}$ is the dynamical time scale], beyond which they finally evolve into a prolate state characterized by $2K_r/K_T \approx 2$ (Fig. 23.3).

In the absence of a clear-cut analytical derivation, interpretation of the mechanism of instability is still open to question.[17] For a one-parameter family of models for which a thermodynamical interpretation has been attempted (i.e., the $f^{(\nu)}$ models; see Chapter 25), a curious coincidence has been noted of the onset of radial-orbit instability with the location in the equilibrium sequence at which a global thermodynamical temperature becomes negative.[18] In any case, the source of free energy for the $\ell = 2$ instability is recognized to be in the pressure anisotropy, in the excessive amount of radial orbits, as indicated earlier. Away from the marginal state, the instabilities are found to be violent; that is, they act on the fast dynamical time scale and are shown by N-body simulations to lead, along with the symmetry break, to quick rearrangements of the mass distribution and orbital structure of the system.[19] In particular, it has been checked that models with $\Psi < 2$ of the sequence described in Section 22.4 are violently unstable and thus uninteresting for modeling observed objects, whereas those found to have realistic profiles ($\Psi \geq 7$) are not too anisotropic and stable with respect to radial-orbit instability. By means of N-body simulations, it also has been found[20] that dark halos probably do not help in changing the preceding instability criterion and thus cannot cure radial-orbit instability.

23.2.2 Constraints on the Intrinsic Structure of Elliptical Galaxies

Observed galaxies are likely to have already evolved away from unstable configurations and thus to be in a state of slow evolution. Within this perception, can radial-orbit instability have had an impact on the initial stages of galaxy evolution? Can the related mechanism have played a role during galaxy collapse so that the current distribution of intrinsic morphologies basically reflects the results of radial-orbit instability? This interesting possibility poses the serious problem that the violent instabilities involved modify the overall density distribution, leading to unrealistic density profiles (i.e., to profiles with dominant cores[21]). Furthermore, the collapse simulations of galaxy formation[22] often lead to prolate and triaxial end products, whereas most observed ellipticals are probably not far from the oblate case.[23] As briefly described at the end of Subsection 7.2.2, one alternative picture for the generation of significantly nonspherical systems is that of merging, among which the possibly rare case of head-on encounters is expected to lead to final products of the prolate type; in contrast, the spiraling in of a captured satellite on a more tangential orbit is likely to generate a final state of the oblate type.

Another conceptually interesting application of stability studies, also beyond the issue of radial-orbit instability, is to exclude a number of models that otherwise might be considered genuine equilibrium states. For example, the negative-temperature part ($\Psi < 0$) of the sequence of f_∞ models[24] is fiercely unstable[25] and thus cannot be used to model galaxies (Figs. 23.4 and 23.5). Some unusual velocity distributions sometimes invoked in conservative analyses as alternatives to the presence of massive black holes in galaxy nuclei[26] or massive dark halos[27] and, in general, many of the models obtained based on the density-priority approach (see Subsection 22.1.2) might simply have to be ruled out by stability considerations. Such a narrowing down of the set of acceptable models is one example of the process that is often called *studying the dynamical window* (see Chapter 5).

Along this line of thinking, another long-term project has been to try to show that elliptical galaxies flatter than E7 cannot exist because all the models that might be imagined for them would be violently unstable.[28] This problem appears to be difficult and not well defined in view of our limitations in constructing nonspherical galaxy models (see Chapter 22).

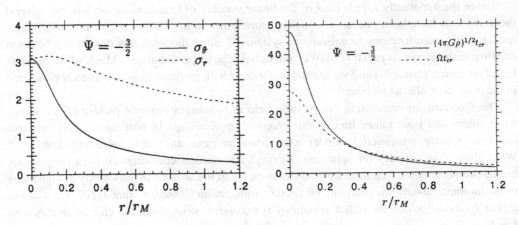

Fig. 23.4. Properties of the basic state of the violently unstable model $\Psi = -3/2$. *Left:* Radial profiles of the radial and tangential velocity dispersions in the natural units of r_M/t_{cr}. *Right:* Radial profiles for two dimensionless frequencies. Here t_{cr} is the crossing time defined as $t_{cr} = GM^{5/2}/(2K)^{3/2}$ (from Bertin, G., Pegoraro, F., Rubini, F., Vesperini, E. 1994. *Astrophys. J.*, **434**, 94).

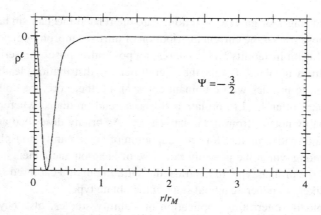

Fig. 23.5. Radial part of the density eigenfunction of the most unstable $\ell = 2$ mode in the $\Psi = -3/2$ model; for this mode, the dimensionless growth rate γt_{cr} is ≈ 8. Note how the mode develops in the vicinity of r_α and is localized in the inner part of the model (from Bertin, G., Pegoraro, F., Rubini, F., Vesperini, E. 1994. *Astrophys. J.*, **434**, 94).

23.2.3   Some Theoretical Issues

Earlier we focused on issues that have a direct impact on the astrophysical context. In the study of the stability of spherical and spheroidal systems, there are many results and open questions of a more theoretical interest that should be mentioned briefly. In particular, it has long been argued, possibly based on experience from the study of collisionless plasmas, that meaningful conditions for stability could be found by considering the derivatives of the underlying distribution function. However, some simple results, by analogy with the conditions for Landau damping, have to be reformulated, as is required by the nontrivial context of a self-consistent inhomogeneous global basic state.

Within the apparently simple case of the linear stability of spherical stellar systems, several theorems thus have been proved that involve conditions on df/dE and $d^3\rho/d\Phi^3$ for isotropic models and generalizations to anisotropic systems;[29] often the theorems distinguish between stability with respect to radial ($\ell = 0$) or nonradial ($\ell \neq 0$) perturbations. Much of the work is based on energy principles and variational methods,[30] with methods basically already pioneered in the context of plasma physics.[31]

The theorems are interesting but of little help in forming a physical picture of the stability problem and have rather limited astrophysical applications. In particular, many intuitive questions remain unanswered: Under what conditions can an $\ell = 0$ instability develop?[32] What about $\ell = 1$ modes for spherical systems? Are models characterized by a tangentially biased pressure tensor free from instabilities even when the amount of anisotropy is high? Can nonmonotonic density distributions be prone to instabilities, and of what type? What about higher-ℓ modes? When the radial anisotropy is excessive, what about the case of modes with $\ell \neq 2$?

Other interesting aspects refer to the spectrum of linear modes within a given value of ℓ. Is it true that modes generally develop with ω^2 real?[33] If so, why? Energy principles may be misleading in this context given the nontrivial role of resonances. An N-body study of marginal stability is very difficult to set up, but even semianalytical studies, based on the equations developed in the preceding section, so far have been unable to provide clear-cut answers. How should we envision the processes of slow evolution as occurring close to marginal stability? Are there well-defined discrete damped modes, or do damped modes come in a continuous spectrum? Obviously, we have a long path ahead toward the goal of understanding nonlinear dynamical evolution.

23.3 Stability of Rotating Stellar Systems

The study of the stability of rotating stellar systems, beyond the case of relatively cool disks described in Part III, is not well developed. Because of the difficulties in setting up a rigorous linear stability analysis, already evident from the spherical case, the study is largely limited to investigations based on numerical experiments.[34] As noted at the beginning of this chapter and in Subsection 23.2.2, a thorough stability analysis would be astrophysically interesting either to interpret the origin of observed morphologies or to identify violently unstable models as unsuitable for application. After the discovery that typically bright ellipticals are not supported by rotation, the study of equilibrium models of rotating stellar systems has made only little progress. Consequently, little attention also has been paid to the related stability problem.

A general underlying theme of the investigations made in this context has been to test the generality of the conjectured criterion, described in Section 20.2, according to which axisymmetry is incompatible with excessive amounts of rotation, as measured by the parameter $t = K_{ord}/|W|$, where the kinetic energy in the form of rotation is compared with the gravitational binding energy.[35] One line of research has focused on the benchmark case of the stellar dynamical analogues of the classical rotating fluids (see Chapter 10), in particular, on the study of the stability of stellar dynamical analogues of the classical Maclaurin spheroids[36] and the uniformly rotating polytropes.[37] Comparison between the possible differences of behavior in fluid and stellar dynamical models has been considered by means of a variety of methods.[38] Study of the stability of rotating self-gravitating fluids has been the focus of renewed attention in a recent line of research, and this has also followed the general theme of the applicability of the Ostriker-Peebles criterion.[39]

A systematic study, by means of N-body simulations, of the stability properties of the family of differentially rotating f_{WT}^d models introduced in Subsection 22.3.4 has been performed.[40] On the one hand, dynamical instabilities are found at relatively low levels of rotation, in terms of the t parameter. On the other hand, in the regime of moderate rotation strength there are equilibrium configurations exhibiting toroidal structure that appear to be free of dynamical instabilities.

Notes

1. See Weinberg, M. D. 1994. *Astrophys. J.*, **421**, 481; Vesperini, E., Weinberg, M. D. 2000. *Astrophys. J.*, **534**, 598. Some of these investigations may have an impact on the dynamics of galaxy disks; for example, the tidal driving on a disk may be mediated by the collective response of a lively spheroidal halo, as discussed by Weinberg, M. D. 1998. *Mon. Not. Roy. Astron. Soc.*, **299**, 499.
2. Fridman, A. M., Polyachenko, V. L. 1984. *Physics of Gravitating Systems*, Springer-Verlag, Berlin; Palmer, P. L. 1994. *Stability of Collisionless Stellar Systems*, Kluwer, Dordrecht, The Netherlands.
3. Polyachenko, V. L., Shukhman, I. G. 1981. *Sov. Astron. AJ*, **25**, 533.
4. In particular, the various theorems that can be derived from energy principles.
5. Here we follow the article by Bertin, G., Pegoraro, F., Rubini, F., Vesperini, E. 1994. *Astrophys. J.*, **434**, 94; in practice, much of the analysis repeats the derivation by Fridman, A. M., Polyachenko, V. L.

1984. *op. cit.*, and has many points in common with the derivation of the matrix equation by Kalnajs, A. J. 1977. *Astrophys. J.*, **212**, 637, for collisionless disks.

6. Vilenkin, N. Ya. 1968. *Special Functions and the Theory of Group Representations*, Vol. 22 of AMS *Translations of Mathematical Monographs*, American Mathematical Society, Providence, RI.

7. Coppi, B., Rewoldt, G. 1976. *Adv. Plasma Phys.*, **6**, 421.

8. Landau, L. D. 1946. *J. Phys. USSR*, **10**, 25.

9. See also Saha, P. 1991. *Mon. Not. Roy. Astron. Soc.*, **248**, 464.

10. See Palmer, P. L., Papaloizou, J. 1987. *Mon. Not. Roy. Astron. Soc.*, **224**, 1043; 1988. *Mon. Not. Roy. Astron. Soc.*, **231**, 935; Weinberg, M. D. 1989. *Mon. Not. Roy. Astron. Soc.*, **239**, 549; 1991. *Astrophys. J.*, **368**, 66; 1994. *Astrophys. J.*, **421**, 481; Saha, P. 1991. *Mon. Not. Roy. Astron. Soc.*, **248**, 464; 1992. *Mon. Not. Roy. Astron. Soc.*, **254**, 132; Bertin, G., Pegoraro, F., Rubini, F., Vesperini, 1994. *op. cit.*

11. A systematic survey of what is known about the stability of stellar systems is given in the monographs by Fridman, A. M., Polyachenko, V. L. 1984. *op. cit.*, and by Palmer, P. L. 1994. *op. cit.*, in which some insight is also gained by considering simpler but less realistic equilibrium configurations.

12. The instability, initially found by Polyachenko, V. L., Shukhman, I. G. 1981. *Sov. Astron. AJ*, **25**, 533, was later confirmed by a number of *N*-body simulations, e.g., by Barnes, J., Goodman, J., Hut, P. 1986. *Astrophys. J.*, **300**, 112.

13. The numbers quoted here correspond to the estimate reported by Fridman, A. M., Polyachenko, V. L. 1984. *op. cit.*

14. Palmer, P. L., Papaloizou, J. 1987. *Mon. Not. Roy. Astron. Soc.*, **224**, 1043; Dejonghe, H., Merritt, D. 1988. *Astrophys. J.*, **328**, 93. For models with density profiles of the form suggested by Dehnen, W. 1993. *Mon. Not. Roy. Astron. Soc.*, **265**, 250, which include Jaffe's and Hernquist's density-potential pairs (see Subsection 22.4.4), and pressure-anisotropy profile of the Osipkov-Merritt type (Osipkov, L. P. 1979. *Sov. Astron. Lett.*, **5**, 42; Merritt, D. 1985. *Astron. J.*, **90**, 1027), it has been argued that the threshold value for the onset of radial-orbit instability should be raised to 2.31 ± 0.27; Meza, A., Zamorano, N. 1997. *Astrophys. J.*, **490**, 136.

15. See Merritt, D. 1988. *Astron. J.*, **95**, 496; Saha, P. 1991. *Mon. Not. Roy. Astron. Soc.*, **248**, 464.

16. Trenti, M., Bertin, G. 2006. *Astrophys. J.*, **637**, 717.

17. See the review articles by Polyachenko, V. L., on p. 301, and by Merritt, D., on p. 315, in 1987. *IAU Symposium*, Vol. 127, ed. P. T. de Zeeuw. Reidel, Dordrecht, The Netherlands; see also Weinberg, M. D. 1991. *op. cit.*

18. Bertin, G., Trenti, M. 2003. *Astrophys. J.*, **584**, 729.

19. See Bertin, G., Stiavelli, M. 1989. *Astrophys. J.*, **338**, 723; Londrillo, P., Messina, A., Stiavelli, M. 1991. *Mon. Not. Roy. Astron. Soc.*, **250**, 54; Bertin, G., Pegoraro, F., Rubini, F., Vesperini, E. 1994. *op. cit.*

20. Stiavelli, M., Sparke, L. S. 1991. *Astrophys. J.*, **382**, 466.

21. See the simulations by Londrillo, P., Messina, A., Stiavelli, M. 1991. *op. cit.*

22. See Aguilar, L., Merritt, D. 1990. *Astrophys. J.*, **354**, 33; Dubinski, J., Carlberg, R. G. 1991. *Astrophys. J.*, **378**, 496; Quinn, P. J. 1993. In *Proceedings of the ESO/EIPC Workshop on Structure, Dynamics, and Chemical Evolution of Early-Type Galaxies*, eds. I. J. Danziger et al. European Southern Observatory Publications, Garching, Germany, p. 371.

23. See Franx, M., Illingworth, G. D., de Zeeuw, P. T. 1991. *Astrophys. J.*, **383**, 112.

24. Considered by Merritt, D., Tremaine, S., Johnstone, D. 1989. *Mon. Not. Roy. Astron. Soc.*, **236**, 829.

25. See Stiavelli, M., Sparke, L. 1991. *op. cit.*; Bertin, G., Pegoraro, F., Rubini, F., Vesperini, E. 1994. *op. cit.*

26. See Binney, J. J., Mamon, G. A. 1982. *Mon. Not. Roy. Astron. Soc.*, **200**, 361.

27. Tonry, J. L. 1983. *Astrophys. J.*, **266**, 58; see the models constructed by the method of Dejonghe, H. 1989. *Astrophys. J.*, **343**, 113.

28. This line of work has been pursued, e.g., by Merritt, D., Stiavelli, M. 1990, *Astrophys. J.*, **358**, 399; Merritt, D., Hernquist, L. 1991. *Astrophys. J.*, **376**, 439. See also Levison, H. F., Duncan, M. J., Smith, B. F. 1990. *Astrophys. J.*, **363**, 66.

29. Antonov, V. A. 1960. *Astr. Zh.*, **37**, 918 (trans. *Sov. Astron.*, **4**, 859); 1962. *Vestn. Leningr. Univ.*, **7**, 135 (trans. in *IAU Symposium*, Vol. 113, eds. J. Goodman, P. Hut. Reidel, Dordrecht, The Netherlands),

and *Vestn. Leningr. Univ.*, **19**, 96 (trans. in *IAU Symposium*, Vol. 127, ed. P. T. de Zeeuw. Reidel, Dordrecht, The Netherlands).

30. See Kulsrud, R., Mark, J. W.-K. 1970. *Astrophys. J.*, **160**, 471; Doremus, J. P., Feix, M. R., Baumann, G. 1971. *Phys. Rev. Lett.*, **26**, 725; Sygnet, J. F., Forets, G. D., et al. 1984. *Astrophys. J.*, **276**, 737; Kandrup, H. E., Sygnet, J. F. 1985. *Astrophys. J.*, **298**, 27; Habib, S., Kandrup, H. E., Yip, P. F. 1986. *Astrophys. J.*, **309**, 176; May, A., Binney, J. J. 1986. *Mon. Not. Roy. Astron. Soc.*, **221**, 13; Goodman, J. 1988. *Astrophys. J.*, **329**, 612; de Zeeuw, P. T., Schwarzschild, M. 1989. *Astrophys. J.*, **345**, 84; 1991. *Astrophys. J.*, **369**, 57; Kandrup, H. E., Morrison, P. J. 1993. *Ann. Phys.*, **225**, 114.

31. See Kruskal, M. D., Oberman, C. R. 1958. *Phys. Fluids*, **1**, 275; Rosenbluth, M. N., Rostoker, N. 1959. *Phys. Fluids*, **2**, 23; Taylor, J. B., Hastie, R. J. 1965. *Phys. Fluids*, **8**, 323; Grad, H. 1966. *Phys. Fluids*, **9**, 225.

32. Hénon, M. 1973. *Astron. Astrophys.*, **24**, 229, found that some systems, generalized polytropes, with $\partial f / \partial E > 0$, are stable and others instead are violently unstable with respect to radial instabilities; the radial-velocity distribution of the unstable models presented two separate maxima and might thus have undergone a kind of fire-hose or two-stream instability. The physical interpretation of these results is still largely unclear.

33. Overstabilities have been claimed to occur in the presence of an external field associated with a massive black hole by Palmer, P. L., Papaloizou, J. 1988. *Mon. Not. Roy. Astron. Soc.*, **231**, 935.

34. For a general review, we may refer again to the monographs by Fridman, A. M., Polyachenko, V. L. 1984. *op. cit.*; Palmer, P. L. 1994. *op. cit.*

35. Ostriker, J. P., Peebles, P. J. E. 1973. *Astrophys. J.*, **186**, 467. By means of *N*-body simulations, Miller, R. H., Smith, B. F. 1980. *Astrophys. J.*, **235**, 793, find that oblate stellar systems can rotate very rapidly and yet exhibit only relatively little flattening; they also identify rotating systems with fairly high values of *t* that appear to violate the Ostriker-Peebles criterion and finally argue, from the class of models explored by their simulations, that flat ellipticals are probably triaxial. A class of inhomogeneous oblate models with density-potential pairs of the Stäckel form has been studied by means of *N*-body simulations by Sellwood, J. A., Valluri, M. 1997. *Mon. Not. Roy. Astron. Soc.*, **287**, 124, and found to be generally consistent with the Ostriker-Peebles criterion.

36. Vandervoort, P. O. 1991. *Astron. J.*, **377**, 49, by means of a linear stability analysis in terms of fluid moment equations, with no explicit use of the supporting distribution function, addresses systematically the stability properties of the main modes associated with these systems. Bending instabilities in homogeneous spheroids, based on the stellar dynamical models suggested by Polyachenko, V. L. 1976. *Sov. Phys. Dokl.*, **21**, 417, are studied by Jessop, C. M., Duncan, M. J., Levison, H. F. 1997. *Astrophys. J.*, **489**, 49, with the general goal of explaining why galaxies flatter than E7 are not observed.

37. By means of *N*-body simulations, Miller, R. H., Vandervoort, P. O., et al. 1989. *Astrophys. J.*, **342**, 105, show that the systems are dynamically unstable with respect to toroidal modes when the ratio of the rotational kinetic energy to the gravitational potential energy exceeds a value of ≈ 0.16, in agreement with theoretical predictions made earlier by Vandervoort, P. O. 1983. *Astrophys. J.*, **273**, 511, with the help of tensor virial methods.

38. Christodoulou, D. M., Shlosman, I., Tohline, J. E. 1995. *Astrophys. J.*, **443**, 551; Christodoulou, D. M., Kazanas, D., et al. 1995. *Astrophys. J.*, **446**, 472, and following papers.

39. Centrella, J. M., New, K. C. B., et al. 2001. *Astrophys. J. Lett.*, **550**, L193; Saijo, M., Yoshida, S. 2006. *Mon. Not. Roy. Astron. Soc.*, **368**, 1429; Ou, S., Tohline, J. E. 2006. *Astrophys. J.*, **651**, 1068, and a number of other papers.

40. Varri, A. L. 2012. Ph.D. dissertation, Università degli Studi di Milano, Milano, Italy.

24 Dark Matter in Elliptical Galaxies

From the discussion of Chapter 20 it should be clear that spiral galaxies possess significant amounts of dark matter in the form of a diffuse halo. Thus we should naturally expect that elliptical galaxies also possess sizable dark halos. Can we prove the validity of this expectation? How can we measure the amount and distribution of dark matter in elliptical galaxies? Note that in contrast to the case of spirals, we may easily imagine here that the halo has the same spatial distribution as the luminous matter because both components can be thought to be the result of collisionless collapse.

In general, elliptical galaxies lack a straightforward, radially extended kinematic tracer, such as the atomic hydrogen emission used to derive the rotation curves of spiral galaxies. Furthermore, although for spirals we can produce convincing tests to judge the overall spatial symmetry of the underlying gravitational potential, for ellipticals, such symmetry is a major open question not only in general but also especially for individual objects; in particular, it is difficult to determine whether the system is basically axisymmetric or triaxial. Thus it is not surprising to find that the data of many objects can in principle be fitted by models that assume a constant mass-to-light ratio. However, in many cases, models of this type are found to be difficult to justify from the physical point of view; for example, if the velocity-dispersion profile is rather flat, as for NGC 4472, a significant bias of the pressure tensor in the tangential direction would be required. Why should ellipticals always be formed with basically an $R^{1/4}$ luminosity profile and yet with a wide variety of distributions of stellar orbits (see discussion in Subsection 22.4.6)? In addition, some of the models obtained when a constant M/L is imposed turn out to require a relatively high mass-to-light ratio anyway, often too high to be justified in terms of current models for a normal stellar population without dark matter. Thus, given the clear indications from the case of spiral galaxies, it is sensible to allow for gradients in M/L if that is going to lead to physically acceptable models.

Therefore, if we follow the arguments presented in Subsection 22.4.6 and interpret the natural consequences of physically acceptable fits, we can state that there are several elliptical galaxies for which sizable halos have been found. Indeed, it turns out that the resulting distribution of matter for those objects is not too different from the behavior found in spirals.[1]

With respect to the initial discussions of the 1980s, the subject has made significant progress in general as a result of the identification and use of a variety of kinematic tracers for mass estimates and in particular as a result of a systematic application of studies of gravitational lensing (to this topic a separate chapter is dedicated in Part V). On the one hand, a curious general trend is observed for large ellipticals in both the nearby and the distant universe; that is, the total gravitational potential in these systems, resulting from the contribution of dark and luminous matter, is found to be very close to being logarithmic (the characteristic behavior of the isothermal sphere; this corresponds directly with the paradigm of flat rotation curves in spiral galaxies). On the other hand, uncertainties remain in relation to the amount and distribution of dark matter in low-luminosity stellar systems; even for the smallest spheroidal stellar systems (in particular, globular clusters and dwarf spheroidals) that, because of their vicinity, can be studied in greatest detail, the apparent variety of behavior in relation to their dark halos remains unexplained. In this chapter some of the main points of interest that have emerged from past and more recent investigations are summarized.

24.1 Stellar Dynamical Diagnostics

Spectroscopic studies of the stellar light (i.e., kinematic information on the distribution of stars that can be derived from the absorption lines detected in the spectrum of starlight) provide the most obvious source of data on the gravitational field in elliptical galaxies. But for a long time this information has been regarded as the least useful diagnostic tool, mainly for two reasons. On the one hand, until recently, the available instruments would not allow us to obtain good kinematic profiles at a sufficiently far distance from the center: The profiles generally extended out to only a fraction of R_e and were subject to sizable errors. On the other hand, general skepticism prevailed because it was noted that the observation of a relatively flat velocity-dispersion profile could be interpreted either as being due to the presence of a gradient of the relevant M/L ratio (i.e., the presence of dark matter) or, alternatively, as an indication that the system is largely populated by stars in tangential orbits.[2]

The situation of stellar dynamical diagnostics has now changed considerably. Accurate kinematic profiles can be measured beyond R_e (an accuracy better than ≈ 20 km s^{-1} out to $2R_e$ can be achieved in some cases). From the modeling point of view, for quasi-spherical systems, a framework has been proposed to interpret the universality of the luminosity profile of elliptical galaxies, following the view that the current state of ellipticals may be understood as if they were formed by an essentially collisionless collapse (see Chapter 22, especially Section 22.4). This framework suggests that without any significant effects on the density distribution of the luminous component, the observed velocity-dispersion profile may change considerably in the presence of dark matter; in particular, it is flatter if a diffuse dark halo is present. A simple relation holds in this description of the dynamical state of ellipticals for the amount of dark matter relative to the amount of luminous matter inside the effective radius:

$$\left(\frac{M_D}{M_L}\right)_{R_e} \approx 3.5 \frac{\sigma^2(R_e)}{\sigma^2(0_+)} - 1, \tag{24.1}$$

where $\sigma(R_e)$ represents the observed velocity dispersion at R_e, and $\sigma(0_+)$ is the central velocity dispersion excluding any peculiar feature that may be associated with the dynamics of the nucleus (in particular, the black-hole cusps mentioned in Subsection 22.2.2).[3] The detailed modeling tools are those described by the two-component models of Subsection 22.4.6, which lead to luminous–dark-matter decompositions of the relevant observed profiles.[4] Some galaxies, such as NGC 3379, can be well modeled without a dark halo, whereas others, such as NGC 4472, are found to require significant amounts of dark matter (Figs. 24.1 and 24.2). It is interesting to find that although relatively wide variations from galaxy to galaxy occur in the total mass-to-light ratio, the ratio M_L/L determined empirically in these decompositions turns out to be relatively

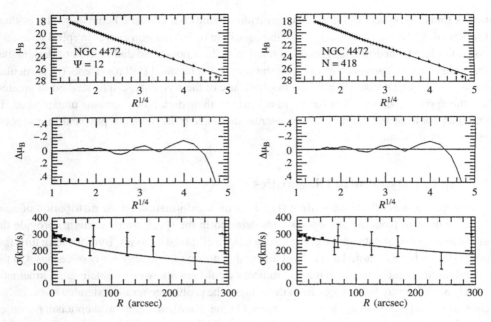

Fig. 24.1. Comparison of best-fit stellar dynamical models for NGC 4472. The fit on the left is performed with one-component constant mass-to-light-ratio f_∞ models, and the fit on the right is with two-component f_∞ models (from Saglia, R. P., Bertin, G., Stiavelli, M. 1992. *Astrophys. J.*, **384**, 433).

stable ($M_L/L_B \approx 5$; this value is based on the long-distance scale, with NGC 4472 at 27 Mpc, a choice often made in the 1990s when these studies were made), consistent with that of reasonable stellar populations.

Some trends with size and X-ray and radio properties of the objects found to possess a dark halo have been noted.[5] From a sample of ten well-studied elliptical galaxies, it appears that evidence for massive dark halos emerges preferentially for systems with large R_e (> 8 kpc, again based on the long-distance scale, with NGC 4472 at 27 Mpc). Furthermore, the cases with a positive determination of a massive dark halo turn out to be usually powerful X-ray and radio sources.

In contrast, if the same objects, such as NGC 4472 and NGC 7796, for which we conclude that dark matter is present are modeled with a constant M/L ratio, then a distribution of stellar orbits is found that is largely populated by tangential orbits (Fig. 24.3). It would be very difficult to imagine a formation process leading to mostly quasi-circular motions with essentially no or little net overall rotation. Furthermore, when the distribution function departs significantly from a Gaussian, collective instabilities may take place on a short time scale, so the related models would not be physically acceptable.

The general picture noted earlier is basically confirmed by a study of twenty-one mostly luminous, slowly rotating, and nearly round ellipticals, for which the relevant line profiles have been measured systematically.[6] The line profiles remained difficult to measure with the desired accuracy.[7] The main conclusions proposed by the study are (1) the circular velocity of the galaxies is often rather flat (see Fig. 22.18), (2) galaxies are characterized by only mild pressure anisotropy,

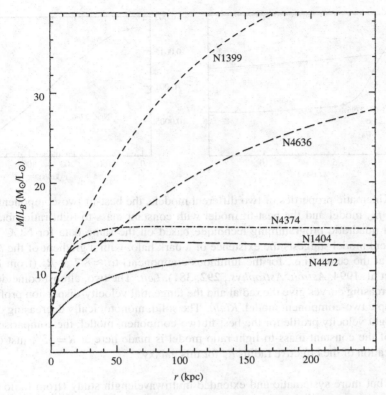

Fig. 24.2. Cumulative mass-to-light ratios (blue band) for the best-fit two-component f_∞ models for a sample of galaxies that turn out to be associated with an X-ray halo (from Bertin, G., Toniazzo, T. 1995. *Astrophys. J.*, **451**, 111; the stellar dynamical analysis had been performed in the paper by Saglia, R. P., Bertin, G., Stiavelli, M. 1992. *Astrophys. J.*, **384**, 433, in which the uncertainties associated with these measured profiles are discussed in detail). The gradients show the amount of dark matter present.

biased in the radial direction, with fairly uniform internal dynamical structure (see the models described in Section 22.4), (3) the dark matter typically contributes ≈ 10 to 40 percent of the mass inside R_e, and (4) some galaxies have no indication for dark matter inside $2R_e$ (see the results on NGC 3379 mentioned earlier and Fig. 22.16).

A first exploratory survey[8] has emphasized the interest in obtaining two-dimensional maps for a combined gas (see also Section 24.2) and stellar dynamics investigation. The scientific goals of this initiative go well beyond the issues related to the amount and distribution of dark matter in early-type galaxies; they include the important issues of galaxy morphology, rotation, stellar populations, presence of a central massive black hole, scaling relations, formation, and evolution. In relation to dark matter, the survey suggests a median dark matter fraction of ≈ 30 percent of the total mass inside R_e, in broad agreement with previous studies. It gives indications that faster-rotating galaxies have lower dark-matter fractions than slower-rotating and generally more massive objects and some evidence against cuspy nuclear dark-matter profiles in galaxies.[9] The stellar dynamical analysis has been carried out by means of two-integral Jeans and three-integral Schwarzschild dynamical models on kinematic data out to $\approx R_e$.

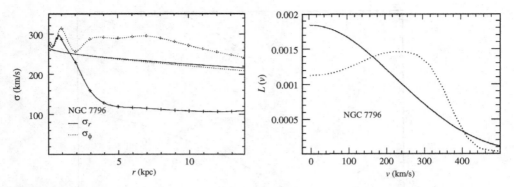

Fig. 24.3. Kinematic properties of two different models, the best-fit two-component (i.e., with dark matter) f_∞ model and the best-fit model with constant mass-to-light ratio obtained from the so-called quadratic programming technique based on the same data for NGC 7796. The two-component model suggests the existence of a dark halo, with a gradient of the M/L_B ratio from ≈ 4.7 (at the center, i.e., for the luminous component) to ≈ 8.7 at R_e (from Bertin, G., Bertola, F., et al. 1994. *Astron. Astrophys.*, **292**, 381). *Left:* The two, almost coincident, monotonically decreasing curves give the radial and the tangential velocity-dispersion profiles for the almost isotropic two-component model. *Right:* The solid, monotonically decreasing curve gives the line-of-sight velocity profile for the best-fit two-component model; the comparison with the line profile of the constant mass-to-light ratio model is made here at $R = 35''$, just outside the estimated location of the effective radius R_e for this galaxy.

A similar but more systematic and extended multiwavelength study (from radio to optical; therefore, the data refer not only to the stars but also to various gas components when present; see also Section 24.2) with two-dimensional kinematics for a volume-limited sample of 260 nearby ($D < 42$ Mpc) early-type galaxies,[10] divided into 68 ellipticals and 192 S0s, has led to a wealth of empirical results. In particular, the study is largely aimed at quantifying the distinct properties of two classes of early-type galaxies, fast rotators (two-thirds of the ellipticals of the sample turn out to be fast rotators) and slow rotators, and at reconstructing their formation and evolution histories. Slow rotators are typically massive. The study leads to the conclusion that the majority of early-type galaxies are consistent with being nearly oblate systems, and only a small fraction of them (less than 12 percent) have central mildly triaxial structures.

24.2 Other Kinematic Tracers

24.2.1 Hot Interstellar Gas

The discovery of radially extended X-ray emission from many ellipticals immediately raised great expectations about the possibility of using these data as independent diagnostics of the potential well in these objects.[11] This X-ray emission points to the presence of a hot, diffuse gas (actually, plasma), at least in the brighter objects. In the simplest model for this hot gas, we assume quasi-hydrostatic equilibrium and spherical symmetry, that is,

$$\left(\frac{1}{\rho}\frac{dp}{dr}\right)_{gas} \approx -\frac{GM(r)}{r^2}, \tag{24.2}$$

and thus hope to derive the gravitational force necessary to bind the galaxy from the observed density and temperature properties of the gas. In the earlier papers on this issue it was concluded that indeed X-rays provide strong evidence, possibly the only reliable proof, that dark halos surround elliptical galaxies. Unfortunately, the extraction of reliable temperature profiles for the hot gas from the spectroscopic information is a formidable task even in the most favorable cases,[12] and the physics of the hot cooling gas is very complex. Thus caution has been urged,[13] especially because the bright X-ray-emitting galaxies are only found in clusters, in which the confinement of the hot gas is due not only to the gravitational field of the individual galaxy but also to the external pressure provided by the intracluster medium.

One interesting case study is that of the galaxy NGC 4472, for which it has been shown that X-ray data available in the 1990s out to $7R_e$ could not discriminate between stellar dynamical models with or without dark matter[14] (Fig. 24.4). The test was performed by taking the best available stellar dynamical diagnostics out to $\approx R_e$ (see Section 24.1) and by considering three potential wells, that of the best-fit model F, which was found to demand the presence of a sizable dark halo; that of the model E, with the most massive halo compatible with the same data; and the potential well of the best-fit model D constructed by imposing a constant mass-to-light ratio (i.e., the best-fit model with no dark matter); for the three models, the predicted X-ray temperature profiles for the hot gas have then been produced, and, surprisingly, they show no significant differences from one another and all in reasonable agreement with the *Einstein* data.[15] Later this type of temperature prediction was checked against much more stringent X-ray temperature profiles from ROSAT (in particular, for NGC 4636.[16])

One interesting fact associated with the hot gas is the possible decoupling of shape between the optical image of an elliptical galaxy (given by the starlight) and its X-ray image (which, it is hoped, traces the gravitational potential well). Such decoupling has been observed in flat objects, such as NGC 720, and suggests the presence of a fairly round halo somewhat tilted with respect to the luminous component[17] (Fig. 24.5).

In the meantime, great progress has taken place in X-ray observations, especially by means of observations taken by *Chandra* and XMM-*Newton*.[18] Some galaxies thus have been restudied in depth. In particular, the E5 galaxy NGC 720 required new investigations because an apparently unexplained twist observed in the position angle of the X-ray axis "cast doubt on whether the X-ray emission indeed traces the potential" in this galaxy.[19] The new *Chandra* data eventually confirmed the presence of the decoupling of the properties of the X-ray isophotes from those of the optical isophotes, although in different quantitative terms (in particular, the observed X-ray ellipticity was found to be significantly smaller than that measured by ROSAT, which was attributed to unresolved point sources present in the ROSAT observations; the position angle of the X-ray axis also was found to have different characteristics). Eventually, the authors concluded that the presence of a triaxial dark halo with geometric properties distinct from those of the stars would rule out explanations of alternative gravity theories, such as Modified Newtonian Dynamics (MOND) (see also the geometric argument offered by the study of the Bullet Cluster, briefly described at the end of this chapter). The galaxy was then reanalyzed in combination with the *Suzaku* data (best suited to probe the faint outer parts), giving general support to a model based on hydrostatic equilibrium conditions.[20] NGC 720 is one of the seven ellipticals considered in an earlier study based on *Chandra* data[21] and one of the fifty-three ellipticals studied by combining *Chandra* and XMM-*Newton* data.[22]

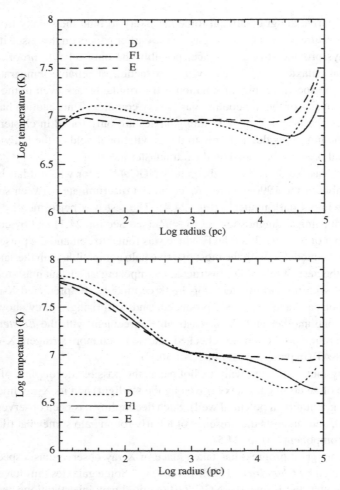

Fig. 24.4. Comparison of the hot-gas temperature profiles for three models with different amounts of dark matter (D is without dark matter; see text) for the X-ray emission of NGC 4472, constrained by the available stellar dynamical data. The upper frame is a simple $q = 0$ cooling-flow model; the lower frame adopts a semiempirical hydrostatic model (see Subsection 25.3.1 for explanations). The emission-weighted temperatures for cases D, E, and F1, in units of 10^6 K, are, respectively, 8.5, 9.5, and 8.4 in the cooling-flow model or 6.5, 9.5, and 7 for the hydrostatic model (from Bertin, G., Pignatelli, E., Saglia, R. P. 1993. *Astron. Astrophys.*, **271**, 381).

Central giant elliptical galaxies, such as M87 and NGC 1399, have been studied to test to what extent the paradigm of a single hot phase in hydrostatic equilibrium leads to reliable mass measurements.[23] The galaxy NGC 3379 also has been the target of intensive investigations. It is one of the fifty-three galaxies mentioned earlier, for which by application of the hydrostatic equilibrium condition of the X-ray-emitting plasma (assumed to be responsible for the observed X-ray emission) the authors measured a total mass that was seven times higher than the total mass deduced from optical observations.[24] In reality, there is very little plasma emitting in

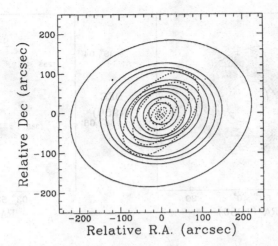

Fig. 24.5. Offset of optical (*dashed contours*; *R* band, spaced by 1 mag arcsec^{-2}) with respect to ROSAT X-ray isophotes (*solid contours*; spaced by 1.2 to 1.7 in intensity) for the galaxy NGC 720, providing geometric evidence for the possible presence of a dark halo (from Buote, D. A., Canizares, C. R. "Geometrical evidence for dark matter: X-ray constraints on the mass of the elliptical galaxy NGC 720," 1994. *Astrophys. J.*, **427**, 86; reproduced by permission of the AAS).

X-rays (with emission down to the level of 4×10^{37} erg s^{-1}, corresponding to a mass of a few $10^5 \, M_\odot$, concentrated in a region of radius smaller than 1 kpc from the galaxy center), and the picture appears to be that of a plasma under conditions of outflow.[25]

Thus, even in the presence of more modern and more reliable X-ray data, it appears that hot interstellar medium diagnostics often can lead to controversial or inconclusive results. On the one hand, even with the more modern telescopes, accurate measurements of density and temperature gradients on the small scale of galaxies are difficult to obtain. On the other hand, the picture of hydrostatic equilibrium of a pure thermal plasma may not be applicable, and a more complex modeling of the hot phase may be required (to include the presence of different phases, unresolved point sources, turbulent and systematic plasma motions, nonspherical geometry, and magnetic fields).[26]

Curiously, up to the 1990s, X-ray-based mass determinations were praised because, in contrast with stellar dynamical methods, they were thought to be simple, accurate, and free from physical complexities (such as those associated with the orbit anisotropy problem in stellar dynamics). Now it seems that the picture is basically reversed because all the complexities associated with the modeling of X-ray emission are becoming more and more evident. The limitations of X-ray diagnostics appear to be more evident in the inner regions, whereas in the outer regions there is better convergence with the results of other kinematic indicators. In conclusion, before a claim is made based on modeling of X-ray data, consistency should be checked with all the other available diagnostic tools.

24.2.2 Cold Atomic Hydrogen

In the first surveys, cold atomic hydrogen was found only rarely around ellipticals, with a generally irregular distribution. Modeling the gas kinematics was not easy. In some cases, evidence

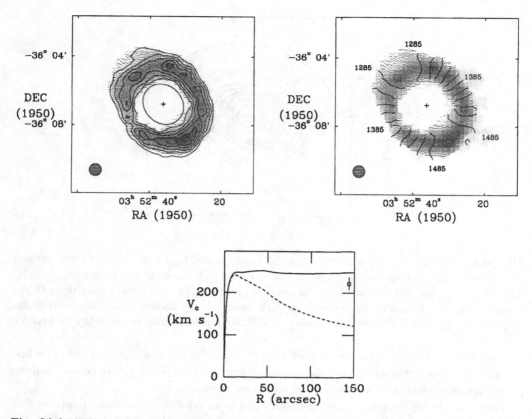

Fig. 24.6. An exceptionally regular HI ring structure is that found around the galaxy IC 2006 (from Franx, M., van Gorkom, J. H., de Zeeuw, P. T. "Evidence for axisymmetric halos: The case of IC 2006," 1994. *Astrophys. J.*, **436**, 642; reproduced by permission of the AAS). The left upper frame shows isodensity contours for the HI distribution (intervals of 3.4×10^{19} atoms cm^{-2}, with the inner ellipse marking an outer isophote of the optical galaxy); the right upper frame shows the observed velocity field (spaced by 20 km s^{-1}). In both frames the cross marks the center of the optical galaxy. The analysis of these data leads to a kinematic point shown in the lower frame (also from Franx, M., van Gorkom, J. H., de Zeeuw, P. T. 1994. *Astrophys. J.*, **436**, 642; reproduced by permission of the AAS), which suggests the presence of a significant dark halo in this galaxy (the dashed curve represents the expected rotation curve under the assumption of a constant mass-to-light ratio).

for dark matter was provided. For the galaxy NGC 4278, it has been argued[27] that the modeling requires that the M/L ratio should increase by a factor of 2 from the center out to $6R_e$.

However, in more recent years, very interesting exceptions have been found, among which the narrow regular ring[28] around IC 2006 (Fig. 24.6). To a large extent, a similar case is that of the E4 elliptical galaxy NGC 2974, for which *Very Large Array* observations also have revealed the presence of a regular ring. The HI kinematic data combined with other diagnostics require the existence of a dark halo, which, together with the visible matter, supports an axisymmetric gravitational field associated with a flat rotation curve; within $5R_e$, at least 55 percent of the total mass is dark.[29]

HI diagnostics for elliptical galaxies have now found renewed attention as a result of the discovery of the existence of regular, extended, low-density HI disks or rings in a number of ellipticals.[30] Studies of NGC 5266 (for which the HI distribution is not so regular[31]) and NGC 3108 (with a more regular distribution of atomic hydrogen[32]) have been followed by several other investigations, which demonstrate the power of such a simple and straightforward kinematic tracer, when present with a regular distribution.[33] It also has been shown that these HI disks are often hosted by radio galaxies. Enormous disks or rings of HI gas (with sizes up to 190 kpc and masses up to $2 \times 10^{10}\ M_\odot$) are detected around the host galaxies of a significant fraction of the compact radio sources.[34]

24.2.3 Planetary Nebulae and Other Discrete Tracers

Another possibility is to probe the gravitational field of ellipticals by studying discrete kinematic tracers (i.e., individual objects), such as satellites, globular clusters, and planetary nebulae. In general, this technique suffers from two problems, that is, the difficulty of collecting a sufficient number of data points (in particular, for satellites) and the fact that the structure of the orbits of the selected class of tracers may be poorly known (this is likely to be a problem in modeling the data relative to globular cluster systems given the fact that globular clusters are subject to dynamical friction, for which the related orbit circularization effects are not fully understood; see Chapter 7). In other words, even if the individual velocities along the line of sight can be measured with good accuracy, the use of these data to infer the underlying total mass distribution is subject to difficult modeling problems.[35] So far, most studies start out by a suitable binning of the individual kinematic data into data points at various radial locations and then address the issue of fitting the observed kinematic profiles thus obtained. More recently, progress is being made on developing a model analysis without binning the available discrete set of line-of-sight velocities.

Among the galaxies studied by means of the kinematics of their globular cluster system we should mention NGC 4486 (M87), NGC 1399, NGC 5128 (Cen A), NGC 4649 (M60), NGC 4472 (M49), NGC 4636, and NGC 3379. In general, when the number of kinematic measurements is significant, the information can be valuable, especially in view of the radial extent to which the globular clusters are detected.[36] However, rather than as an independent diagnostic for mass determinations, the data are best used in combination with other tracers to infer the internal kinematics of the globular cluster system.

Planetary nebulae are a more promising class of tracers of the gravitational field in ellipticals. They are bright and easy to identify, are associated with strong emission lines, sample the entire radial range of the host stellar systems, and are likely to be well representative of the kinematics of the galaxy of which they are part.[37] This type of diagnostic also has been applied to the study of spiral and S0 galaxies.[38]

Studies based on the collection of line-of-sight velocity measurements of planetary nebulae are now at a well-advanced stage for relatively nearby early-type galaxies, such as NGC 5128, M87, NGC 4472, and others mentioned earlier. They often reveal the presence of significant rotation in the outer regions.[39] The studies also confirm (see Section 24.1) that a few ellipticals, such as NGC 3379, NGC 4494, and NGC 821, are associated with rapidly declining velocity-dispersion profiles and thus appear to lack prominent dark-matter halos.[40] Others have flatter profiles and show clear evidence of a dark halo.[41]

24.2.4   Warm Ionized Gas

Warm ionized gas (observed through optical emission lines) is present more frequently, but it usually has only a modest radial extent.[42] Here, even more than for cold atomic hydrogen, it is hard to make a case that the gas has settled into a quiet, nonturbulent equilibrium state, as required by the use of closed orbits to model the gas flow. A detailed study of NGC 5077,[43] based on cold closed orbits for the gas in a triaxial potential, indicates that the observed gas kinematics can be fitted with a constant M/L out to R_e, thus showing no evidence for dark matter on that scale.

At present it appears that data about the kinematics of warm ionized gas provide valuable information on the gravitational field of the host galaxy if combined with the use of other diagnostic tools.

24.3   Globular Clusters versus Dwarf Spheroidal Galaxies

At the very low end of the luminosity distribution of ellipsoidal stellar systems, globular clusters and dwarf spheroidal galaxies populate different regions of the luminosity-diameter plane.[44] They are generally hosted by larger systems. Many are found in our Galaxy and thus can be studied in great detail. Typically they display several distinctive morphological characteristics, but they also exhibit several elements of continuity, even in relation to galaxies.[45]

In relation to the problem of dark matter, globular clusters and dwarf spheroidals appear to be characterized by opposite trends. On the one hand, globular clusters are traditionally thought to be free of dark matter, although the detailed kinematic information that would be required to make a strong statement in this direction is generally not available.[46] On the other hand, dwarf spheroidals are commonly thought to be the stellar systems most dominated by dark matter.[47]

We note that for globular clusters, a satisfactory modeling framework appears to be within reach (see Section 22.3). In contrast, the dynamical modeling of dwarf spheroidals, because of their more irregular morphology, is not as straightforward as desired.[48]

24.4   Early-Type Galaxies at Cosmological Distances

Gravitational lensing has made it possible to weigh galaxies and clusters of galaxies at cosmological distances. A separate chapter (Chapter 26) is devoted to this important phenomenon and to its role as an independent diagnostic tool. Here we briefly summarize some of the results relative to the internal structure and dark-matter content of early-type galaxies. Clearly, these results refer mainly to relatively massive galaxies, which are the galaxies most likely to exhibit significant effects as gravitational lenses.

An impressive long-term project aimed at combining stellar-dynamical and gravitational lensing measurements for early-type galaxies in the distant Universe has led to a wealth of data and to relatively simple, although to a large extent puzzling, results. In particular, we refer to the Lenses Structure and Dynamics (LSD) survey[49] and the following Sloan Lens ACS (SLACS) survey.[50]

These investigations lead to the conclusion that the total (luminous plus dark) density profile in these distant galaxies, if modeled as a single power law, behaves typically as $\rho \sim 1/r^2$, which is the signature of the isothermal sphere.[51] Attempts have been made to determine whether the power-law index changes significantly with redshift, but only little, if any, evolution has been found.

The database also has been used to study the properties of the stellar populations and the relation between total mass and stellar mass in distant galaxies.[52] Another investigation has tried to see whether the innermost regions are characterized by a different power-law index.[53]

24.5   Dark Matter in Groups and Clusters

Because, with Chapter 20 and the preceding discussion, we have covered much of the current state of the problem of dark matter in normal galaxies, we may now take a quick look at some studies that refer to the larger scale of groups and clusters of galaxies to see how the picture may match with that of the cosmological context.

One historically important argument, which has been very recently revisited by means of a superb measurement of the velocity vector of M31, refers to the dynamics of galaxy orbits in the vicinity of the Milky Way Galaxy (i.e., our Galaxy). The Local Group is a set of approximately thirty galaxies that includes the Galaxy and some well-known objects, such as the Magellanic Clouds, Andromeda (M31), and M33. In a simplified model, we may assume that the dynamics of the group is dominated by the two brightest members, that is, the Milky Way and M31, with the current separation of ≈ 750 kpc decreasing at a rate of ≈ 125 km s^{-1}. Because the distance scale obtained by multiplying this speed by the Hubble time is of the order of 1 Mpc, it seems unlikely that we are witnessing a chance encounter between two galaxies; rather, the system is to be considered as a gravitationally bound system that has decoupled from the Hubble expansion. In this case, we obtain a lower limit to the total mass of M31 and the Milky Way by arguing that initially the two galaxies were close to each other, that their separation then increased with the expansion of the Universe, and that they are now falling back toward each other as a result of their mutual gravitational attraction.[54] For a present time of $\approx 15 \times 10^9$ years, this simple model leads to a total mass of $4 \times 10^{12} M_\odot$. Combined with the known values of total luminosities for the two galaxies, this implies a global value of the mass-to-light ratio close to 100 $M_\odot/L_\odot$ or, in other words, that the halo inferred for our own Galaxy (see Chapters 14 and 20) extends out at least to ≈ 100 kpc. This result is essentially confirmed by more recent and detailed studies of the Local Group, also based on the assumption that the system is gravitationally bound. In particular, by means of a sophisticated observational strategy with the use of HST, based on the statistical analysis of the proper motions of thousands of M31 stars relative to the positions of hundreds of background galaxies, it has been possible to measure the three-dimensional velocity vector of the galaxy M31 relative to the Milky Way, which, combined with other modern accurate measurements available (of distances and line-of-sight velocities), has led to a much more realistic and detailed picture of the dynamics of the Local Group.[55]

The analysis just outlined can be extended by consideration of the dynamics of galaxy pairs outside our Local Group. In contrast with the case of binary stars, for which the relevant orbits can be sampled over significant time intervals, here we can measure only the projected separation between the two objects and the line-of-sight components of the orbital velocity at one given time. Any selected set of objects is going to be contaminated by nonphysical pairs. In addition, even the pairs that are genuine may suffer from the effects of nearby galaxies and groups (as is true for M31 and the Milky Way). Furthermore, assumptions have to be made on the kind of gravitational interaction that occurs between the two members of a given pair (especially if extended halos are invoked, for then the interaction is not that of simple point masses) and on the geometry of the orbits. Thus it is no wonder that such a statistical study of galaxy binaries has so far yielded rather inconclusive evidence for the presence of large amounts of dark matter,[56] even though the study of pairs with high-quality kinematic data[57] suggests that dark halos are needed to explain the observations and pose significant constraints on their radial extent. One emerging puzzle is that the extended halos that are invoked may come into direct contact, with subsequent orbital decay, by dynamical friction, at a rather fast rate.

Similar trends and similar, and even more severe, problems are encountered in the study of galaxy groups, with the added complication that the cosmological effects, as a result of the Hubble expansion and the matter distributed outside the groups, may become important factors. Despite all the difficulties that may be raised, masses of groups are derived from application of the virial theorem or from related mass estimators.[58] Wide ranges in mass-to-light ratios are found, with typical values similar to those reported earlier for the Local Group.[59] In relation to the presence and distribution of dark matter in galaxy groups, some excitement has been generated by the finding by ROSAT of diffuse X-ray emission in a few cases on the scale of about 200 kpc. Of particular interest is the small group of NGC 2300, which apparently contains only three galaxies.[60] Study of the X-ray data for the spherically symmetric group of NGC 5044, composed of a single giant elliptical surrounded by a large number of dwarfs, is supported by a detailed cooling-flow model and yields[61] a mass-to-light ratio of $\approx 130 \, M_\odot/L_\odot$ (Fig. 24.7); the same group was reobserved by XMM-*Newton* and *Chandra* and modeled in better detail.[62]

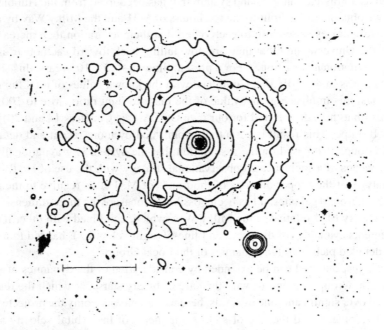

Fig. 24.7. ROSAT X-ray isophotes superimposed on an optical image of the regular group associated with NGC 5044 (from David, L. P., Jones, C., Forman, W., Daines, S. "Mapping the dark matter in the NGC 5044 group with ROSAT: Evidence for a nearly homogeneous cooling flow with a cooling wake," 1994. *Astrophys. J.*, **428**, 544; reproduced by permission of the AAS).

Similar results[63] have been obtained from the study of the Hickson compact group HCG 62. If the X-ray emission is tracing the underlying potential well, the evidence is that the light coming from the galaxies that belong to the group is not tracing the overall mass distribution. However, if groups are pervaded by large amounts of dark matter, it is not clear how long their members can survive (or how they have been able to survive so far) from merging, given the presence of dynamical friction.[64]

As mentioned in Chapter 20, studies of the motions of galaxies in clusters, starting with those of Zwicky and Smith in the 1930s, gave the first clear evidence that large amounts of nonluminous matter are present in the Universe. More recent studies of the kinematics of galaxies inside clusters have confirmed the general conclusions of the earlier work. Under the assumption that dark matter and luminous galaxies have similar spatial distributions, typical mass-to-light ratios of several hundred were found. One well-investigated case is that of the Coma cluster (a cluster of more than 1,000 bright galaxies, primarily ellipticals and lenticulars, located in the nearby Universe). Models that leave open the possibility that dark matter is not distributed as luminous matter[65] show a wide range of equally acceptable options, with values of the mass-to-light ratio up to 500, which definitely would be interesting from the cosmological point of view (see Chapter 4).[66]

Rich clusters of galaxies are usually strong X-ray sources; the X-rays are produced by a hot intracluster plasma. A simple hydrostatic model for this intracluster medium (ICM), as an isothermal plasma coexisting with a distribution of galaxies in a common spherical potential well,[67] has been quite successful in providing a way to fit X-ray and optical data. The model predicts a simple relation

$$\rho_{gas} \propto \rho_{galaxies}^{\beta} \tag{24.3}$$

between the galaxy and the hot-plasma density-distribution profiles depending on a parameter (β) that is the square of the ratio of the galaxy velocity dispersion to the plasma thermal speed.

Starting with the 1970s, initial discussions of the mass distribution inside clusters focused on the use of the model just introduced; on the need to measure the temperature profile of the ICM accurately, with a controversy on whether temperature gradients are present (in particular, on whether the outer temperature profile is generally declining or not); on the relative share between galactic and intergalactic material in the mass budget of the visible matter; and in general on the relative distribution of visible and dark matter. Typical line-of-sight galaxy velocity dispersions[68] in rich clusters are close to 900 km s^{-1}, and typical plasma temperatures are ≈ 7 keV. The mean value of β derived from fits to density distributions[69] is 2/3, whereas that obtained from direct measurements of plasma temperatures and galaxy velocity dispersions[70] is 0.9 (this somewhat reduces a discrepancy noted earlier[71]). Because the X-ray emission appears to be due to thermal *bremsstrahlung*, in principle, the mass of the hot intracluster gas can be derived directly from the observed X-ray surface-brightness profiles. The mass in the ICM thus was estimated to often exceed the total mass associated with the individual galaxies in the cluster. Uncertainties on the amount of mass contained in the outer regions suggested that even up to 30 to 60 percent of the total virial mass could be accounted for by the intracluster plasma.[72] One of the most striking clusters[73] was first identified by ROSAT/ASCA data and then optically confirmed by observations at the *New Technology Telescope*, which is a cluster at $z = 0.45$ with $L_X \approx 2 \times 10^{46}$ erg s^{-1}, $T \approx 9$ keV, and total estimated mass $\approx 10^{15}$ $M_\odot$ within a radius of ≈ 2 Mpc; the first analysis of its X-ray properties led to the picture of a cluster with a very substantial cooling flow (see Chapter 25), of more than $3 \times 10^3 \, M_\odot \, \mathrm{yr}^{-1}$, and ICM mass fraction estimated to be larger than 30 percent. The issue of whether luminous galaxies in clusters trace the overall mass distribution also was studied by comparing the shape of optical and X-ray isophotes; a study of five Abell clusters shows significantly rounder X-ray isophotes with respect to the underlying galaxy distribution.[74]

In practice, after these initial discussions, enormous progress in understanding the internal structure of clusters and the relative distribution of dark and luminous matter has been made as a result of two major advances in the field. On the one hand, X-ray observations have made use of the power of the new generation of telescopes after the pioneering work by ROSAT, in particular, of XMM-*Newton* and *Chandra*.[75] It is interesting to note that the X-ray astrophysical community has generally referred to the long cosmological distance scale (i.e., a Hubble constant set to $H_0 = 50 \, \mathrm{km \, s^{-1} \, Mpc^{-1}}$), and only in the very last few years has it adapted to the now commonly accepted value of $H_0 \approx 70 \, \mathrm{km \, s^{-1} \, Mpc^{-1}}$; this has significant effects on the overall quantitative picture of the mass distribution inside clusters because the determination of the relevant quantities involves nontrivial scalings with the Hubble constant.[76] On the other hand, the study of the effects of strong and weak[77] gravitational lensing has made it possible to make completely independent mass measurements on different scales, with a diagnostic tool that does not require the assumption of dynamical equilibrium. The term *strong lensing* refers to the observations of giant luminous arcs (see also Fig. 1.2) that are interpreted as due to gravitational lensing by the intervening cluster of very distant sources in the background. Spectral analysis confirms the plausibility of the interpretation in terms of gravitational lensing.[78] In turn, the term *weak lensing* refers to the observed distortion of a large number of distant source galaxies behind the cluster.[79] In the context of galaxies, the topic of gravitational lensing will be treated separately in Chapter 26. In general, the mass-to-light ratios derived from the gravitational lens models turn out to be reasonably consistent with those of X-ray studies.

The general picture is that in contrast to the trends characteristic of individual galaxies, in clusters of galaxies the dark component is more concentrated than the visible component. The mass of the visible component, especially outside the first central 100 kpc, is dominated by that of the ICM, which can be below 10 percent of the total mass when we refer to a sphere of radius 0.5 Mpc. Cosmological models predict that when referring to larger and larger radii, the ratio of visible to dark mass in clusters eventually should approach ≈ 15 percent.[80]

Combined X-ray and gravitational lensing studies of two clusters of galaxies have led to especially interesting results, claimed to provide decisive evidence for the very existence of dark matter, against alternative theories of modified gravity (such as MOND; see Chapter 20). In particular, here we refer to the study of the so-called Bullet Cluster.[81] For this cluster of galaxies, optical and X-ray data indicate that the center of mass of the visible components (especially the ICM) is offset with respect to the center of mass of the total mass distribution, as determined by weak lensing analysis. The presence of such an offset is explained in terms of a collision between two clusters, during which dissipative (ICM) and collisionless (dark matter and stars) components are subject to different dynamical processes. In practice, no matter what the origin of the observed offset actually is, such a displacement appears to exclude an interpretation of the available data by a change in the laws of gravity without invoking the existence of a separate dark-matter component.

Notes

1. Bertin, G., Stiavelli, M. 1993. *Rep. Prog. Phys.*, **56**, 493; see also Bertin, G. 1993. In *Proceedings of the ESO/EIPC Workshop on Structure, Dynamics, and Chemical Evolution of Early-Type Galaxies*, eds. I. J. Danziger, et al. European Southern Observatory Publications, Garching, Germany, p. 243.

2. Tonry, J. L. 1983. *Astrophys. J.*, **266**, 58.

3. It is reassuring to find that indeed the steepest velocity-dispersion profiles are those that in this physical picture correspond to $M_D = 0$; if galaxies were formed with arbitrary orbital distributions, as is advocated by those who insist on assuming a constant mass-to-light ratio, steeper profiles also would be expected.

4. The fit is thus performed in terms of spherical models; for this purpose, it is appropriate to work on E0 to E2 objects. The limitations of the spherical assumption have been estimated and found to be not too severe by means of simulated observations of intrinsically nonspherical objects generated by N-body codes (see Fig. 22.17 of Chapter 22, and Saglia, R. P., Bertin, G., Stiavelli, M. 1992. *op. cit.*).

5. Bertin, G., Bertola, F., et al. 1994. *Astron. Astrophys.*, **292**, 381.

6. Gerhard, O., Kronawitter, A., et al. 2001. *Astron. J.*, **121**, 1936.

7. As described by the authors in section 2.2 of their paper; in particular, see their fig. 5.

8. The SAURON survey; Bacon, R., Copin, Y., et al. 2001. *Mon. Not. Roy. Astron. Soc.*, **326**, 23; de Zeeuw, P. T., Bureau, M., et al. 2002. *Mon. Not. Roy. Astron. Soc.*, **329**, 513; these are the first two of a long series of articles. The survey focuses on twenty-four ellipticals and twenty-four S0s representative galaxies, equally divided into cluster and field galaxies. SAURON is basically an integral-field spectrograph with a 33×41 arcsec field of view.

9. Cappellari, M., Bacon, R., et al. 2006. *Mon. Not. Roy. Astron. Soc.*, **366**, 1126.

10. Cappellari, M., Emsellem, E., et al. 2011. *Mon. Not. Roy. Astron. Soc.*, **413**, 813; Emsellem, E., Cappellari, M., et al. 2011. *Mon. Not. Roy. Astron. Soc.*, **414**, 888, which are the first and third of a long series of articles.

11. Forman, W., Schwarz, J., et al. 1979. *Astrophys. J. Lett.*, **234**, L27.

12. See Trinchieri, G., Kim, D.-W., et al. 1994. *Astrophys. J.*, **428**, 555.

13. See Fabbiano, G. 1989. *Annu. Rev. Astron. Astrophys.*, **27**, 87, and references therein.

14. Bertin, G., Pignatelli, E., Saglia, R. P. 1993. *Astron. Astrophys.*, **271**, 381.

15. Forman, W., Jones, C., Tucker, W. 1985. *Astrophys. J.*, **293**, 102.

16. Bertin, G., Toniazzo, T. 1995. *Astrophys. J.*, **451**, 111; the accurate ROSAT X-ray temperature profile presented by Trinchieri, G., Kim, D.-W., et al. 1994. *op. cit.*, for NGC 4636 reaches out to $\approx 1,000$".

17. Buote, D. A., Canizares, C. R. 1994. *Astrophys. J.*, **427**, 86.

18. Kim, D.-W., Pellegrini, S. eds. 2012. *Hot Interstellar Matter in Elliptical Galaxies*, Vol. 378, Astrophysics and Space Science Library, Springer Science and Business Media, LLC, Heidelberg; in particular, see the chapters written by G. Fabbiano, p. 1, by T. S. Statler, p. 207, and by D. A. Buote and P. J. Humphrey, p. 235.

19. *Chandra* data were studied by Buote, D. A., Jeltema, T. E., et al. 2002. *Astrophys. J.*, **577**, 183.

20. Humphrey, P. J., Buote, D. A., et al. 2011. *Astrophys. J.*, **729**, 53.

21. Humphrey, P. J., Buote, D. A., et al. 2006. *Astrophys. J.*, **646**, 899.

22. Fukazawa, Y., Botoya-Nonesa, J. G., et al. 2006. *Astrophys. J.*, **636**, 698.

23. For an XMM-*Newton* study of M87, see Matsushita, K., Belsole, E., et al. 2002. *Astron. Astrophys.*, **386**, 77. For a study of *Chandra* data relative to M87 and NGC 1399, see Churazov, E., Forman, W., et al. 2008. *Mon. Not. Roy. Astron. Soc.*, **388**, 1062. Das, P., Gerhard, O., et al. 2010. *Mon. Not. Roy. Astron. Soc.*, **409**, 1362, study a sample of six galaxies that includes the above-mentioned objects. Murphy, J. D., Gebhardt, K., Adams, J. J. 2011. *Astrophys. J.*, **729**, 129, find discrepancies between the mass model for M87 derived from SAURON integral-field stellar dynamical data and that derived from X-ray data.

24. See table 5 in Fukazawa, Y., Botoya-Nonesa, J. G., et al. 2006. *op. cit.*

25. Trinchieri, G., Pellegrini, S., et al. 2008. *Astrophys. J.*, **688**, 1000.

26. Ciotti, L., Pellegrini, S. 2004. *Mon. Not. Roy. Astron. Soc.*, **350**, 609, focused on the case of NGC 4472; Pellegrini, S., Ciotti, L. 2006. *Mon. Not. Roy. Astron. Soc.*, **370**, 1797, focused on NGC 3379.

See also Humphrey, P. J., Buote, D. A., et al. 2013. *Mon. Not. Roy. Astron. Soc.*, **430**, 1516, and the many references included in these papers. For a study of systematic errors in the application of X-ray diagnostics to clusters of galaxies, see Rasia, E., Ettori, S., et al. 2006. *Mon. Not. Roy. Astron. Soc.*, **369**, 2013, and many following papers.

27. Lees, J. F. 1991. In *Warped Disks and Inclined Rings around Galaxies*, eds. S. Casertano et al. Cambridge University Press, Cambridge, UK, p. 50.
28. Franx, M., van Gorkom, J. H., de Zeeuw, P. T. 1994. *Astrophys. J.*, **436**, 642.
29. Weijmans, A.-M., Krajnović, D., et al. 2008. *Mon. Not. Roy. Astron. Soc.*, **383**, 1343.
30. Morganti, R., Sadler, E., Oosterloo, T. 1997. In *The Nature of Elliptical Galaxies*, eds. M. Arnaboldi, G. S. Da Costa, P. Saha, Vol. 116 of the Astronomical Society of Pacific Series, p. 354.
31. Morganti, R., Sadler, E., et al. 1997. *Astron. J.*, **113**, 937.
32. Oosterloo, T. A., Morganti, R., et al. 2002. *Astron. J.*, **123**, 729.
33. Oosterloo, T. A., Morganti, R., et al. 2007. *Astron. Astrophys.*, **465**, 787.
34. Emonts, B. H. C., Morganti, R., et al. 2010. *Mon. Not. Roy. Astron. Soc.*, **406**, 987; see also Oosterloo, T., Morganti, R., et al. 2010. *Mon. Not. Roy. Astron. Soc.*, **409**, 500. For a general discussion of HI in early-type galaxies, see also Serra, P., Oosterloo, T., et al. 2012. *Mon. Not. Roy. Astron. Soc.*, **422**, 1835.
35. See also Bahcall, J. N., Tremaine, S. 1981. *Astrophys. J.*, **244**, 805.
36. Here we just mention a partial list of relevant contributions; many other references are given in the cited papers (see also Deason, A. J., Belokurov, V., et al. 2012. *Astrophys. J.*, **748**, 2). When the article by Schuberth, Y., Richtler, T., et al. 2010. *Astron. Astrophys.*, **513**, A52, appeared, NGC 1399 was the galaxy with the richest data set of kinematically measured globular clusters (see also Richtler, T., Dirsch, B., et al. 2004. *Astron. J.*, **127**, 2094; and Samurović, S., Danziger, I. J. 2006. *Astron. Astrophys.*, **458**, 79). For NGC 5128, Woodley, K. A., Gómez, M., et al. 2010. *Astron. J.*, **139**, 1871, bring the number of globular clusters with measured line-of-sight velocity to 563. For NGC 4636, Lee, M. G., Park, H. S., et al. 2010. *Astrophys. J.*, **709**, 1083, consider a kinematic data set of 368 globular clusters. For NGC 4472, Côté, P., McLaughlin, D. E., et al. 2003. *Astrophys. J.*, **591**, 850, study a sample of 263 globular clusters. Other investigations rely on much smaller numbers of kinematic data points; for M87, see the analysis by Wu, X., Tremaine, S. 2006. *Astrophys. J.* **643**, 210. This is natural for the first investigations based on this diagnostics, such as Mould, J. R., Oke, J. B., et al. 1990. *Astron. J.*, **99**, 1823. But there are more modern studies based on small data sets. For example, using thirty-eight globular clusters in NGC 4649, Bridges, T., Gebhardt, K., et al. 2006. *Mon. Not. Roy. Astron. Soc.*, **373**, 157, argue about the fact that globular clusters are characterized by isotropic orbits in the inner regions and tangential orbits outside (see also Hwang, H. S., Lee, M. G., et al. 2008. *Astrophys. J.*, **674**, 869). Working on a sample of thirty-six globular clusters, Pierce, M., Beasley, M. A., et al. 2006. *Mon. Not. Roy. Astron. Soc.*, **366**, 1253, suggest evidence of a dark halo in NGC 3379. Norris, M. A., Gebhardt, K., et al. 2012. *Mon. Not. Roy. Astron. Soc.*, **421**, 1485, work on a sample of seventy-nine globular clusters in NGC 3923.
37. In the beginning, the number of kinematic data points secured for a given object was very small. In NGC 3379, twenty-nine planetary nebulae were studied by Ciardullo, R., Jacoby, G. H., Dejonghe, H. B. 1993. *Astrophys. J.*, **414**, 454. For a study of nineteen planetary nebulae in NGC 4406, see Arnaboldi, M., Freeman, K. C., et al. 1996. *Astrophys. J.*, **472**, 145. Arnaboldi, M., Freeman, K. C., et al. 1998. *Astrophys. J.*, **507**, 759, study forty-three planetary nebulae in NGC 1316; Napolitano, N. R., Arnaboldi, M., Capaccioli, M. 2002. *Astron. Astrophys.*, **383**, 791, discuss the dynamics of NGC 1399 on the basis of kinematic data for fifty-seven planetary nebulae. For a study of the outer regions of M87, see Doherty, M., Arnaboldi, M., et al. 2009. *Astron. Astrophys.*, **502**, 771. Instruments, in particular the planetary nebula spectrograph (Douglas, N. G., Arnaboldi, M., et al. 2002. *Publ. Astron. Soc. Pacific*, **114**, 1234), and major observing campaigns have been dedicated to this diagnostic.
38. Merrett, H. R., Merrifield, M. R., et al. 2006. *Mon. Not. Roy. Astron. Soc.*, **369**, 120; Noordermeer, E., Merrifield, M. R., et al. 2008. *Mon. Not. Roy. Astron. Soc.*, **384**, 943.

39. For example, in NGC 5128, based on the line-of-sight velocities of 433 planetary nebulae; Hui, X., Ford, H., et al. 1995. *Astrophys. J.*, **449**, 592; see also Peng, E. W., Ford, H. C., Freeman, K. C. 2004. *Astrophys. J.*, **602**, 685.

40. Romanowsky, A. J., Douglas, N. G., et al. 2003, *Science*, **301**, 1696; Douglas, N. G., Napolitano, N. R., et al. 2007. *Astrophys. J.*, **664**, 257; Napolitano, N. R., Romanowsky, A. J., et al. 2009. *Mon. Not. Roy. Astron. Soc.*, **393**, 329.

41. For NGC 1344, see Teodorescu, A. M., Méndez, R. H., et al. 2005. *Astrophys. J.*, **635**, 290. For NGC 4374, see Napolitano, N. R., Romanowsky, A. J., et al. 2011. *Mon. Not. Roy. Astron. Soc.*, **411**, 2035. See also Coccato, L., Gerhard, O., et al. 2009. *Mon. Not. Roy. Astron. Soc.*, **394**, 1249.

42. See Buson, L. M., Sadler, E. M., et al. 1993. *Astron. Astrophys.*, **280**, 409, and references therein.

43. Bertola, F., Bettoni, D., et al. 1991. *Astrophys. J.*, **373**, 369.

44. van den Bergh, S. 2008. *Mon. Not. Roy. Astron. Soc.*, **390**, L51.

45. van den Bergh, S. 2008. *Mon. Not. Roy. Astron. Soc.*, **385**, L20.

46. See Zocchi, A., Bertin, G., Varri, A. L. 2012. *Astron. Astrophys.*, **539**, A65; Sollima, A., Bellazzini, M., Lee, J.-W. 2012. *Astrophys. J.*, **755**, id.156. Sollima, A., Nipoti, C. 2010. *Mon. Not. Roy. Astron. Soc.*, **401**, 131, and Ibata, R., Sollima, A., et al. 2011. *Astrophys. J.*, **743**, 43, make a test of the MOND paradigm on a few globular clusters, which turns out to be favorable to the use of Newtonian dynamics.

47. Mateo, M. L. 1998. *Annu. Rev. Astron. Astrophys.*, **36**, 435; Walker, M. G., Mateo, M., et al. 2006. *Astron. J.*, **131**, 2114; Gilmore, G., Wilkinson, M. I., et al. 2007. *Astrophys. J.*, **663**, 948; Walker, M. G., Mateo, M., et al. 2007. *Astrophys. J. Lett.*, **667**, L53; Walker, M. G., Mateo, M., et al. 2009. *Astron. J.*, **137**, 3109. The MOND paradigm also has been tested on these systems: Gerhard, O. E., Spergel, D. N. 1992. *Astrophys. J.*, **397**, 38; Łokas, E. L. 2001. *Mon. Not. Roy. Astron. Soc.*, **327**, L21; Łokas, E. L. 2002. *Mon. Not. Roy. Astron. Soc.*, **333**, 697; see also Łokas, E. L. 2009. *Mon. Not. Roy. Astron. Soc.*, **394**, L102.

48. See Amorisco, N. A., Evans, N. W. 2011. *Mon. Not. Roy. Astron. Soc.*, **411**, 2118, and references therein.

49. Koopmans, L. V. E., Treu, T. 2002. *Astrophys. J. Lett.*, **568**, L5; Treu, T., Koopmans, L. V. E. 2002. *Astrophys. J.*, **575**, 87; and many following papers.

50. Bolton, A. S., Burles, S., et al. 2006. *Astrophys. J.*, **638**, 703; Treu, T., Koopmans, L. V. E., et al. 2006. *Astrophys. J.*, **640**, 662; and many following papers, among which Newton, E. R., Marshall, P. J., et al. 2011. *Astrophys. J.*, **734**, 104. A new project, the Sloan WFC Edge-on Late-type Lens Survey (SWELLS) has been started recently, using a similar strategy for spiral galaxies; see Treu, T., Dutton, A. A., et al. 2011. *Mon. Not. Roy. Astron. Soc.*, **417**, 1601, and following papers. Another important project with results about the internal structure of early-type galaxies is the Cosmic Lens All-Sky (CLASS) Survey; see Myers, S. T., Jackson, N. J., et al. 2003. *Mon. Not. Roy. Astron. Soc.*, **341**, 1.

51. See also Rusin, D., Kochanek, C. S., Keeton, C. R. 2003. *Astrophys. J.*, **595**, 29.

52. Grillo, C., Gobat, R., et al. 2008. *Astron. Astrophys. Lett.*, **477**, L25; Grillo, C., Gobat, R., et al. 2009. *Astron. Astrophys.*, **501**, 461.

53. Grillo, C. 2012. *Astrophys. J. Lett.*, **747**, 15.

54. Kahn, F. D., Woltjer, L. 1959. *Astrophys. J.*, **130**, 705.

55. Sohn, S. T., Anderson, J., van der Marel, R. P. 2012. *Astrophys. J.*, **753**, id.7; van der Marel, R. P., Fardal, M., et al. 2012. *Astrophys. J.*, **753**, id.8; van der Marel, R. P., Besla, G., et al. 2012. *Astrophys. J.*, **753**, id.9. Similar measurements have been made for the dwarf spheroidal galaxy Leo I, a member of the Local Group.

56. White, S. D. M., Huchra, J., et al. 1983. *Mon. Not. Roy. Astron. Soc.*, **203**, 701; Karachentsev, I. D. 1985. *Sov. Astron.*, **29**, 243.

57. van Moorsel, G. A. 1987. *Astron. Astrophys.*, **176**, 13. See also Erickson, L. K., Gottesman, S. T., and Hunter, J. H., Jr. 1999. *Astrophys. J.*, **515**, 153; Honma, M. 1999. *Astrophys. J.*, **516**, 693.

58. Bahcall, J. N., Tremaine, S. 1981. *op. cit.*

59. Tully, R. B. 1987. *Astrophys. J.*, **321**, 280.

60. Mulchaey, J. S., Davis, D. S., et al. 1993. *Astrophys. J. Lett.*, **404**, L9.
61. David, L. P., Jones, C., et al. 1994. *Astrophys. J.*, **428**, 544.
62. Buote, D. A., Lewis, A. D., et al. 2003. *Astrophys. J.*, **594**, 741.
63. Ponman, T. J., Bertram, D. 1993. *Nature (London)*, **363**, 51. The compact group HCG 62 also was observed by XMM-*Newton* and *Chandra*; see Morita, U., Ishisaki, Y., et al. 2006. *Publ. Astron. Soc. Jpn.*, **58**, 719.
64. For a study of the NGC 1407 group based on different diagnostic tools, see Romanowsky, A. J., Strader, J., et al. 2009. *Astron. J.*, **137**, 4956.
65. Merritt, D. 1987. *Astrophys. J.*, **313**, 121.
66. However, a detailed study of the combined optical and X-ray results available for the Coma cluster (Hughes, J. P. 1989. *Astrophys. J.*, **337**, 21) restricted the allowed range for the mass-to-light ratio considerably, down to 90 to 250 $M_\odot/L_\odot$; actually, the best-fit model was found to be the one for which the mass distribution follows that of the light and is characterized by a range 140 to 190 $M_\odot/L_\odot$.
67. Cavaliere, A., Fusco-Femiano, R. 1976. *Astron. Astrophys.*, **49**, 137.
68. Struble, M. F., Rood, H. J. 1991. *Astrophys. J. Suppl.*, **77**, 363.
69. Jones, C., Forman, W. 1984. *Astrophys. J.*, **276**, 38.
70. Edge, A. C., Stewart, G. C. 1991. *Mon. Not. Roy. Astron. Soc.*, **252**, 428.
71. Sarazin, C. L. 1988. *X-Ray Emission from Clusters of Galaxies*, Cambridge University Press, Cambridge, UK.
72. See Sarazin, C. L. 1988. *op. cit.*; Hughes, J. P. 1989. *op. cit.*
73. Schindler, S., Hattori, M., et al. 1997. *Astron. Astrophys.*, **317**, 646.
74. Buote, D. A., Canizares, C. R. 1992. *Astrophys. J.*, **400**, 385.
75. Rosati, P., Borgani, S., Norman, C. 2002. *Annu. Rev. Astron. Astrophys.*, **40**, 539.
76. For example, see De Boni, C., Bertin, G. 2008. *Il Nuovo Cimento B*, **123**, 31.
77. Mellier, Y. 1999. *Annu. Rev. Astron. Astrophys.*, **37**, 127; Bartelmann, M., Schneider, P. 2001. *Phys. Rep.*, **340**, 291.
78. See Soucail, G., Fort, B. 1991. *Astron. Astrophys.*, **243**, 23, and references therein.
79. See Tyson, J. A., Valdes, F., Wenk, R. A. 1990. *Astrophys. J. Lett.*, **349**, L1; Fahlman, G. G., Kaiser, N., et al. 1994. *Astrophys. J.*, **437**, 56; Smail, I., Ellis, R. S., et al. 1994. *Mon. Not. Roy. Astron. Soc.*, **237**, 871; Hoekstra, H., Franx, M., et al. 1998. *Astrophys. J.*, **504**, 636.
80. For a detailed discussion of two well-studied nearby clusters, A496 and Coma, based on simple models for the X-ray emitting ICM, see De Boni, C., Bertin, G. 2008. *op. cit.*
81. The cluster 1E 0657-558; Clowe, D., Bradač, M., et al. 2006. *Astrophys. J. Lett.*, **648**, L109. Among the latest analyses of the Bullet Cluster, see Paraficz, D., Kneib, J.-P., et al. 2012, arXiv:1209.0384. A somewhat similar case with different morphological aspects has been made for the Ring Cluster, Cl 0024+17; Jee, M. J., Ford, H. C., et al. 2007. *Astrophys. J.*, **661**, 728.

Part V
In Perspective

25 Selected Aspects of Formation and Evolution

The galaxies we see today owe their current structure to a combination of initial conditions, set by the processes that led to their formation, and a number of mechanisms that have shaped their evolution. Some regularities that we observe, in their morphology (such as the regularities captured by the Hubble classification scheme), in their overall luminosity profiles (exponential or $R^{1/4}$), or in the existence of well-defined scaling laws (such as the luminosity-velocity relation for spirals and the fundamental plane for elliptical galaxies), demand a physical explanation in terms of formation and evolution.

The discussion here becomes necessarily speculative, especially because we lack strong and direct empirical constraints. It is true that distant quasars and gas-rich absorption systems detected at relatively high redshifts provide clues about the conditions under which galaxies were formed, and relatively normal galaxies have been observed out to $z \approx 5$ (i.e., up to a lookback time greater than 90 percent of the age of the Universe) and beyond.[1] Yet we lack the type of detailed quantitative structural information that has allowed us to develop a satisfactory picture of the dynamics of normal galaxies, as observed at $z \approx 0$. Furthermore, even when we manage to obtain some data on their internal constitution, for example, in galaxies at $z \approx 1$, we do not have direct information on the way such objects have evolved from their progenitors and are going to evolve into the systems that we see in the nearby Universe.

In addition, the study of formation and evolution processes is intrinsically difficult because it should face the dynamics of nonequilibrium configurations; this practically excludes the possibility of semianalytical treatments, except for highly idealized scenarios. Once key factors that determine the general characteristics of the products of formation, such as the ratio of the cooling time to the dynamical time in the early stages of collapse,[2] are recognized, many ideas are often gathered through numerical simulations, which may inspire, support, or rule out some astrophysical scenarios. Collapse will proceed mainly in a collisionless fashion if star formation is initially rapid, so most stars are formed before quasi-equilibrium is reached. Otherwise the collapse will be dissipative, with the possibility of generating a disk if the initial configuration has a significant angular momentum to begin with. In practice, we still do not know what determines the origin of the two main broad categories of objects, that is, spirals and ellipticals.

In other physical contexts, the tools of statistical mechanics have been able to bypass many apparent difficulties and explain the basic properties of observed equilibrium configurations. The great success of these studies has been to show that if the system has a chance to relax, the properties of the final state can be anticipated and are essentially independent of the initial conditions. We may wonder whether similar arguments can be pursued for the case of galaxies. In this respect, we are at a clear disadvantage here. Galaxies, being primarily collisionless stellar systems, have had no time to relax fully by means of star-star encounters. Even the collisionless relaxation processes that can be invoked (see Section 7.3) have only led to partial relaxation. A posteriori, this is demonstrated by the pressure anisotropy observed directly, for example, in the solar neighborhood. A similar situation is familiar in many systems studied theoretically and experimentally in plasma physics.

The purpose of this chapter is not to review and evaluate the formation and evolution scenarios that have been investigated so far, in view of the available astrophysical evidence; instead, in line with the main goals of this book, only a few interesting ideas that form the background to many aspects of the structure and dynamics of normal galaxies described in the previous parts of this book are summarized. The ideas selected have the advantage that they can be made quantitative in terms of a semianalytical discussion. Thus the models involved are somewhat oversimplified, but precisely because of their simplicity, they serve as a useful framework for many of the current thoughts in this important research area. Many questions remain without a proper answer. The topic of dissipative accretion, which is at the basis of the formation of disks, will be briefly discussed separately in Chapter 27.

25.1   Gravothermal Catastrophe

In the 1950s, the study of self-gravitating isothermal spheres led to the discovery that gravity can have a major effect in changing the Boyle law of ordinary gases. Such spheres are characterized by two dimensional scales (e.g., the central density ρ_0 and the temperature T) and one dimensionless parameter [e.g., the dimensionless radius $\Xi = R/\lambda$ or the density contrast $\rho_0/\rho(R)$]; here R is the finite radius of the sphere, $\lambda = \sqrt{kT/(m4\pi G\rho_0)}$ is a length scale that is reminiscent of the Jeans length (see Chapter 12), k is the Boltzmann constant, and m is the gas-particle mass. In particular, it was found that in the presence of self-gravity, in the (volume, boundary-pressure) plane (V, p_{ext}) for low values of the total volume V, the hyperbola characteristic of the Boyle law changes into a curve that spirals into a point; equilibrium is possible only up to a maximum value of p_{ext}. All the points of the spiral along such a line of isothermal self-gravitating equilibrium configurations beyond the point of maximum external pressure turn out to be unstable toward collapse (Fig. 25.1).[3] These results have been generalized to the case in which the boundary surface has no specific shape and symmetry.[4]

This investigation, initially prompted by interest in the processes that lead to star formation, was then extended to cover the case of truncated isothermal spheres in the context of stellar dynamics[5] and soon gained popularity in the study of globular clusters.[6] The interest in the case of globular clusters is twofold. On the one hand, some of these relatively small stellar systems possess the desired degree of collisionality that is commonly thought to be required for the phenomenon to occur.[7] The reason is that the catastrophe is generally interpreted as the result of a heat flux from the innermost regions, dominated by gravity and associated with a negative specific heat,[8] to the tenuous outer regions, akin to an ordinary gas; strictly speaking, purely collisionless stellar systems should be stable. [We recall that other phenomena associated with collisionality that are known to affect the evolution of globular clusters are mass segregation and star evaporation (see Chapter 7).] On the other hand, some globular clusters, with relatively short relaxation times, are indeed observed to possess a central cusp, distinct from the core structure expected in King models (see Chapter 22). These clusters are naturally interpreted as

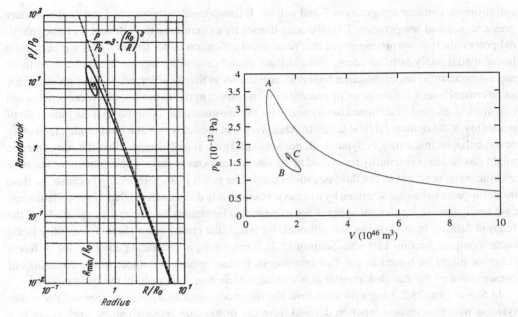

Fig. 25.1. Illustration of the external pressure-size relation characteristic of a self-gravitating isothermal cloud and its deviation from the Boyle law. *Left:* The diagram from the original article by Ebert, R. 1955. *Zeit. Astrophys.*, **37**, 217. *Right:* The pressure-volume relation for an isothermal spherical cloud of 1 $M_\odot$ made of molecular hydrogen at a temperature of 10 K. The points along the curve on the right of the point of maximum external pressure, denoted by A, correspond to stable equilibria; the points on the curve beyond A that spiral into the critical point C are all unstable (after Lombardi, M., Bertin, G. 2001. *Astron. Astrophys.*, **375**, 1091).

systems that have suffered core collapse as a result of the gravothermal catastrophe. Indeed, direct N-body simulations show that on the collisional time scale, core collapse can develop, and this is eventually halted by the effects of binaries, which are present in a real system and alter the energy budget; the postcollapse case is thought to be characterized by large-amplitude gravothermal oscillations.[9] The process appears to be accelerated in systems characterized by finite total angular momentum (gyrogravothermal catastrophe).[10]

Of course, globular clusters are truncated by tides in a way that is physically different from the picture of a sphere bounded by a reflecting wall, as envisioned in the original discussion of the phenomenon. That discussion was centered on the use of the Boltzmann entropy, the application of which to self-gravitating, inhomogeneous systems is rather subtle. In addition, a collisionless or quasi-collisionless (see Section 8.2) or fluid (e.g., for the case of a disk, see Subsection 8.3.2) description should be regarded as only a highly idealized model description when applied to study the dynamical evolution of a finite number of particles in mutual gravitational interaction.

The original discussion, centered on the use of the Boltzmann entropy in the study of a self-gravitating stellar system bounded by a spherical reflecting wall, led to the identification of different points for the onset of instability toward collapse, in particular, occurring at $\Xi \approx 34.4$, $\Xi \approx 8.99$, and $\Xi \approx 6.45$ (corresponding to a density contrast of ≈ 709, 32.1, and 14, respectively) for perturbations at constant total energy E_{tot} and volume V (gravothermal catastrophe), constant temperature T and volume V (isothermal collapse), and constant boundary pressure p_{ext} and temperature T (isobaric collapse) by extremizing the relevant thermodynamical potentials.[11] A surprising result that requires clarification is the following. If we perform a linear modal analysis of an ideal, inviscid fluid model (assuming infinite thermal conductivity and isothermality) under the same boundary conditions as listed earlier, we recover precisely the same critical points for the onset of instability with respect to radial modes; eigenfrequencies and eigenfunctions can be calculated (analytically for the marginal case), proving that the time scale of instability is the dynamical time scale. In other words, in the context of the fluid model, the nature of the collapse instability is dynamical, not secular. These results suggest that the phenomenon might just be Jeans instability recovered in the inhomogeneous context. The argument of negative specific heats is absent in the fluid equations used. The result is also surprising because the fluid model appears to be characterized by a behavior completely different from that of the collisionless case (collisionless isothermal spheres are expected to be stable), in contrast to the study of the Jeans instability in the homogeneous model, for which the criterion for (Jeans) instability is the same (compare Section 12.1 with Section 12.2). Apparently, in spherical geometry, the different behavior might be traced to the fact that a collisionless sphere is subject to the constraint of conservation of the detailed angular momentum, which does not apply in the fluid model.[12]

In Subsection 25.2.3 we will show how the thermodynamical argument proposed for stellar systems with distribution function derived from the Boltzmann entropy can be worked out in a relatively straightforward manner for systems with finite mass and infinite radial extent.

25.2 Collisionless Collapse

One major challenge to dynamicists in the past five decades has been to give a proper statistical categorization of collisionless collapse. In Part IV, several indications have been given that for elliptical galaxies this framework indeed appears to provide a sensible way of describing how the current equilibrium configurations have been reached (see Section 22.4). The same dynamical scenario can be the proper framework not only for the case that ellipticals were formed in the distant past from primordial clumps of stars[13] but also, to some extent, if they result from more recent mergers.[14] (Of course, mergers under different conditions[15] can produce a variety of effects; in fact, one problem of the merging paradigm is the difficulty in making quantitative predictions.[16])

Here the expansive set of numerical simulations,[17] some in the cosmological context, that have provided insights on many aspects of the formation picture of collisionless collapse is not reviewed. Instead, we focus on some points in the context of statistical mechanics. There have been several attempts in this area, and these broadly follow three lines of thought.

One general direction has been to investigate whether it is possible to develop a sensible statistical mechanics for a system taken to be collisionless a priori. The natural relaxation process that is considered is that of violent relaxation, acting efficiently when the system is initially significantly far from equilibrium. In practice, this is equivalent to taking a system of N stars in the continuum limit $N \to \infty$ for which the collisionless Boltzmann equation provides the appropriate evolution equation in six-dimensional phase space. This is a well-posed mathematical problem and has led to curious findings, in particular, the possibility of a dynamical exclusion principle,[18] which, in turn, has led to a sort of Fermi-Dirac statistic for a classical system made of mutually interacting stars. Memory of the initial conditions remains by means of the constraint that the relevant distribution function cannot exceed, during evolution, the maximum value initially available.[19] This line of thought raises several questions.[20] In particular, one focus of interest is a proper definition and the problem of uniqueness of the entropy functional.[21] These studies do recognize the difference between fine-grained and coarse-grained distribution functions (in particular, when issues such as phase mixing are addressed), but in practice, they are fully devoted to evolution of the fine-grained distribution function.

A different approach, more directly aimed at the description of galaxies, recognizes that in problems of physical interest, a clear limit is posed by the finite number of stars that populate the relevant phase space, which then naturally emphasizes the use of the coarse-grained distribution function.[22] The analysis based on such a discrete approach appears to indicate that the non-Boltzmann behavior associated with degeneracy (resulting from the dynamical exclusion principle) may be of only little physical interest because the systems for which degeneracy might be significant would turn out to fall under conditions for which star-star encounters cannot be ignored.

These lines of thought address some very general questions, such as whether a stellar system under collisionless relaxation would tend to become close to isothermal in the thermodynamical sense (with velocity dispersions related to masses) or instead would evolve in the direction of a common velocity dispersion independent of mass (following the picture of phase mixing and violent relaxation). In practice, we may even wonder whether some commonly assumed modeling aspects are actually justified. For example, a common procedure is to impose the existence of a closed box, with total number of particles and total energy fixed. Note that the interesting study of collisional relaxation and the finding of the gravothermal catastrophe resulting from complete relaxation are based on the device of enclosing the stellar system in a box of given volume with reflecting walls (see Section 25.1). In contrast, simulations show that as a result of collapse, many stars escape, even in the collisionless case. Nevertheless, some of these concepts turn out to be useful for the explanation of a number of numerical simulations. Note that all studies of collisionless collapse stress the fact that it would be very important to better clarify the role and the ways of incomplete relaxation.[23]

A third line of thought aims at more limited and practical goals. Can we rationalize in the framework of statistical mechanics the success of some dynamical arguments in constructing some realistic anisotropic models? Can we do so by proceeding naively as if a relatively standard statistical analysis were applicable? If not, what do we learn from these attempts in relation to the possibility of categorizing the statistical mechanics of collisionless collapse? Some studies have brought attention to the integrated energy distribution $N(E)$. In particular, we may recall here the finding that the asymptotic behavior at large radii $\rho \sim r^{-4}$, which appears to be characteristic of the structure of elliptical galaxies (see Chapter 22, especially Subsection 22.4.4), can be traced[24] to the existence of a nonvanishing finite value for the integrated energy distribution $N(E)$ at $E \sim 0$. In Subsection 22.4.3 it was remarked that the wide range of exponential behavior[25] in the $N(E)$ profile often noted in realistic galaxy models turns out to be indicative of core concentration rather than of the overall density profile.[26] Before a summary of some other clues on the statistical mechanics of incomplete violent relaxation is given, a short digression is in order.

When referring to the end products of collisionless collapse, it should be emphasized once more that in the study of the dynamics of galaxies we are working with idealized models that provide only an approximate description of the real systems. For example, when we consider stellar dynamical models of elliptical galaxies, even in the absence of violent instabilities, significant tidal interactions, or merger events, we should be ready to recognize that the actual basic state is not frozen, as implied by the equilibrium solution to the collisionless Boltzmann equation, but that it is likely to be in a process of gradual evolution. In fact, a number of dissipative effects (the presence of gas in various phases, the presence of a globular cluster system, and other gentle interactions of various types) may well contribute to establish such a process of secular evolution. In a sense, much as for globular clusters, we might have to consider that elliptical galaxies also evolve along sequences of collisionless equilibrium configurations.

25.2.1   Analogous Themes in Hydrodynamics and Magnetohydrodynamics

To some extent, these discussions present analogies with research in hydrodynamics and plasma physics. These are studies of the evolution of either a nonviscous hydrodynamical flow or a perfectly conductive magnetohydrodynamic (MHD) plasma. In the dissipationless case, we can identify important invariants. Some well-known invariants are the vorticity integral in fluid dynamics (after Helmholtz and Kelvin[27]) and the conservation law that guarantees the freezing of the magnetic-field lines in ideal MHD flow.[28] In reality, a dissipationless flow has a rich variety of conservation laws. One interesting invariant in magnetohydrodynamics is the helicity integral,[29] which is related to the linkage of vortex lines, a topological invariant in hydrodynamics.[30] Actually, we can prove the existence of an infinite number of invariants.[31] All this can be traced to the Hamiltonian structure of the underlying set of equations.[32] Note that the state of minimum magnetic energy of a plasma configuration is the one called *force free*, in which the currents are parallel to the magnetic-field lines; here the conservation of all the topological invariants is such that the proportionality constant can be different for different field lines, each of which possesses a well-defined identity.

The natural step that follows for physical applications is the study of the evolution of the plasma or hydrodynamical flow in the presence of small amounts of dissipation that are due to either finite resistivity or finite viscosity. Thus we would like to know whether some of the invariants found in the dissipationless analysis can be used as additional constraints to find energetically favored configurations. From some physical arguments we may try to isolate which invariant is likely to be best conserved in the presence of small amounts of dissipation and use it to find unique minimum energy states. In magnetohydrodynamics, this has led to identification of the so-called reversed-field pinch configurations, with the magnetic field proportional to the current,[33] as particularly attractive from this general point of view.

25.2.2   Statistical Mechanics of Partially Relaxed Stellar Systems

Let us focus on elliptical galaxies. The fact that mechanisms of violent relaxation and phase mixing have had only a limited amount of time to act in the process of collisionless collapse and the fact that there is a sharp difference in the dynamical time scale between the inner and outer regions of a galaxy make it clear that galaxies should be considered as only partially relaxed. There are two ways that can be used to try to describe quantitatively the results of partial relaxation.[34]

The exponential behavior in a Maxwellian distribution function $f \propto \exp(-aE)$ is associated with the constraint of fixing the total energy of the system, which for a spherical equilibrium can be written as

$$E_{\text{tot}} = \frac{1}{3} \int E f(E, J^2) \, d^3x d^3v. \tag{25.1}$$

The constant a is the Lagrange multiplier used in the process of maximizing the Boltzmann entropy of the system. The single-particle energy E is not conserved during violent relaxation, but the related global quantity is. The ansatz of standard statistical mechanics is that besides the total number of particles, this is the only memory of the initial conditions that a system keeps under complete relaxation.

Because it is recognized that realistic models of elliptical galaxies cannot be given in terms of a simple isothermal (or truncated isothermal; see Sections 22.2 through 22.4), we may wonder whether the dynamical requirements that have led to the construction of the f_∞ models (see Subsection 22.4.2) can be explained by the presence of an (approximate) additional constraint, in which a single-particle quantity of the form $q = J^\nu(-E)^{\nu_1}$ is not conserved during collapse, but the related global quantity is. We find indeed that only a specific choice of ν_1 for a given value of ν guarantees that the additional constraint meets the dynamical requirements. In particular, the additional constraint of fixing the global quantity

$$Q = \int J^\nu |E|^{-3\nu/4} f \, d^3v d^3x \tag{25.2}$$

leads to the distribution function

$$f^{(\nu)} = A \exp \left[-aE - d \left(\frac{J^2}{|E|^{3/2}} \right)^{\nu/2} \right], \tag{25.3}$$

which can be shown to satisfy the dynamical selection criterion of Subsection 22.4.2 and, in particular, to be characterized by $\rho \sim r^{-4}$ at large radii, as desired. In addition, it has been shown that for $1/2 \leq \nu \leq 1$, sufficiently concentrated models are indeed associated with small departures from the $R^{1/4}$ profile over a wide radial range. The interesting aspect of this line of reasoning is that indeed numerical simulations show that the global quantity Q of Eq. (25.2) is reasonably well conserved during collisionless collapse. Unfortunately, no simple physical interpretation for this quantity is available.[35]

A different way to incorporate the incompleteness of relaxation is the following. Consider the natural partition in E and J^2 suitable for a spherical system. The distribution $N(E, J^2)$ is related to the distribution function f by means of the Jacobian of the transformation from the ordinary partition $d^3v d^3x$ to $dEdJ^2$. The entropy to be extremized now becomes

$$S = - \int N(E, J^2) \ln N(E, J^2) \, dEdJ^2 = - \int f(E, J^2) \ln \left[\frac{2\pi f(E, J^2)}{\Omega_r(E, J^2)} \right] d^3v d^3x. \tag{25.4}$$

Here Ω_r is the radial orbital frequency discussed in Chapter 21, which, for a given potential, is a well-defined function of energy and angular momentum. Then we argue that violent relaxation is ineffective at large values of angular momentum, so the distribution

$$N_J(J^2) = 4\pi^2 \int \frac{2\pi f(E, J^2)}{\Omega_r(E, J^2)} \, dE \tag{25.5}$$

is conserved for $J > J_0$ (with $J_0^2 \approx G^2 M^3 / |E_{\text{tot}}|$). Then, extremizing the entropy function under the constraint of a fixed total number of particles, total energy, and detailed angular-momentum distribution $N_J(J^2)$ leads to

$$f(E,J^2) \propto \Omega_r(E,J^2)\exp(-aE-c_J). \tag{25.6}$$

Here c_J is the Lagrange multiplier associated with the continuous set of constraints $N_J(J^2)$. If we impose such detailed conservation of angular momentum with respect to an initial homogeneous sphere with a Maxwellian distribution function and note that in the outer regions $\Omega_r \sim |E|^{3/2}$ (see Subsection 21.1.1), we find $c_J \sim cJ^2/2$, so for large values of the angular momentum,

$$f(E,J^2) \sim f_\infty \propto |E|^{3/2}\exp(-aE-cJ^2/2). \tag{25.7}$$

This is the distribution function identified by means of dynamical arguments in Subsection 22.4.2. These arguments confirm and quantify the suggestion[36] of using the dynamical time as a weight factor for microstates.

25.2.3 Some Interesting Properties of the f_∞ and $f^{(\nu)}$ Models

Some interesting properties of the self-consistent models associated with the distribution functions identified in Subsection 25.2.2 have been noted well after the models were constructed originally. In Part IV we saw that the f_∞ models prove to be a useful framework for interpreting photometric and spectroscopic observations of bright elliptical galaxies. One curious aspect of the structure of the models was noted much later, under the influence of current discussions on a quantity that is often called *pseudo-phase-space density* or *proxy for the phase-space density* for a collisionless system of particles. In search of clues on the phase-space structure of dark halos, an examination of cosmological simulations of the formation of cold dark-matter halos has led to the finding that the quantity ρ/σ^3, where ρ is the mass density and σ is the dark-matter velocity dispersion, follows a power-law decline over more than two decades in radius.[37] Initially, the relevant quantity was referred to the isotropic case so that σ did not require additional specifications. As a simple test, Fig. 25.2 shows the behavior of the quantity ρ/σ^3, where $\sigma^3 = \sigma_r\sigma_T^2$ (σ_r and σ_T represent radial and tangential velocity dispersion) for high-Ψ f_∞ models.[38] The quantity R_{scale} used in the figure is approximately the half-mass radius; obviously, the asymptotic power-law behavior at large radii is guaranteed by the asymptotic behavior recorded in Eqs. (22.53) and (22.54). Because the f_∞ and $f^{(\nu)}$ models represent the products of incomplete violent relaxation, based on the dynamical arguments given in Section 22.4 and the statistical arguments of Subsection 25.2.2 they are characterized by similar general properties; thus a plot similar to that illustrated in Fig. 25.2 but based on the $f^{(\nu)}$ models would have a similar appearance.

The study of the $f^{(\nu)}$ models was reconsidered a few years ago for two main reasons. On the one hand, their derivation from statistical arguments based on the use of the Boltzmann entropy is straightforward and thus makes them suitable for a rigorous discussion of the paradigm of the gravothermal catastrophe. On the other hand, improved simulations of collisionless collapse (Fig. 25.3) opened the way to a better test on the conservation of the quantity Q mentioned in Subsection 25.2.2 and a closer inspection of the quality of the models in fitting the results of the simulations. In particular, there remained one difficulty with the f_∞ models: At the time when the f_∞ models were constructed and investigated, it was immediately clear[39] that their pressure anisotropy distribution is qualitatively in line with the simulations of collisionless collapse but

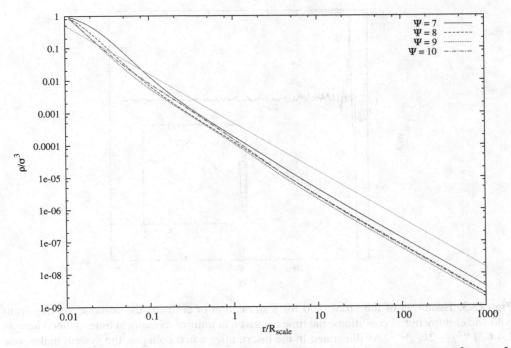

Fig. 25.2. The behavior of the quantity ρ/σ^3, where ρ is the mass density and $\sigma^3 = \sigma_r \sigma_T^2$, represents the third power of the effective velocity dispersion for concentrated f_∞ models (from Zocchi, A. 2010. Tesi di Laurea, Università degli Studi di Milano, Milano, Italy).

quantitatively different because high-Ψ models are too isotropic (with $r_\alpha/r_M \approx 3$) with respect to the end products of the simulations (with $r_\alpha/r_M \approx 1$).

The $f^{(\nu)}$ distribution function (25.3) has been found by extremizing the Boltzmann entropy S under the assumption that the quantity Q introduced in Eq. (25.2) is fixed, together with the total mass M and the total energy E_{tot}. Therefore, for the family of the $f^{(\nu)}$ models, the functions $S = S(M, Q, \Psi)$ and $E_{\text{tot}} = E_{\text{tot}}(M, Q, \Psi)$ can be calculated in a relatively straightforward way.[40] We recall that in the notation of Chapter 22, the quantity $\Psi = -a\Phi(r = 0)$ represents the relevant dimensionless concentration parameter, which parameterizes (for given ν) the family of the $f^{(\nu)}$ models. The specific entropy $\sigma = S/M$ and specific total energy $\epsilon = E_{\text{tot}}/M$ as a function of Ψ are illustrated in the left panel of Fig. 25.4 for the case $\nu = 1$ at fixed M and fixed Q. The curves exhibit the oscillations that contain the information that high-Ψ models can be subject to the gravothermal catastrophe.[41] Indeed, we can calculate the global temperature T from the relation $1/T = \partial S/\partial E_{\text{tot}}$ and show that in the plane $(\epsilon, 1/T)$ the equilibrium sequence exhibits the spiral characteristic of the gravothermal catastrophe. In the right panel of Fig. 25.4 the spiral is compared with the one obtained by using the (positive-definite) quantity a as a proxy for the global temperature.[42] As briefly mentioned in Subsection 23.2.1, a curious coincidence has been found by inspection of the plots shown in Fig. 25.4. This is the fact that the global temperature defined from the relation $1/T = \partial S/\partial E_{\text{tot}}$ becomes negative at low values of Ψ, at the point at which the equilibrium sequence becomes unstable against the radial-orbit instability.

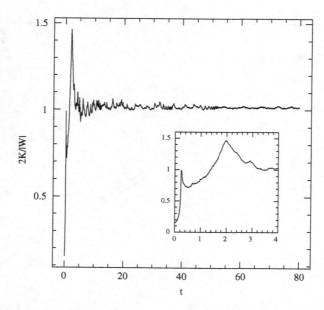

Fig. 25.3. Evolution of the virial ratio for a simulation of collisionless collapse starting from cold and clumpy initial conditions; the time t is given in units of dynamical time, defined here as $1.4 \, GM^{5/2}/(-2E_{\text{tot}})^{3/2}$. As illustrated in the insert, after a first collapse, the system makes one bounce and then quickly settles into conditions of quasi-equilibrium (around a value of the virial ratio slightly above unity because of the presence of some mass loss; from Trenti, M., Bertin, G., van Albada, T. S. 2005. *Astron. Astrophys.*, **433**, 57).

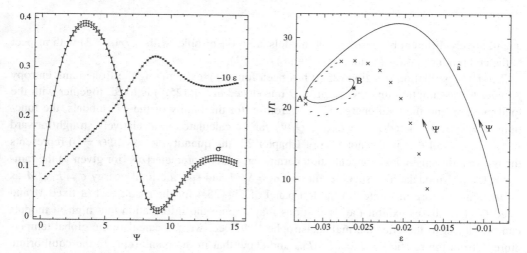

Fig. 25.4. *Left:* Specific Boltzmann entropy σ and total energy ϵ for the $f^{(\nu)}$ models ($\nu = 1$), at constant total mass M and Q, as a function of the concentration parameter Ψ; note that at low concentrations, below a value of $\Psi \approx 3.5$, the models are characterized by a negative global temperature. *Right:* Temperature versus specific total energy for the family of $f^{(\nu)}$ models ($\nu = 1$) showing the structure of the instability spiral characteristic of the gravothermal catastrophe; crosses and other symbols refer to the computed global temperature defined from $\partial S/\partial E_{\text{tot}}$, whereas the line describes the curve based on the local central temperature defined from the coefficient a (from Bertin, G., Trenti, M. 2003. *Astrophys. J.*, **584**, 729).

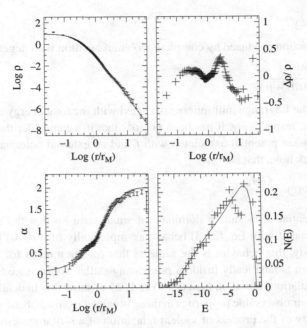

Fig. 25.5. Comparison between the end products of a simulation of collisionless collapse (*symbols with error bars*) with a best-fit $f^{(\nu)}$ model (*lines*). The top left panel represents the density, the top right panel gives the density residuals from the fit, the bottom left frame illustrates the anisotropy profile, and the bottom right frame shows the energy-density distribution $N(E)$ in code units (from Trenti, M., Bertin, G., van Albada, T. S. 2005. *Astron. Astrophys.*, **433**, 57).

Figure 25.5 gives an example of the quality of the $f^{(\nu)}$ models in matching the properties of the end products of collisionless collapse obtained from numerical simulations.[43] The models are able to match the simulated density distributions over nine orders of magnitude. In addition, with no free parameters left once the velocity and length scales and the dimensionless concentration parameter are fixed to identify the best-fit model for a given simulation, they predict a pressure anisotropy profile that is in excellent agreement with the anisotropy profile observed in the simulations (see the lower-left panel of Fig. 25.5), and in general, they provide a good representation of the overall structure in phase space; in particular, the high-Ψ models are characterized by $r_\alpha/r_M \approx 1$.

At present, a physical interpretation of the approximately conserved quantity Q is still missing. In addition, studies of how well the quantity is conserved during collapse simulations do not identify a preferred value of ν in the range $1/2$ to 1; in this range of ν, the dynamical and structural characteristics of the models do not change significantly.

Finally, the spherical, nontruncated $f^{(\nu)}$ models have been shown to perform reasonably well in matching the structural properties of some less relaxed globular clusters.[44] For a better assessment of this issue and a fair comparison with the fits obtained by application of King models (see Section 22.3), it would be desirable to construct truncated $f^{(\nu)}$ models, but so far these models have not been studied systematically.

25.2.4   Degeneracy?

The distribution function produced by complete violent relaxation with degeneracy is[45]

$$f = \frac{A}{1 + \exp(aE - \mu)}, \tag{25.8}$$

where a and μ are the Lagrange multipliers associated with the total energy and total number of particles constraints, respectively. It has been noted[46] that if we consider the particle energy to be the sum of the mean potential associated with f and an external potential associated with a central massive black hole, that is,

$$E = \frac{1}{2}v^2 + \Phi(r) - \frac{GM_\bullet}{r}, \tag{25.9}$$

the black hole potential turns out to dominate at small radii so that the density and velocity dispersion associated with Eq. (25.8) behave asymptotically (at $r \to 0$) like $\rho \sim r^{-3/2}$ and $\langle v^2 \rangle \sim r^{-1}$. Curiously, this behavior is the same as that for the models for which the massive black hole has grown adiabatically inside a preexisting stellar system (see Subsection 22.2.2). Self-consistent calculations based on Eqs. (25.8) and (25.9) show that in detail these models can be very similar in their observable properties to those obtained from adiabatic growth. Obviously, a more realistic study of the process of violent relaxation of a stellar system in the presence of a massive black hole should allow the black hole to move, but this would require nontrivial calculations.

It has been argued that the picture of adiabatic growth of massive black holes inside isothermal cores can well represent the currently available photometric data for the stars in galaxy nuclei.[47] Now the preceding results on degenerate models produced by violent relaxation appear to revalue precisely the opposite picture (i.e., that in which the massive black hole preexists). At least the conclusion would be that observations may be unable to discriminate between the two formation pictures.

25.3   **Dissipative Collapse**

There have been systematic numerical studies of the process of dissipative collapse.[48] Most of the ideas developed in this context, among which similarity solutions have attracted much interest, are not described here.[49] Only a brief outline is given of one framework in which the mechanism of gas infall, which is one element of the process of dissipative collapse, can be followed in detail, with many interesting connections with the observations. The case for which the analysis can be carried out in a relatively simple manner is that in which the process proceeds in a quasi-stationary manner. The steady-state solutions sketched later provide many useful applications. Still, they should not be taken literally as implying that evolution does not occur. As with many stationary solutions, these should be taken as possible "snapshots" to be matched as part of a more complex "movie."

In particular, we will consider spherical infall. The framework we wish to describe is that often called *cooling flow*.[50] Here we refer to a configuration in which the gas would be supported by hydrostatic equilibrium. The support can be lost and the gas flows inward if the loss of energy by means of emission of radiation is sufficiently rapid. This picture appears to be applicable to many cases in which X-rays are observed from clusters of galaxies and some bright elliptical galaxies. The intracluster hot gas thus can generate significant inflows, whereas for individual galaxies the estimated accretion rates rarely exceed one solar mass per year. However, even for clusters of galaxies, it should be stressed that the interesting volume at which the actual flow becomes significant is rather small and often coincides with the location of a central dominant galaxy. In some extreme cases, accretion rates beyond 1,000 solar masses per year have been claimed to occur.

A second paradigm that also would fall in the general context of dissipative collapse is that of accretion disks.[51] Here we usually consider the situation in which a relatively thin disk is formed, with gas in approximate circular motion. Friction within the disk may allow the gas to slowly accrete toward the center. This picture is often used to explain the energy release from very compact objects (such as X-ray binaries or active galactic nuclei). It is also relevant for the formation of protostars. However, the same general picture also may play a role in the formation of galaxy disks. This topic will be addressed separately in Chapter 27.

25.3.1 Cooling Flows

For a hot, optically thin gas, thermal radiation is emitted[52] at a rate basically proportional to the square of the gas density $\rho^2 \Lambda(T)$. The emission function $\Lambda(T)$ is dominated by free-free emission $\propto T^{1/2}$ at high temperatures. At lower temperatures, below 3 keV, the contribution of line emission is important and must be computed carefully for the chemical composition of the gas under investigation.[53] For a gas at temperature T, pressure p, and density ρ, a cooling time associated with radiation losses thus is given by

$$\tau_{\text{cool}} = \frac{5}{2} \frac{p}{\rho^2 \Lambda(T)}. \tag{25.10}$$

A cooling flow may be generated where this cooling time is shorter than the Hubble time. In subsequent paragraphs we focus on conditions best applicable to the hot coronal gas around an individual bright elliptical galaxy for which the expected gas temperature is in the range 0.5 to 1.5 keV (for intracluster gas, the material is sitting in a deeper potential well, and the natural temperature is of a few kiloelectronvolts, up to 10 keV). Note that the hot interstellar medium in a galaxy is expected to be regenerated and heated by fresh gas injected by the stellar component.

Neglecting the contribution of magnetic fields and various transport processes in the gas, we can describe a simple steady-state, spherically symmetric galactic cooling flow by the following equations:[54]

$$\frac{1}{r^2} \frac{d}{dr}(r^2 \rho u) = \alpha \rho_\star - q \frac{\rho}{\tau_{\text{dec}}}, \tag{25.11}$$

$$\rho u \frac{du}{dr} + \frac{dp}{dr} = -\rho \frac{d\Phi}{dr} - \alpha \rho_\star u, \tag{25.12}$$

$$\frac{1}{r^2} \frac{d}{dr} \left[r^2 \rho u \left(\frac{1}{2} u^2 + \frac{5}{2} \frac{p}{\rho} + \Phi \right) \right]$$
$$= -\rho^2 \Lambda(T) + \alpha \rho_\star(\epsilon_{\text{inj}} + \Phi) - q \frac{\rho}{\tau_{\text{dec}}} \left(\frac{1}{2} u^2 + \frac{5}{2} \frac{p}{\rho} + \Phi \right), \tag{25.13}$$

where a decoupling time $\tau_{\text{dec}} = \tau_{\text{cool}}/c(T)$ is introduced, and $c(T) = 2 - d \ln \Lambda / d \ln T$. Equation (25.11) represents the continuity equation, where u is the radial velocity ($u < 0$ indicates inflow), the term $\alpha \rho_*$ indicates mass injection taken to be proportional to the local star density, and the last term is a phenomenological term describing the decoupling of cold gas from the flow as a result of thermal-instability-related[55] processes (thus q is a parameter that is expected to be of the order of unity). Equation (25.12) is the radial momentum balance equation, where Φ represents the gravitational potential. When u is small, as it turns out to be in many cases, the system is in approximate hydrostatic equilibrium over much of the volume involved. Equation (25.13) describes energy transport; the energy-injection term $\alpha \rho_* \epsilon_{inj}$ is expected to be due mainly to the energy released in the gas by supernova explosions (especially those of type SNIa). Pressure is related to temperature by the equation of state $p = \rho kT/\mu m_p$, where μ is the mean mass per particle in units of the proton mass ($\mu \approx 0.63$ for a cosmic chemical composition).

The preceding system of equations includes a set of gas variables (ρ, p, and u; the temperature T is then determined by the equation of state), a set of input parameters for the gas (chemical composition of the gas, the parameters α and ϵ_{inj} that determine the mass and energy injection rate; these can be related to the distribution of stars, the mass-to-light ratio for the stellar component, and the relevant supernova rate), and one phenomenological parameter q for the gas that we may constrain empirically. The remaining two profiles involved, the stellar density ρ_* and the gravitational potential Φ, may be taken as decoupled in a first approximation by considering the contribution of the gas mass to gravity to be negligible; these functions define the properties of the galaxy as a stellar system, with the gravitational potential possibly including the contribution from a dark halo (see Part IV).

The equations for the gas variables are to be solved by the imposition of appropriate boundary conditions. These are nontrivial, and different options are available.[56] We may consider the outer boundary located at r_{ext}, where the dynamics of the coronal gas matches that of the intra-cluster gas. The matching conditions are specified by two parameters (i.e., the mass inflow rate $\dot{M}_{ext}$ and the pressure p_{ext}). We may then treat the sonic point $r = r_s$ as a free boundary at which the flow is assumed to be regular. The values of r_s and the mass flow $4\pi r^2 u$ at r_s are guessed and changed until the integration makes the adopted values compatible with the outer boundary conditions at r_{ext}. In some cases the value of r_s becomes too small, and the problem reduces to the determination of a fully subsonic flow. This can be integrated from the outside, starting with different values of the density ρ_{ext} until the central-mass flux becomes vanishingly small.

The resulting models turn out to be rather sensitive to the values of the phenomenological parameter q and the outer boundary parameter $\dot{M}_{ext}$, but they are less sensitive to p_{ext}. The use of a finite value of q resolves one of the major difficulties with the standard steady-state cooling-flow models[57] for the hot gas around elliptical galaxies: that of the X-ray overbrightness in the inner regions.

There are different ways of approaching the observations within the previously described general framework. In the community of X-ray astrophysics, it is common practice to use the X-ray data to reconstruct the properties of the global potential well [i.e., of $\Phi(r)$]. This has initially been quite appealing given the fact that X-ray data sample a large volume (the hot gas is rather diffuse) and the fact that kinematic data for the stellar component were inaccurate and radially restricted (see Chapter 24). In practice, such a reconstruction technique suffers from the fact that the X-ray determinations of the temperature profile of the gas are rather problematic. This has led to some controversial conclusions about the detection of dark halos around elliptical galaxies.

In the 1990s, accurate measurements of the kinematic data for the stars have become possible even beyond R_e, and overall stellar dynamical modeling has been significantly improved. Under these circumstances, it is possible to select the best stellar dynamical models for some elliptical galaxies with firm confidence in their accuracy. For a sample of well-studied elliptical galaxies (NGC 1399, NGC 1404, NGC 4374, NGC 4472, and NGC 4636), the quantities $\rho_\star$ and Φ (these models include the presence of a dark halo) thus have been used as inputs to the cooling-flow equations, and it has been possible to show that rather accurate models for the available X-ray data can be provided with moderate amounts of accretion (up to 4 $M_\odot$ yr^{-1}) with a reasonable value of the outer pressure and a moderate value of the supernova rate;[58] the best models have $q \approx 0.5$ (Fig. 25.6). A particularly encouraging case is that of NGC 4636, for which a rather accurate temperature profile out to 1,000 arcsec has been measured by ROSAT[59] and is well explained by the cooling-flow model based on the potential well determined from stellar dynamical observations.

As mentioned earlier, the inflow velocity becomes appreciable only in the innermost regions, whereas in most of the volume the resulting gas distribution turns out to be in approximate hydrostatic equilibrium. Thus a conceptually simpler test on X-ray observations can be done in the following way. Because the emission profile $\rho^2 \Lambda(T)$ is largely determined by the gas density distribution, we can make a rough guess on a value of T and use the observed X-ray brightness profile to derive an empirical $\rho(r)$ (instead, in the model described earlier, the density profile was an unknown function to be computed from the cooling-flow equations). Then, for a given form of the potential well, we can produce the expected temperature profile from the hydrostatic equilibrium equation

$$T(r) = \frac{\mu m_p}{\rho(r)} \left[G \int_r^{r_{\text{ext}}} \frac{M(r')\rho(r')}{r'^2} \, dr' + p_{\text{ext}} \right], \tag{25.14}$$

where $M(r')$ is the total mass enclosed in a sphere of radius r'. In general, this would be sufficient, but we could make a more accurate determination through iteration by inserting the profile $T(r)$ just calculated in the emission function $\rho^2 \Lambda(T)$ to obtain a more accurate empirical density distribution of the gas. This rather straightforward modeling procedure has been used (see lower frame of Fig. 24.4) to prove that the final temperature profiles are not very sensitive to how the stellar dynamical models are extrapolated to large radii (within a class of physically plausible models), thus indicating that X-ray observations are not very helpful in discriminating between models with a dark halo and others without a dark halo. This is in sharp contrast to what is commonly believed.[60]

Other studies[61] have discussed the time evolution of spherical models (Fig. 25.7) with similar physical ingredients but with the possibility of incorporating the effects of a time-dependent energy and mass injection in the gas. In particular, it has been argued that different stages of flow evolution (which may include, in addition to inflow, periods of outflow and wind) may be responsible for the large scatter observed around the relation $L_X \sim L_B^s$ (with $s = 1.4-2.2$) between X-ray and optical total luminosity of elliptical galaxies[62] (Fig. 25.8). In fact, the preceding steady-state cooling-flow solutions are probably relevant only for the X-ray-brightest elliptical galaxies. Other investigations have focused on the specific issue of the conditions for the applicability of a steady-state description as a snapshot of a time-evolving process.[63]

The simplified description provided in this subsection refers mainly to one interesting modeling tool for the observed diffuse X-ray emission. This has sometimes been taken as the physical basis for broader pictures of dissipative formation. In fact, it has been suggested that significant

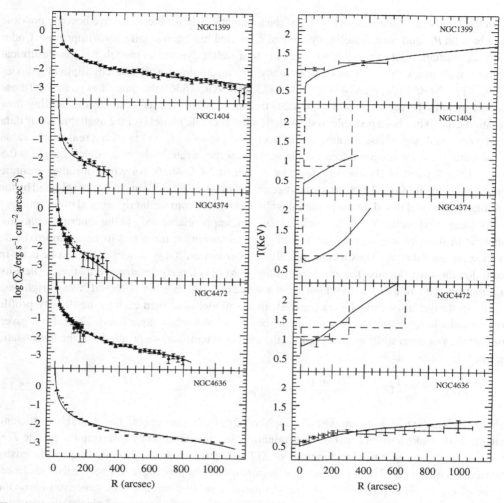

Fig. 25.6. A cooling-flow fit to X-ray emission (*left*) and temperature (*right*) profiles based on models with gas loss for a set of galaxies for which the underlying potential well has been taken to be the one provided by the best-fit models of stellar dynamical data (usually available out to R_e); the models are the ones illustrated in Fig. 24.2. Note the rather accurate prediction of the extended X-ray temperature profile available for NGC 4636 (from Bertin, G., Toniazzo, T. 1995. *Astrophys. J.*, **451**, 111).

amounts of cold gas decouple from the cooling flow and that stars are thus formed in the central regions.[64] The possibility of considering this as the basis for a mechanism for the dissipative formation of some galaxies, especially the giant cDs that are often seen at the center of massive cluster cooling flows, is in principle supported by the values of some estimates of the relevant cluster accretion rates, which would indeed easily be able to build up a galactic mass over a Hubble time. The picture then might be applicable in explaining the origin of the stellar halo observed around many cDs.

In practice, progress in the study of the X-ray emission from galaxies and clusters of galaxies has identified a number of difficulties with the simple cooling-flow picture proposed originally. In

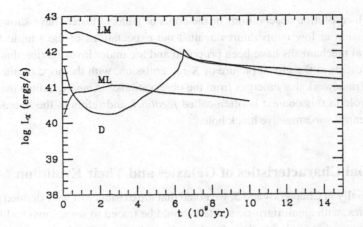

Fig. 25.7. Variation of the X-ray luminosity with time in three evolution models. (Selected models are from Mathews, W. G., Loewenstein, M. 1986. *Astrophys. J. Lett.*, **306**, L7; Loewenstein, M., Mathews, W. G. 1987. *Astrophys. J.*, **319**, 614; D'Ercole, A., Renzini, A., Ciotti, L., Pellegrini, S. 1989. *Astrophys. J. Lett.*, **341**, L9. The figure is from Sarazin, C. L. 1990. In *The Interstellar Medium in Galaxies*, eds. H. A. Thronson, J. M. Shull. Kluwer, Dordrecht, The Netherlands, p. 230; ©1990 Kluwer Academic Publishers, with kind permission from Kluwer Academic Publishers.)

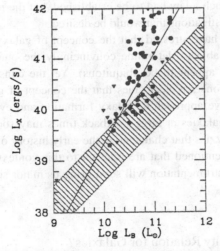

Fig. 25.8. Correlation of X-ray luminosity and optical luminosity for early-type galaxies. Filled circles are detections; inverted triangles are upper limits (from Canizares, C. R., Fabbiano, G., Trinchieri, G. 1987. *Astrophys. J.*, **312**, 503). The hatched area gives the range of estimated stellar X-ray luminosities (from Sarazin, C. L. 1990. In *The Interstellar Medium in Galaxies*, eds. H. A. Thronson, J. M. Shull. Kluwer, Dordrecht, The Netherlands, p. 204; ©1990 Kluwer Academic Publishers, with kind permission from Kluwer Academic Publishers).

particular, high-resolution observations of the cores of galaxy clusters have shown a significant deficit of emission at low temperatures against the expectations of the simplest cooling-flow models. Several mechanisms have been proposed and are under investigation that include more physical ingredients in the global picture of X-ray emission, with the hope of matching the rich astrophysical framework that emerges from the observations.[65] One ingredient that is thought to play a major role in this context is often called *feedback* and refers to the interaction with the activity of a central supermassive black hole.[66]

25.4 Global Characteristics of Galaxies and Their Evolution

In Part I, especially in Chapters 4 and 5, we noted that the existence of well-defined global scaling laws for galaxies, with small intrinsic scatter, should be traced to some universal formation and evolution processes that are left to the astrophysicist to discover. Purely dynamical and collision-less processes are scale-free, and, indeed, the stellar dynamical equilibrium models discussed in Part IV can be normalized to satisfy the empirical scaling relations. Consequently, little help is expected from stellar dynamical arguments alone. Similarly, the discussion of Section 25.2 may help us to understand the universality of the $R^{1/4}$ profile, but it is of little use for clarifying the existence of the fundamental plane of elliptical galaxies. The study of dissipative collapse, briefly touched on in Section 25.3 and to be expanded in Chapter 27, is richer in this respect but still keeps us far from some striking facts, such as the luminosity-velocity relation for spiral galaxies. The latter empirical relation obviously suggests that a fine-tuning has taken place between the disk and the dark halo, if such a law had to be established, but the mechanisms involved so far escape the quantitative investigations that would be desired.

On many occasions we have stressed that the concept of galaxy mass is not well defined simply because we are unable to determine convincingly the outer edge of the dark halos that are invoked (and they appear to be ubiquitous). Yet the conspiracy effects brought out by the global scaling relations are clear clues that the concept of galaxy mass is useful after all. Even though a deductive approach to galaxy formation and evolution may be out of the question, observations of galaxies at large lookback times may soon allow us to disentangle empirically many of the puzzles that characterize the early history of galaxies. In the next subsection a few points are mentioned that are relevant to this context, with the confidence that new telescopes and new instrumentation will soon bring us major surprises and some decisive answers.

25.4.1 A Mass-Luminosity Relation for Galaxies?

Let us define[67] the mass M of an elliptical galaxy as a quantity proportional to the square of the observed central velocity dispersion σ_0 and the observed half-luminosity radius R_e. This is a purely empirical definition, where no position is taken on the galaxy model or on the fact that part of the mass may be luminous and part may be dark, with possibly different origins altogether. In a sense, this definition recognizes the role of an empirical virial relation. Then suppose that elliptical galaxies are homologous so that, for given M, the total luminosity L can be taken to be proportional to the square of the observed half-luminosity radius and to the surface brightness I_e at that location. Then a mass-luminosity relation for elliptical galaxies in the form $L = f(M)$

would imply the existence of a surface in the natural space of the observed parameters:

$$I_e R_e^2 \propto f(\sigma_0^2 R_e). \tag{25.15}$$

Note that the homology required here is weak, in the sense that the surface-brightness profiles are assumed to have a universal shape (e.g., the $R^{1/4}$ law, but a different law would also work) at given values of L; in principle, different luminosity profiles (e.g., different values of the index n in a generalized $R^{1/n}$ law) at different values of L also would be allowed. If the function f is approximated by a power law, say, $f(M) \propto M^a$, then we have $I_e \propto \sigma_0^{2a} R_e^{a-2}$. Curiously, for $a = 0.807$ we derive the same expression $R_e \propto \sigma_0^{1.35} I_e^{-0.84}$ of the fundamental plane as that given by Eq. (4.14). A similar use of the homology argument shows that the often-considered $D_n - \sigma$ relation needs a surface-brightness correction.[68]

Why should a mass-luminosity relation hold at all? Can we set up a similar relation for spiral galaxies based on the existence of the velocity-luminosity relation $L \propto V^4$ so neatly established in the near infrared?[69] If we replace σ_0 with V and R_e with the disk horizontal scale $h_\star$ in the preceding argument, we find that a constant mass-to-light ratio for spiral galaxies would give $M \propto h_\star^2$ [see Section 4.3, relation (4.10)]. Whatever the final empirical relations turn out to be, can we hope to find some physical justification for them? These open questions leave us at a disadvantage with respect to other fields, such as stellar evolution, for which clear answers can be provided.

Progress in these directions will depend on two basic steps. On the one hand, it is of primary importance to identify the set of optimal observables that guarantee the smallest scatter in global relations. We note that the operational definitions of the relevant spectroscopic quantities (e.g., the notion of central velocity dispersion for ellipticals and the notion of circular velocity for spiral galaxies) and photometric quantities (e.g., the use of near-infrared images for spiral galaxies) have now become rather sharp. On the other hand, we should address the issue of connecting these empirical parameters to physically interesting intrinsic quantities. In particular, we should eventually find a way to assess the relative share between dark and luminous matter in determining the mass variable of the fundamental plane relation. Even in the absence of recent mergers and recent major star-formation events, the luminous component is expected to be subject to a passive evolution (thus reflected in the total luminosity L) as a result of the natural aging of a set of stars formed in the distant past.

25.4.2 Weak Homology of Elliptical Galaxies

The suggestion provided in Subsection 25.4.1 can be analyzed further and reformulated in the following way:[70] Consider the fundamental plane in the form given by Eq. (4.15), with the notation introduced after that equation, in Chapter 4, and write the virial theorem as

$$\frac{GL}{R_e} \times \left(\frac{M}{L}\right)_\star = K_V \sigma_0^2. \tag{25.16}$$

Here L is the total luminosity, $(M/L)_\star$ is the mass-to-light ratio in the relevant wave band for the stellar component, σ_0 is the central velocity dispersion of the stars, and R_e is the effective radius, measured and defined as in Subsection 4.3.3. The dimensionless virial coefficient K_V takes into account projection effects, the structural properties of the underlying distribution function of the stellar component, and effects related to the presence of the dark-matter component. If we eliminate σ_0 from Eqs. (4.15) and (25.16), we obtain

$$\frac{(M/L)_\star}{K_V} = R_e^{(2-10\beta+\alpha)/\alpha} L^{(5\beta-\alpha)/\alpha}. \tag{25.17}$$

Empirically, the exponent of R_e is very small [in the blue band, $(2-10\beta+\alpha)/\alpha \approx 0.04$]. Therefore, if we argue that elliptical galaxies are homologous, as we did in Subsection 25.4.1, the virial coefficient should be considered to be approximately constant, and we thus recover a mass-luminosity relation of the form

$$\left(\frac{M}{L}\right)_\star \propto L^\delta, \tag{25.18}$$

where $\delta \approx 0.30 \pm 0.064$. An alternative interpretation would be to assume an approximately constant stellar mass-to-light ratio and thus invoke a case of weak homology, that is,

$$K_V \propto L^{-\delta}. \tag{25.19}$$

In this picture, all the dynamical factors that contribute to the virial coefficient should cooperate and determine a systematic variation of K_V as a function of the total luminosity L. Note that a factor of 20 in L would require a change by a factor of ≈ 2.45 in K_V.

One argument against strict homology derives from the fact that the $R^{1/4}$ law has been found to be violated (Fig. 25.9), so preference is often given to an $R^{1/n}$ description, with the index n free to change from galaxy to galaxy (see Subsection 4.1.2). A fit to the NGC 1379 photometric profile (in the wide radial range considered in the left frame of Fig. 25.9) by means of the $R^{1/4}$ law would exhibit deviations by up to 1 magnitude (in the outer regions); similarly, a fit to NGC 4552 by the $R^{1/4}$ law would show residuals exceeding 0.2 magnitude in the inner regions and even larger than 1 magnitude (in the outer regions). In practice, for bright ellipticals, the $R^{1/4}$ law gives a reasonable description of the observed photometric profiles, with few exceptions, whereas a law of the form $R^{1/n}$ with $n \approx 2$ is generally representative of the low-luminosity end of the distribution of elliptical galaxies. We thus might be left to conclude that homology is well justified for the bright end of the luminosity distribution of elliptical galaxies.

However, there is a more subtle argument that suggests that caution should be used in applying strict homology and that some sort of weak homology must be at work. This point is illustrated in Fig. 25.10. Even if we restrict our attention to galaxies with an $R^{1/4}$ photometric profile, their internal structure can vary significantly so that the virial coefficient K_V can be subject to appreciable variations. Note that isotropic models constructed to have a projected density profile characterized by $n = 4$ at all radii are associated with a virial coefficient $K_V \approx 4.5$. In turn, if we consider f_∞ models with Ψ in the range 7 to 10, which are all consistent with an $R^{1/4}$ profile, they are associated with a virial coefficient that changes by a factor of 50 percent, from $K_V \approx 2.5$ to $K_V \approx 3.25$. In other words, systems that are photometrically similar could have significantly different virial coefficients.

In addition, by checking the values of K_V on galaxies for which a dark–luminous-matter decomposition is available (the set of ten galaxies[71] mentioned in Section 24.1), no clear trends emerge in terms of either strict homology (i.e., constant K_V) or constant $(M/L)_\star$. We are thus

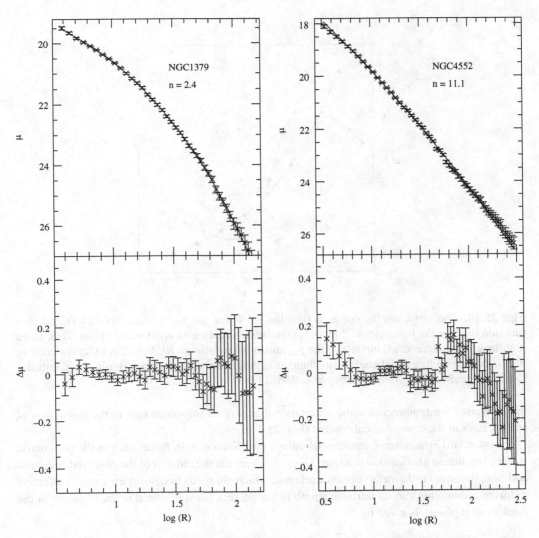

Fig. 25.9. *Left:* Best fit to the photometric profile of the galaxy NGC 1379 (*top*), by means of the $R^{1/n}$ law, and the corresponding residuals (*bottom*). *Right:* Best fit to the photometric profile of the galaxy NGC 4552, by means of the $R^{1/n}$ law, in a similar format (from Bertin, G., Ciotti, L., Del Principe, M. 2002. *Astron. Astrophys.*, **386**, 149).

forced to conclude that the fundamental plane reflects a tuning of changes in K_V and $(M/L)_\star$ that remains unexplained.

25.4.3 The Universe at Large Redshifts

As mentioned earlier, observations of galaxies at large distances bring us back to the initial stages of galaxy evolution and may just provide directly some of the answers we are looking for. Here we recall that information coming from cosmological distances is mediated by the geometry of

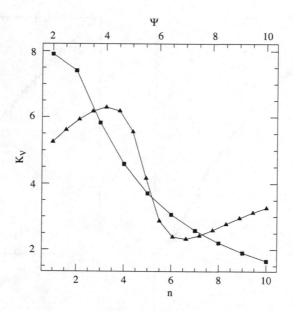

Fig. 25.10. The virial coefficient as a function of Ψ for the f_∞ models (*triangles*) and as a function of n for the isotropic $R^{1/n}$ models (*squares*), based on an aperture of radius $R_e/8$. Even if sufficiently concentrated models of the f_∞ family all exhibit an $R^{1/4}$ profile, independently of Ψ, the corresponding virial coefficient changes significantly with Ψ (from Bertin, G., Ciotti, L., Del Principe, M. 2002. *Astron. Astrophys.*, **386**, 149).

the Universe. We thus record some relations[72] that play an important role in the discussion of scaling laws in the cosmological context (Fig. 25.11).

In a standard Friedmann–Lemaître cosmology (see Section 4.4), fluxes such as the bolometric surface brightness are dimmed at a rate $(1+z)^{-4}$, where z is the redshift of the observed object. In addition, the angular-diameter–distance relation, which allows us to convert a measured effective radius in arcseconds into an intrinsic length in kiloparsecs (as is intended in the equation of the fundamental plane), is given by

$$D(z) = \frac{c}{H_0(1+z)\sqrt{\Omega_R}} \sinh\left[\sqrt{\Omega_R} \int_0^z \frac{dz'}{\sqrt{\Omega_m(1+z')^3 + \Omega_R(1+z')^2 + \Omega_\Lambda}}\right]. \qquad (25.20)$$

Here we recall that $\Omega_m + \Omega_R + \Omega_\Lambda = 1$; the expression with the hyperbolic function is suited for an open Universe $\Omega_R > 0$. For the Einstein–de Sitter case ($\Omega_m = 1$, $\Omega_\Lambda = 0$), we have

$$D(z) = \frac{2c}{H_0(1+z)}\left(1 - \frac{1}{\sqrt{1+z}}\right). \qquad (25.21)$$

Thus the (bolometric) distance modulus (i.e., the difference between apparent and absolute magnitude of an object at redshift z) is given (within a constant) by $2.5\log[(1+z)^4 D(z)^2]$. Note that the distance $D(z)$ is a nonmonotonic function of redshift. Because $H_0(10\ \mathrm{kpc})/c \approx 0.7\ h$ arcsec (where h is the reduced Hubble constant; see Chapter 4), we see that a galaxy at redshift $z \approx 1$ will appear to have a size of the order of 1 arcsec, and at redshifts larger than ≈ 2, such angular size will grow with z (Fig. 25.12).

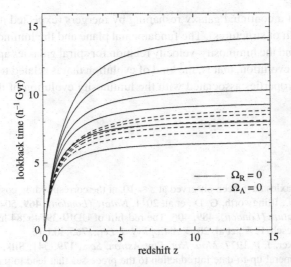

Fig. 25.11. Lookback time as a function of redshift for different values of the density parameter Ω_m *(from top to bottom:* $\Omega_m = 0.05, 0.1, 0.2, 0.3, 0.5, 1$). Solid curves refer to a flat Universe, and dashed curves refer to models with a vanishing cosmological constant; the $\Omega_m = 1$ solid curve coincides with the corresponding dashed curve. The quantity h is the reduced Hubble constant.

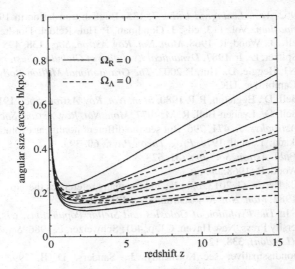

Fig. 25.12. Angular size as a function of redshift for different values of the density parameter Ω_m *(from bottom to top:* $\Omega_m = 0.05, 0.1, 0.2, 0.3, 0.5, 1$); solid and dashed lines are as in Fig. 25.11. The quantity h is the reduced Hubble constant.

In Chapter 4 a few references were given about a field in which enormous progress has been made in the last few years, that is, the study of the relevant global scaling laws for galaxies at cosmological distances ($z \approx 1$ and farther out). Describing this progress would bring us well beyond the scope of this book. Here we only mention that, in general, very little evidence is obtained for significant dynamical evolution of galaxies in the past ≈ 8 to 10 billion years, which goes against the picture of a continual galaxy reshaping by mergers (expected in the standard model of hierarchical growth of structures). The fundamental plane and the luminosity-size relation for early-type galaxies and the luminosity-velocity relation for spiral galaxies appear to be consistent with a rather passive evolution, that is, the kind of evolution that is related to the gradual changes in the photometric properties associated with the luminosity evolution of the underlying stellar populations.[73]

Notes

1. Two candidate galaxies have been observed at $z \approx 10$, at the corresponding cosmic age of ≈ 500 Myr. See Bouwens, R. J., Illingworth, G. D., et al. 2011. *Nature (London)*, **469**, 504; Zheng, W., Postman, M., et al. 2012. *Nature (London)*, **489**, 406. The redshift of UDFJ-39546284 has been confirmed; see Bouwens, R. J., Oesch, P. A., et al. 2013. *Astrophys. J. Lett.*, **765**, id.L16.

2. Rees, M. J., Ostriker, J. P. 1977. *Mon. Not. Roy. Astron. Soc.*, **179**, 541; Silk, J. 1977. *Astrophys. J.*, **211**, 638. For a general up-to-date introduction to the processes that lead to the formation of the first stars and galaxies, see Stiavelli, M. 2009. *From First Light to Reionization: The End of the Dark Ages*, Wiley, New York.

3. Ebert, R. 1955. *Zeit. Astrophys.*, **37**, 217; Bonnor, W. B. 1956. *Mon. Not. Roy. Astron. Soc.*, **116**, 351.

4. Lombardi, M., Bertin, G. 2001. *Astron. Astrophys.*, **375**, 1091. Part of the discussion given in this section follows the arguments presented by Sormani, M. C., Bertin, G. 2013. *Astron. Astrophys.*, **552**, A37.

5. Antonov, V. A. 1962. *Vest. Leningrad Univ.*, **7**, 135; English translation in 1985. *Dynamics of Star Clusters, IAU Symposium*, Vol. 113, eds. J. Goodman, P. Hut. Reidel, Dordrecht, The Netherlands, p. 525; Lynden-Bell, D., Wood, R. 1968. *Mon. Not. Roy. Astron. Soc.*, **138**, 495.

6. For example, see Spitzer, L., Jr. 1987. *Dynamical Evolution of Globular Clusters*, Princeton University Press, Princeton, NJ; Heggie, D., Hut, P. 2003. *The Gravitational Million-Body Problem*, Cambridge University Press, Cambridge, UK.

7. See also Lynden-Bell, D., Eggleton, P. P. 1980. *Mon. Not. Roy. Astron. Soc.*, **191**, 483.

8. See also Lynden-Bell, D., Lynden-Bell, R. M. 1977. *Mon. Not. Roy. Astron. Soc.*, **181**, 405.

9. Makino, J. 1996. *Astrophys. J.*, **471**, 796. But see the different nonlinear evolution scenario discussed by Hachisu, I., Nakada, Y., et al. 1978. *Prog. Theor. Phys.*, **60**, 393.

10. Hachisu, I. 1979. *Publ. Astron. Soc. Jpn.*, **31**, 523.

11. Lynden-Bell, D., Wood, R. 1968. *op. cit.*

12. See Sormani, M. C., Bertin, G. 2013. *op. cit.*, and the many references therein.

13. van Albada, T. S. 1982. *Mon. Not. Roy. Astron. Soc.*, **201**, 939.

14. Toomre, A. 1977. In *The Evolution of Galaxies and Stellar Populations*, eds. B. T. Tinsley, R. B. Larson. Yale University Press, New Haven, CT, p. 401; Schweizer, F. 1986. *Science*, **231**, 193; Barnes, J. F. 1989. *Nature (London)*, **338**, 123.

15. Dissipative and nondissipative; see Kormendy, J., Sanders, D. B. 1992. *Astrophys. J. Lett.*, **390**, L53.

16. See Schechter, P. L. 1990. In *Dynamics and Interactions of Galaxies*, ed. R. Wielen, Springer-Verlag, Berlin, p. 508. See also Governato, F., Reduzzi, L., Rampazzo, R. 1993. *Mon. Not. Astron. Soc.*, **261**, 379; Quinn, P. J., Hernquist, L., Fullagar, D. P. 1993. *Astrophys. J.*, **403**, 74.

17. Among others, in addition to the article by van Albada, T. S. 1982. *op. cit.*, we may recall Hénon, M. 1964. *Ann. Astrophys.*, **27**, 83; Melott, A. L. 1983. *Astrophys. J.*, **264**, 59; McGlynn, T. A. 1984. *Astrophys. J.*, **281**, 13; Villumsen, J. V. 1984. *Astrophys. J.*, **284**, 75; May, A., van Albada, T. S. 1984. *Mon. Not. Roy. Astron. Soc.*, **209**, 15; Aguilar, L., White, S. D. M. 1986. *Astrophys. J.*, **307**, 97; Stiavelli, M., Bertin, G. 1987. *Mon. Not. Roy. Astron. Soc.*, **229**, 61; Aguilar, L., Merritt, D. 1990. *Astrophys. J.*, **354**, 33; McGlynn, T. A. 1990. *Astrophys. J.*, **348**, 515; Londrillo, P., Messina, A., Stiavelli, M. 1991. *Mon. Not. Roy. Astron. Soc.*, **250**, 54; Stiavelli, M., Londrillo, P., Messina, A.

1991. *Mon. Not. Roy. Astron. Soc.*, **251**, 57. See also Trenti, M., Bertin, G., van Albada, T. S. 2005. *Astron. Astrophys.*, **433**, 57, and references therein.

18. Lynden-Bell, D. 1967. *Mon. Not. Roy. Astron. Soc.*, **136**, 101.

19. See also Tremaine, S., Gunn, J. E. 1979. *Phys. Rev. Lett.*, **42**, 407; Madsen, J., Epstein, R. I. 1984. *Astrophys. J.*, **282**, 11; Madsen, J., Epstein, R. I. 1985. *Phys. Rev. Lett.*, **54**, 2720.

20. Starting with the "water-bag" simulations by Hohl, F., Feix, M. R. 1967. *Astrophys. J.*, **147**, 1164. See Rephaeli, Y. 1982. *Phys. Rev. D*, **26**, 770; Rephaeli, Y. 1983. *Astron. Astrophys.*, **123**, 98; Luwel, M., Severne, G. 1985. *Astron. Astrophys.*, **152**, 305; Severne, G., Luwel, M. 1986. *Astrophys. Sp. Sci.*, **122**, 299; Madsen, J. 1987. *Astrophys. J.*, **316**, 497; Nozakura, T. 1992. *Mon. Not. Roy. Astron. Soc.*, **257**, 455; Kull, A., Treumann, R. A., Boehringer, H. 1997. *Astrophys. J.*, **484**, 58. See also Colombi, S., Touma, J. 2008. *Comm. Nonlin. Sci. Num. Sim.*, **13**, 46.

21. Tremaine, S., Hénon, M., Lynden-Bell, D. 1986. *Mon. Not. Roy. Astron. Soc.*, **219**, 285; Wiechen, H., Ziegler, H. J., Schindler, K. 1988. *Mon. Not. Roy. Astron. Soc.*, **232**, 623; Ziegler, H. J., Wiechen, H. 1989. *Mon. Not. Roy. Astron. Soc.*, **238**, 1261; Spergel, D., Hernquist, L. 1992. *Astrophys. J. Lett.*, **397**, L75.

22. Shu, F. H. 1978. *Astrophys. J.*, **225**, 83; Shu, F. H. 1987. *Astrophys. J.*, **316**, 502.

23. See Lynden-Bell, D. 1967. *op. cit.*; Shu, F. H. 1978. *op. cit*; 1987. *op. cit.*

24. Jaffe, W. 1987. In *Structure and Dynamics of Elliptical Galaxies*, ed. P. T. de Zeeuw. Reidel, Dordrecht, The Netherlands, p. 511; see also Hjorth, J., Madsen, J. 1991. *Mon. Not. Roy. Astron. Soc.*, **253**, 703.

25. Binney, J. J. 1982. *Mon. Not. Roy. Astron. Soc.*, **200**, 951.

26. Bertin, G., Stiavelli, M. 1989. *Astrophys. J.*, **338**, 723.

27. Kelvin, Lord 1910. *Mathematical and Physical Papers*, Vol. IV, Cambridge University Press, Cambridge, UK, pp. 101, 172; see also Lynden-Bell, D., Katz, J. 1981. *Proc. Roy. Soc. London A*, **378**, 179.

28. Alfvén, H. 1950. *Cosmic Electrodynamics*, Oxford University Press, Oxford, UK.

29. Woltjer, L. 1958. *Proc. Natl. Acad. Sci. USA*, **44**, 489. This paper provides a simpler interpretation to the theorem on force-free fields described earlier by Chandrasekhar, S., Woltjer, L. 1958. *Proc. Natl. Acad. Sci. USA*, **44**, 285.

30. Moffatt, H. K. 1969. *J. Fluid Mech.*, **35**, 117.

31. See the review by Tur, A. V., Yanovsky, V. V. 1993. *J. Fluid Mech.*, **248**, 67.

32. Arnol'd, V. I. 1965. *J. Appl. Math. Mech.*, **29**, 1002; Arnol'd, V. I. 1989. *Mathematical Methods of Classical Mechanics*, Springer-Verlag, Berlin; Kuvshnikov, B. N., Pegoraro, F., Schep, T. J. 1994. *Phys. Lett. A*, **191**, 296; see also Morrison, P. J., Greene, J. M. 1980. *Phys. Rev. Lett.*, **45**, 790.

33. Taylor, J. B. 1974. *Phys. Rev. Lett.*, **33**, 1139.

34. The following is based on the article by Stiavelli, M., Bertin, G. 1987. *op. cit.*; the article also includes the numerical simulations that demonstrate the reasonable conservation of the quantity Q of Eq. (25.2).

35. This topic was studied extensively by Trenti, M. 2005. Ph.D. dissertation, Scuola Normale Superiore, Pisa, Italy. See Trenti, M., Bertin, G., van Albada, T. S. 2005. *Astron. Astrophys.*, **433**, 57, and Subsection 25.2.3. A systematic study of the dynamical properties of the $f^{(\nu)}$ models is given by Trenti, M., Bertin, G. 2005. *Astron. Astrophys.*, **429**, 161.

36. Lynden-Bell, D. 1967. *op. cit.*; Tremaine, S. 1987. In *Structure and Dynamics of Elliptical Galaxies*, ed. P. T. de Zeeuw. Reidel, Dordrecht, The Netherlands, p. 367; see also Jaynes, E. T. 1973. *Found. Phys.*, **3**, 477.

37. For example, see Taylor, J. E., Navarro, J. F. 2001. *Astrophys. J.*, **563**, 483; Dehnen, W., McLaughlin, D. E. 2005. *Mon. Not. Roy. Astron. Soc.*, **363**, 1057.

38. The realization of this test was encouraged by Luca Ciotti and performed by Zocchi, A. 2010. Tesi di Laurea, Università degli Studi di Milano, Milano, Italy.

39. Stiavelli, M., Bertin, G. 1985. *Mon. Not. Roy. Astron. Soc.*, **217**, 735. For the definition of the anisotropy radius r_α, see Subsection 22.4.1.

40. Here we follow Bertin, G., Trenti, M. 2003. *Astrophys. J.*, **584**, 729.

41. As described by Lynden-Bell, D., Wood, R. 1968. *op. cit.*

42. The use of a proxy for the global temperature was argued in order to explore the possibility of a gravothermal catastrophe in the context of stellar dynamics for systems for which the distribution function could not be derived by extremizing the Boltzmann entropy. See appendix V in Lynden-Bell, D., Wood, R. 1968. *op. cit.*; Katz, J. 1980. *Mon. Not. Roy. Astron. Soc.*, **190**, 497; Magliocchetti, M., Pucacco, G., Vesperini, E. 1998. *Mon. Not. Roy. Astron. Soc.*, **301**, 25.

43. This short summary is based on the article by Trenti, M., Bertin, G., van Albada, T. S. 2005. *Astron. Astrophys.*, **433**, 57.

44. Zocchi, A., Bertin, G., Varri, A. L. 2012. *Astron. Astrophys.*, **539**, A65.

45. Lynden-Bell, D. 1967. *op. cit.*

46. Stiavelli, M. 1998. *Astrophys. J. Lett.*, **495**, L91.

47. van der Marel, R. P. 1999. *Astron. J.*, **117**, 744; van der Marel, R. P., Cretton, N., et al. 1998. *Astrophys. J.*, **493**, 613.

48. See Larson, R. B. 1969. *Mon. Not. Roy. Astron. Soc.*, **145**, 505; 1974. *Mon. Not. Roy. Astron. Soc.*, **166**, 385; 1975. *Mon. Not. Roy. Astron. Soc.*, **173**, 671; Carlberg, R. C. 1984. *Astrophys. J.*, **286**, 403.

49. For a spherically symmetric isothermal gas, similarity solutions were found and discussed by Penston, M. V. 1969. *Mon. Not. Roy. Astron. Soc.*, **144**, 425; Larson, R. B. 1969. *Mon. Not. Roy. Astron. Soc.*, **145**, 271; Shu, F. H. 1977. *Astrophys. J.*, **214**, 488; Hunter, C. 1977. *Astrophys. J.*, **218**, 834. An analysis of wider classes of models of collapsing radiative gas is given by Chevalier, R. A. 1987. *Astrophys. J.*, **318**, 66; Bertschinger, E. 1989. *Astrophys. J.*, **340**, 666; Boily, C. M., Lynden-Bell, D. 1995. *Mon. Not. Roy. Astron. Soc.*, **276**, 133. A characteristic feature of these studies is that they include sonic points where the gas flow changes from subsonic to supersonic. The general perception, except for the role of self-gravity, is very similar to that of the spherical stellar wind problem; see Holzer, T. E., Axford, W. I. 1970. *Annu. Rev. Astron. Astrophys.*, **8**, 31.

50. A review of the properties of this general framework can be found in the monograph by Sarazin, C. L. 1988. *X-Ray Emissions from Clusters of Galaxies*, Cambridge University Press, Cambridge, UK, and in the article by Fabian, A. C. 1994. *Annu. Rev. Astron. Astrophys.*, **32**, 277.

51. See Frank, J., King, A., Raine, D. 2002. *Accretion Power in Astrophysics*, 3rd ed., Cambridge University Press, Cambridge, UK.

52. See Rybicki, G. B., Lightman, A. P. 1979. *Radiative Processes in Astrophysics*, Wiley, New York.

53. Raymond, J. C., Cox, D. P., Smith, B. W. 1976. *Astrophys. J.*, **204**, 290; see also Sarazin, C. L., White, R. E., III 1987. *Astrophys. J.*, **320**, 32. For the general problem in a variety of contexts, see also Smith, B., Sigurdsson, S., Abel, T. 2008. *Mon. Not. Roy. Astron. Soc.*, **385**, 1443; Wiersma, R. P. C., Schaye, J., Smith, B. D. 2009. *Mon. Not. Roy. Astron. Soc.*, **393**, 99; Gnedin, N. Y., Hollon, N. 2012. *Astrophys. J. Suppl.*, **202**, id.13, and references therein.

54. Sarazin, C. L., Ashe, G. A. 1989. *Astrophys. J.*, **345**, 22; the discussion of this section follows the article by Bertin, G., Toniazzo, T. 1995. *Astrophys. J.*, **451**, 111; see also Bertin, G., Pignatelli, E., Saglia, R. P. 1993. *Astron. Astrophys.*, **271**, 381.

55. See Mathews, W. G., Bregman, J. N. 1978. *Astrophys. J.*, **224**, 308.

56. Here we follow the choice made by Bertin, G., Toniazzo, T. 1995. *op. cit.* A different integration scheme, described by Vedder, P. W., Trester, J. J., Canizares, C. R. 1988. *Astrophys. J.*, **332**, 725, was adopted by Bertin, G., Pignatelli, E., Saglia, R. P. 1993. *op. cit.*

57. See Sarazin, C. L., White, R. E., III 1988. *Astrophys. J.*, **331**, 102; Bertin, G., Pignatelli, E., Saglia, R. P. 1993. *op. cit.*; different remedies have been considered by Thomas, P. A. 1986. *Mon. Not. Roy. Astron. Soc.*, **220**, 949; Thomas, P. A., Fabian, A. C., et al. 1986. *Mon. Not. Roy. Astron. Soc.*, **222**, 655.

58. Cappellaro, E., Turatto, M., et al. 1993. *Astron. Astrophys.*, **268**, 472.

59. Trinchieri, G., Kim, D.-W., et al. 1994. *Astrophys. J.*, **428**, 555.

60. See Bertin, G., Pignatelli, E., Saglia, R. P. 1993. *op. cit.* Some cautionary notes on the use of X-ray data to diagnose the distribution of dark halos are also expressed by Fabbiano, G. 1989. *Annu. Rev. Astron. Astrophys.*, **27**, 87.

61. Loewenstein, M., Mathews, W. G. 1987. *Astrophys. J.*, **319**, 614; David, L. P., Forman, W., Jones, C. 1990. *Astrophys. J.*, **359**, 29; Ciotti, L., D'Ercole, A., et al. 1991. *Astrophys. J.*, **376**, 380.

62. The values of *s* reported bracket the different estimates provided by Canizares, C. R., Fabbiano, G., Trinchieri, G. 1987. *Astrophys. J.*, **312**, 503; Donnelly, R. H., Faber, S. M., O'Connell, R. M. 1990. *Astrophys. J.*, **354**, 52; White, R. E., III, Sarazin, C. L. 1991. *Astrophys. J.*, **367**, 476.

63. Murray, S. D., Balbus, S. A. 1992. *Astrophys. J.*, **395**, 99.

64. See the volume Fabian, A. C., ed. 1987. *Cooling Flows in Clusters and Galaxies*, Kluwer, Dordrecht, The Netherlands.

65. For a relatively recent review of the main issues that require clarification, see Peterson, J. R., Fabian, A. C. 2006. *Phys. Rep.*, **427**, 1.

66. Ciotti, L., Ostriker, J. P. 1997. *Astrophys. J. Lett.*, **487**, L105, and many following papers: see Ciotti, L. 2009. *La Rivista del Nuovo Cimento*, **32**, 1; Pellegrini, S., Ciotti, L., Ostriker, J. P. 2012. *Astrophys. J.*, **744**, id.21, and references therein.

67. This discussion follows the article by van Albada, T. S., Bertin, G., Stiavelli, M. 1995. *Mon. Not. Roy. Astron. Soc.*, **276**, 1255; see also Faber, S. M., Dressler, A., et al. 1987. In *Nearly Normal Galaxies, from the Planck Time to the Present*, ed. S. M. Faber. Springer-Verlag, New York, p. 175; Bender, R., Burstein, D., Faber, S. M. 1992. *Astrophys. J.*, **399**, 462; Renzini, A., Ciotti, L. 1993. *Astrophys. J. Lett.*, **416**, L49.

68. van Albada, T. S., Bertin, G., Stiavelli, M. 1993. *Mon. Not. Roy. Astron. Soc.*, **265**, 627.

69. See Verheijen, M. A. W. 1997. Ph.D. dissertation, University of Groningen, Groningen, The Netherlands.

70. The material presented here follows the paper by Bertin, G., Ciotti, L., Del Principe, M. 2002. *Astron. Astrophys.*, **386**, 149. See also the references cited there.

71. See table 5 in the article by Bertin, G., Bertola, F., et al. 1994. *Astron. Astrophys.*, **292**, 381.

72. Peebles, P. J. E. 1993. *Principles of Physical Cosmology*, Princeton University Press, Princeton, NJ.

73. See also the discussion given by Saracco, P., Gargiulo, A., Longhetti, M. 2012. *Mon. Not. Roy. Astron. Soc.*, **422**, 3107, and references therein.

26 Galaxies and Gravitational Lensing

This chapter is devoted to a very simple summary of the main concepts at the basis of gravitational lensing and its astrophysical applications, with special attention to the frequent case in which galaxies act either as lenses or as sources affected by the phenomenon of lensing. After the discovery of the first clear macroscopic evidence of gravitational lensing, that is, the observation of a double image of a single quasar produced by an intervening galaxy acting as a lens,[1] enormous progress has been made in terms of observations and theoretical results, which has produced a vast literature dealing with a number of distinct and extremely interesting related topics.[2] As we will see, the physical basis of gravitational lenses is such that the most spectacular phenomena are produced by distant lenses on very distant sources; this is why the rapid progress and growing interest in this subject largely coincide with the systematic exploration of the distant Universe at redshifts of cosmological interest started at the end of the past century.

The main reason to present such a summary here is that one of the most interesting themes developed in the past ten to fifteen years in relation to the general problem of mass diagnostics for galaxies and clusters of galaxies is that of the combined use of gravitational lensing and dynamics. The use of dynamics as a diagnostic tool to probe the mass distribution in galaxies and other systems is generally based on the assumption that the object under investigation has reached a state of approximate dynamical equilibrium. The use of gravitational lensing thus is not only interesting per se, as a completely independent diagnostic tool, but also especially because it can deal with systems caught in a state of relatively rapid evolution or with systems for which the assumption of a specific type of equilibrium (e.g., hydrostatic equilibrium) might be questionable.

Another reason to give some room to gravitational lensing in the last part of this book, in which galaxies are looked at in a more general perspective, is that on the large scale (of clusters of galaxies or large-scale structure of the Universe), galaxies are used as tracers of the overall gravitational field, with important implications for cosmology; often a good knowledge of galactic structure is a necessary prerequisite for a proper use of this kind.

For the galactic context, applications were initially limited to the study of the halo of our Galaxy (microlensing) and to the study of the mass distribution in elliptical galaxies at intermediate redshifts; now extensions to the study of spiral galaxies are also considered. In this respect, this chapter is to be regarded as a necessary completion of Chapters 20 and 24. In addition, the general phenomenon has some interesting applications in apparently unrelated fields (e.g., the discovery of extrasolar planets).

26.1 Elements of Gravitational Lensing

Consider[3] a geometric configuration of the type illustrated in Fig. 26.1. An observer O looks at a distant point source, the angular position of which in the sky is given by the vector θ. The light ray coming from the source would have been seen at the position θ^s, but a mass distribution ρ (denoted by D in the figure) along the line of sight deflects the light rays along the relevant geodesics according to the laws of general relativity. In the limit of weak gravitational fields, it can be shown that a mass distribution associated with a (position-dependent) mean potential Φ acts as a lens characterized by (position-dependent) refraction index

$$n \sim 1 - \frac{2\Phi}{c^2}, \tag{26.1}$$

where c represents the speed of light; here the gravitational potential is taken to vanish at large radii, away from the intervening mass distribution. Thus a mass distribution along the line of sight acts as a converging achromatic lens.

For most astrophysical applications, the deflections induced on distant sources by a galaxy or a cluster of galaxies acting as a lens are small (of the order of few arcseconds), and the typical size of the lens is small with respect to the distances D_{od} and D_{ds} to the observer and to the source, respectively. For simplicity, in a first analysis, the presence of additional material (i.e., of additional lenses in the foreground or the background with respect to the lens under investigation) is ignored. Under these circumstances, the intervening galaxy (or cluster of galaxies) can be described as a thin gravitational lens. In other words, the impact of the mass distribution on the light rays that pass through it depends on only the density distribution Σ obtained from ρ by projection along the line of sight. The situation thus can be depicted schematically as in Fig. 26.2. The ray-tracing equation relates the source true location θ^s to the observed location θ by means of a vector function $\beta(\theta)$:

$$\theta^s(\theta) = \theta - \beta(\theta). \tag{26.2}$$

Note that θ, θ^s, and β are all two-dimensional vectors. For a given vector θ we have a unique value of the vector θ^s, but the application is not always one to one (in other words, under appropriate circumstances, a given source may appear to the observer as a multiple image[4]).

It can be proved that the vector function $\beta(\theta)$ can be expressed in terms of the dimensionless projected density along the line of sight

$$\kappa \equiv \frac{\Sigma}{\Sigma_c} \tag{26.3}$$

as

$$\beta(\theta) = \frac{1}{\pi} \int \frac{\kappa(\theta')(\theta - \theta')}{||\theta - \theta'||^2} \, d^2\theta', \tag{26.4}$$

where the critical density is defined as

$$\Sigma_c = \frac{c^2}{4\pi G} \frac{D_{os}}{D_{ds} D_{od}}. \tag{26.5}$$

The critical density is a purely geometric quantity. Its expression explains why, in general, given astrophysical systems, such as galaxies or clusters of galaxies, that have their own characteristic projected densities (e.g., a cluster of galaxies may have a typical peak density on the

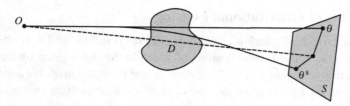

Fig. 26.1. Observer, gravitational lens, and source (from Lombardi, M. 2000. Ph.D. dissertation, Scuola Normale Superiore, Pisa, Italy).

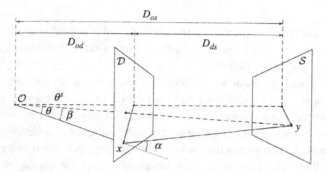

Fig. 26.2. The definition of the relevant angles in a gravitational lens configuration (from Lombardi, M. 2000. Ph.D. dissertation, Scuola Normale Superiore, Pisa, Italy).

order of 1 kg m$^{-2} \approx 478\, M_\odot$ pc^{-2}) generate stronger effects when located at redshifts of cosmological interest. For such distant lenses, the critical density of possible geometric configurations can be quite small, so κ may exceed unity. (Note that at redshifts of cosmological interest, the distances that appear in the definition of the critical density should be replaced by the appropriate relativistic angular-diameter distances.)

By virtue of the superposition principle, Eq. (26.4) generalizes the notion of the deflection angle α of light rays by a point-mass lens of mass M, which can be written as

$$\alpha \approx \frac{4GM}{c^2 r},\tag{26.6}$$

where r is the minimum distance of the light ray passing close to the point mass. Equation (26.6) has been tested with greater and greater precision as one of the main empirical pieces of evidence in support of general relativity.

We note that the deflection vector angle $\beta(\theta)$ can be expressed as the gradient of a scalar $\psi(\theta)$ often called the *deflection potential*:

$$\psi(\theta) = \frac{1}{\pi} \int \kappa(\theta') \ln \|\theta - \theta'\| \, d^2\theta',\tag{26.7}$$

so the function θ^s of the ray-tracing equation can be expressed as the gradient of a potential:

$$\theta^s = \nabla \left[\frac{\|\theta\|^2}{2} - \psi(\theta) \right].\tag{26.8}$$

The dimensionless density κ is related to the deflection potential ψ by means of a two-dimensional Poisson equation:

$$\nabla^2 \psi(\theta) = 2\kappa(\theta). \tag{26.9}$$

26.1.1 Einstein Rings, Multiple Images, Time Delays

One interesting phenomenon associated with small extended sources, when they happen to be well aligned with the observer and the center of an intervening axisymmetric lens, is the generation of an Einstein ring. The ring is the image of the small extended source that occurs at the angular radius at which the average projected density inside such radius equals the critical density of the geometric configuration. One extremely rare and particularly curious example is that shown in Fig. 26.3. The system actually exhibits a double Einstein ring because the lens, at $z \approx 0.222$, is well aligned with two distant sources, one at redshift $z \approx 0.609$ (responsible for the observed inner ring) and the other at redshift estimated to be somewhat below $z = 6.9$ (responsible for the outer ring).

When the geometric requirements for the production of a ring are met only approximately, distant sources are often split into multiple images around the foreground lens (Fig. 26.4). Here the interest in the phenomenon is related not only to the fact that the configuration provides information on the density distribution of the lens (as we will briefly describe later in this chapter) but also to the possibility of detecting variability in the distant source, which generally occurs at different times in different images. Such time delays are due to the different distances traveled by photons along the different paths related to the various images. Under these circumstances,

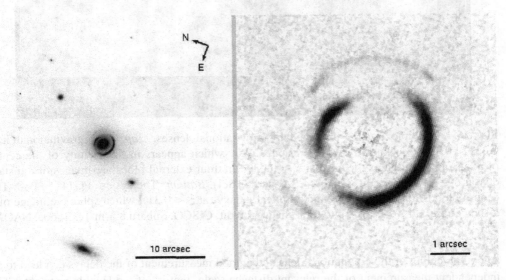

Fig. 26.3. The system SDSS J0946+1006, a double Einstein ring; the right panel shows a zoomed-in version of the image shown on the left, with the surface brightness of the lens galaxy subtracted out (from Gavazzi, R., Treu, T., et al. "The Sloan Lens ACS Survey. VI: Discovery and analysis of a double Einstein ring," 2008. *Astrophys. J.*, **677**, 1046; reproduced by permission of the AAS).

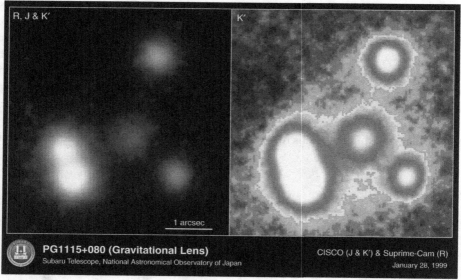

Fig. 26.4. Multiple images produced by gravitational lenses. *Top:* The gravitational lens G2237+0305, also known as the Einstein Cross, which appears in the vicinity of the center of a lensing spiral galaxy located at $z \approx 0.039$; the four external blobs are images of a distant quasar at $z \approx 1.69$ (credit: NASA, ESA, and STScI). *Bottom:* The system PG1115+080. The central blob is a lensing foreground elliptical galaxy (at $z \approx 0.31$), which splits the image of a distant quasar (at $z \approx 1.722$) into four images (credit: CISCO, Subaru 8.3-m Telescope, NAOJ).

a good modeling of the gravitational lens and a good measurement of the delays can lead to an independent measurement of the relevant distance scale, that is, of the Hubble constant.[5] The basic point is that the measurement of a time delay breaks the dimensionless nature of the underlying geometry of a gravitational lens (which involves only angles) and leads to determination of an intrinsic length scale, thus setting the basis for determination of the cosmological distance scale.

In addition to the case of a point-mass lens, simple analytical properties of the ray-tracing equations can be given for the so-called singular isothermal sphere (see Chapter 22), for which the projected dimensionless density is $\kappa \propto 1/\|\theta\|$, and other regularized pseudoisothermal spheres from which the divergence at the origin is removed. The use of these simple lens models is also instructive, because the behavior of their density at large values of $\|\theta\|$ mimics the behavior of dark-matter halos. The relevant gravitational lens analysis is often discussed in the literature[6] and will not be repeated here.

26.1.2 Extended Sources

Distant galaxies act as small but extended sources (see Fig. 25.12). It is thus natural to inspect the properties of the (symmetric) Jacobian matrix $A(\theta)$, which can be written as the sum of a matrix proportional to the identity matrix and a symmetric matrix γ_{ij} with vanishing trace, called the *shear matrix*:

$$A(\theta) = \frac{\partial \theta^s}{\partial \theta} = \begin{pmatrix} 1 - \kappa(\theta) - \gamma_1(\theta) & -\gamma_2(\theta) \\ -\gamma_2(\theta) & 1 - \kappa(\theta) + \gamma_1(\theta) \end{pmatrix} \tag{26.10}$$

The matrix γ_{ij} is defined implicitly by Eq. (26.10). Note that this notation is consistent with Eqs. (26.8) and (26.9). Thus the distortion induced on the image of a finite source includes a magnification by the factor

$$\mu(\theta) = \frac{1}{|\det A(\theta)|} = \frac{1}{(1 - \kappa)^2 - (\gamma_1^2 + \gamma_2^2)} \tag{26.11}$$

and a stretching associated with the shear. The situation is illustrated in Fig. 26.5. The magnification corresponds to an increase in the area of the source. It can be proved that in contrast to the action of the point-spread function on optical images of extended sources (caused by atmospheric seeing or the imperfect optics of a telescope), gravitational lensing conserves the surface brightness of an extended source, that is, $I(\theta) = I^s(\theta^s)$. Therefore, magnification corresponds to a brightening in terms of the total flux received by the observer from a lensed object.

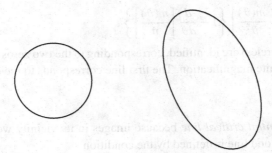

Fig. 26.5. Shear and magnification (*right*) induced by a lens on an extended circular source (*left*).

26.2 Kinds of Gravitational Lensing

26.2.1 Microlensing

Microlensing is the term often used to describe the situation in which lens, source, and source image are not resolved in the sky, and gravitational lensing manifests itself as a temporary brightening of a point source when the proper alignment conditions with an intervening lensing object are temporarily met. Initially, great interest in this phenomenon was raised by the possibility of measuring properties of possible constituents of the dark halo of our Galaxy. Some of the projects and the results obtained were described in Section 20.3. Revived interest in this kind of gravitational lensing has been raised relatively recently by application to the detection of extrasolar planets.[7] Curiously, about forty years ago,[8] one application of the effect was studied in the context of the now very hot topic of the search for intermediate-mass black holes at the center of globular clusters.

26.2.2 Strong Lensing

The term *strong lensing* generally refers to very prominent and easily detectable macroscopic phenomena produced by gravitational lensing, such as multiple images, arcs, arclets, and rings. For the occurrence of these phenomena, the projected density of the lens Σ must be close to the critical density Σ_c, that is, $\kappa \approx 1$. Note that this property is unrelated to the magnitude of the gravitational potential Φ associated with the lens density distribution ρ. In other words, the quasi-Newtonian Eq. (26.1) is generally applicable when strong lensing phenomena are observed.

To understand some of the underlying conditions for the occurrence of strong lensing, it is instructive to refer to axisymmetric lenses. If we call $\vartheta = ||\theta||$ the magnitude of the vector θ, with the center of the lens located at the origin, it can be shown that the relevant ray-tracing equation can be written as

$$\vartheta^s = \vartheta - \frac{m(\vartheta)}{\vartheta}, \tag{26.12}$$

where $m(\vartheta)$ represents the dimensionless integrated mass, so

$$\frac{dm(\vartheta)}{d\vartheta} = 2\vartheta\kappa(\vartheta). \tag{26.13}$$

The determinant of the Jacobian matrix is

$$\det A(\vartheta) = \left[1 - \frac{m(\vartheta)}{\vartheta^2}\right]\left\{1 - \frac{d}{d\vartheta}\left[\frac{m(\vartheta)}{\vartheta}\right]\right\}. \tag{26.14}$$

Two critical lines (circles) are identified, corresponding to the two zeros of $\det A$, which would be associated with infinite magnification. The first line corresponds to the condition

$$\frac{m(\vartheta)}{\vartheta^2} = 1 \tag{26.15}$$

and is called the *tangential critical line* because images in its vicinity would be stretched into tangential arcs. The second line is defined by the condition

$$\frac{d}{d\vartheta}\left[\frac{m(\vartheta)}{\vartheta}\right] = 1 \tag{26.16}$$

Fig. 26.6. Arcs and arc features in the cluster Abell 370 (credit: NASA, ESA, the Hubble SM4 ERO Team, and ST-ECF).

and is called the *radial critical line* because images in its vicinity would be stretched into radial arcs.

These concepts and definitions can be generalized to the situation of nonaxisymmetric lenses. Critical lines are determined by the vanishing of the determinant of the Jacobian matrix in the plane of the lens, that is, by det $A(\theta) = 0$; the corresponding lines in the source plane, obtained from the ray-tracing equation, are called *caustics*.

Figures 26.3 and 26.4 are examples of strong lensing. Additional examples are shown in Fig. 26.6. Here we wish to mention that the modeling of strong lensing systems with multiple images involves two separate aspects. In general, a satisfactory model for the observed location of the multiple images produced by the lens is found in a relatively straightforward way, even in complex systems.[9] A much more difficult part of the study is the interpretation of the observed fluxes associated with the different images, which in the ideal case should follow well-defined rules, such as the vanishing of the so-called cusp ratio.[10] Here the modeling often fails. It is commonly believed and argued[11] that the discrepancies between predicted and observed fluxes are due to substructures present in the lens. Because of this and of the difficulties in finding the substructures predicted by cosmological simulations, these discrepancies are actually considered to be very interesting and are currently studied with great attention.

26.2.3 Weak or Statistical Lensing

The study of *weak gravitational lensing*[12] refers to a statistical analysis of the weak distortions produced by a lens with $\kappa < 1$ on distant individual extended sources. The main assumption is that the shapes (quadrupoles) of distant sources have random orientation on the source plane,[13] so we can detect the polarization induced locally by the shear term associated with gravitational lensing. Typically we consider a relatively large field (a cluster of galaxies at intermediate redshifts $z \approx 0.5$ has a typical size of a few arcminutes) in the sky and subdivide it into tiles (subfields) of suitable size. Deep observations may detect a galaxy density of the order of 100 galaxies arcmin^{-2}. In each tile we measure the shapes of many distant sources and thus determine the polarization locally. Qualitatively, the observed polarization is such that the minor axes of the distant source galaxies point preferentially in the direction of the center of the lens. Once these data points are collected for the entire field of view, we can proceed to reconstruct a map for the distribution of $\kappa(\theta)$. The process can be carried out in different ways but is always subject to a degeneracy, called the *mass-sheet degeneracy*: The shape of the density distribution (the density gradients) is measured, but its overall scale is a priori unknown. Only a direct measurement of magnification could break such a degeneracy.

In a simplified description, the shape of a source galaxy is associated with an observed quadrupole matrix Q that corresponds to a source quadrupole matrix

$$Q^s = AQA,$$
(26.17)

where the symmetric matrix A is the Jacobian matrix defined in Eq. (26.10). Different definitions of Q have been considered; we may take the natural definition in which the dimensions of Q are those of an (angular) area.[14]

An alternative way of approaching the problem is based on the introduction of a complex observed ellipticity parameter defined as[15]

$$\varepsilon = \frac{Q_{11} - Q_{22} + 2iQ_{12}}{Q_{11} + Q_{22} + 2\sqrt{Q_{11}Q_{22} - Q_{12}^2}}.$$
(26.18)

After defining the complex shear $\gamma \equiv \gamma_1 + i\gamma_2$, where γ_i are the quantities introduced in Eq. (26.10), and the complex reduced shear

$$g \equiv \frac{\gamma}{1 - \kappa},$$
(26.19)

simple relations in terms of g can be obtained between the source complex ellipticity ε^s and the observed complex ellipticity ε.

The isotropy assumption on the orientation of distant sources corresponds to the statement $\langle Q^s \rangle = M$ Id, where M is a positive constant, and Id is the identity matrix, or $\langle \varepsilon^s \rangle = 0$. Thus, locally, the measurement of many individual quadrupole moments $Q^{(n)}$ leads to an estimate[16] of the matrix A^2, up to a constant, or the measurement of $\langle \varepsilon \rangle$ is used as an estimator of the reduced shear g.[17]

In the weak lensing limit $\kappa \ll 1$, the following relations hold: $g(\theta) \sim \gamma(\theta)$, $\varepsilon \sim \varepsilon^s + \gamma - \gamma(\varepsilon^s)^2$, and for the magnification factor defined in Eq. (26.11), $\mu(\theta) \sim 1 + 2\kappa(\theta)$.

To proceed to the global mass reconstruction, the information gathered for the various tiles (imagined in the preceding description) has to be smoothed out by the use of proper weight functions so as to obtain a map of the shear over the field under investigation. It can be shown that a vector u defined in terms of the derivatives of the shear matrix (i.e., $u_i \equiv \gamma_{ij,j}$) satisfies the relation

$$\nabla \kappa = u.$$
(26.20)

[A similar relation can be written for the function $\tilde{\kappa} \equiv \ln(1 - \kappa)$ with a vector $\tilde{u}$ defined in terms of the reduced shear g.] Note that the u field is curl-free.

In the method based on measurement of the various $Q^{(n)}$, the mass-sheet degeneracy corresponds to the fact that the matrix A is determined up to an unknown constant M. For the methods based on estimation of the shear field, the degeneracy is clear from Eq. (26.20) (or the analogous equation involving $\tilde{\kappa}$ and $\tilde{u}$), which indicates that the density distribution is determined up to an unknown constant. The statement of the mass-sheet degeneracy that follows from the equation relating $\tilde{\kappa}$ to $\tilde{u}$ is as follows: The projected dimensionless mass density map κ can be determined only up to transformations of the form $\kappa \to \kappa' = (1 - C)\kappa + C$, where C is an unknown constant.

A number of factors further complicate the entire analysis, but they have been addressed and discussed in great detail in a number of dedicated articles. We just list them as a set of important items: object detection, elimination of foreground objects, photometry and shape determination, corrections for seeing and instrumental point-spread function, estimate of the source redshift and inclusion of effects associated with the fact that different sources are located on different source planes, cosmological factors in the definition of the various distances, and effects of the finite field sampled by the data. In conclusion, very deep multiband observations are required for good weak lensing measurements; the mass reconstruction suffers from a degeneracy that would be broken if a direct magnification measurement could be made.

26.3 Some Interesting Applications

In this section some applications of strong and weak gravitational lensing are briefly described. The main use of a gravitational lensing model to interpret a set of data on distant sources is to reconstruct properties of the density distribution of the intervening lens. Clearly, strong lensing is best suited to probe the inner regions of the lens, whereas weak lensing analysis best probes the outer parts of the density distribution of the deflecting material. In some cases, both types of investigations can be combined, with robust determinations of the total mass distribution of the lens.[18] Most applications are thus in the general context of mass diagnostics. Other applications address some cosmological objectives (e.g., we already mentioned, in Subsection 26.1.1, the use of time delays to measure the Hubble constant, or at the end of Chapter 24 we referred to the study of the Bullet Cluster as a serious case in favor of the existence of dark matter and against alternative theories of gravity, such as MOND). At the end of this section we will describe two less standard applications that appear to be feasible but have so far not yet found adequate observational confirmation.

26.3.1 Combined Use of Stellar Dynamics and Gravitational Lensing

In Section 24.4 we briefly reported on the interesting results of the combined use of stellar dynamics and gravitational lensing in probing the total mass-density distribution of early-type galaxies (in short, these studies find for the total density distribution a profile of the form $\rho \sim 1/r^2$; Fig. 26.7). In this respect, because of its multiple internal constraints, the case of the double Einstein ring (see Fig. 26.3) has proved to be a very important system.

26.3.2 Galaxy-Galaxy Lensing

By considering large sets of galaxies close to each other in the sky but otherwise dynamically unrelated, of which the foreground objects act as a lens on the more distant galaxies, we may aim at carrying out a weak lensing statistical analysis that in principle should be able to provide information on the radial size and shape of galactic halos.[19] In general, a parametric model

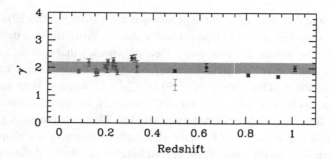

Fig. 26.7. Logarithmic total-mass density slopes of field early-type galaxies as a function of redshift, derived from the combined use of stellar dynamics and gravitational lensing; the case $\gamma' = 2$ corresponds to the slope characteristic of the singular isothermal sphere (from Koopmans, L. V. E., Treu, T., Bolton, A. S., Burles, S., Moustakas, L. A. "The Sloan Lens ACS Survey. III: The structure and formation of early-type galaxies and their evolution since $z \sim 1$," 2006. *Astrophys. J.*, **649**, 599; reproduced by permission of the AAS).

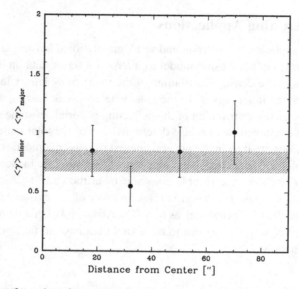

Fig. 26.8. Example of results of a study of galaxy-galaxy lensing based on early data from the Canada-France-Hawaii Telescope Legacy Survey. Ratio of mean shear for sources closest to the minor axes of a foreground lens to that for sources closest to the major axes; the measured weighted-average shear ratio favors a halo ellipticity of 0.3 (from Parker, L. C., Hoekstra, H., Hudson, M. J., van Waerbeke, L., Mellier, Y. "The masses and shapes of dark matter halos from galaxy-galaxy lensing in the CFHT Legacy Survey," 2007. *Astrophys. J.*, **669**, 21; reproduced by permission of the AAS).

for the density distribution of the dark-matter halo is adopted, and the data (and the very weak polarization signal associated with them) are analyzed in the sense of an ensemble average by assuming that galaxies and their halos are characterized by some sort of homology. Typical detected values are ≈ 0.3 for the halo average ellipticity (Fig. 26.8) and ≈ 500 kpc (less well constrained) for the halo truncation radius.

26.3.3 Measurement of the Cosmological Parameters and the Cosmic Shear

There are two general applications of gravitational lensing that are at the interface with observational cosmology: the possibility of independent measurements of the cosmological parameters and study of the so-called cosmic shear.[20] Separately from the method based on time delays (briefly mentioned in Subsection 26.1.1), one general idea at the basis of determination of the cosmological parameters from gravitational lensing consists of recognizing that lensing analyses are centered around three items: the mass reconstruction of the lens, the redshifts of the distant source galaxies, and the large-scale geometry of the Universe. Therefore, if we make assumptions on (or if we measure separately) one of the preceding three items, we may put significant constraints on the determination of the other two. In the weak lensing context, starting from a good estimate of the redshift distribution of the distant source galaxies (in particular, by application of the technique of photometric redshifts), it has been shown that the cosmological parameter that could be better constrained is the dimensionless mass-density parameter Ω_m.[21] In contrast, a strong lensing study based on the assumption that the total density distribution of the lens galaxies is $\rho \sim 1/r^2$ leads to significant constraints on the other cosmological parameter Ω_Λ.[22]

To a large extent, studies of the cosmic shear are similar to those of galaxy-galaxy lensing. The general goal is to determine the very weak polarization signal (even weaker than that characteristic of galaxy-galaxy lensing) associated with the large-scale structure of the Universe. By these studies we may determine the properties of the primordial structure formation in the Universe, as evolved at $z \approx 1$, which we believe reflects the initial inhomogeneities (anisotropies) detected in the distribution of the cosmic microwave background radiation. This technique requires extensive and accurate data acquisition (Fig. 26.9) and a very deep statistical analysis. For its importance, it has been and is the target of a number of projects.[23]

26.3.4 A Direct Measurement of Magnification Based on Observation of the Fundamental Plane

In Chapter 4 (with further discussion in Section 25.4) we mentioned that early-type galaxies populate a plane, the fundamental plane, in the natural three-dimensional space of global parameters $(\log R_e, \log \sigma_0, SB_e)$. Equation (4.15), which defines such a plane, establishes a standard rod. Similarly, the luminosity-velocity relation described in Subsection 4.3.1 establishes a standard candle for spiral galaxies. In fact, both scaling laws are commonly used as distance estimators. Here we briefly describe how the standard rod provided by the fundamental plane can be used to make a direct measurement of magnification and thus break the mass-sheet degeneracy.[24]

If we take a sample of early-type galaxies behind a lensing cluster, that is, looked at through a gravitational lens, we expect that the basic fundamental plane relation

$$\log R_e = \log r_e + \log D_A(z) = \alpha \log \sigma_0 + \beta SB_e + \gamma, \tag{26.21}$$

where R_e is the intrinsic effective radius (measured in kiloparsecs), r_e is the apparent (angular) effective radius, and $D_A(z)$ is the angular-diameter distance to the galaxy observed at redshift z, will be changed into

$$\frac{1}{2}\log|\det A| = \alpha \log \sigma_0 + \beta SB_e + \gamma - \log r_e - \log D_A(z). \tag{26.22}$$

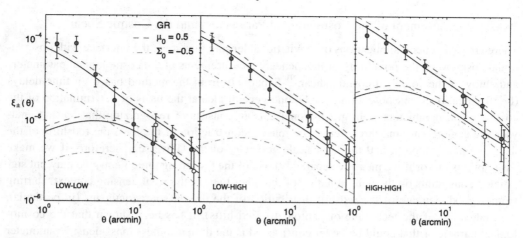

Fig. 26.9. Example of a cosmic shear study. Data points for two lensing correlation functions ξ_+ and ξ_- obtained from the Canada-France-Hawaii Telescope Lensing Survey are plotted and compared with theoretical curves constructed from the best-fit ΛCDM cosmology; the source galaxies used are divided in two redshift bins, a low-redshift group with $0.5 < z \leq 0.85$ and a high-redshift group with $0.85 < z \leq 1.3$ (from Simpson, F., Heymans, C., et al. "CFHTLenS: testing the laws of gravity with tomographic weak lensing and redshift-space distortions," 2013. *Mon. Not. Roy. Astron. Soc.*, **429**, 2249, where the relevant definitions and interpretations are provided; by permission of Oxford University Press, on behalf of the Royal Astronomical Society).

(We recall that the intrinsic surface brightness SB_e is obtained from the data by taking into account a number of factors, in particular, cosmological dimming, the so-called k-correction, and Galactic extinction, but is otherwise unaffected by lensing.) Thus the existence of the fundamental plane relation allows us to obtain a direct measurement of the magnification because all the quantities appearing on the right-hand side of Eq. (26.22) can be measured from the observations. From Subsection 26.1.2 and Eq. (26.19), we recall that $\det A = (1 - \kappa)^2 (1 - g)^2 \sim 1 - 2\kappa$. The scatter in the fundamental plane relation, of ≈ 15 percent in R_e in the local Universe, with apparently only a small increase with redshift,[25] implies a corresponding error of ≈ 30 percent in $\det A$ and thus of ≈ 15 percent in κ. Obviously, a source of uncertainty to be considered is related to the evolution (with z) of the fundamental plane relation, but fortunately, this is reasonably well under control, as described briefly in Subsection 4.3.3.

Therefore, a measurement of the fundamental plane relation on a set of distant galaxies behind a cluster can break the mass-sheet degeneracy completely (for weak lensing analyses of the cluster). In principle, the technique leads to an absolute measurement of the mass distribution of the lens on the small pencil beams that characterize the size of background galaxies, and this might lead to interesting tests for current scenarios of structure formation; however, it has been shown that within a realistic cosmological scenario, substructures do not contribute much to the magnification signal that is looked for but only add a modest amount of scatter.

A study has been made of what would be the best strategy to follow in this context and has shown that the optimal choice is that of measuring the fundamental plane properties for a sample of early-type galaxies behind the cluster distributed approximately uniformly in the sky. The role of the redshift distribution of the source galaxies, in relation to the redshift of

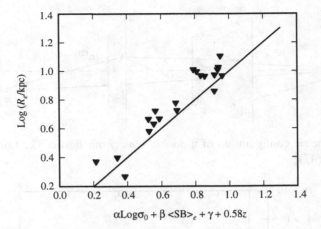

Fig. 26.10. Shift of the fundamental plane resulting from the magnification by an intervening cluster of galaxies. The fundamental plane measurements for twenty simulated galaxies are plotted and compared with the straight line, which is the fundamental plane expected in the absence of lensing viewed edge-on. The positions in the fundamental plane space have been corrected for evolution of the plane with redshift (from Sonnenfeld, A., Bertin, G., Lombardi, M. 2011. *Astron. Astrophys.*, **532**, id.A37).

the lensing cluster and to the limitations of fundamental plane measurements, also has been discussed. Simple simulations have been carried out for clusters with intrinsic properties similar to those of the Coma cluster. For a massive cluster ($M > 10^{15} M_\odot$) located at redshift 0.3 ± 0.1, a set of about twenty fundamental plane measurements, combined with a good weak lensing analysis, should be able to lead to a mass determination with a precision of 20 percent or better.[26] So far an actual observational test of these ideas has not been performed. Simulated observations (Fig. 26.10) illustrate the size of the expected effect.

26.3.5 Double Lenses

The chance of having two clusters aligned along the line of sight is not too small; some cases for which a similar spatial configuration takes place have indeed been noted.[27] In addition, astronomers have often looked for the possible presence of mass structures without a significant visible counterpart (dark galaxies or dark clusters). Therefore, it is interesting to consider a situation, as illustrated in Fig. 26.11, in which two gravitational lenses are present along the same line of sight. This problem has received some attention in cases of strong lensing[28] but has been considered only little in the weak lensing context.[29]

The main qualitative change that takes place in a configuration of this kind is that the ray-tracing Jacobian matrix of the combined system is no longer symmetric. In different words, the u field, introduced in Subsection 26.2.3, is no longer curl-free. To detect the presence of a double lens, we can thus aim at detecting an asymmetry in the Jacobian matrix or, correspondingly, the small deviations from the standard relation $\nabla \wedge u = 0$. Because the deviations are small, only particularly favorable conditions are expected to lead to detectable effects. Based on the cluster density distribution known empirically,[30] and by assuming that the first lens is located at $z \approx 0.1$

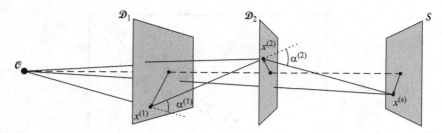

Fig. 26.11. Geometric configuration of a double lens (from Bertin, G., Lombardi, M. 2001. *Astrophys. J.*, **546**, 47).

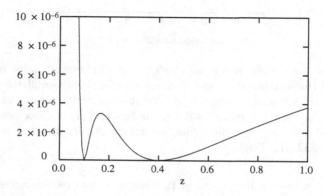

Fig. 26.12. Detection of a dark cluster at redshift $z = 0.4$ in a simulated observation. The value of a suitably defined integral S is shown as a function of the redshift z. The luminous cluster is assumed to be at a known redshift, $z = 0.1$ (from Bertin, G., Lombardi, M. 2001. *Astrophys. J.*, **546**, 47).

and the second at $z \approx 0.4$, a statistical analysis suggests that a few cases of double lenses with detectable effects actually may be present in the sky.

The weak lensing analysis and the corresponding statistical analysis to quantify the various sources of error and the limitations of a study under realistic conditions are rather complex but have been carried out. In particular, it has been shown that the redshift distribution of the source galaxies, which also might introduce asymmetries that would be confused with the effects of double lensing, are actually associated with systematic effects typically two orders of magnitude smaller than the double-lensing effects that are looked for. An integral function $S(z)$ can be introduced in such a way that its zeros (or minima) occur in correspondence of the redshifts of the two lenses. In principle, measurements of weak lensing on a luminous cluster may lead to the detection of a second cluster and to the determination of its redshift.[31] An example based on simulated observations is shown in Fig. 26.12.

Notes

1. Walsh, D., Carswell, R. F., Weymann, R. J. 1979. *Nature (London)*, **279**, 381. A lensing galaxy at redshift $z \approx 0.36$ splits the image of a QSO at redshift $z \approx 1.4$ into two images separated by $\approx 6''$.
2. A classical reference is the monograph by Schneider, P., Ehlers, J., Falco, E. E. 1992. *Gravitational Lenses*, Springer-Verlag, Heidelberg. Many other important reviews have been written on the subject,

among which we wish to mention the articles by Wambsganss, J. 1998. *Living Reviews in Relativity*, **1**, 12; and Bartelmann, M. 2010. *Classical and Quantum Gravity*, **27**, 233001.

3. Much of this short introduction to gravitational lensing follows Lombardi, M. 2000. Ph.D. dissertation, Scuola Normale Superiore, Pisa, Italy.

4. An elegant proof that a transparent, nonsingular gravitational lens generally leads to an odd number of images of a point source is given by Lombardi, M. 2009. *Il Nuovo Cimento B*, **124**, 363.

5. The general ideas governing this application of gravitational lensing are surprisingly anticipated in an old but very modern article by Refsdal, S. 1964. *Mon. Not. Roy. Astron. Soc.*, **128**, 307. There is a vast literature on this topic; in one of the most recent studies, the possibility has been claimed of measuring the Hubble constant by this method with ≈ 5 percent precision; see Suyu, S. H., Auger, M. W., et al. 2013. *Astrophys. J.*, **766**, id.70. Note that some significant discrepancies in the determination of the Hubble constant appear to emerge from the newly released data from the *Planck* mission, which point to a relative low value of the Hubble constant $H_0 = 67.3 \pm 1.2$ km s^{-1} Mpc^{-1}; see Ade, P. A. R., Aghanim, N., et al. 2013. arXiv:1303.5076, and comments and other references given in Section 4.4.

6. In particular, see chapter 8 in the monograph by Schneider, P., Ehlers, J., Falco, E. E. 1992. *op. cit.*

7. There are several reviews dedicated to microlensing. A recent one is by Mao, S. 2012. *Res. Astron. Astrophys.*, **12**, 947; see also the expository article by Mao, S. 2008. arXiv:0811.0441, an unpublished contribution to the Manchester Microlensing Conference held in January 2008.

8. Silvestro, G. 1974. *Astron. Astrophys.*, **36**, 41.

9. For example, see the twelve-image system found in the $z = 0.84$ cluster Cl J0152.7-1357; Grillo, C., Lombardi, M., et al. 2008. *Astron. Astrophys.*, **486**, 45.

10. See Schneider, P., Ehlers, J., Falco, E. E. 1992. *op. cit.*

11. Mao, S., Schneider, P. 1998. *Mon. Not. Roy. Astron. Soc.*, **295**, 587; and many following papers.

12. For a review of weak gravitational lensing, see Bartelmann, M., Schneider, P. 2001. *Phys. Rep.*, **340**, 291.

13. This assumption is currently under scrutiny; in the context of galaxy-galaxy lensing, see Blazek, J., Mandelbaum, R., et al. 2012. *J. Cosmol. Astrop. Phys.*, issue 5, id.041, and references therein.

14. But see Kaiser, N., Squires, G. 1993. Astrophys. J., **404**, 441.

15. Seitz, C., Schneider, P. 1997. *Astron. Astrophys.*, **318**, 687; here the definition of complex ellipticity is changed in its normalization with respect to previous articles (e.g., Seitz, C., Schneider, P. 1995. *Astron. Astrophys.*, **294**, 411; **297**, 287), in which the normalization was limited to the trace of the quadrupole Q.

16. Lombardi, M., Bertin, G. 1998. *Astron. Astrophys.*, **330**, 791.

17. Seitz, C., Schneider, P. 1997. *op. cit.*

18. For example, see Bradač, M., Schneider, P., et al. 2005. *Astron. Astrophys.*, **437**, 39; Bradač, M., Erben, T., et al. 2005. *Astron. Astrophys.*, **437**, 49.

19. Griffiths, R. E., Casertano, S., et al. 1996. *Mon. Not. Roy. Astron. Soc.*, **282**, 1159; Brainerd, T. G., Blandford, R. D., Smail, I. 1996. *Astrophys. J.*, **466**, 623. Among many more recent articles on this general topic, we wish to mention Hoekstra, H., Franx, M., et al. 2003. *Mon. Not. Roy. Astron. Soc.*, **340**, 609; Kleinheinrich, M., Rix, H.-W., et al. 2005. *Astron. Astrophys.*, **439**, 513; Mandelbaum, R., Seljak, U., et al. 2006. *Mon. Not. Roy. Astron. Soc.*, **372**, 758; Parker, L. C., Hoekstra, H., et al. 2007. *Astrophys. J.*, **669**, 21. See also van Uitert, E., Hoekstra, H., et al. 2012. *Astron. Astrophys.*, **545**, id.A71, and references therein.

20. For a review focusing on the cosmological applications of weak gravitational lensing, see Hoekstra, H., Jain, B. 2008. *Annu. Rev. Nucl. Part. Syst.*, **58**, 99; for a review on cosmic shear, see Refregier, A. 2003. *Annu. Rev. Astron. Astrophys.*, **41**, 645.

21. Lombardi, M., Bertin, G. 1999. *Astron. Astrophys.*, **342**, 337.

22. Grillo, C., Lombardi, M., Bertin, G. 2008. *Astron. Astrophys.*, **477**, 397.

23. For example, a series of papers has been published on the results from the Canada-France-Hawaii Telescope Lensing Survey (CFHTLenS); see Simpson, F., Heymans, C., et al. 2013. *Mon. Not. Roy. Astron. Soc.*, **429**, 2249.

24. The ideas presented in this subsection follow the exploratory article by Bertin, G., Lombardi, M. 2006. *Astrophys. J. Lett.*, **648**, L17. A more thorough investigation, with attention to several statistical issues

related to the proposed method, is carried out by Sonnenfeld, A., Bertin, G., Lombardi, M. 2011. *Astron. Astrophys.*, **532**, id.A37.

25. The work of Treu, T., Ellis, R. S., et al. 2005. *Astrophys. J.*, **633**, 174, suggests an increase of the scatter up to ≈ 23 percent at $z = 1$; the work of Auger, M. W., Treu, T., et al. 2010. *Astrophys. J.*, **724**, 511, suggests an intrinsic scatter of only ≈ 11 percent.

26. Sonnenfeld, A., Bertin, G., Lombardi, M. 2011. *op. cit.*

27. In particular, the cases of the clusters Cl 0317+15 and A1758; see Molinari, E., Buzzoni, A., Chincarini, G. 1996. *Astron. Astrophys. Suppl.*, **119**, 391; Wang, Q. D., Ulmer, M. P. 1997. *Mon. Not. Roy, Astron. Soc.*, **292**, 920.

28. For example, see Crawford, C. S., Fabian, A. C., Rees, M. J. 1986. *Nature (London)*, **323**, 514; Seitz, S., Schneider, P. 1992. *Astron. Astrophys.*, **265**, 1; 1994. *Astron. Astrophys.*, **287**, 349.

29. Bertin, G., Lombardi, M. 2001. *Astrophys. J.*, **546**, 47.

30. See Girardi, M., Borgani, S., et al. 1998. *Astrophys. J.*, **506**, 45; Borgani, S., Rosati, P., et al. 1999. *Astrophys. J.*, **517**, 40.

31. Bertin, G., Lombardi, M. 2001. *op. cit.*

27   Self-Gravitating Accretion Disks

This final chapter is meant to be a bridge between the dynamics of disk galaxies and the dynamics of disks on much smaller scales. The paradigm of mass accretion on a central object by means of a dissipative disk has long been studied in various contexts. The general framework was initially considered as a tool to describe the formation stages of the solar nebula to clarify the mechanisms that generate stars and planets.[1] Later, the picture was reexplored in the context of high-energy astrophysics.[2] There are thus two interesting aspects: the formation and dynamical evolution of a disk during collapse and the possibility of extracting gravitational energy from systems dominated by rotation. In the general picture, key roles are played by viscous dissipation and angular-momentum transport.

These concepts bring us to consider some of the topics that are currently at the frontier of astrophysical investigations. At the smallest scale, for masses of the order of a few solar masses and lengths of the order of a few astronomical units, much of the current interest addresses the processes of star and planet formation; here great progress comes from the ongoing discovery and study of extrasolar planets[3] and detailed observations of star-forming regions with dedicated telescopes from the ground and from space. At similarly small scales, high-energy astrophysics phenomena are produced when accretion occurs onto compact stellar objects and stellar-mass black holes; these processes generally affect binary systems, in which material from a relatively normal star is captured by the compact star. In addition, at the scale of active galactic nuclei, that is, for masses in the range of millions to a few billions solar masses and lengths of the order of 1 parsec or less, accretion takes place onto central massive black holes. Here one important issue is the long-term evolution of the nuclei, possibly originating from primordial central black holes of much smaller mass, for which we would like to understand how the central objects have grown to the sizes observed at recent epochs. We also may consider the possibility of accretion onto intermediate-mass black holes (masses of the order of a few thousand solar masses), but the evidence for their existence (possibly at the center of some globular clusters) is still under discussion. Finally, we may interpret gaseous disks on the galactic scale (especially the light HI disks discovered in some elliptical galaxies; see Subsection 24.2.2) as a result of accretion from the intergalactic medium, and by this we are brought back to the scales of galaxies, that is, to the general topics of this book.

A vast literature is available on this subject.[4] Obviously, even a superficial review of the relevant theoretical and observational aspects related to it would require a very long digression that would bring us well beyond the scope of this book. Here we wish only to provide a general picture of some of the mechanisms that are applicable to relatively cold, gaseous accretion disks, for which the disk self-gravity becomes important, because in practice the mechanisms involve in rescaled form the properties of density waves, as described in Chapters 15 and 16 for spiral galaxies. We recall that a line of research that has greatly profited from the study of density waves is that of the dynamics of planetary rings, which became very popular especially after interplanetary probes, such as the *Voyager* missions, managed to gather direct information on ring structure and morphology.[5]

27.1   The Traditional Paradigm of Accretion Disks

In the standard treatment of the dynamics of an accretion disk,[6] the gravitational potential is taken to be dominated by the central object, which may be a protostar, a compact star, or a black hole, with mass $M_\bullet$ and very small characteristic radius $r_\bullet$. In the nonrelativistic case, the gravitational field is thus Keplerian. Accretion can take place because some viscosity is present and allows the rotating gas to lose angular momentum. Most analyses focus on the case of rather thin disks; this makes it possible to decouple the study of the vertical from the horizontal structure and dynamics.[7]

27.1.1   Viscous Diffusion

One effect of viscosity is to make gas diffuse. Consider an axisymmetric thin disk characterized by density distribution σ and velocity field $(u, r\Omega)$. By eliminating the radial velocity u from the continuity equation

$$\frac{\partial \sigma}{\partial t} + \frac{1}{r}\frac{\partial}{\partial r}(r\sigma u) = 0 \tag{27.1}$$

and the angular-momentum balance equation

$$\frac{\partial}{\partial t}(\sigma r^2 \Omega) + \frac{1}{r}\frac{\partial}{\partial r}(r^3 \sigma \Omega u) = \frac{1}{r}\frac{\partial}{\partial r}[\nu \sigma r^3 (d\Omega/dr)], \tag{27.2}$$

where ν represents the kinematic viscosity so that the right-hand-side term describes the result of the viscous torque, we obtain the diffusion equation

$$\frac{\partial \sigma}{\partial t} = \frac{1}{r}\frac{\partial}{\partial r}\left\{ \frac{1}{[d(r^2\Omega)/dr]}\frac{\partial}{\partial r}\left[\nu \sigma r^3(-d\Omega/dr)\right] \right\}. \tag{27.3}$$

Here it is interesting to note that the viscous transport described by the right-hand side of Eq. (27.2) depends on the gradient of Ω and not, as might have seemed more natural, on the gradient of the angular momentum.[8] In standard analyses, the disk is considered to rotate with the Keplerian angular velocity $\Omega = \sqrt{GM_\bullet/r^3}$, so the diffusion equation becomes

$$\frac{\partial \sigma}{\partial t} = \frac{3}{r}\frac{\partial}{\partial r}\left[r^{1/2}\frac{\partial}{\partial r}\left(\nu \sigma r^{1/2}\right) \right]. \tag{27.4}$$

If ν is constant or a power of radius, the equation can be solved analytically, and the related studies of density evolution, starting from assigned initial conditions for σ, such as those of a narrow ring, are relatively simple to carry out and instructive.[9]

27.1.2 Luminosity by Viscous Dissipation for a Steady-State Accretion Disk

As indicated in the introductory remarks of this chapter, the other aspect of interest is the possibility of extracting energy by means of matter infall. Consider the equations of mass and angular-momentum conservation for a thin axisymmetric disk in steady state [which follow from Eqs. (27.1) and (27.2), reduced to the time-independent case by integration over the radial coordinate]:

$$\dot{M} = -2\pi r\sigma u, \tag{27.5}$$

$$\dot{J} = \dot{M}r^2\Omega + 2\pi\nu\sigma r^3\frac{d\Omega}{dr}. \tag{27.6}$$

The disk density σ is associated with a constant accretion rate $\dot{M}$, taken to be positive for an inflow velocity $u < 0$. The quantity Ω is the local angular velocity of the disk, which is typically decreasing with radius. Thus some angular momentum is convected inward, whereas viscosity induces an outward flux of angular momentum, which combines in a constant angular-momentum flux $\dot{J}$.

A general argument often considered to estimate the constant $\dot{J}$ is to refer to the vicinity of the central star, where the disk rotation should depart from Keplerian to become sub-Keplerian and match the rotation of the central object. Therefore, at some point close to the radius $r_\bullet$ we should have $d\Omega/dr = 0$ so that $\dot{J} = \dot{M}r_\bullet^2\Omega(r_\bullet)$. In other words, away from the boundary layer close to the surface of the central object, for a Keplerian disk, we expect that Eq. (27.6) could be written as

$$\nu\sigma = \frac{\dot{M}}{2\pi|d\ln\Omega/d\ln r|}\left[1 - \frac{r_\bullet^2\Omega(r_\bullet)}{r^2\Omega(r)}\right] = \frac{\dot{M}}{3\pi}\left[1 - \left(\frac{r_\bullet}{r}\right)^{1/2}\right]. \tag{27.7}$$

The dissipation rate per unit area (for one of the two sides of a thin disk) associated with the viscous term in the fluid equations is

$$D(r) = \frac{1}{2}\nu\sigma\left[r\left(\frac{d\Omega}{dr}\right)\right]^2. \tag{27.8}$$

The quantity $\nu\sigma$ can be eliminated from Eqs. (27.7) and (27.8), which shows that the available energy reservoir is gravitational, and the presence of viscosity is only required to release it, but in a sense is irrelevant to the final energy budget. The disk-accretion luminosity L_{acc} can be obtained by integration

$$L_{\text{acc}} = 2\pi\int_{r_\bullet}^{\infty} D(r)r\,dr. \tag{27.9}$$

Therefore, in a steady-state model, the gravitational energy of material accreting at a rate $\dot{M}$ is released by the disk (possibly in the form of radiation), with $L_{\text{acc}} \approx GM_\bullet\dot{M}/2r_\bullet$, within the general picture that a sort of boundary layer forms in the vicinity of the surface of the central object.

One major concern in the theory of accretion disks is determination of the viscosity in the accreting gas or plasma. It soon became clear that a derivation from first principles of gas dynamics or plasma physics would lead to insufficient amounts of viscosity, that is, accretion rates that would be too small to interpret the observed luminosity in terms of accretion luminosity.

Therefore, the theory has resorted to the concept of turbulent or anomalous viscosity. A whole research area has thus developed that is aimed at finding convincing mechanisms able to produce the desired levels of anomalous viscosity.[10]

27.1.3   A Useful Semiempirical Framework: The α-Disks

Given the complexity of the problems involved, it is common practice to refer to the presence of a turbulent viscosity in terms of a physically intuitive prescription:[11]

$$\nu = \alpha c z_0, \tag{27.10}$$

where c is an effective thermal speed in the gas, z_0 is the thickness of the disk, and α is a phenomenological parameter likely to be smaller than unity, where we hide our ignorance of the detailed mechanisms that are responsible for the viscosity present. For a Keplerian thin disk, the thickness can be written as $z_0 = c/\Omega$. Equation (27.10) basically results from dimensional analysis under the condition that the eddies associated with turbulence is smaller than the disk thickness. The parameter α is often considered to be a constant, but some studies have argued that it should be allowed to vary with radius.[12]

27.1.4   Energy Budget: Different Types of Accretion Disks

Although the momentum transport problem is often simplified by the preceding treatment but universally considered an issue of great concern, the energy equations are generally accepted in their ideal form. In some cases we consider the various radiation processes that may be involved and the related energy transport, much as is done for stellar atmospheres.

Two hypotheses are often made in the construction of the models of accretion disks. One is that viscosity is the only source of heating [see Eq. (27.8)], with a possible additional contribution coming from external irradiation; such heating rate per unit area is often denoted by Q^+. The second hypothesis is that the system cools radiatively at a rate per unit area indicated by Q^-, which requires a discussion of the vertical energy radiative transport. Thus we can set up a horizontal energy-transport equation of the form

$$\sigma u T_0 \frac{ds}{dr} = Q^+ - Q^-, \tag{27.11}$$

where T_0 represents the midplane temperature of the disk, and s is the entropy per unit mass of the disk material; here horizontal energy transport by radiation is generally ignored.

Standard-disk solutions are those for which $Q^+ \approx Q^-$ and the disk is taken to be optically thick. The relevant set of equations is then closed in terms of disk quantities and leads to the construction of models that have been successfully compared with the observations for a variety of interesting astrophysical objects (such as active galactic nuclei or X-ray binaries or cataclysmic variables).[13] These standard-disk solutions become colder and colder at large radii so that eventually the disk is expected to become self-gravitating in its outer parts (see discussion in the next section).

Different models have been constructed under the assumption that the disk is optically thin.[14] If we argue instead that most of the energy remains stored in the disk and only little energy is radiated away, we find a class of solutions called *advection-dominated accretion flows*,[15] for which Eq. (27.11) becomes

$$\sigma u T_0 \frac{ds}{dr} = fD, \tag{27.12}$$

and f is a parameter (or, rather, a function of radius) considered to be ≈ 1.

27.2 Steady-State Self-Gravitating Accretion Disks

Given the general framework introduced earlier, we may ask under what conditions and with what consequences would the disk self-gravity affect the picture of disk accretion.[16] The role of self-gravity associated with the disk can be taken into account at three levels. At the first stage, we consider the effects of self-gravity on the vertical structure of the disk.[17] In the second step, we consider the role of self-gravity on waves and instabilities in the disk,[18] much as in the discussion of the dynamics of planetary rings; these waves and instabilities may have a beneficial impact on the problem of viscosity.[19] At the third level, we consider disks for which the rotation curve is not Keplerian, as generally assumed.[20]

In this section, a curious result is summarized for the case in which the disk gravity is fully incorporated[21] that ties up the problem of accretion disks with the possibility of generating flat rotation curves. Some general problems remain open; for example, it would be interesting to discuss diffusion Eq. (27.3) in the case in which $\Omega(r)$ is not an assigned (external) quantity but is (partly) determined by the function $\sigma(r)$ appearing in the equation.[22]

27.2.1 Thickness

For a disk dominated by self-gravity, we may adopt the prescription (see Section 14.1)

$$z_0 = \frac{c^2}{\pi G \sigma}. \tag{27.13}$$

More refined analyses have been performed to produce an expression for the thickness beyond that of the simple slab model and to incorporate the effects of an external spherical field.[23] Note that to judge whether self-gravity is unimportant, we may simply impose that the thickness as determined by the Keplerian restoring force c/Ω is much smaller than the thickness as determined by the disk self-gravity, given by Eq. (27.13). If we do so, we obtain the condition $Q \gg 1$, where $Q = c\kappa/(\pi G \sigma)$ is the standard axisymmetric stability parameter. In other words, by moving from the inner, hotter regions out to the regions where the disk becomes cooler, in the sense that the related Q parameter decreases and then approaches unity, the disk self-gravity eventually must take over and become important.

27.2.2 Momentum and Energy Transport by Density Waves: Self-Regulation

If we refer back to Eq. (27.11), we can argue that the disk self-gravity contributes an additional heating term, which we take to be of the form[24]

$$H_{\text{grav}} = g(Q)\sigma c^2 \Omega, \tag{27.14}$$

where Q is the standard axisymmetric stability parameter and, for values of Q below a certain threshold $\bar{Q} \approx 1$, g is a fast-increasing function of Q; an example of such function could be $g(Q) = (\bar{Q}/Q)^n$, with $n \gg 1$. This additional heating could be interpreted as the result of the presence of a modified viscosity generated by gravitational instabilities that are excited if the disk is too cold. In fact, by making use of Eq. (27.13), we could write Eq. (27.14) as

$$H_{\mathrm{grav}} = (\alpha_{\mathrm{grav}} c z_0) \sigma \left(r \frac{d\Omega}{dr} \right)^2 , \qquad (27.15)$$

and refer to a modified viscosity characterized by $\alpha \rightarrow \alpha + \alpha_{\mathrm{grav}}$.[25] Actually, there is no need to tie the effects of gravitational instabilities on momentum to those on energy transport; it is only important to recognize that anomalous viscosity and anomalous energy transport are naturally expected in the presence of gravitational instabilities.[26] In practice, for a nonadvective self-gravitating accretion disk, the conjectured steep function of Q in the heating term is bound to enforce a self-regulation process (see Section 16.3), so the energy balance equation can be replaced by the simple condition

$$\frac{c\kappa}{\pi G \sigma} \approx \bar{Q}, \qquad (27.16)$$

with $\bar{Q}$ not far from unity. Note that this condition might allow for the presence of vigorous nonaxisymmetric instabilities, which are beyond the reach of the present simplified axisymmetric model. The hope is that these nonaxisymmetric effects can be, at least in a first approximation, included to justify the viscosity prescription and the level of $\bar{Q}$ determining self-regulation. Note also that the self-regulation prescription considered here is local, whereas the process probably should be treated at the global level. Unfortunately, such a global self-regulation mechanism is difficult to model. Numerical simulations confirm the general validity of these arguments.[27]

27.2.3 Rotation Curves

For a cool, slowly accreting disk, the rotation is determined by the contributions of self-gravity and the Keplerian field of the central object:

$$\Omega^2 \sim \frac{1}{r} \frac{d\Phi_D}{dr} + \frac{GM_\bullet}{r^3}, \qquad (27.17)$$

where $\Phi_D(r)$ is the potential generated by the disk [see Subsection 14.3.1, in particular, Eq. (14.26)].

27.2.4 Self-Regulated Accretion Disks

By combining Eqs. (27.6), (27.10), and (27.13), we obtain

$$G\dot{j} = GM r^2 \Omega + 2\alpha c^3 r^3 \frac{d\Omega}{dr}. \qquad (27.18)$$

One qualitative new aspect involved in the description just developed is the fact that we now have a closed set of equations [Eqs. (27.18), (27.17), (14.26), and (27.16)] in which the self-regulation constraint replaces the energy equation. Thus energy transport is dealt with at a phenomeno-logical, physically justified level of description, more in line with the way angular-momentum transport is treated by the prescription of Eq. (27.10).

At sufficiently large radii (i.e., $r \gg r_s$, $r \gg r_J$; these scale radii will be better defined later in this subsection), we expect that the effects associated with $M_\bullet$ and $\dot{J}$ become negligible. Therefore, we may look for solutions where, for simplicity, $M_\bullet$ and $\dot{J}$ vanish exactly. We can easily eliminate c and obtain one equation for σ imposed by self-regulation:

$$\pi G \sigma = \frac{2\Omega}{\bar{Q}} \left(1 + \frac{1}{2} \frac{d \ln \Omega}{d \ln r}\right)^{1/2} \left(\frac{G\dot{M}}{2\alpha}\right)^{1/3} \left|\frac{d \ln \Omega}{d \ln r}\right|^{-1/3}. \tag{27.19}$$

The other equation relating Ω to σ is obviously Eq. (14.26), defining a self-gravitating disk. Curiously, the natural solution to this problem is readily provided by the self-similar disk (see Subsection 14.3.1) with

$$2\pi G \sigma r = r^2 \Omega^2 = V^2 = \text{constant}, \tag{27.20}$$

$$c = \left(\frac{G\dot{M}}{2\alpha}\right)^{1/3} \approx \left(\frac{27\dot{M}}{4\pi\alpha}\right)^{1/3} 10 \text{ km s}^{-1}, \tag{27.21}$$

and $V = 2\sqrt{2}c/\bar{Q}$, $u = -\alpha\bar{Q}^2 c/4$, $z_0/r = \bar{Q}^2/4$, and $\nu\sigma = \dot{M}/2\pi$; in the last expression of Eq. (27.21), $\dot{M}$ is given in units of $M_\odot$ yr^{-1}. Thus the solution is fully determined by the two parameters $\dot{M}$ and α; $\bar{Q}$ is expected to be close to unity, well constrained by dynamics. These models can be generalized to the case of finite $M_\bullet$, $\dot{J}$, and truncation radius, to situations for which the self-regulation constraint fails to operate in parts of the disk, and to the framework within which a spheroidal mass component is also present.[28]

By construction, the scale radius r_s marks the transition from Keplerian to non-Keplerian rotation. A straightforward asymptotic analysis suggests that

$$r_s = GM_\bullet \left(\frac{\bar{Q}^2}{8}\right) \left(\frac{G\dot{M}}{2\alpha}\right)^{-2/3}. \tag{27.22}$$

Similarly, it is easily shown that a sensible choice for r_J is

$$r_J = \frac{\sqrt{2}|\dot{J}|}{\dot{M}} \left(\frac{\bar{Q}}{4}\right) \left(\frac{G\dot{M}}{2\alpha}\right)^{-1/3}. \tag{27.23}$$

For $r \ll r_s$, the rotation curve is expected to be Keplerian. We can thus envision a general situation in which, if we move out to larger radii but still below r_s, we meet an entire transition region that connects the innermost regions with $Q \gg 1$, where the effects of self-gravity are negligible,[29] to an intermediate region in which self-gravity effects show up by modifying the standard Keplerian thickness prescription, and self-regulation (based on the competition between cooling and gravitational instabilities) is established, out to the region where the rotation curve exhibits significant deviations from a Keplerian behavior. The detailed structure of the transition region and the outer parts depends on the type of cooling mechanisms that are involved.[30]

In the discussion of the shape of dark halos around spiral galaxies (see comments in Sections 14.5 and 20.1), it is often noted that a round halo is associated with the density profile $\rho \sim r^{-2}$ characteristic of an isothermal sphere (see Subsection 22.2.1), which is physically appealing, whereas a flat halo would require a mass density of the form $\sigma \sim r^{-1}$, which would have no simple physical justification. The study of the cool self-regulated accretion disk presented in this section surprisingly leads to a disk mass distribution with a flat rotation curve.

27.3 Some Applications

Applications of the preceding framework basically refer to disks, or their outer parts, that can be considered to be sufficiently cold. Self-gravity generally changes the picture in important qualitative ways. For example, it has been shown that if cooling is dominated by *bremsstrahlung*, optically thin accretion disks,[31] which are usually considered to be thermally unstable, are stabilized, even in the nonadvective case.[32]

In the context of protostellar disks, self-gravity is generally expected to affect evolution in two different ways. If the cooling time is much smaller than the dynamical time scale, gravitational instabilities are expected to generate fragments that eventually may originate planets. When the cooling rate is slower, a steady-state, self-regulated accretion can be established. These general expectations have been confirmed by a number of numerical simulations aimed at studying the nonlinear evolution of self-gravitating disks and quantitatively determining the threshold condition that discriminates between the two types of behavior.[33] Obviously, the actual occurrence of the two different scenarios depends on the specific physical conditions of the disk under investigation (from the macroscopic scales down to the level of the relevant microscopic physical processes that participate in determining the cooling rate). The two types of evolution define two lines of research characterized by many interesting results and open questions related to separate astrophysical problems.

Other applications are prompted by the advent of new telescopes and instrumentation. For example, direct evidence of spiral structure in accretion disks can be looked for by means of interferometric measurements that will soon be made available by ALMA (see Chapter 2).[34]

Instead of providing a long list of items of interest and related references, in the following subsections we focus briefly on three specific applications: the first related to the emission of some protostellar objects, the second to the measurement of deviations from Keplerian rotation in accretion disks on very different scales, and the last bringing us back to the general problem of the dynamics of galaxies.

27.3.1 Spectral Energy Distribution of Protostellar Disks

The presence of circumstellar disks around pre-main-sequence stars (in particular, the T Tauri and FU Orionis systems) has long been inferred from the properties of the spectral energy distribution associated with protostellar objects in the infrared and millimetric parts of the spectrum.[35] It is now commonly believed that the excess luminosity at long wavelengths around many objects (especially T Tauri stars, for which the estimated accretion rate is very low) is largely determined only passively by the disk, by irradiation of light received from the central object. In turn, other objects (such as FU Orionis objects or high-mass protostars, for which the accretion rate is estimated to be higher) are likely to be associated with a more active disk, able to dissipate sufficient amounts of accretion energy as a contribution to the observed infrared luminosity.[36]

Let us now consider an active accretion disk, that is, a disk for which the observed luminosity is generated by the energy dissipated by accretion. For an assumed power-law dependence of the temperature profile of an optically thick disk $T \propto r^{-q}$, the standard-disk solution mentioned in Subsection 27.1.4 [see comments after Eq. (27.11)] predicts a spectral index $n \equiv d\log(\nu F_\nu)/d\log\nu \approx 4 - 2/q$ (the quantity νF_ν is the standard quantity used in the representation of the spectral-energy distribution of a source; obviously, the standard symbol ν for the frequency should not be confused with the kinematic viscosity) in the infrared part of the spectrum. If we combine this statement with the expression for the accretion luminosity given by Eq. (27.8), we see that an active Keplerian accretion disk would imply $q = 3/4$, that is, $n = 4/3$, whereas T Tauri stars are typically characterized by $n = 2/3$ or even flatter spectral-energy distributions ($n \approx 0$). Curiously, if the rotation curve of the disk were flat, the temperature profile would be associated with $q \approx 1/2$, and the expected spectral index would be $n \approx 0$.[37] In some cases, for example, for the young star TW Hydrae,[38] evidence has been found for $q = 1/2$ and for a power-law surface density distribution $\sigma \propto r^{-1}$, as would be natural in a self-gravitating disk, but it is not clear to what extent the properties of the relevant profiles refer to the disk material or only to the dust component of the disk.

Instead of proceeding further with general arguments, we simply take the task of calculating in detail the expected spectral energy distribution for a full model of a partially self-regulated and fully active self-gravitating disk. By partially self-regulated, we mean a disk characterized by a Q profile so that, in relation to a scale radius r_Q to be considered as a free parameter, the disk is nongravitating in its inner regions and self-regulated, self-gravitating in its outer parts. The study can be performed by including as additional free parameters an inner truncation radius r_{in} and an outer truncation radius r_{out}. The final expected spectral-energy distribution depends on one geometric parameter (the inclination of the disk with respect to the line of sight), three scale parameters (a luminosity scale, a temperature scale, and the length scale r_s, all defined in terms of the accretion quantities $\dot{M}$, α, and $M_\bullet$), and three dimensionless length parameters (r_Q/r_s, r_{in}/r_s, and r_{out}/r_s). The disk surface temperature profile, which is required to produce the final spectral-energy distribution by integrating the local Planck spectrum in the radial coordinate, is set by the local condition

$$\sigma_B T^4 = D(r) + H_{grav}, \tag{27.24}$$

where σ_B is the Stefan-Boltzmann constant, and $D(r)$ and H_{grav} are defined by Eqs. (27.8) and (27.14), respectively. Note that a test based on this framework, as will be briefly summarized in the following paragraph, turns out to provide numbers significantly different from earlier estimates of the role of self-gravity[39] that did not include the important role of self-regulation, as incorporated by Eq. (27.24); in particular, we find that lighter disks are required but still able to contribute significantly to the spectral-energy distribution through their self-gravity.

This procedure has been applied to fit the spectral-energy distribution of two FU Orionis objects and four T Tauri stars. For the FU Orionis objects, for which the interpretation in terms of an active self-gravitating accretion disk is better justified, some results are illustrated in Figs. 27.1 and 27.2. The infrared excess at long wavelengths $\lambda > 10 \ \mu m$ for the FU Orionis objects can be described by means of a self-gravitating accretion-disk model, with general characteristics (in particular, $\dot{M}$ in the range 10^{-5} to $10^{-4} \ M_\odot \ yr^{-1}$) consistent with models that attribute the observed excess to other physical processes; the required accretion parameter α is also found to

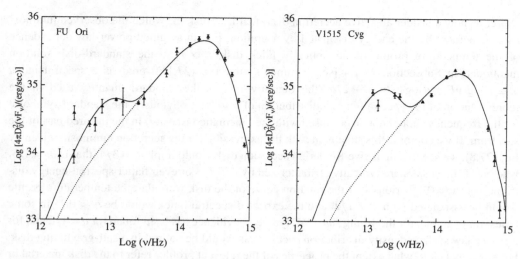

Fig. 27.1. Best fit to the spectral-energy distributions for the FU Orionis sample (*solid curves*) compared with the spectra resulting from Keplerian disks with the same parameters ($\dot{M}$, r_{in}/r_s, r_{out}/r_s, and disk inclination; *thin curves*). Data from Kenyon, S. J., Hartmann, L. 1991. *Astrophys. J.*, **382**, 664, and from Weaver, W. B., Jones, G. 1992. *Astrophys. J. Suppl.*, **78**, 239 (from Lodato, G., Bertin, G. 2001. *Astron. Astrophys.*, **375**, 455).

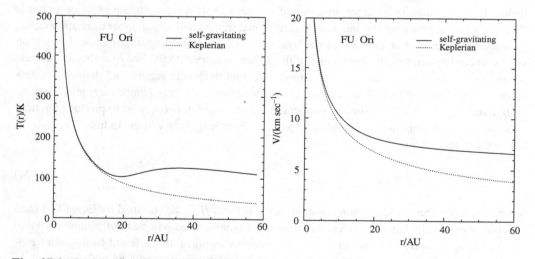

Fig. 27.2. Surface-temperature profile (*left*) and rotation curve (*right*) for the best-fit model to FU Ori (with $r_Q/r_s = 0.4$) compared with the corresponding Keplerian model (from Lodato, G., Bertin, G. 2001. *Astron. Astrophys.*, **375**, 455).

be in a reasonable range, 10^{-3} to 10^{-2}, and the disk mass required is not too large. For the T Tauri stars, the fits are in principle acceptable but would require accretion rates ($\approx 10^{-6}$ $M_\odot$ yr^{-1}) that are judged to be unrealistically large for these objects; in this latter case, the contribution to the emission from a self-gravitating disk may help to alleviate the difficulties that the more standard interpretation, in terms of disk irradiation, sometimes encounters (these difficulties are often resolved by invoking a suitable flaring of the disk); in other words, irradiation and some contributions from a partially active self-gravitating disk may turn out to provide the best picture of what is observed.

27.3.2 Non-Keplerian Rotation

In Subsection 22.2.2 we briefly referred to a beautiful study of the central region of the galaxy NGC 4258 (= M106), for which the measurement of the properties of water-vapor emission has been interpreted as an indication of a disk in Keplerian rotation.[40] This, together with the study of the motion of stars in the vicinity of the Galactic center, mentioned in Chapter 1, has led to the best evidence for the existence of supermassive black holes in galaxies. Since then, a number of active galactic nuclei, in which maser emission has been observed and interpreted as tracing the presence of a central cold rotating gaseous disk, have been investigated.[41]

One interesting case is that of the galaxy NGC 1068, for which discrepancies have been noted in the observed rotation-curve profile from a pure Keplerian decline with radius.[42] Such discrepancies may signal the presence of a disk of nonnegligible mass, in the framework of the self-gravitating accretion-disk models described in this chapter.[43] In this interpretation, a full model can be constructed (Fig. 27.3). A good fit is obtained, leading to a mass of the black hole $M_\bullet \approx (8.0 \pm 0.3) \times 10^6 \, M_\odot$ comparable with that of the surrounding cold accretion disk; the implied viscosity parameter is $\alpha \approx 10^{-2}$, in line with some theoretical expectations. The result is interesting for at least two reasons. On the one hand, with respect to other estimates based on a purely Keplerian model, it suggests a significantly lighter (by a factor of 2) mass for the central black hole, which may be relevant to the ongoing discussions about the scaling laws that are obeyed by supermassive black holes in galaxies (see Subsection 4.3.4). On the other hand, the modeling confirms that the water maser emission data come from a region where indeed deviations from Keplerian rotation are expected a priori because for this case the radius r_s (see Subsection 27.2.4) is of the order of a few parsecs. Instead, for the galaxy NGC 4258, which was found to be characterized by a Keplerian rotation curve, the estimate of r_s indicates that the data sample only the innermost parts of the disk, which are indeed expected a priori to be characterized by Keplerian rotation.

Another very interesting case is that of the water maser emission from the center of NGC 3079.[44] Here the evidence is for a flat rotation curve (Fig. 27.4); therefore, as noted by the authors of this study, "the mass of the parsec-scale disk is significant with respect to the central mass." It thus appears that the paradigm of cold self-regulated, self-gravitating accretion disks leads to predictions that are confirmed by observations of some active galactic nuclei.[45]

Finally, we may wonder whether signs of deviations from Keplerian rotation can be found in protostellar disks. A study in this direction has been proposed by application of long-wavelength spectroscopy to FU Orionis objects.[46] We can take advantage of the fact that because the disk surface temperature is expected to decrease with radius, different lines are expected to diagnose different parts of the disk. A high-resolution study of the profiles of the far-infrared molecular hydrogen rotational lines and submillimeter CO lines may be able to probe the rotation curve of the outer accretion disks in these objects.[47] Some results of a first exploratory work are illustrated in Fig. 27.5.

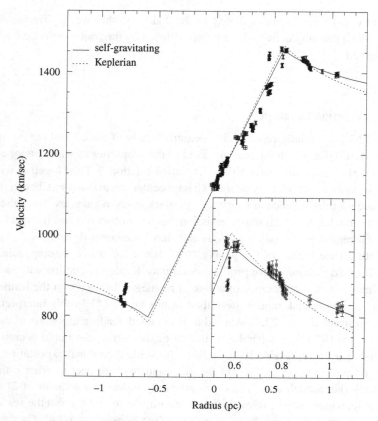

Fig. 27.3. Fit by a self-gravitating accretion-disk model to the central parsec of the rotation curve of NGC 1068 derived from water maser emission. The data are from Greenhill, L., Gwinn, C. 1997. *Astrophys. Sp. Sci.*, **248**, 261. The inset shows the declining part of the rotation curve together with the best fit obtained by assuming Keplerian rotation (from Lodato, G., Bertin, G. 2003. *Astron. Astrophys.*, **398**, 517).

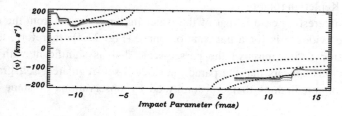

Fig. 27.4. Average rotation velocity as a function of distance from the location of the dynamical center along the disk major axis, that is, impact parameter, for the central region of NGC 3079. The dotted lines show the analytical results for a Keplerian edge-on disk of negligible mass around a $(0.5, 2, 4) \times 10^6 \, M_\odot$ point mass computed using an averaging aperture of 4 mas (from Kondratko, P. T., Greenhill, L. J., Moran, J. M. "Evidence for a geometrically thick self-gravitating accretion disk in NGC 3079," 2005. *Astrophys. J.*, **618**, 618; reproduced by permission of the AAS).

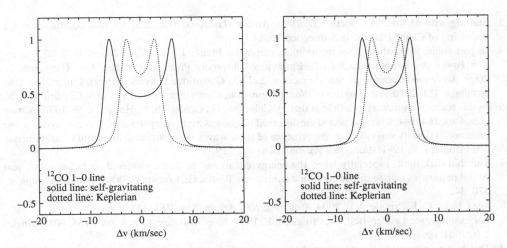

Fig. 27.5. ^{12}CO 1–0 line profile at 110 GHz for a model applicable to FU Ori with a non-Keplerian rotation curve associated with a self-gravitating accretion disk (*solid line*) and for a strictly Keplerian model (*dotted line*). The results are shown for $r_s \approx 18$ AU (*left*) and $r_s \approx 36$ AU (*right*) (from Lodato, G., Bertin, G. 2003. *Astron. Astrophys.*, **408**, 1015).

27.3.3 HI Disks

Here we would like to close the chapter by mentioning two additional research areas, in the context of galaxies, in which the model of self-gravitating accretion disks may find application. If the conditions are such that star formation does not occur on the fast dynamical scale, we could consider the problem of the dissipative collapse (within a dark halo made of collisionless matter) of a rotating gas cloud into a protogalactic disk. After the initial rapid collapse, we can expect that a slow contraction by accretion might take place while stars begin to form and populate the disk. A study of this type should face the challenging task of interpreting the origin of exponential disks in differential rotation, characterized by a flat rotation curve. Some investigations along these lines have been made,[48] but to our knowledge, the role of self-gravitating accretion, of the type briefly described in this chapter, has not been fully taken into account.[49]

The other research area in the context of galaxies, inspired by observations, is that of the structure and dynamics of the light HI disks discovered in early-type galaxies (see Subsection 24.2.2 and references therein). It is likely that these disks formed by accretion, possibly at later epochs with respect to the formation of the host galaxies. Because data on the galactic scale are acquired in a relatively straightforward manner, it would be interesting to check whether these light HI disks conform to the general picture of self-regulated, self-gravitating accretion disks, even though their dynamics is dominated by the (collisionless) matter associated with the host galaxy. To some extent, this work would be analogous to the study of self-gravity effects in planetary rings, for which the gravitational field is obviously dominated by the host planet.

Notes

1. See von Weizsäcker, C. F. 1948. *Z. Naturforsch.*, **3A**, 524; Lüst, R. 1952. *Z. Naturforsch.*, **7A**, 87.
2. Prendergast, K. H., Burbidge, G. R. 1968. *Astrophys. J. Lett.*, **151**, L83; Lynden-Bell, D. 1969. *Nature (London)*, **223**, 690.

3. Starting with Mayor, M., Queloz, D. 1995. *Nature (London)*, **378**, 355; see also comments at the beginning of Chapter 20 and in Subsection 26.2.1.

4. In particular, we wish to refer to the monographs by Frank, J., King, A., Raine, D. 2002. *Accretion Power in Astrophysics*, 3rd ed., Cambridge University Press, Cambridge, UK; Hartmann, L. 2009. *Accretion Processes in Star Formation*, 2nd ed., Cambridge University Press, Cambridge, UK; Armitage, P. J. 2010. *Astrophysics of Planet Formation*, Cambridge University Press, Cambridge, UK.

5. In this respect, a pioneering article is that by Colombo, G., Goldreich, P., Harris, A. W. 1976. *Nature (London)*, **264**, 344, who interpreted the observed brightness asymmetry in Saturn's A ring as due to the presence of density waves before the presence of these waves was confirmed by in situ measurements.

6. See Pringle, J. E. 1981. *Annu. Rev. Astron. Astrophys.*, **19**, 307.

7. The thin disk limit, especially when the energy equations are also considered, is rather subtle and would require a separate discussion; for example, see Bertin, G., Lodato, G. 2001. *Astron. Astrophys.*, **370**, 342.

8. See Clarke, C. J., Pringle, J. E. 2004. *Mon. Not. Roy. Astron. Soc.*, **351**, 1187.

9. For example, see Lynden-Bell, D., Pringle, J. E. 1974. *Mon. Not. Roy. Astron. Soc.*, **168**, 603; Pringle, J. E. 1981. *op. cit.*

10. This is not unexpected if we look at the experience in laboratory and space plasmas; see Coppi, B. 1980. *Comm. Plasma Phys. Cont. Fusion*, **5**, 261. For plasma disks, attention has often been addressed to mechanisms involving the presence of magnetic fields; see Blandford, R. D. 1976. *Mon. Not. Roy. Astron. Soc.*, **176**, 465; Lovelace, R. V. E. 1976. *Nature (London)*, **262**, 649; Balbus, S. A., Hawley, J. F. 1991. *Astrophys. J.*, **376**, 214; Coppi, B., Coppi, P. S. 1998. *Phys. Lett., A*, **239**, 261; see also Velikhov, E. 1959. *Sov. Phys. JETP*, **36**, 1938; Chandrasekhar, S. 1960. *Proc. Natl. Acad. Sci. USA*, **46**, 253.

11. Shakura, N. I., Sunyaev, R. A. 1973. *Astron. Astrophys.*, **24**, 337; see also Lynden-Bell, D., Pringle, J. E. 1974. *op. cit.*

12. For example, see Laughlin, G., Różyczka, M. 1996. *Astrophys. J.*, **456**, 279.

13. Shakura, N. I., Sunyaev, R. A. 1973. *op. cit.*

14. Shapiro, S. L., Lightman, A. P., Eardley, D. M. 1976. *Astrophys. J.*, **204**, 187.

15. Narayan, R., Yi, I. 1994. *Astrophys. J. Lett.*, **428**, L13. Issues related to the stability of these configurations have led to the development of different models; in particular, see the adiabatic inflow-outflow solutions by Blandford, R. D., Begelman, M. C. 1999. *Mon. Not. Roy. Astron. Soc.*, **303**, L1, and 2004. *Mon. Not. Roy. Astron. Soc.*, **349**, 68.

16. A review of this topic is given by Lodato, G. 2007. *La Rivista del Nuovo Cimento*, **30**, 293.

17. Paczyński, B. 1978. *Acta Astron.*, **28**, 91; see also Bardou, A., Heyvaerts, J., Duschl, W. J. 1998. *Astron. Astrophys.*, **337**, 966.

18. Lin, D. N. C., Pringle, J. E. 1987. *Mon. Not. Roy. Astron. Soc.*, **225**, 607; Adams, F. C., Ruden, S. P., Shu, F. H. 1989. *Astrophys. J.*, **347**, 959.

19. Lin, D. N. C., Pringle, J. E. 1990. *Astrophys. J.*, **358**, 515; Laughlin, G., Bodenheimer, P. 1994. *Astrophys. J.*, **436**, 335.

20. This topic was addressed by Bodo, G., Curir, A. 1992. *Astron. Astrophys.*, **253**, 318.

21. Bertin, G. 1997. *Astrophys. J. Lett.*, **478**, L71; see also Bertin, G., Lodato, G. 1999. *Astron. Astrophys.*, **350**, 694.

22. Aspects of this general problem, within a simplified model, have been considered by Lin, D. N. C., Pringle, J. E. 1990. *op. cit.*

23. The interpolation formula derived in appendix A of the article by Bertin, G., Lodato, G. 1999. *op. cit.*, improves on earlier expressions; see Sakimoto P. J., Coroniti F. 1981. *Astrophys. J.*, **247**, 19; and Bardou, A., Heyvaerts, J., Duschl, W. J. 1998. *op. cit.*

24. Here we follow arguments presented in detail by Bertin, G., Lodato, G. 2001. *op. cit.*

25. The argument is similar to that considered by Lin, D. N. C., Pringle, J. E. 1987. *op. cit.*

26. The conditions under which anomalous viscosity and anomalous energy transport can be tied to a single diffusion coefficient have been studied and clarified by Lodato, G., Rice, W. K. M. 2004. *Mon. Not. Roy. Astron. Soc.*, **351**, 630; Lodato, G., Rice, W. K. M. 2005. *Mon. Not. Roy. Astron. Soc.*, **358**, 1489; Cossins, P., Lodato, G., Clarke, C. J. 2009. *Mon. Not. Roy. Astron. Soc.*, **393**, 1157.

27. See Lodato, G., Rice, W. K. M. 2004. *op. cit.*, especially their fig. 3; Lodato, G., Rice, W. K. M. 2005. *op. cit.*, especially their fig. 6.

28. Bertin, G., Lodato, G. 1999. *op. cit.*

29. See also Bardou, A., Heyvaerts, J., Duschl, W. J. 1998. *op. cit.*

30. A preliminary discussion of this physically complex multiscale problem is given by Bertin, G., Lodato, G. 2001. *op. cit.*

31. Of the kind studied by Shapiro, S. L., Lightman, A. P., Eardley, D. M. 1976. *op. cit.*

32. Bertin, G., Lodato, G. 2001. *op. cit.*

33. For example, see Gammie, C. F. 2001. *Astrophys. J.*, **553**, 174; Rice, W. K. M., Armitage, P. J., et al. 2003. *Mon. Not. Roy. Astron. Soc.*, **339**, 1025. In the context of quasi-stellar objects (QSOs), the general concepts have been revisited by Goodman, J. 2003. *Mon. Not. Roy. Astron. Soc.*, **339**, 93.

34. Cossins, P., Lodato, G., Testi, L. 2010. *Mon. Not. Roy. Astron. Soc.*, **407**, 181.

35. See Rydgren, A. E., Strom, S. E., Strom, K. M. 1976. *Astrophys. J. Suppl.*, **30**, 307; Beckwith, S. V. W., Sargent, A. I., et al. 1990. *Astron. J.*, **99**, 924.

36. In this subsection we mainly follow the article by Lodato, G., Bertin, G. 2001. *Astron. Astrophys.*, **375**, 455. In the context of QSOs, similar general concepts have been revisited by Sirko, E., Goodman, J. 2003. *Mon. Not. Roy. Astron. Soc.*, **341**, 50.

37. The arguments given here basically summarize those by Adams, F. C., Lada, C., Shu, F. H. 1988. *Astrophys. J.*, **326**, 865. Given that a massive disk with flat rotation curve appeared to be implausible, these authors proposed other ways, based on the role of gravitational instabilities, to justify the unexpected temperature profile of the disk in the cases of an observed flat spectrum.

38. Wilner, D. J., Ho, P. T. P., et al. 2000. *Astrophys. J. Lett.*, **534**, L101.

39. Kenyon, S. J., Hartmann, L. 1987. *Astrophys. J.*, **323**, 714; Shu, F. H., Adams, F. C., Lizano, S. 1987. *Annu. Rev. Astron. Astrophys.*, **25**, 23.

40. Miyoshi, M., Moran, J., et al. 1995. *Nature (London)*, **373**, 127.

41. In particular, see Kuo, C. Y., Braatz, J. A., et al. 2011. *Astrophys. J.*, **727**, 20.

42. Greenhill, L., Gwinn, C. 1997. *Astophys. Sp. Sci.*, **248**, 261; Kumar, P. 1999. *Astrophys. J.*, **519**, 599.

43. Here we follow Lodato, G., Bertin, G. 2003. *Astron. Astrophys.*, **398**, 517.

44. Kondratko, P. T., Greenhill, L. J., Moran, J. M. 2005. *Astrophys. J.*, **618**, 618.

45. Unfortunately, the authors of the study of NGC 3079 estimate the value of the parameter Q by adopting in its definition the thermodynamical value of c ("With these parameters we obtain $Q = 0.01 - 0.02$, where the range reflects only the uncertainty in temperature, etc."), which is obviously bound to be severely underestimated with respect to the effective value determined by the presence of turbulence, as is well known, for example, for the atomic hydrogen on the scale of spiral structure in galaxies. In practice, the disk is expected to be effectively characterized by $Q \approx 1$.

46. Lodato, G., Bertin, G. 2003. *Astron. Astrophys.*, **408**, 1015.

47. Double peaked line profiles have been observed in the near infrared and in the optical, probing the inner parts of the disk; Kenyon, S., Hartmann, L., Hewett, R. 1988. *Astrophys. J.*, **325**, 231.

48. See Lin, D. N. C., Pringle, J. E. 1987. *Astrophys. J. Lett.*, **320**, L87.

49. In the context of supermassive black-hole formation, some interesting results along these lines have been obtained by Lodato, G., Natarajan, P. 2006. *Mon. Not. Roy. Astron. Soc.*, **371**, 1813.

Bibliography

Aarseth, S. J. 2003. *Gravitational N-Body Simulations*. Cambridge University Press, Cambridge, UK.

Abramowitz, M., Stegun, I. A. 1965. *Handbook of Mathematical Functions*. Dover, New York.

Alfvén, H. 1950. *Cosmic Electrodynamics*. Oxford University Press, Oxford, UK.

Alfvén, H., Fälthammar, C.-G. 1963. *Cosmic Electrodynamics*. Oxford University Press, Oxford, UK.

Armitage, P. J. 2010. *Astrophysics of Planet Formation*. Cambridge University Press, Cambridge, UK.

Arnol'd, V. I. 1976. *Les Méthodes Mathématiques de la Mécanique Classique*. Mir, Moscow.

Arnol'd, V. I. 1989. *Mathematical Methods of Classical Mechanics*. Springer-Verlag, Berlin.

Ashman, K., Zepf, S. E. 1998, *Globular Cluster Systems*. Cambridge University Press, Cambridge, UK.

Batchelor, G. K. 1967. *Fluid Dynamics*. Cambridge University Press, Cambridge, UK.

Bender, C. M., Orszag, S. A. 1978. *Advanced Mathematical Methods for Scientists and Engineers*. McGraw-Hill, New York.

Bertin, G., Lin, C. C. 1996. *Spiral Structure in Galaxies: A Density Wave Theory*. MIT Press, Cambridge, MA.

Bertin, G., Farina, D., Pozzoli, R., eds. 2004. *Plasmas in the Laboratory and in the Universe: New Insights and New Challenges*. AIP Conference Proceedings, Vol. 703, Melville, NY.

Bertin, G., Pozzoli, R., et al. eds. 2007. *Collective Phenomena in Macroscopic Systems*. World Scientific, Singapore.

Bertin, G., De Luca, F., et al. eds. 2010. *Plasmas in the Laboratory and in the Universe: Interactions, Patterns, and Turbulence*. AIP Conference Proceedings, Vol. 1242, Melville, NY.

Binney, J., Merrifield, M. 1998. *Galactic Astronomy*. Princeton University Press, Princeton, NJ.

Binney, J., Tremaine, S. 1987. *Galactic Dynamics*. Princeton University Press, Princeton, NJ.

Binney, J., Tremaine, S. 2008. *Galactic Dynamics*, 2nd ed. Princeton University Press, Princeton, NJ.

Birdsall, C. K., Langdon, A. B. 1985. *Plasma Physics via Computer Simulation*. McGraw-Hill, New York.

Biskamp, D. 1993. *Nonlinear Magnetohydrodynamics*. Cambridge University Press, Cambridge, UK.

Blaauw, A., Schmidt, M., eds. 1965. *Galactic Structure*. University of Chicago Press, Chicago.

Block, D. L., Greenberg, J. M., eds. 1996. *New Extragalactic Perspectives in the New South Africa*. Kluwer, Dordrecht, The Netherlands.

Boccaletti, D., Pucacco, G. 1996. *Theory of Orbits*, Vol. 1: *Integrable Systems and Nonperturbative Methods*. Springer-Verlag, Berlin.

Boccaletti, D., Pucacco, G. 1999. *Theory of Orbits*, Vol. 2: *Perturbative and Geometrical Methods*. Springer-Verlag, Berlin.

Bok, B. J., Bok, P. F. 1974. *The Milky Way*, 4th ed. Harvard University Press, Cambridge, MA.

Budden, K. G. 1985. *The Propagation of Radio Waves*. Cambridge University Press, Cambridge, UK.

Buta, R. J., Corwin, H. G., Oderwahn, S. C. 2007. *The de Vaucouleurs Atlas of Galaxies*. Cambridge University Press, Cambridge, UK.

Cairns, R. A. 1991. *Radiofrequency Heating of Plasmas*. Hilger, Bristol, UK.

Casertano, S., Sackett, P., Briggs, F., eds. 1991. *Warped Disks and Inclined Rings around Galaxies*. Cambridge University Press, Cambridge, UK.

Chandrasekhar, S. 1939. *An Introduction to the Theory of Stellar Structure*. University of Chicago, Chicago (reprinted in 1958 by Dover, New York).

Chandrasekhar, S. 1942. *Principles of Stellar Dynamics*. University of Chicago Press, Chicago (reprinted in 1960 by Dover, New York).

Chandrasekhar, S. 1961. *Hydrodynamic and Hydromagnetic Stability*. Oxford University Press, Oxford, UK (reprinted in 1981 by Dover, New York).

Chandrasekhar, S. 1969. *Ellipsoidal Figures of Equilibrium*. Yale University Press, New Haven, CT.

Chandrasekhar, S. 1983. *The Mathematical Theory of Black Holes*. Clarendon Press, Oxford, UK.

Chapman, S., Cowling, T. G. 1952. *The Mathematical Theory of Non-Uniform Gases*. Cambridge University Press, Cambridge, UK.

Chen, F. F. 1984. *Introduction to Plasma Physics and Controlled Fusion*, Vol. 1: *Plasma Physics*, 2nd ed. Plenum, New York.

Ciotti, L. *An Introduction to Stellar Dynamics*. Cambridge University Press, Cambridge, UK (in preparation).

Combes, F., Casoli, F., eds. 1991. *Dynamics of Galaxies and Their Molecular Cloud Distributions*. Kluwer, Dordrecht, The Netherlands.

Contopoulos, G. 2002. *Order and Chaos in Dynamical Astronomy*. Springer-Verlag, Berlin.

Cox, J. P. 1980. *Theory of Stellar Pulsation*. Princeton University Press, Princeton, NJ.

De Bruijn, N. G. 1970. *Asymptotic Methods in Analysis*. North-Holland, Amsterdam.

de Vaucouleurs, G., de Vaucouleurs, A., et al. 1991. *Third Reference Catalogue of Bright Galaxies*. Springer-Verlag, New York.

Dyson, F. J. 1971. *Neutron Stars and Pulsars*. Fermi Lectures 1970. Accademia Nazionale dei Lincei, Rome.

Eddington, A. S. 1926. *Internal Constitution of Stars*. Cambridge University Press, Cambridge, UK.

Erdélyi, A. 1956. *Asymptotic Expansions*. Dover, New York.

Fabbiano, G., Gallagher, J. S., Renzini, A., eds. 1990. *Windows on Galaxies*. Kluwer, Dordrecht, The Netherlands.

Fabian, A. C., ed. 1987. *Cooling Flows in Clusters and Galaxies*. Kluwer, Dordrecht, The Netherlands.

Feynman, R. P., Leighton, R. B., Sands, M. 1963. *The Feynman Lectures on Physics*. Addison-Wesley, Reading, MA.

Frank, J., King, A., Raine, D. 2002. *Accretion Power in Astrophysics*, 3rd ed. Cambridge University Press, Cambridge, UK.

Fridman, A. M., Polyachenko, V. L. 1984. *Physics of Gravitating Systems*. Springer-Verlag, Berlin.

Fried, B. F., Conte, S. 1961. *The Plasma Dispersion Function*. Academic, New York.

Goldstein, H. 1950. *Classical Mechanics*. Addison-Wesley, Reading, MA.

Hartmann, L. 2009. *Accretion Processes in Star Formation*, 2nd ed. Cambridge University Press, Cambridge, UK.

Hartmann, D., Burton, W. B. 1997. *Atlas of Galactic Neutral Hydrogen*. Cambridge University Press, Cambridge, UK.

Harwit, M. 2006. *Astrophysical Concepts*, 4th ed. Springer, New York.

Heading, J. 1962. *An Introduction to Phase-Integral Methods*. Wiley, New York.

Heggie, D., Hut, P. 2003. *The Gravitational Million-Body Problem: A Multidisciplinary Approach to Star Cluster Dynamics*. Cambridge University Press, Cambridge, UK.

Hockney, R. W., Eastwood, J. W. 1981. *Computer Simulations Using Particles*. McGraw-Hill, New York.

Hodge, P. 1992. *The Andromeda Galaxy*. Kluwer, Dordrecht, The Netherlands.

Huba, J. D. 2002. *NRL Plasma Formulary*. Naval Research Laboratory, Washington, DC.

Hubble, E. 1936. *The Realm of the Nebulae*. Yale University Press, New Haven, CT.

Ichimaru, S. 1973. *Basic Principles of Plasma Physics: A Statistical Approach*. Benjamin, Reading, MA.

Jackson, J. D. 1962. *Classical Electrodynamics*. Wiley, New York.

Jeans, J. 1929. *Astronomy and Cosmogony*. Cambridge University Press, Cambridge, UK (reprinted in 1961 by Dover, New York).

Kelvin, Lord 1910. *Mathematical and Physical Papers*. Cambridge University Press, Cambridge, UK.

Kim, D.-W., Pellegrini, S., eds. 2012. *Hot Interstellar Matter in Elliptical Galaxies*. Astrophysics and Space Science Library, Vol. 378, Springer Science and Business Media, Heidelberg.

Kitchin, C. R. 1984. *Astrophysical Techniques*. Adam Hilger, Bristol, UK (5th ed. published in 2009 by CRC Press, Taylor & Francis Group, Boca Raton, FL).

Klimontovich, Y. L. 1967. *The Statistical Theory of Nonequilibrium Processes in a Plasma.* MIT Press, Cambridge, MA.

Kolb, E. W., Turner, M. S. 1990. *The Early Universe.* Addison-Wesley, Redwood City, CA.

Krall, N. A., Trivelpiece, A. W. 1973. *Principles of Plasma Physics.* McGraw-Hill, New York.

Landau, L., Lifchitz, E. 1967. *Physique Statistique.* Mir, Moscow.

Lecar, M., ed. 1972. *Gravitational N-Body Problem. IAU Colloquium,* Vol. 10. Reidel, Dordrecht, The Netherlands.

Lin, C. C., Siegel, L.A. 1988. *Mathematics Applied to Deterministic Problems in the Natural Sciences.* Society for Industrial and Applied Mathematics, Philadelphia, PA.

Linde, A. D. 1990. *Particle Physics and Inflationary Cosmology.* Harwood, New York.

MacMillan, W. D. 1958. *Theoretical Mechanics: The Theory of the Potential.* Academic, New York.

Martinet, M., Mayor, M., eds. 1973. *Dynamical Structure and Evolution of Stellar Systems.* Geneva Observatory, Sauverny, Switzerland.

Maslov, V. P., Fedoriuk, M. V. 1981. *Semi-Classical Approximation in Quantum Mechanics.* Reidel, Dordrecht, The Netherlands.

Matteucci, F. 2001. *The Chemical Evolution of the Galaxy.* Kluwer, Dordrecht, The Netherlands.

Mayer, J. E., Mayer, M. G. 1940. *Statistical Mechanics of Fluids.* Wiley, New York.

Messiah, A. 1962. *Quantum Mechanics.* Wiley, New York.

Mihalas, D., Binney, J. J. 1981. *Galactic Astronomy.* Freeman, San Francisco.

Nayfeh, A. H. 1973. *Perturbation Methods.* Wiley, New York.

Nezlin, M. V., Snezhkin, E. N. 1993. *Rossby Vortices, Spiral Structure, Solitons.* Springer-Verlag, Heidelberg.

Noether, E. 1983. *Collected Papers,* ed. N. Jacobson. Springer-Verlag, Berlin.

Northrop, T. G. 1963. *The Adiabatic Motion of Charged Particles.* Wiley Interscience, New York.

Ogorodnikov, K. F. 1965. *Dynamics of Stellar Systems.* Pergamon, New York.

Pagel, B. E. J. 1998. *Nucleosynthesis and Chemical Evolution of Galaxies.* Cambridge University Press, Cambridge, UK.

Palmer, P. L. 1994. *Stability of Collisionless Stellar Systems.* Kluwer, Dordrecht, The Netherlands.

Peebles, P. J. E. 1980. *The Large-Scale Structure of the Universe.* Princeton University Press, Princeton, NJ.

Peebles, P. J. E. 1993. *Principles of Physical Cosmology.* Princeton University Press, Princeton, NJ.

Peterson, B. M. 1997. *An Introduction to Active Galactic Nuclei.* Cambridge University Press, Cambridge, UK.

Poincaré, H. 1908. *Science et Méthode.* Trans. as *Science and Method,* published in 1914 by Courier Dover Publications, New York.

Poston, T., Stewart, I. 1978. *Catastrophe Theory and Its Applications.* Pitman, London.

Potter, D. 1973. *An Introduction to Computational Physics.* Wiley, London.

Rybicki, G. B., Lightman, A. P. 1979. *Radiative Processes in Astrophysics.* Wiley, New York.

Sandage, A. 1961. *The Hubble Atlas of Galaxies.* Publication 618, Carnegie Institution of Washington, Washington, DC.

Sandage, A., Bedke, J. 1988. *Atlas of Galaxies Useful for Measuring the Cosmological Distance Scale.* NASA SP-496, Washington, DC.

Sandage, A., Bedke, J. 1994. *The Carnegie Atlas of Galaxies.* Publication 638, Carnegie Institution of Washington, Washington, DC.

Sandage, A., Sandage, M., Kristian, J., eds. 1975. *Galaxies and the Universe.* University of Chicago Press, Chicago.

Sandage, A., Tammann, G. A. 1987. *A Revised Shapley-Ames Catalog of Bright Galaxies,* 2nd ed. Publication 635, Carnegie Institution of Washington, Washington, DC.

Sanders, R. H. 2010. *The Dark Matter Problem: A Historical Perspective.* Cambridge University Press, Cambridge, UK.

Sarazin, C. L. 1988. *X-Ray Emissions from Clusters of Galaxies.* Cambridge University Press, Cambridge, UK.

Saslaw, W. C. 1985. *Gravitational Physics of Stellar and Galactic Systems.* Cambridge University Press, Cambridge, UK.

Schmidt, G. 1979. *Physics of High Temperature Plasmas,* 2nd ed. Academic, New York.

Schneider, P., Ehlers, J., Falco, E. E. 1992. *Gravitational Lenses.* Springer-Verlag, Heidelberg.

Sciama, D. W. 1993. *Modern Cosmology and the Dark Matter Problem.* Cambridge University Press, Cambridge, UK.

Sérsic, J.-L. 1968. *Atlas de Galaxias Australes.* Observatorio Astronomico de Córdoba, Córdoba, Argentina.

Sommerfeld, A. 1949. *Partial Differential Equations in Physics.* Academic, New York.

Spitzer, L., Jr. 1962. *Physics of Fully Ionized Gases.* Wiley, New York.

Spitzer, L., Jr. 1968. *Diffuse Matter in Space.* Wiley, New York.

Spitzer, L., Jr. 1978. *Physical Processes in the Interstellar Medium.* Wiley, New York.

Spitzer, L., Jr. 1987. *Dynamical Evolution of Globular Clusters.* Princeton University Press, Princeton, NJ.

Stiavelli, M. 2009. *From First Light to Reionization: The End of the Dark Ages.* Wiley, New York.

Stix, T. H. 1962. *The Theory of Plasma Waves.* McGraw-Hill, New York.

Stoker, J. J. 1957. *Water Waves.* Wiley Interscience, New York.

Tajima, T. 1989. *Computational Plasma Physics, with Applications to Fusion and Astrophysics.* Addison-Wesley, Reading, MA.

Tassoul, J.-L. 1978. *Theory of Rotating Stars.* Princeton University Press, Princeton, NJ.

Thirring, W. 1980. *Quantum Mechanics of Large Systems, A Course in Mathematical Physics*, Vol. IV. Springer-Verlag, New York.

Tricomi, F. G. 1985. *Integral Equations.* Dover, New York.

Van Dyke, M. 1975. *Perturbation Methods in Fluid Mechanics.* Parabolic, Stanford, CA.

Vilenkin, N. Ya. 1968. *Special Functions and the Theory of Group Representations.* AMS Translations of Mathematical Monographs Series, Vol. 22. American Mathematical Society, Providence, RI.

Whitham, G. B. 1974. *Linear and Nonlinear Waves.* Wiley, New York.

Whittet, D. C. B. 1992. *Dust in the Galactic Environment.* Institute of Physics Publishing, Bristol, UK.

Zel'dovich, Ya. B., Novikov, I. D. 1971. *Relativistic Astrophysics*, Vol. 1: *Stars and Relativity.* University of Chicago Press, Chicago.

Zwicky, F. 1957. *Morphological Astronomy.* Springer-Verlag, Berlin.

Index of objects

Galaxies

NGC 221 (= M32), 20, 321, 350
NGC 224 (= M31), 4–5, 48, 80, 282–3, 289,
    321, 350, 381
NGC 309, 258, 265, 288
NGC 598 (= M33), 223–4, 381
NGC 628 (= M74), 23, 174, 259, 261, 265
NGC 720, 375, 377
NGC 801, 36, 284
NGC 821, 379
NGC 891, 24, 177, 180, 182
NGC 936, 264
NGC 1035, 284
NGC 1058, 259
NGC 1068, 445–6
NGC 1300, 13, 225, 254, 256
NGC 1316, 386
NGC 1344, 387
NGC 1365, 225
NGC 1379, 410–1
NGC 1399, 351, 373, 376, 379, 385–6,
    405–6
NGC 1404, 373, 405–6
NGC 1407, 388
NGC 1549, 22
NGC 1566, 21
NGC 1637, 265, 288
NGC 2300, 382
NGC 2403, 28, 31, 177
NGC 2599 (= UGC 4458), 36
NGC 2663, 37
NGC 2683, 36
NGC 2685, 159
NGC 2841, 22
NGC 2859, 253
NGC 2903,174, 293
NGC 2915, 259, 261
NGC 2974, 378
NGC 2997, 23, 224, 258, 264–5
NGC 3031 (= M81), 17–8, 22–3, 163, 186, 249,
    252, 257, 262–4
NGC 3079, 445–6, 449
NGC 3108, 379
NGC 3115, 49, 321

NGC 3198, 31, 36, 175–6, 178, 255, 276, 283,
    285, 287, 293
NGC 3344, 264
NGC 3377, 321
NGC 3379, 32–3, 314, 342, 344, 347, 371, 373,
    376, 379, 385–6
NGC 3521, 36
NGC 3741, 259
NGC 3923, 386
NGC 3938, 264
NGC 3998, 159
NGC 4013, 276
NGC 4038, 75
NGC 4039, 75
NGC 4244, 27
NGC 4254 (= M99), 223–4, 256, 258, 261, 264
NGC 4258 (= M106), 294, 321–2, 350, 445
NGC 4278, 378
NGC 4321 (= M100), 23, 224, 254, 264–5
NGC 4374, 33, 373, 387, 405–6
NGC 4406, 386
NGC 4472 (= M49), 347, 349–50, 370–3, 375–6,
    379, 385–6, 405–6
NGC 4486 (= M87), 5, 8, 20, 72, 80, 321, 350–1,
    376, 379, 385–6
NGC 4494, 379
NGC 4550, 37
NGC 4552, 33, 410–1
NGC 4559, 177
NGC 4565, 27
NGC 4594 (= M104), 321
NGC 4596, 264
NGC 4622, 223, 254, 256–7, 288
NGC 4636, 5, 373, 375, 379, 385–6, 405–6
NGC 4649 (= M60), 379, 386
NGC 4650, 159
NGC 4725, 174
NGC 4762, 27
NGC 5018, 37
NGC 5044, 382
NGC 5055 (= M63), 267
NGC 5077, 380
NGC 5128 (= Cen A), 352, 379, 386–7
NGC 5194 (= M51), 12, 27, 224, 252, 255–6, 263–4
NGC 5195, 224

Globular clusters

Clusters of galaxies

Index